CONTEMPORARY ENDOCRINOLOGY

Series Editor:
P. Michael Conn, PhD
Oregon Health & Science University
Beaverton, OR, USA

For further volumes:
http://www.springer.com/series/7680

Ken Ho
Editor

Growth Hormone Related Diseases and Therapy

A Molecular and Physiological Perspective for the Clinician

Humana Press

Editor
Ken Ho
Centres for Health Research
Princess Alexandra Hospital
The University of Queensland
Brisbane
Australia
k.ho@uq.edu.au

ISBN 978-1-61779-740-8 ISBN 978-1-60761-317-6 (eBook)
DOI 10.1007/978-1-60761-317-6
Springer New York Dordrecht Heidelberg London

Softcover reprint of the hardcover 1st edition 2011

Printed on acid-free paper

Humana Press is part of Springer Science+Business Media (www.springer.com)

Preface

The molecular era ushered in the cloning of the growth hormone (GH) gene and the production of unlimited amounts of GH through recombinant technology. The continuing momentum of research from basic science to clinical evaluation has brought unprecedented advances to the understanding of GH biology for the clinical endocrinologist. This book endeavours to distill the new information of relevance to the endocrinologist spanning the last 20 years. It contains five sections covering physiology, molecular genetics, GH deficiency, acromegaly and pharmacotherapy.

The first section on physiology focuses on GH action. A review of the structure and function of the GH receptor is followed by a perspective on the regulatory role of ghrelin on GH secretion. Attention is drawn to the pattern of GH secretion as an important determinant of tissue action. The metabolic actions of GH are diverse affecting fat, carbohydrate and protein homeostasis in humans.

The second section on genetics covers pituitary function and adenomas. Transcription factors in pituitary cell type development and the disease phenotypes resulting from loss of function mutations causing isolated or combined GH deficiency are complemented by a timely review of associated structural abnormalities identifiable by modern day imaging. This section also presents new and fascinating information on familial pituitary adenomas, their genotype and phenotype.

The section on adult GH deficiency spans the epidemiology and diagnosis of GH deficiency with a strong reminder for the clinician that the transition period represents a critical time of somatic maturation, occurring years after cessation of liner growth. Long-term global experience in replacement therapy has reconfirmed the safety and efficacy of GH in restoring body composition and fitness, with scant evidence for malignancy risk.

The section on acromegaly focuses on management, giving practical guides to the value of GH and IGF-1 measurements, the place of somatostatin analogues and of radiotherapy while reminding the reader as to why evaluating the quality of life is an important part of management. Compelling evidence is provided for clinicians to strive for tight control based on epidemiological evidence that mortality is returned to that of the general population when this is achieved.

The section on GH pharmacology takes the reader through innovative developments of long-acting GH formulations with some products on the threshold of clinical use. While there is much abuse of GH in the community, this section provides a balanced review of the effects of GH supplementation in ageing and in sports where recent data indicate an enhancing effect on a selective aspect of performance.

This book integrates a wealth of information for the paediatric endocrinologists, adult endocrinologists, endocrine scientists and internists interested in the human biology of GH.

Brisbane, Australia Ken Ho

Contents

Part V Use of Growth Hormone

Contributors

Xavier Badia Health Economics and Outcomes Research, IMS Health, and CIBERER (Centro de Investigación Biomédica en Red en Enfermedades Raras), Barcelona, Spain

Matteo Baldi Department of Internal Medicine, Division of Endocrinology, Diabetology and Metabolism, University of Turin, Torino, Italy

Ariel L. Barkan Division of Metabolism, Endocrinology and Diabetes, Department of Neurosurgery, University of Michigan, Ann Arbor, MI, USA

Albert Beckers Department of Endocrinology, C.H.U. de Liège, University of Liège, Domaine Universitaire du Sart-Tilman, Liège, Belgium

Vita Birzniece Department of Endocrinology, Garvan Institute of Medical Research and St Vincent's Hospital, Darlinghurst, NSW, Australia

Morton G. Burt Southern Adelaide Diabetes and Endocrine Services, Repatriation General Hospital and Flinders University, Adelaide, SA, Australia

Jens Sandahl Christiansen Department of Internal Medicine and Endocrinology, Aarhus University Hospital, Aarhus, Denmark

Adrian F. Daly Department of Endocrinology, University of Liège, Domaine Universitaire du Sart-Tilman, Liège, Belgium

Natascia Di Iorgi Department of Pediatrics, IRCCS, Giannina Gaslini – University of Genova, Genova, Italy

Helen Freeman Academic Unit of Diabetes, Endocrinology & Metabolism, Department of Human Metabolism, University of Sheffield, Royal Hallamshire Hospital, Sheffield, UK

Pamela U. Freda Department of Medicine, Columbia University, College of Physicians and Surgeons, New York, NY, USA

Valentina Gasco Department of Internal Medicine, Division of Endocrinology, Diabetology and Metabolism, University of Turin, Torino, Italy

Ezio Ghigo Department of Internal Medicine, Division of Endocrinology, Diabetology and Metabolism, University of Turin, Torino, Italy

James Gibney Department of Endocrinology and Diabetes, Adelaide and Meath Hospital, Dublin, Ireland

Helena Gleeson Department of Endocrinology, Leicester Royal Infirmary, Leicester, UK

Haliza Haniff Department of Endocrinology, Leeds Teaching Hospitals NHS Trust, Leeds, UK

Claire E. Higham Department of Endocrinology, Christie Hospital, Manchester, UK

Ken Ho Centres for Health Research, Princess Alexandra Hospital and The University of Queensland, Brisbane, Australia

Ian M. Holdaway Department of Endocrinology, Greenlane Clinical Centre and Auckland Hospital, Auckland, New Zealand

Vivian Hwa Department of Pediatrics, Oregon Health & Science University, Portland, OR, USA

Fabio Lanfranco Department of Internal Medicine, Division of Endocrinology, Diabetology and Metabolism, University of Turin, Torino, Italy

Mohamad Maghnie Department of Pediatrics, IRCCS, Giannina Gaslini – University of Genova, Genova, Italy

Filippa Marotta Department of Internal Medicine, Division of Endocrinology, Diabetology and Metabolism, University of Turin, Torino, Italy

Anne McGowan Department of Endocrinology and Diabetes, Adelaide and Meath Hospital, Dublin, Ireland

Marco Alessandro Minetto Department of Internal Medicine, Division of Endocrinology, Diabetology and Metabolism, University of Turin, Torino, Italy

Giovanna Motta Department of Internal Medicine, Division of Endocrinology, Diabetology and Metabolism, University of Turin, Torino, Italy

Robert D. Murray Department of Endocrinology, Leeds Teaching Hospitals NHS Trust, Leeds, UK

Ralf Nass Division of Endocrinology and Metabolism, University of Virginia, Charlottesville, VA, USA

Anne E. Nelson Pituitary Research Unit, Garvan Institute of Medical Research, Darlinghurst, NSW, Australia

Jennifer Park Division of Diabetes, Endocrinology and Metabolism, University of California, San Francisco, CA, USA

Sandra Pekic Neuroendocrine Unit, Institute of Endocrinology, University Clinical Center, Belgrade, Serbia

Elyse Pine-Twaddell Department of Pediatrics, Division of Endocrinology, Johns Hopkins University School of Medicine, Baltimore, MD, USA

Bruce E. Pollock Department of Neurological Surgery, Mayo Clinic College of Medicine, Rochester, MN, USA

Department of Radiation Oncology, Mayo Clinic College of Medicine, Rochester, MN, USA

Vera Popovic Neuroendocrine Unit, Institute of Endocrinology, University Clinical Center, Belgrade, Serbia

Sarbendra L. Pradhananga Academic Unit of Diabetes, Endocrinology & Metabolism, Department of Human Metabolism, University of Sheffield, Sheffield, UK

Sally Radovick Department of Pediatrics, Division of Endocrinology, Johns Hopkins University School of Medicine, Baltimore, MD, USA

Eugenia Resmini Department of Endocrinology Medicine, Hospital Sant Pau, Universitat Autònoma de Barcelona and CIBERER (Centro de Investigación Biomédica en Red en Enfermedades Raras), Barcelona, Spain

Antonio Ribeiro-Oliveira Jr. Federal University of Minas Gerais, Belo Horizonte, Minas Gerais, Brazil

Christopher J. Romero Department of Pediatrics, Division of Endocrinology, Johns Hopkins University School of Medicine, Baltimore, MD, USA

Richard J.M. Ross Academic Unit of Diabetes, Endocrinology & Metabolism, Department of Human Metabolism, University of Sheffield, Royal Hallamshire Hospital, Sheffield, UK

Alicia Santos Department of Endocrinology Medicine, Hospital Sant Pau, Universitat Autònoma de Barcelona and CIBERER (Centro de Investigación Biomédica en Red en Enfermedades Raras), Barcelona, Spain

Andrea Secco Department of Pediatrics, IRCCS, Giannina Gaslini – University of Genova, Genova, Italy

Kirstine Stochholm Department of Internal Medicine and Endocrinology, Aarhus University Hospital, Aarhus, Denmark

Alice Thorpe Academic Unit of Diabetes, Endocrinology & Metabolism, Department of Human Metabolism, University of Sheffield, Royal Hallamshire Hospital, Sheffield, UK

Peter J. Trainer Department of Endocrinology, Christie Hospital, Manchester, UK

Silvia Vandeva Department of Endocrinology, C.H.U. de Liège, University of Liège, Domaine Universitaire du Sart-Tilman, Liège, Belgium

Clinical Center of Endocrinology and Gerontology, Medical University Sofia, Sofia, Bulgaria

Susan M. Webb Department of Endocrinology Medicine, Hospital Sant Pau, Universitat Autònoma de Barcelona and CIBERER (Centro de Investigación Biomédica en Red en Enfermedades Raras), Barcelona, Spain

Ian R. Wilkinson Academic Unit of Diabetes, Endocrinology & Metabolism, Department of Human Metabolism, University of Sheffield, Royal Hallamshire Hospital, Sheffield, UK

Sabina Zacharieva Clinical Center of Endocrinology and Gerontology, Medical University Sofia, Sofia, Bulgaria

Part I
Physiology

Chapter 1
Growth Hormone Receptor in Growth

Vivian Hwa

Abstract It has been approximately 20 years since the cloning and characterization of the human growth hormone (GH) receptor, *GHR*, gene. Cell-surface GHR binds circulating GH, which promotes postnatal growth by regulating the expression of insulin-like growth factor (IGF)-I. Mutations in the *GHR* gene cause GH insensitivity (GHI) syndrome, also known as Laron syndrome, a syndrome characterized by severe postnatal growth retardation and low serum IGF-I concentrations in the presence of normal or elevated GH levels. Over 70 *GHR* mutations have been reported, with majority of the mutations found in exons encoding for the extracellular domain of the GHR. Inheritance of *GHR* mutations is predominantly autosomal recessive. Evaluating the impact of identified mutations on GHR structure and function is important to understand the pathophysiology of the disease. Therapeutic options for patients carrying mutations in the *GHR* gene have recently expanded to include recombinant IGF-I therapy.

Keywords Growth hormone insensitivity • Growth hormone receptor • IGF-I deficiency

Introduction

The growth-promoting effects of growth hormone (GH) are mediated primarily through regulating expression of insulin-like growth factor (IGF)-I, both circulating and peripheral, as demonstrated in rodent models and in case studies in humans. The critical importance of IGF-I for growth is highlighted by *Igf1* −/− null mice who

V. Hwa (✉)
Department of Pediatrics, Oregon Health & Science University,
3181 SW Sam Jackson Park Road, Portland, OR 97239-3098, USA
e-mail: hwav@ohsu.edu

K. Ho (ed.), *Growth Hormone Related Diseases and Therapy: A Molecular and Physiological Perspective for the Clinician*, Contemporary Endocrinology,
DOI 10.1007/978-1-60761-317-6_1,

are severely growth retarded with most dying soon after birth from the consequences of muscular hypoplasia [1, 2]. In humans, homozygous *IGF1* mutations are extremely rare, with only three convincing cases reported [3–5]. The observed intrauterine growth retardation (IUGR) and severe postnatal growth failure (height SDS, HtSDS, below –4.9) in each case supported the importance of IGF-I for growth both in utero as well as postnatal. Microcephaly and mental retardation associated with the *IGF1* mutation, furthermore, suggested IGF-I is critical for brain development, and sensorineural deafness was reported for two of the three cases [3, 4].

While it remains unclear how intrauterine IGF-I production is regulated, although both nutrition and insulin appear to play some role [6], postnatal production of circulating IGF-I is most dependent on GH. The clinical syndrome of GHI associated with an IGF-I deficiency (IGFD) and accompanied by severe growth retardation [7], in particular, has focused much attention on the pivotal role of the GH receptor (GHR) in this process.

Growth Hormone Insensitivity Syndrome

The clinical conditions of GHI and congenital GH deficiency (GHD) are characterized by minimal growth retardation in utero, profound postnatal growth retardation, infantile facial appearance, and markedly reduced serum concentrations of IGF-I. GHI is distinguished from GHD by demonstrated resistance to endogenous GH and exogenous (recombinant) GH in terms of growth, metabolic changes, or significant elevation of serum IGF-I [8, 9].

The condition of GHI was first reported in 1966 by Laron et al. [10], who described three siblings with clinical features of GHD (frontal bossing, hypoplasia of the midfacies and the nasal bridge, sparse hair, high-pitched voices, and blue scleria), but who had abnormally high levels of GH. The lack of response to GH was subsequently shown to be due to an absence of appropriate functional receptors for GH [11]. The molecular basis for this condition came with the eventual identification and cloning of the growth hormone receptor (*GHR*) gene [12, 13]. The patients in the reported study, who presented with severe growth failure (height standard deviations, Ht SDS, of –7.3 and –4.2), were shown to carry partial deletions in the *GHR* gene [13]. Since these first reports, the more than 70 *GHR* mutations identified in over 250 reported cases of GHI indicate a broader spectrum of phenotypic and biochemical abnormalities associated with GHI [14, 15].

The GH-IGF-I Axis in Postnatal Growth

The activation of the GH-IGF-I axis is initiated upon the interaction of pituitary-derived, circulating, GH with cell-surface GHR, a homodimeric transmembrane protein that belongs to the Type I class of the cytokine receptor superfamily. Like other members of the Type I family, which includes the prolactin receptor, the erythropoietin receptor, and a number of interleukin receptors, the GHR lacks the intrinsic

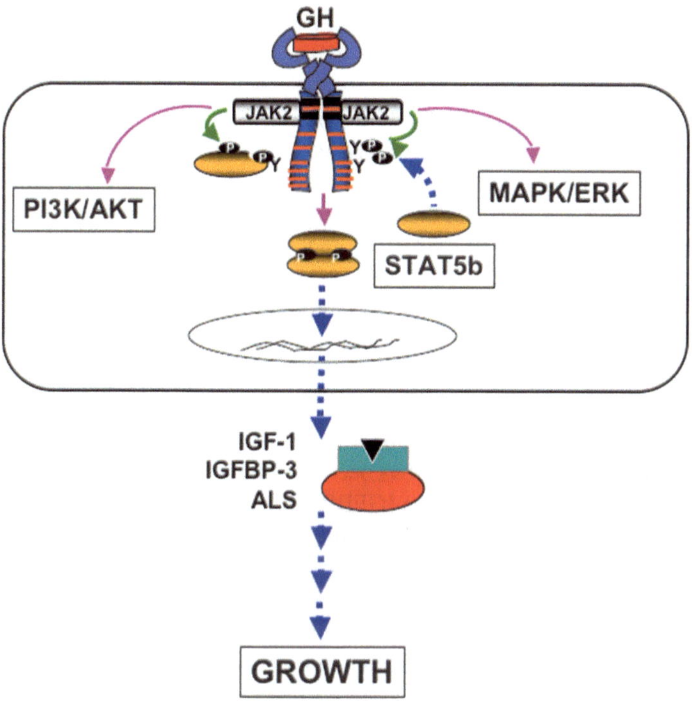

Fig. 1.1 Schematic of the growth hormone (GH) – insulin-like growth factor (IGF)-I axis. The association of GH with the homodimeric GH receptor (GHR) complex induces JAK2 recruitment, transphosphorylation of JAK2, and subsequent activation of the MAPK/ERK, PI3K/AKT, and STAT5b pathways. JAK2 phosphorylates (P) intracellular GHR tyrosines (the seven tyrosines, Y, are depicted by *purple lines*), which leads to the recruitment and docking of STAT5b to the GHR phosphorylated tyrosines. The recruited STAT5b is phosphorylated by JAK2, homodimerizes, and translocates to the nucleus where it binds to DNA and transcriptionally activates genes encoding for proteins such as IGF-I, IGF-binding protein (IGFBP)-3, and the acid labile subunits (ALS). IGF-I, IGFBP-3 and ALS are secreted and circulate in serum as a 150 kDa ternary complex, and exert endocrine effects on growth. *Solid purple arrows*, signaling pathways activated by GHR-JAK2; *solid green arrows*, tyrosyl phosphorylation by JAK2; *dashed blue arrows*, translocation of indicated protein molecules. *AKT* AKT8 virus oncogene cellular homolog; *ALS* acid labile subunit; *ERK* extracellular signal-related kinase; *GH* growth hormone; *IGF-I* insulin-like growth factor-I; *IGFBP-3* IGF-binding protein-3; *JAK* Janus-family tyrosine kinase; *MAPK* mitogen-activated protein kinase; *PI3K* phosphoinositide 3 kinase; *STAT5b* signal transducer and activator of transcription 5b

kinase activity necessary to initiate signal transduction. Instead, the GHR associates preferentially with cytosolic Janus kinase 2 (JAK2), and upon binding of one molecule of GH to the dimeric GHR, conformational changes in the GHR induce the activation of JAK2 by auto-transphosphorylation [16]. The activated JAK2 subsequently phosphorylates multiple tyrosines located on the intracellular domain of the GHR, which can serve as docking sites for cytosolic components of at least three distinct signaling pathways: the STAT (signal transducer and activator of transcription), the MAPK (mitogen-activated protein kinase), and the PI3K (phosphoinositide-3 kinase) pathways (Fig. 1.1). The signaling cascades culminate in the regulation of multiple genes, including genes encoding for IGF-I, IGF-binding protein (IGFBP)-3, and acid labile subunit (ALS). Also upregulated are negative regulators, such as the SOCS family of proteins [17], which act in a feedback loop to dampen GH-GHR signal transduction, thereby modulating IGF-I production.

The GH-induced, circulating IGF-I, IGFBP-3, and ALS are predominantly liver-derived and are found in the circulation as a ternary complex of 150 kDa. Reported aberrancies in the *GHR* [14] and the *STAT5b* genes [14, 18] significantly reduce the concentrations of serum IGF-I, IGFBP-3, and ALS, while defects in the *IGF1* gene resulted in a deficiency in IGF-I only [3–5]. Mutations in the *IGFALS* gene, encoding for the ALS protein, were recently identified in a subset of clinical GHI cases with mild short stature [19]. The abnormally low serum IGF-I and IGFBP-3 levels associated with homozygous and compound heterozygous *IGFALS* mutations are incongruous with the mild short stature phenotype but are consistent with the critical role of ALS in prolonging the half-life of both IGF-I and IGFBP-3. The mild effect(s) of ALS defects on growth has been attributed to a rapid clearance of IGF-I from the circulation [19]. Defects in IGFBP-3 have yet to be identified, suggesting that the other members of the IGFBP family may compensate for lack of a functional IGFBP-3, as has been demonstrated in rodent models for growth [20]. Indeed, circulating IGFBP-5 is also known to form a complex with IGF-I and ALS [21].

When serum GH levels are normal or elevated, but serum IGF-I, IGFBP-3, and ALS concentrations are below normal, and remain abnormally low upon GH therapy or in an IGF-I generation test (5–7 days of daily injections of recombinant GH, [22]), GHI is indicated and an aberrancy in the GH-IGF axis suggested. Measurements of serum levels of GH-binding protein (GHBP), which is the circulating extracellular domain of GHR (see below), furthermore, can indicate whether the defect is in the *GHR* gene.

The GHR Gene: Organization and Expression

The human GHR protein is encoded by a single *GHR* gene that spans 297.9 kilobases (kb), located on chromosome 5p13-p12. The gene consists of ten exons, from which a 4.4 kb mRNA is transcribed. Exons 2–10 encode for a prepeptide of 638 amino acid residues with the first 18 amino acids, the signal peptide, proteolytically removed upon the insertion of the receptor into the plasma membrane (Fig. 1.2). The mature GHR protein, 620 residues in length, is comprised of three domains: an extracellular, GH-binding domain encoded by exons 2–7 (246 amino acid residues), a short transmembrane domain encoded by exon 8 (24 residues), and the intracellular portion of the GHR encoded by exons 9 and exon 10 (350 residues). Exon 10 also carries 2.4 kb of the 3′ untranslated region (3′UTR), a region that is usually necessary for stabilizing mRNA expression and can, therefore, potentially harbor detrimental mutations [23]. Posttranslational modification of the mature GHR produces a monomer of approximately 125 kDa in molecular weight, and the final GHR product translocates to the cell surface as a preformed homo-dimer [24].

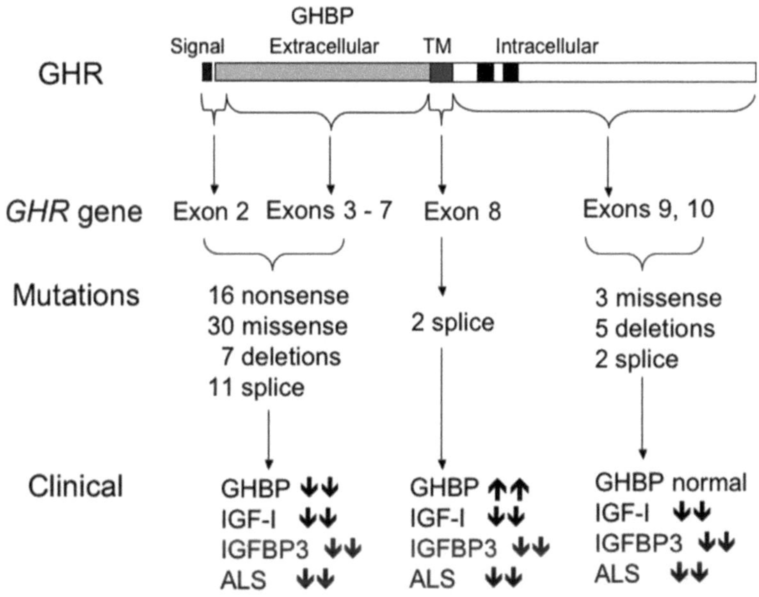

Fig. 1.2 Schematic of GHR primary structure. The exons encoding for the GHR domains are indicated. Box 1 and 2 in the intracellular domain are indicated as *filled boxes*. Mutations/variants identified within each domain are summarized [14, 15] and include more recently identified mutations (see text). The general effects of mutations on serum GHBP, IGF-I, IGFBP-3, and ALS concentrations are as indicated

The expression of full-length *GHR* (*GHRfl*) mRNA has been found to be widely distributed in human tissues with highest expression in the liver, fat, muscle, kidney, and heart [25]. Two *GHR* mRNA variants that utilize alternative splice sites located in exon 9 have also been detected in most tissues tested [26, 27]. Expression, however, appeared to be considerably reduced compared to that of *GHRfl* mRNA [25]. Reconstitution studies have shown that the predicted peptides generated from these smaller mRNA isoforms could have serious biological consequences as the resultant truncated GHR 1–277 and 1–279 peptides lack the majority of the intracellular domain and are capable of exerting dominant-negative effects on GH signaling when coexpressed with GHRfl [25, 27]. Indeed, unique *GHR* splice site mutations that spliced out the entire exon 9 have been identified in heterozygous state and proved to be dominant-negative in two nonrelated, severely short-statured patients who were GHI and IGFD [28, 29].

GHR mutations associated with GHI have been identified in all coding exons (see below), except within exon 8 and encompass the normal spectrum of genetic variations: nonsense (single base change that alters an amino acid residue to a stop codon), missense (single base change that results in a substitution of amino acid residue for the normal residue), and nucleotide(s) insertions and deletions. Intronic polymorphisms are more common, but only those variants that directly impact splicing events have been characterized [14]. The deletion of exon 3 (d3), reported in approximately 50% of the population, has turned out to be a common polymorphism whose biological impact has remained controversial (see below).

It should be clarified at this time that reference to the position of the amino acid residues in the primary structure of the GHR protein differs between the established GHR literature and current, on-line, genetic and protein databases. The numbering of amino acid residues traditionally did not include the 18 residues of the GHR signal peptide, with only residues in the final, processed, mature GHR protein counted. However, the numeration now standardized in all databases includes the signal peptide and all amino acids that are translated from the mRNA. In this chapter, the amino acids' positions indicated will be as they were published (i.e., mature peptide); where appropriate, the GHR prepeptide position will be indicated.

The Extracellular Domain of GHR Is Also the GH-Binding Protein

The extracellular domain of the dimeric, cell surface, GHR has been crystallized [30] and is considered to contain two functional subdomains of which subdomain 1, which comprised of the first 123 residues of the mature protein (exons 2–5), is involved in GH–GHR interaction, and subdomain 2, consisting of 6 beta-sheet regions that encompass residues 128–246 (exons 6–7), is involved in receptor dimerization and GH-induced rotation [16, 31]. Proteolytic cleavage at residues 242–244 releases the extracellular domain to circulate in plasma as GHBP, where it is believed to bind about 50% of circulating GH, thereby prolonging the half-life of GH, as well as modulating its bioactivity [32, 33]. Mutations identified in the extracellular domain can result in loss of detectable serum GHBP or render the GHR dysfunctional.

To date, almost all reported mutations are recessively inherited and are identified either in homozygous (the majority) or compound heterozygous forms. Nonsense and missense mutations predominate, with splicing and deletions (gross and small deletions) found with less frequency [14, 15].

GHBP-Negative, Classical GHI

The preponderance of mutations identified in GHI cases, including the first described cases, had abnormally low levels of circulating GHBP. These patients have been referred to as having classical GHI or Laron syndrome and present features typical of the first described cases [10].

The largest cohort of GHI patients who were GHBP-negative, IGFD, and presented with typical Laron features was identified in an inbred population in Ecuador [34]. All the affected individuals were homozygous for a sense mutation in exon 6 of the *GHR* gene that created a cryptic splice site [35]. The single nucleotide change did not alter the amino acid involved (E180), but induced the in-frame deletion of eight residues (181–189). The resultant mutant GHR protein was originally predicted to be unstably expressed [36]. More recent evidence, based on reconstitution studies, suggested that, on the contrary, the aberrant GHR(E180sp) was stably expressed and had retained the ability to homodimerize independently of ligand binding [37], a process that occurs in the endoplasmic reticulum [16, 24]. The mutant protein, however, could not localize appropriately to the cell membrane [37].

The highly inbred Ecuador cohort ultimately numbered approximately 100 individuals, and despite genetic homogeneity, displayed variable severity of GHI with final heights ranging from −12 to −5.3 SDS. Obesity was uniform in this group, typically 50% body fat, but diabetes mellitus was not observed [8]. The same mutation has since been identified in an Israeli patient of Moroccan origin [38], in 8 Brazilians [39], and in a Chilean family [40].

Mutations in the GHR Extracellular Domain That Are GHBP-Positive

Only a handful of defects in the extracellular domain that did not affect circulating GHBP concentrations but disrupted GHR functions have been described. These included a homozygous D152H missense mutation in exon 6 [41], a homozygous missense change in intron 6 that introduced a cryptic splice site and resulted in an insertion of a pseudoexon 6 [42], and C94S/H150Q compound heterozygous mutations identified in exon 5 and 6, respectively [43]. The D152H defect disrupted the structure of the extracellular domain sufficiently to inhibit the ability of the extracellular domain to dimerize [41]. The clinical features of the siblings who carried the C94S/H150Q compound heterozygous mutation were somewhat milder and atypical than classical Laron syndrome [43]: the elder child had normal hair, mild midfacial hypoplasia, a depressed nasal bridge, moderate frontal bossing, and no high-pitched voice; the younger child displayed a lack of Laron features. The small cohort of unrelated subjects (with overlapping ethnic background) who carried the homozygous insertion of a pseudoexon 6 (36 amino acid sequence inserted after residue 189) also displayed varying degrees of GHI, IGFD, and typical Laron features [44]. Interestingly, although the pseudoexon 6 and E180sp are both splicing mutations that cause alterations in the same region of the GHR protein, the deletion of eight amino acids (E180sp) resulted in more consistent clinical GHI features than the insertion of 36 residues [37].

Rare cases of partial GHI have been described in which serum GHBP concentrations have been abnormally high, up to 100-fold greater than normal [28, 45–50]. The unusual biochemistry has been attributed to splicing defects identified in intron 7 and intron 8 of the *GHR* gene (see below). The predicted end result of these mutations was the stable expression of mutant GHR variants, which could not anchor to the cell surface due to the splicing out of exon 8 [49]. The consequence of the massive serum concentrations of GHBP was hypothesized to compete with membrane-attached GHRs for physiologically secreted GH and led to the observed partial GHI.

The genetic-phenotype correlations that are perhaps the most difficult to reconcile are reports where the mutations identified in the GHR extracellular domain are heterozygous and are not obviously dominant-negative, yet appears to correlate to idiopathic short stature (ISS) and some degree of GHI [51]. For example, the heterozygous Arg211His, R211H (or R229H of the GHR prepeptide) was believed to be the etiology for the severe growth retardation in a subject who had a height SDS of −5.1 [51]. Regeneration of the R211H variant in the extracellular portion of the GHR only [51] demonstrated poor expression and the hypotheses proposed was that R211H functioned as a dominant-negative or that the second allele was poorly expressed, therefore accounting for the short stature and undetectable serum GHBP levels in the subject [51]. An in silico program that predicts possible impact of an amino acid substitution on protein structure and function of a human protein (Polymorphism Phenotyping: http://genetics.bwh.harvard.edu/pph2/) suggests that the R229H (prepeptide) substitution was probably damaging. However, in the most recent SNP database (http://www.ncbi.nlm.nih.gov/projects/SNP/snp_ref.cgi?locusId=2690), heterozygous R211H (prepeptide designation, R229H, in the database) was found with a frequency of 0.023, suggesting this is a polymorphism that is more common than should be associated with severe growth retardation. Indeed, we have identified the same heterozygous variant in family members who are of normal stature as well as in respective probands with growth retardation (unpublished; [52]). The conundrum of heterozygous GHR variants on growth phenotype awaits better understanding of the GHR structure and function. The possibility of a coexisting defect(s) should also be considered.

Mutations in the Transmembrane and GHR Intracellular Domain

Unlike the extracellular domain, no crystal structures of the GHR transmembrane or intracellular domains are available, although several regions critical to GHR functions have been characterized. Two regions, a proline-rich motif-designated Box 1 (residues 279–287, ILPPVPVPK, encoded by exon 9) and a region designated Box 2 (residues 325–338), are critical for interacting with JAK2, and a 10-amino acid motif (residues 322–331, DSWVEFIELD, exon 10), which overlaps with Box 2, has been demonstrated in vitro to be necessary for ubiquitin-dependent endocytosis of GH-bound GHR [53]. The human GHR also contains seven tyrosines which are believed to be phosphorylated by JAK2 upon ligand binding.

Mutations have not been identified within the short transmembrane domain of GHR, encoded by exon 8, although three splicing mutations (two homozygous and one heterozygous) which effectively excise exon 8 have been described [46, 47, 49]. The transcriptional fusion of exon 7 to exon 9 upon excision of exon 8 resulted in a frameshift and premature termination of protein expression. These nonfunctional GHR variants were unable to anchor to the cell surface, and as a consequence, serum GHBP levels were dramatically elevated by up to 110-fold (see above). The two subjects carrying homozygous splicing mutations had features typical of Laron syndrome, with height SDS below −5 and low serum levels of IGF-I and IGFBP-3 consistent with a nonfunctioning GHR [46, 47]. Subjects who were heterozygous for the splicing mutations also had supranormal levels of GHBP, but serum IGF-I and IGFBP-3 concentrations were normal, and subjects were either modestly short [49] or had heights within low-normal range [47, 49]. Altogether, these observations suggested that the presence of one wild-type *GHR* allele was sufficient to permit some normality in GH-induced actions.

Intracellular Domain Mutations

Surprisingly, few *GHR* mutations identified in GHI patients are located in the intracellular domain. Five of these mutations involved deletions – a gross chromosomal deletion in one allele that included loss of the *GHR* exons 4–10 [54] or small *GHR* deletions [55–58]. Two splice site mutations [28, 29] leading to the excision of exon 9 have also been described, and recently, a heterozygous single nucleotide duplication in Box 1 (*c.899dupC*) was identified in a young boy, age 2.8 years, with a height SDS of −4.07 [52]. With the exception of the gross deletion [54], the remaining seven reported mutations induced frameshifts with subsequent premature protein termination. The truncated GHR variants were expressed, as serum GHBP levels were relatively normal in the respective patients. The two splice site mutations, *IVS8-1G>C* and *IVS9+1G>A*, generated truncated GHRs that exerted dominant-negative effects on normal GHR [28, 29]. Phenotypically, the majority of the subjects resembled those with Laron syndrome.

It is of note that neither homozygous nonsense nor homozygous missense mutations in the GHR intracellular domain have been identified to date. Only three heterozygous missense variants are reported to be associated with short stature and an IGF deficiency: C422F [59], A478T [60], and P561T [59, 61]. However, evaluation of C422F (prepeptide, C440F) in reconstitution systems [62] and P561T (prepeptide, P579T) in a population study [63] indicated that neither variant appeared to be responsible for short stature, and each variant, in fact, has been found with a heterozygous frequency of 0.098 and 0.093, respectively, according to the SNP database (http://www.ncbi.nlm.nih.gov/projects/SNP/snp_ref.cgi?locusId=2690). Further, these two SNPs appear to be found only in those of Asian decent, with recent analysis suggesting that the polymorphisms were correlated to mandibular height [64, 65]. Polymorphisms associated with codon A478 (prepeptide, A496), to date, have not been reported. It remains unclear whether an A478T change has biological significance, although the in silico Polymorphism Phenotyping program (http://genetics.bwh.harvard.edu/pph2/) predicted that an A496T (prepeptide) substitution would likely to be damaging to GHR structure/function.

The activation of the GHR tyrosines on the intracellular domain is crucial to the recruitment and activation of the STAT5b signaling pathway, a pathway that is responsible for regulating IGF-I production. The handful of mutations identified in the intracellular domain of GHR that are clearly implicated in IGFD and GHI (see above) involve frameshifts due to deletions, duplications, or splicing mutations, all of which resulted in either premature protein truncations that abrogated most of the tyrosines, or destabilizes the entire GHR protein structure. For the human GHR, recent investigations suggested that the critical tyrosines appear to be Y534, Y566, and Y627, and the inactivation of all three tyrosines is necessary to abrogate, or significantly reduce, STAT5b signal transduction [66]. This redundancy of tyrosine usage by STAT5b, which is consistent among all mammalian GH receptors analyzed, could explain why a simple, homozygous, missense mutation, even within one of the critical tyrosines, has yet to be identified in GHI subjects. Such a mutation would be predicted to have minimal impact on STAT5b signaling unless the mutation significantly compromised the structure of the GHR.

Exon 3-Deleted GHR

In addition to identified, specific, GHR mutations, it has been recently reported that a common polymorphism of the GHR is associated with increased responsiveness to GH [67]. Approximately half of western Europeans are hetero- or homozygous with respect to an allele encoding an isoform of the GHR gene that is lacking exon 3 (d3-GHR). Evaluation of two cohorts of children of European descent with heights < −2SD, carrying a diagnosis of either ISS or small for gestational age (SGA) and receiving treatment with GH, suggested that children carrying one or both d3-GHR alleles grew almost twice as well as children who were homozygous for the full-length isoform. These observations have been confirmed in some, but not all, studies, leading to continued controversy as to whether the d3-GHR isoform actually confers increased GH responsiveness. In an attempt to put all such studies into perspective, a recent meta-analysis of prepubertal children with short stature concluded that the d3-GHR does appear to be associated with an increased baseline height in response to 1 year of GH therapy only in GHD, but not in non-GHD children with short stature [68]. The magnitude of this growth response, however, was only 0.5 cm/year, suggesting that genotyping for d3-GHR may have limited value as a predictor of therapeutic results.

Perspective and Summary

It has been approximately 20 years since the cloning and characterization of the human *GHR* gene. Mutations in the *GHR* gene are the cause of GHI syndrome, with inheritance predominantly autosomal recessive. Evaluating the impact of identified mutations on GHR structure and function is important to understanding the pathophysiology of the disease. Therapeutic options for patients carrying mutations in the *GHR* gene have recently expanded to include recombinant IGF-I therapy; Chernausek et al., for example, reported an average baseline of 2.8–8.0 cm/year during the first year of IGF-I treatment, and lower height velocities, but above baseline, during subsequent years [69]. Based on studies demonstrating the efficacy and safety of IGF-I therapy, the US Food and Drug Administration has approved rhIGF-I therapy for children with height SD score of −3 or less, serum IGF-I score of −3 or less (<2.5th percentile in the European Union), and normal or elevated GH [70]. Optimizations of doses are currently under investigations [70].

References

1. Baker J, Liu JP, Robertson EJ, Efstratiadis A. Role of insulin-like growth factors in embryonic and postnatal growth. Cell. 1993;75:73–82.
2. Liu JP, Baker J, Perkins AS, Robertson EJ, Efstratiadis A. Mice carrying null mutations of the genes encoding insulin-like growth factor I (Igf-I) and type I IGF receptor (Igflr). Cell. 1993;75:59–72.
3. Woods KA, Camacho-Hubner C, Savage MO, Clark AJ. Intrauterine growth retardation and postnatal growth failure associated with deletion of the insulin-like growth factor I gene. New Engl J Med. 1996;335:1363–7.
4. Walenkamp MJE, Karperien M, Pereira AM, et al. Homozygous and heterozygous expression of a novel insulin-like growth factor-I mutation. J Clin Endocrinol Metab. 2005;90:2855–64.
5. Netchine I, Azzi S, Houang M, et al. Partial primary deficiency of insulin-like growth factor (IGF)-I activity associated with IGF-1 mutation demonstrates its critical role in growth and brain development. J Clin Endocrinol Metab. 2009;94(10):3913–21.
6. Saenger P, Czernichow P, Hughes I, Reiter EO. Small for gestional age: short stature and beyond. Endocr Rev. 2007;28(2):219–51.
7. Rosenfeld RG. Molecular mechanisms of IGF-I deficiency. Horm Res. 2006;65 Suppl 1:15–20.
8. Rosenfeld RG, Rosenbloom AL, Guevara-Aguirre J. Growth hormone (GH) insensitivity due to primary GH receptor deficiency. Endocr Rev. 1994;15(3):369–90.
9. Rosenbloom AL. Physiology and disorders of the growth hormone receptor (GHR) and GH-GHR signal transduction. Endocrine. 2000;12(2):107–19.
10. Laron Z, Pertzelan A, Mannheimer S. Genetic pituitary dwarfism with high serum concentation of growth hormone – a new inborn error of metabolism? Isr J Med Sci. 1966;2(2):152–5.
11. Eshet R, Laron Z. Defects of human growth hormone receptors in the liver of two patients with Laron-type dwarfism. Isr J Med Sci. 1984;20:8–11.
12. Leung DW, Spencer SA, Cachianes G, et al. Growth hormone receptor and serum binding protein: purification, cloning and expression. Nature. 1987;330(6148):537–43.
13. Godowski PJ, Leung DW, Meacham LR, et al. Characterization of the human growth hormone receptor gene and demonstration of a partial gene deletion in two patients with Laron-type dwarfism. Proc Natl Acad Sci USA. 1989;86:8083–7.
14. Savage MO, Attie KM, David A, Metherell LA, Clark AJ, Camacho-Hubner C. Endocrine assessment, molecular characterization and treatment of growth hormone insensitivity disorders. Nat Clin Pract Endocrinol Metab. 2006;2:395–407.
15. Diniz ET, Jorge AA, Arnhold IJ, Rosenbloom AL, Bandeira F. Novel nonsense mutation (p.Y113X) in the human growth hormone receptor gene in a Brazilian patient with Laron syndrome. Arq Bras Endocrinol Metab. 2008;52(8):1263–70.
16. Brown RJ, Adams JJ, Pelekanos RA, et al. Model for growth hormone receptor activation based on subunit rotation within a receptor dimer. Nat Struct Mol Biol. 2005;12(9):814–21.
17. Flores-Morales A, Greenhalgh CJ, Norstedt G, Rico-Bautista E. Negative regulation of growth hormone receptor signaling. Mol Endocrinol. 2006;20(2):241–53.
18. Rosenfeld RG, Hwa V. The growth hormone cascade and its role in mammalian growth. Horm Res. 2009;71 Suppl 2:36–40.

19. Domene HM, Hwa V, Argente J, et al. Human acid-labile subunit deficiency: clinical, endocrine and metabolic consequences. Horm Res. 2009;72(3):129–41.
20. Ning Y, Schuller AG, Bradshaw S, et al. Diminished growth and enhanced glusoce metabolism in triple knockout mice containing mutations of the insulin-like growth factor binding protein-3, -4, and -5. Mol Endocrinol. 2006;20(9):2173–86.
21. Twigg SM, Baxter RC. Insulin-like growth factor binding protein 5 forms an alternative ternary complex with IGFs and the acid-labile subunit. J Biol Chem. 1998;273:6074–9.
22. Buckway CK, Guevara-Aguirre J, Pratt KL, Burren CP, Rosenfeld RG. The IGF-I generation test revisited: a marker of GH sensitivity. J Clin Endocrinol Metab. 2001;86(11):5176–83.
23. Chatterjee S, Pal JK. Role of 5′- and 3′-untranslated regions of mRNAs in human diseases. Biol Cell. 2009;101(5):251–62.
24. Gent J, van Kerkhof P, Roza M, Bu G, Strous GJ. Ligand-independent growth hormone receptor dimerization occurs in the endoplastic reticulum and is required for ubiquitin system-dependent endocytosis. Proc Natl Acad Sci USA. 2002;99(15):9858–63.
25. Ballesteros M, Leung K-C, Ross RMJ, Iismaa TP, Ho KKY. Dostribution and abundance of messenger ribonucleic acid for growth hormone receptor isoforms in human tissues. J Clin Endocrinol Metab. 2000;85(8):2865–71.
26. Dastot F, Sobrier M-L, Duquesnoy P, Duriez B, Goossens M, Amselem S. Alternative spliced forms in the cytoplasmic domain of the human growth hormone (GH) receptor regulate its ability to generate a soluble GH-binding protein. Proc Natl Acad Sci USA. 1996;93: 10723–8.
27. Ross RJM, Esposito N, Shen XY, et al. A short isoform of the human growth hormone receptor functions as a dominant negative inhibitor of the full-length receptor and generates large amounts of binding protein. Mol Endocrinol. 1997;11:265–73.
28. Ayling RM, Ross R, Towner P, et al. A dominant-negative mutation of the growth hormone receptor causes familial short stature. Nat Genet. 1997;16(1):13–4.
29. Iida K, Takahashi Y, Kaji H, et al. Growth hormone (GH) insensitivity syndrome with high serum GH-binding protein levels caused by a heterozygous splice site mutation of the GH receptor gene producing a lack of intracellular domain. J Clin Endocrinol Metab. 1998;83(2):531–7.
30. de Vos AM, Ultsch M, Kossiakoff AA. Human growth hormone and extracellular domain of its receptor: crystal structure of the complex. Science. 1992;255:306–12.
31. Behncken SN, Waters MJ. Molecular recognition events involved in the activation of the growth hormone receptor by growth hormone. J Mol Recognit. 1999;12:355–62.
32. Amit T, Youdim MB, Hochberg Z. Clinical review 112: does serum growth hormone (GH) binding protein reflect human GH receptor function? J Clin Endocrinol Metab. 2000;85(3):927–32.
33. Fisker S. Physiology and pathophysiology of growth hormone-binding protein: methodological and clinical aspects. Growth Horm IGF Res. 2006;16:1–28.
34. Rosenbloom AL, Guevara-Aguirre J, Rosenfeld RG, Fielder PJ. The little women of Loja-growth hormone receptor deficiency in an inbred population of southern Ecuador. N Engl J Med. 1990;323(20):1367–74.
35. Berg MA, Guevara-Aguirre J, Rosenbloom AL, Rosenfeld RG, Francke U. Mutation creating a new splice site in the growth hormone receptro genes of 37 Ecuadorean patients with Laron syndrome. Hum Mutat. 1992;1:24–32.
36. Rosenbloom AL, Guevara-Aguirre J, Rosenfeld RG, Francke U. Growth hormone receptor deficiency in Ecuador. J Clin Endocrinol Metab. 1999;84:4436–43.
37. Fang P, Girgis R, Little BM, et al. Growth hormone (GH) insensitivity and insulin-like growth factor-I deficiency in Inuit subjects and in Ecuadorian cohort: functional stufied of two codon 180 GH receptor gene mutations. J Clin Endocrinol Metab. 2008;93:1030–7.
38. Berg MA, Peoples R, Pérez-Jurado L, et al. Receptor mutations and haplotypes in growth hormone receptor deficiency: a global survey and identification of the Ecuadorean D180splice mutation in an oriental Jewish patient. Acta Paediatr Scand Suppl. 1994;399:112–4.
39. Jorge AA, Menezes Filho HC, Lins TS, et al. Founder effect of E180splice mutation in growth hormone receptor gene (GHR) identified in Brazilian patients with GH insensitivity. Arq Bras Endocrinol Metab. 2005;49(3):384–9.

40. Espinosa C, Sjoberg M, Salazar T, et al. E180splice mutation in the growth hormone receptor gene in a Chilean family with growth hormone insensitivity: a probable common Mediterranean ancestor. J Pediatr Endocrinol Metab. 2008;21(12):1119–27.
41. Duquesnoy P, Sobrier ML, Duriez B, et al. A single amino acid substitution in the exoplasmic domain of the human growth hormone (GH) receptor confers familial GH resistance (Laron syndrome) with positive GH-binding activity by abolishing receptor homodimerization. EMBO J. 1994;13(6):1386–95.
42. Metherell LA, Akker SA, Munroe PB, et al. Pseudoexon activation as a novel mechanism for disease resulting in atypical growth-hormone insensitivity. Am J Hum Genet. 2001;69(3):641–6.
43. Fang P, Riedl S, Amselem S, et al. Primary growth hormone (GH) insensitivity insulin-like growth factor deficiency caused by novel compound heterozygous mutations of the GH receptor gene: genetic and functional studies of simple and compound heterozygous state. J Clin Endocrinol Metab. 2007;92(6):2223–31.
44. David A, Camacho-Hubner C, Bhangoo A, et al. An intronic growth hormone receptor mutation causing activation of a pseudoexon is associated with a broad spectrum of growth hormone insensitivity phenotypes. J Clin Endocrinol Metab. 2007;92(2):655–9.
45. Rieu M, Le Bouc Y, Villares SM, Postel-Vinay M-C. Familial short stature with very high levels of growth hormone binding protein. J Clin Endocrinol Metab. 1993;76(4):857–60.
46. Woods KA, Fraser NC, Postel-Vinay MC, Savage MO, Clark AJ. A homozygous splice site mutation affecting the intracellular domain of the growth hormone (GH) receptor resulting in Laron syndrome with elevated GH-binding protein. J Clin Endocrinol Metab. 1996;81(5):1686–90.
47. Silbergeld A, Dastot F, Klinger B, et al. Intronic mutation in the growth homone (GH) receptor gene from a girl with Laron syndrome and extremely high serum GH binding protein: extended phenotypic study in a very large pedigree. J Pediatr Endocrinol Metab. 1997;10:265–74.
48. Iida K, Takahashi Y, Kaji H, et al. Functional characterization of truncated growth hormone (GH) receptor- (1-277) causing partial GH insensitivity syndrome with high GH-binding protein. J Clin Endocrinol Metab. 1999;84(3):1011–6.
49. Aalbers AM, Chin D, Pratt KL, et al. Extreme elevation of serum growth hormone-binding protein concentrations resulting from a novel heterozygous splice site mutation of the growth hormone receptor gene. Horm Res. 2009;71:276–84.
50. David A, Miraki-Moud F, Shaw NJ, Savage MO, Clark AJ, Metherell LA. Identification and characterisation of a novel GHR defect disrupting the polypyrimidine tract and resulting in GH insensitivity. Eur J Endocrinol. 2010;162(1):37–42.
51. Goddard AD, Covello R, Luoh SM, et al. Mutations of the growth hormone receptor in children with idiopathis short stature. The Growth Hormone Insensitivity Study Group. N Engl J Med. 1995;333(17):1093–8.
52. Aisenberg J, Auyeung V, Pedro JF, et al. Atypical growth hormone insensitivity syndrome (GHIS) and severe insulin-like growth factor-I deficiency (IGFD) resulting from compound heterozygous mutations of the GH receptor (GHR), including a novel frameshift mutation affecting the intracellular domain. Horm Res Paediatr. 2010;74(6):406–11.
53. Strous GJ, Gent J. Dimerization, ubiquitylation and endocytosis go together in growth hormone receptor function. FEBS Lett. 2002;529:102–9.
54. Yamamoto H, Kouhara H, Iida K, Chihara K, Kasayama S. A novel growth hormone receptor gene deletion mutation in a patient with primary growth hormone insensitivity syndrome (Laron syndrome). Growth Horm IGF Res. 2008;18(2):136–42.
55. Kaji H, Nose O, Tajiri H, et al. Novel compound heterozygous mutations of growth hormone (GH) receptor gene in a patient with GH insensitivity syndrome. J Clin Endocrinol Metab. 1997;82(11):3705–9.
56. Gastier JM, Berg MA, Vesterhus P, Reiter EO, Francke U. Diverse deletions in the growth hormone receptor gene cause growth hormone insensitivity syndrome. Hum Mutat. 2000;16(4):323–33.
57. Milward A, Metherell L, Maamra M, et al. Growth hormone (GH) insensitivity syndrome due to a GH receptor truncated after Box1, resulting in isolated failure of STAT5 signal transduction. J Clin Endocrinol Metab. 2004;89:1259–66.

58. Tiulpakov A, Rubtsov P, Peterkova V, Bezlepkina O, Chrousos GP, Hochberg Z. A novel C-terminal growth hormone receptor (GHR) mutation results in impaired GHR-STAT5 but normal STAT3 signaling. J Clin Endocrinol Metab. 2005;90(1):542–7.
59. Kou K, Lajara R, Rotwein P. Amino acid substitutions in the intracellular part of the growth hormone receptor in a patient with the Laron syndrome. J Clin Endocrinol Metab. 1993;76(1):54–9.
60. Goddard AD, Dowd P, Chernausek S, et al. Partial growth-hormone insensitivity: the role of growth-hormone receptor mutations in idiopathic short stature. J Pediatr. 1997;131:S51–5.
61. Baumbach L, Schiavi A, Bartlett R, et al. Clinical, biochemical, and molecular investigations of a genetic isolate of growth hormone insensitivity (Laron's syndrome). J Clin Endocrinol Metab. 1997;82(2):444–51.
62. Iida K, Takahashi Y, Kaji H, et al. The C422F mutation of the growth hormone receptor gene is not responsible for short stature. J Clin Endocrinol Metab. 1999;84(11):4214–9.
63. Chujo S, Kaji H, Takahashi Y, Okimura Y, Abe H, Chihara K. No correlation of growth hormone receptor gene mutation P561T with body height. Eur J Endocrinol. 1996;134(5):560–2.
64. Tomoyasu Y, Yamaguchi T, Tajima A, Nakajima T, Inoue I, Maki K. Further evidence for an association between mandibular height and the growth hormone receptor gene in a Japanese population. Am J Orthod Dentofacial Orthop. 2009;136(4):536–41.
65. Kang EH, Yamaguchi T, Tajima A, et al. Association of the growth hormone receptor gene polymorphisms with mandibular height in a Korean population. Arch Oral Biol. 2009;54(6): 556–62.
66. Derr MA, Fang P, Sinha SK, Ten S, Hwa V, Rosenfeld RG. A novel Y332C missense mutation in the intracellular domain of the human growth hormone receptor does not alter STAT5b signaling: redundancy of GHR intracellular tyrosines involved in STAT5b signaling. Horm Res Paediatr. 2011;75(3):187–99.
67. Dos Santos C, Essioux L, Teinturier C, Tauber M, Goffin V, Bougneres P. A common polymorphism of the growth hormone receptor is associated with increased responsiveness to growth hormone. Nat Genet. 2004;36:720–4.
68. Wassenaar MJ, Dekkers OM, Pereira AM, et al. Impact of the exon 3-deleted growth hormone (GH) receptor polymorphism on baseline height and the growth response to recombinant human GH therapy in GH-deficient (GFD) and non-GHD children with short stature: a systematic review and meta-analysis. J Clin Endocrinol Metab. 2009;94(10):3721–30.
69. Chernausek SD, Backeljauw PF, Frane J, Kuntze J, Underwood LE, Group GISC. Long-term treatment with recombinant insulin-like growth factor (IGF)-I in children with severe IGF-I deficiency due to growth hormone insensitivity. J Clin Endocrinol Metab. 2007;92:902–10.
70. Bright GM, Mendoza JR, Rosenfeld RG. Recombinant human insulin-like growth factor-1 treatment: ready for primetime. Endocrinol Metab Clin North Am. 2009;38(3):625–38.

Chapter 2
Ghrelin in the Regulation of GH Secretion and Other Pituitary Hormones

Fabio Lanfranco, Matteo Baldi, Giovanna Motta, Marco Alessandro Minetto, Filippa Marotta, Valentina Gasco, and Ezio Ghigo

Abstract Ghrelin, a 28 amino acid octanoylated peptide predominantly produced by the stomach, was discovered to be the natural ligand of the type 1a growth hormone (GH) secretagogue receptor (GHS-R1a). Thus, it was considered as a natural GHS additional to growth hormone-releasing hormone (GHRH), although later on ghrelin has mostly been considered a major orexigenic factor. Ghrelin activity at the pituitary level is not fully specific for GH, because it also includes stimulatory effects on both the lactotroph and corticotroph system. In fact, ghrelin in humans significantly stimulates prolactin (PRL) secretion, independently of both gender and age and probably involving a direct action on somatomammotroph cells, and possesses an acute stimulatory effect on the activity of the hypothalamus-pituitary-adrenal axis, which is similar to that of the opioid antagonist naloxone, arginine-vasopressin (AVP) and even corticotropin-releasing hormone (CRH). Finally, ghrelin plays a relevant role in the modulation of the hypothalamus-pituitary-gonadal axis function, with a predominantly central nervous system (CNS)-mediated inhibitory effect upon the gonadotropin pulsatility both in animals and in humans.

Overall, ghrelin is a pleiotropic hormone with a wide spectrum of biological actions. Further studies are required to gain insights into the exact mechanisms involved in ghrelin physiology and pathophysiology and to define the potential therapeutic roles, if any, of ghrelin and its analogs.

Keywords Ghrelin • Pituitary • GH • PRL • ACTH • LH • FSH

F. Lanfranco (✉)
Department of Internal Medicine, Division of Endocrinology, Diabetology and Metabolism, University of Turin, Corso Dogliotti 14, 10126 Torino, Italy
e-mail: fabio.lanfranco@unito.it

K. Ho (ed.), *Growth Hormone Related Diseases and Therapy: A Molecular and Physiological Perspective for the Clinician*, Contemporary Endocrinology,
DOI 10.1007/978-1-60761-317-6_2,

Introduction

Ghrelin is a 28-amino acid peptide initially isolated from human and rat stomach as an endogenous ligand for the growth hormone secretagogue receptor type 1a (GHS-R1a) [1]. Apart from a potent growth hormone (GH)-releasing effect, ghrelin has other actions including stimulation of lactotroph and corticotroph function, inhibition of the gonadal axis at both the central and peripheral level, stimulation of appetite, control of energy balance, influence on sleep and behavior, control of gastric motility and acid secretion, influence on exocrine and endocrine pancreatic function and on glucose metabolism, cardiovascular actions, and modulation of proliferation of neoplastic cells, as well as of the immune system [2, 3].

Production and Structure of Ghrelin

Ghrelin derives from the 117 amino acid precursor preproghrelin encoded by the gene GHRL located on chromosome 3 (3p25-26) [4–6].

Ghrelin peptides exist in two major molecular forms, acylated ghrelin and unacylated ghrelin. The acylation occurs on the third residue (Ser) and is essential for binding to GHS-R1a, which is responsible for ghrelin GH-releasing and orexigenic central activities [7, 8]. Apart from the stomach, ghrelin protein has also been identified in several peripheral tissues, such as the gastrointestinal tract, adrenal gland, thyroid, breast, ovary, placenta, fallopian tube, testis, prostate, liver, gallbladder, fat tissue, human lymphocytes, spleen, kidney, lung, skeletal muscle, myocardium, vein, and skin [1, 9–14]. In the brain, ghrelin-producing neurones have been identified in the pituitary, in the hypothalamic arcuate nucleus, and in a group of neurones adjacent to the third ventricle between the dorsal, ventral, paraventricular, and arcuate hypothalamic nuclei [1, 15–17]. Ghrelin produced in tissues other than the gastrointestinal tract may have a range of still unidentified physiological autocrine or paracrine effects, since the expression of specific receptors is also detected in many of these tissues [18].

The Growth Hormone Secretagogue Receptor, GHS-R1a

The gastric hormone ghrelin was identified as an endogenous ligand for the former orphan receptor GHS-R 1a [1, 7, 19]. The discovery of this receptor followed by 20 years that of synthetic GHS, which specifically binds it [1, 19–22]. This makes the discovery of ghrelin an example of reverse pharmacology.

Synthetic GHS are a family of peptidyl and nonpeptidyl molecules synthesized for the first time in the 1970s as met-enkephalin derivatives devoid of any opioid activity [22–25]. GHRP-6 was the first hexapeptide to actively release GH in vivo, in humans even more than in animals. One of its most remarkable properties was that GHRP-6 showed strong GH-releasing activity even after oral administration, although with low bioavailability and short-lasting effects [19, 22, 23, 26]. Further research that aimed to select orally active molecules with better bioavailability and longer half-lives led to the synthesis of other GHRPs as well as the discovery of orally active nonpeptidyl molecules. The most representative of these nonpeptidyl GHS that was studied in humans was the spiroindoline L-163,191 (MK-0677) [3, 19]. MK-0677 has been shown to possess a high bioavailability and is able to enhance 24-h GH secretion after a single oral administration [3, 19]. MK-0677 resulted in the discovery and cloning of the GHS-R, the existence of which had been previously indicated by binding studies [3, 19, 25].

Studies focusing on the distribution of the identified GHS-Rs showed a particular concentration of these receptors in the hypothalamus-pituitary area. In fact, in the central nervous system (CNS), GHS-R1a expression is highest in several hypothalamic nuclei, including the anterior and lateral hypothalamic areas, and the ventromedial and arcuate nuclei [27, 28]. Additionally, GHS-R mRNA is expressed in several extrahypothalamic regions, including the dorsal motor nucleus of the vagus and many parasympathetic preganglionic neurones [27, 28]. In the arcuate nuclei, GHS-R mRNA is coexpressed with both neuropeptide Y and GHRH [29, 30] and is thereby able to induce orexigenesis and facilitate GH secretion. In peripheral tissues, GHS-R1a is expressed in the anterior pituitary, pancreas [27], thyroid, spleen, myocardium, and adrenal glands [11]. GHS binding has also been observed in several other peripheral tissues [31].

Control of Ghrelin Secretion

Spontaneous ghrelin secretion in rats is pulsatile and displays an ultradian rhythmicity with the major peak preceding the onset of the dark phase [32]. This pattern is not related to the underlying pattern of GH [33] or IGF-I [34] secretion, or to photic cues, but peaks of circulating ghrelin approximately coincide with the commencement of feeding at the beginning of the dark period [33]. In the absence of feeding, basal and episodic ghrelin secretion continues to rise, fasting also upregulating gastric ghrelin mRNA expression and hypothalamic GHS-R mRNA expression [35].

In humans, a diurnal and nocturnal rhythmicity of ghrelin levels has also been observed by some [36, 37] but not by other authors [38].

It is unclear whether aging is a determinant of serum ghrelin concentrations. Ghrelin secretion is reported to be sexually dimorphic in humans, with women in the late follicular stage having higher levels than men [38].

Circulating ghrelin is modulated by energy intake, increasing in fasting states and declining 60–120 min after meals [36, 39], suggesting that ghrelin may act as an initiation signal for food intake and that its secretion may be controlled by blood levels of some nutritional factors [6, 36, 39–41]. Among determinants of ghrelin secretion, the most important appear to be insulin, glucose, and SS [3]. Possibly, GH, leptin, melatonin, thyroid hormones, glucagon, and the parasympathetic nervous system also play a role in ghrelin metabolism [3].

Unacylated Ghrelin and Obestatin

The nonacylated form of ghrelin is present in circulation in far greater amount than its acylated form, does not bind GHS-R1a, and is devoid of any neuroendocrine action. However, several recent studies have shown that unacylated ghrelin exhibits biological activities on cell proliferation and metabolism and binds to cell membranes of cardiomyocytes, adipocytes, prostatic, and skeletal muscle cells [42–45]. These effects are likely mediated through different GHS-receptor subtypes or completely different, unknown, ghrelin receptors [9, 42, 43, 46–48].

The 23-amino acid amidated peptide obestatin is a novel ghrelin gene product, which was identified as the G-protein-coupled receptor 39 (GPR39) ligand and claimed to be a physiological opponent of acylated ghrelin [49, 50]. However, these findings have lately been questioned and obestatin physiological relevance remains unclear [51]. Obestatin is mainly produced in the stomach by the same endocrine cells as ghrelin, and at lower level in the pancreas [52, 53]. Central activities have been reported for obestatin, i.e., inhibition of thirst, modulation of mnemonic functions, of anxiety and sleep, but also peripheral effects. At the cellular level, obestatin has been shown to regulate cell proliferation and survival [51, 52].

Physiological Actions of Ghrelin

In addition to its GH-releasing action, ghrelin exerts multiple endocrine and nonendocrine effects such as stimulation of prolactin (PRL) and adrenocorticotropic hormone (ACTH) secretion, inhibition of the gonadal axis at both the central and peripheral level, stimulation of appetite and of a positive energy balance, and influence on sleep and behavior, on gastric motility and acid secretion and on pancreatic exocrine and endocrine function as well as on glucose levels [3, 41, 46, 54].

In the following paragraphs, the role of ghrelin in the regulation of GH and other pituitary hormone secretion will be reviewed.

Growth Hormone-Releasing Action

The GH-releasing property was the first recognized effect of ghrelin [1]. Ghrelin possesses a strong and dose-related GH-releasing effect, both in vitro and in vivo, in humans and animals [1, 3, 46, 54, 55]. Natural and synthetic GHS stimulate GH release from somatotroph cells in vitro, probably by depolarizing the somatotroph membrane and by increasing the amount of GH secreted per cell [56, 57]. At the hypothalamic level, ghrelin and GHS act via mediation of growth hormone-releasing hormone (GHRH)-secreting neurons as indicated by evidence that passive immunization against GHRH, as well as pretreatment with GHRH antagonists, reduces their stimulatory effect on GH secretion [26, 38, 58]. At the hypothalamic level, ghrelin and GHS do not inhibit somatostatin secretion in vitro in rats; however, some inhibition of hypothalamic somatostatin secretion after exposure to GHS was observed in vivo in pigs [59–61]. On the other hand, there is evidence, both in humans and in animals, that ghrelin may act as a functional somatostatin antagonist at both the pituitary and the hypothalamic level [3, 46].

The GH-releasing activity of GHS is clearly greater in hypothalamic-pituitary preparations than in pituitary preparations, in agreement with evidence that their stimulatory effect on GH secretion is greater in vivo than in vitro [26, 58]. Indeed, in vivo, GHS show synergistic effects on GHRH-stimulated GH release [26, 62] and prevent the normal cyclic refractoriness to GHRH [58]. To confirm that the most important action of ghrelin and synthetic GHS to release GH takes place at the hypothalamic level, the GH-releasing effect of GHS is markedly reduced in animals with lesions of the pituitary stalk [3].

The circulating levels and patterns of ghrelin and GH appear weakly related [33], whereas many studies have established strong correlations between ghrelin variations and food intake episodes. Ghrelin gene deletion in mice impairs neither growth nor appetite [63, 64], although deleting the GHS-R gene does abolish both ghrelin and synthetic GHS effects on these two functions [8]. Zizzari et al. [65] have demonstrated that a novel analog of ghrelin, BIM-28163, developed as a full competitive antagonist of the GHS-R1a decreases spontaneous GH secretion without causing major changes in food intake. Moreover, BIM-28163 blunts the GH-releasing effect of ghrelin, but not the GHRH-induced GH rise [66]. These results indicate that ghrelin, acting through the GHS-R1a, appears to be an endogenous regulator of spontaneous GH secretion, but not necessarily of food intake. Interestingly, antagonism of the GHS-R1a in freely moving male rats does not impair the pulsatile pattern of GH secretion; however, it significantly lowers pulse amplitude, suggesting that endogenous ghrelin acts to amplify the basic pulsatile pattern of GH established by the interplay of hypothalamic GHRH and somatostatin [61].

Iwakura et al. [67] have recently generated a mouse model of ghrelinoma, which allows to investigate the chronic effects of ghrelin excess: adult mice showed elevated plasma ghrelin levels with preserved physiological regulation; IGF-I levels were elevated despite poor nutrition; basal GH levels were not changed while those after GHRH injection tended to be higher. These data indicate that chronic elevation of ghrelin activates the GH/IGF-I axis [67].

The GH-releasing effect of GHS undergoes marked age-related variations, increasing at puberty, reaching a plateau in adulthood, and decreasing during further aging. The mechanisms underlying these variations differ by age. The enhanced GH-releasing effect of GHS at puberty, for instance, is caused by the positive influence of increased serum estrogen levels, which increase GHS-R expression [68–70]. However, estrogen insufficiency does not fully explain the reduced GH response to GHS in postmenopausal women [25, 71–73]. The most important mechanism accounting for reduced GH-releasing activity of GHS in aging is probably represented by age-related variations in neural control of somatotroph function, including GHRH hypoactivity and somatostatinergic hyperactivity [25, 74].

As with GHS, the GH-releasing effect of ghrelin is independent of gender. Moreover, as a reduced expression of the hypothalamic GHS receptors has been demonstrated in the aged human brain, an impairment of the ghrelin system could have a role in the age-related decrease of GH secretion [3].

Some diagnostic and therapeutic implications based on the strong and reproducible GH-releasing effects have been proposed for acylated ghrelin. Particularly when combined with GHRH, ghrelin and GHS could be used as a potent and reliable provocative test to evaluate the capacity of the pituitary to release GH for the diagnosis of GH deficiency [3, 75]. Veldhuis et al. [76] have recently shown that continuous sc ghrelin infusion elevates IGF-I concentrations and sustains physiologically pulsatile 24-h GH secretion in a cohort of adults of different ages and BMIs. These authors suggest the potential utility of prolonged ghrelin administration to amplify GH production in conditions of reversible hyposomatotropism, such as aging or obesity. On the other hand, long-acting and orally active ghrelin analogs might represent an anabolic treatment in frail elderly subjects or in catabolic patients. At present, however, there is no definite evidence that shows the therapeutic efficacy of ghrelin analogs as GH/IGF-I axis-mediated anabolic agents in humans.

Prolactin and Adrenocorticotropic Hormone-Releasing Actions

Ghrelin activity at the pituitary level is not fully specific for GH, because it also includes stimulatory effects on both the lactotroph and corticotroph system [3, 46].

Acylated ghrelin significantly stimulates PRL secretion in vitro from pituitary cell cultures [77] probably acting on somatomammotroph cells [78]. Ghrelin significantly stimulates PRL secretion in humans and this effect is far less age- and gender-dependent than the effect on GH secretion [79, 80]. On the contrary, an inhibitory effect on PRL secretion acting primarily at the hypothalamus has been shown in prepubertal male and female rats [81]. Although the reason for this discrepancy is unclear, the inhibition might be limited to the rat prepubertal period.

More recent data in mice have shown that different parts of the brain, including the hypothalamus, contain neurons that coexpress the dopamine receptor subtype 1 and GHS-R [82]. In such neurons, ghrelin had the capacity to amplify dopamine-induced cyclic adenosine monophosphate accumulation, thus providing temporal control over the magnitude of dopamine signaling [82]. Additional data have shown that ghrelin can activate the mesoaccumbal dopamine system in the ventral tegmental area of mice and rats [83–85] and protect nigral dopaminergic neurons by reducing apoptosis [86].

Messini et al. [87] have recently showed that the dopaminergic agent bromocriptine blocked the stimulating effect of ghrelin on PRL release and attenuated the GH response to the same stimulus in women. The same authors also evaluated the effect of exogenous thyrotropin-releasing hormone (TRH) on ghrelin-induced PRL release in women. They showed that ghrelin induced a smaller PRL increase than TRH and that the stimulating effect of ghrelin on PRL secretion is not additive to that of TRH [88].

The exact mechanism of ghrelin action on PRL secretion requires further investigation.

On the other hand, the stimulatory effect of ghrelin and synthetic GHS on the hypothalamus-pituitary-adrenal axis in humans is remarkable and similar to that of the administration of naloxone, vasopressin, and even corticotropin-releasing hormone (CRH) [3, 46, 79]. Interestingly, the effect of ghrelin on ACTH secretion is even more pronounced than that elicited by synthetic GHS [3, 25, 79, 89–91].

The ACTH-releasing effect of GHS is acute, being attenuated during prolonged treatment, is independent of gender, and shows peculiar age-related variations, increasing at puberty, then reduction in adulthood followed by a trend toward an increase in aging, when the GH-releasing activity of GHS is clearly reduced [25, 71, 80, 92, 93]. The age-related dissociation between the stimulatory effect of ghrelin on somatotroph cells on one hand and on lactotroph and corticotroph cells on the other hand suggests that ghrelin acts at different levels and/or on different receptor subtypes to modulate these hormones [80].

Under physiological conditions, the ACTH-releasing activity of GHS is entirely mediated via the CNS [25, 92, 94]. These mechanisms via the CNS not only include CRH and/or vasopressin-mediated actions [25, 92, 95], but also via neuropeptide Y and/or γ-aminobutyric acid (GABA) [60, 96, 97]. The ACTH response to natural and synthetic GHS is generally sensitive to the negative cortisol feedback mechanism [25, 92, 97]. However, the stimulatory effect of ghrelin and GHS on corticotroph secretion is exaggerated and higher than that of human CRH in patients with pituitary ACTH-dependent Cushing's disease, probably reflecting a direct action of ghrelin and GHS on the pituitary ACTH-secreting tumor cells [25, 71, 98–101]. Interestingly, the administration of CRH to humans does not induce any significant increase in ghrelin secretion [102]. In agreement with the presence of ghrelin and GHS-R expression in ectopic ACTH-secreting tumors, exaggerated ACTH and cortisol response to GHS has also been observed in patients with ectopic ACTH-dependent Cushing's syndrome [92, 103]. These observations, however, reduce the potential use of GHS in testing ACTH secretion to distinguish patients with pituitary from ectopic ACTH-dependent hypercortisolism.

It seems unlikely that ghrelin plays a role in the regulation of corticotroph function in physiological conditions. In fact, twofold increments of plasma ghrelin, which reflect physiological fluctuations in healthy subjects, do not elicit ACTH levels in humans, whereas they stimulate GH secretion [104]. At least threefold increase in circulating ghrelin is required to stimulate corticotroph function [104]. Such a magnitude of variation has been observed in pathological conditions associated with severe malnutrition and weight loss, such as anorexia nervosa, liver cirrhosis, cancer, cardiac cachexia, and end-stage renal failure [104].

Inhibitory Action of Ghrelin on Gonadotropin Secretion

A mounting body of evidence indicates that ghrelin participates in the regulation of reproductive physiology by two actions: (i) through systemic release of the stomach-derived peptide, which acts at different levels of the reproductive system, and (ii) through biological actions on reproductive organs by locally expressed ghrelin [6, 105, 106].

Ghrelin and its receptor are expressed in various components of the reproductive system, such as placenta, testis Leydig cells, rat ovary, mouse embryo, and endometrium [6, 12, 107–110] and available data suggest that ghrelin regulates several aspects of reproductive physiology, at least partially, in a paracrine-autocrine manner [6, 18].

With regard to the effect of systemic ghrelin on gonadotropin secretion, several in vitro and in vivo animal studies indicate that the ghrelin system negatively influences the gonadal axis [105, 107, 109, 111–115]. Acylated ghrelin suppresses luteinizing hormone (LH) pulsatility in rodent, ovine, and primate models [105, 109, 111–115]. Ghrelin decreases pituitary LH responsiveness to gonadotropin-releasing hormone (GnRH). However, ghrelin infusion decreased LH pulse frequency, but not pulse amplitude in adult ovariectomized rhesus primates, suggesting that ghrelin could inhibit the GnRH pulse activity [113].

In a recent study, we showed that a prolonged infusion of acylated ghrelin quantitatively and qualitatively inhibited LH secretion in healthy young males; in fact, the infusion of the peptide was associated with clear inhibition of LH mean concentration and pulsatility [116]. Our data agree with the observation by Kluge et al. [117], who reported that night-time spontaneous LH pulsatility in healthy males is delayed and inhibited after consecutive i.v. acylated ghrelin boluses.

In contrast with in vitro data showing that ghrelin reduces the LH response to GnRH in rodents [114], the LH response to this neurohormone is not modified by the exposure to acylated ghrelin in humans [116]. These findings are therefore against the hypothesis that ghrelin plays any direct inhibitory role on pituitary gonadotropic cells. As acylated ghrelin inhibits the gonadotropin response to naloxone in humans, this clearly points toward a CNS-mediated inhibitory action through the opioid system on the human gonadal axis [116]. Furthermore, ghrelin and its receptor are expressed within the CNS acting as an orexigenic signal through neuropeptides such as neuropeptide Y, agouti-related protein, and orexin [109] that, in turn, are known to play an inhibitory role in the central control of the gonadal axis [118]. Thus, the inhibitory influence of acylated ghrelin on LH secretion could be mediated by these peptides.

On the other hand, ghrelin could, at least partially, affect reproductive capacity by negatively regulating hypothalamic Kiss1 and/or Kiss1r mRNA expression, having been previously demonstrated to have a countervailing effect on kisspeptin's ability to induce LH secretion [119]. Forbes et al. [120] have recently demonstrated in rats that Kiss1 mRNA levels were significantly lowered in the medial preoptical area after i.v. ghrelin administration, fasting, or their combination, while Kiss1r mRNA was not affected. LH pulse frequency was significantly lowered by ghrelin in fed rats, an effect significantly enhanced by food deprivation. Thus, considering the pivotal role for kisspeptin signaling in the activation of the gonadal axis, the ability of ghrelin to downregulate Kiss1 expression may be a contributing factor in ghrelin-related suppression of pulsatile LH secretion.

Finally, it has recently been demonstrated that ghrelin's inhibition of LH secretion can also be mediated by CRH [121]. Administration of a CRH antagonist prevents the suppressive effect of ghrelin on LH pulse frequency in adult rhesus monkeys [121] and as the distribution of Kiss1 and Kiss1r mRNA overlaps with CRH and its receptors, CRH-R1 and CRH-R2, in the medial preoptical area and the arcuate nucleus [122], it is possible that CRH could mediate this suppression by modulating kisspeptin signaling [120].

The actions of ghrelin on follicle-stimulating hormone (FSH) secretion in vivo remain less well characterized. A dissociation in LH and FSH secretion and response to GnRH during ghrelin administration has been reported in vivo and in vitro in animals [112]. In humans, we did not find a quantitative or qualitative effect of acylated ghrelin infusion on both basal and stimulated FSH secretion. We hypothesized that LH secretion is more sensitive to ghrelin inhibitory effect and that FSH might be sensitive to ghrelin modulation at different doses or different administration patterns than those used until now [116].

Finally, the central effects of ghrelin on the reproductive system also involve modulation of secretion of PRL that inhibits gonadotropin secretion. However, as previously mentioned, while stimulation of PRL secretion has been documented in adult humans [3], an inhibitory effect acting primarily at the hypothalamus has been shown in prepubertal male and female rats [81].

Whatever the mechanism may be, the inhibitory effect of a gastroenteropancreatic hormone, like ghrelin, on the gonadal axis fits well with clinical data in pathophysiological conditions. Anorexia nervosa, malnutrition, and cachexia are generally associated with hypogonadism that reflects a functional impairment of neuroendocrine mechanisms [123]. Metabolic factors have a major impact on ghrelin secretion regulation, and the pathophysiological conditions mentioned above are not, by chance, associated with ghrelin hypersecretion [124, 125]. Since the reproductive axis is highly dependent on body energy status, ghrelin, by acting both at the central and the peripheral level, could be one of the signals linking the nutritional status to the hypothalamus-pituitary-gonadal axis [6].

Conclusions

Since its discovery in 1999, there has been a tremendous interest on ghrelin, so that it has become one of the most important subjects of scientific research. A mounting body of evidence shows that ghrelin is involved in GH physiology, participates in PRL and ACTH secretion, and regulates several aspects of reproductive physiology and pathology. Moreover, ghrelin is involved in the regulation of energy fluxes in situations of food deprivation and affects gastrointestinal, cardiovascular, pulmonary and immune function, cell proliferation and differentiation. Overall, ghrelin is a pleiotropic hormone with a wide spectrum of biological activities. It is likely that the peripheral actions of ghrelin play a relevant role in the modulation of the central effects of this peptide, particularly its neuroendocrine actions, which are tightly related to energy balance and nutritional status.

Further studies are required to gain insights into the exact mechanisms involved in ghrelin physiology and pathophysiology and to define the potential therapeutic roles of ghrelin and its analogs.

Acknowledgments This work was supported by a grant from the Italian Ministry of University and Scientific Research (PRIN 2007, Protocol 2007RFFFFN_004) to E. Ghigo.

References

1. Kojima M, Hosoda H, Date Y, Nakazato M, Matsuo H, Kangawa K. Ghrelin is a growth-hormone acylated peptide from stomach. Nature. 1999;402:656–60.
2. Broglio F, Gottero C, Arvat E, Ghigo E. Endocrine and non-endocrine actions of ghrelin. Horm Res. 2003;59:109–17.
3. van der Lely AJ, Tschöp M, Heiman ML, Ghigo E. Biological, physiological, pathophysiological, and pharmacological aspects of ghrelin. Endocr Rev. 2004;25:426–57.
4. Ukkola O, Ravussin E, Jacobson P, et al. Mutations in the preproghrelin/ghrelin gene associated with obesity in humans. J Clin Endocrinol Metab. 2001;86:3996–9.
5. Ueno H, Yamaguchi H, Kangawa K, Nakazato M. Ghrelin: a gastric peptide that regulates food intake and energy homeostasis. Regul Pept. 2005;126:11–9.
6. Lorenzi T, Meli R, Marzioni D, et al. Ghrelin: a metabolic signal affecting the reproductive system. Cytokine Growth Factor Rev. 2009;20:137–52.
7. Howard AD, Feighner SD, Cully DF, et al. A receptor in pituitary and hypothalamus that functions in growth hormone release. Science. 1996;273:974–7.
8. Sun Y, Wang P, Zheng H, Smith RG. Ghrelin stimulation of growth hormone release and appetite is mediated through the growth hormone secretagogue receptor. Proc Natl Acad Sci USA. 2004;101:4679–84.
9. Date Y, Kojima M, Hosoda H, et al. Ghrelin, a novel growth hormone-releasing acylated peptide, is synthesized in a distinct endocrine cell type in the gastrointestinal tracts of rats and humans. Endocrinology. 2000;141:4255–61.
10. Date Y, Nakazato M, Hashiguchi S, et al. Ghrelin is present in pancreatic alpha-cells of humans and rats and stimulates insulin secretion. Diabetes. 2002;51:124–9.
11. Gnanapavan S, Kola B, Bustin SA, et al. The tissue distribution of the mRNA of ghrelin and subtypes of its receptor, GHS-R, in humans. J Clin Endocrinol Metab. 2002;87:2988–91.
12. Gaytan F, Barreiro ML, Chopin LK, et al. Immunolocalization of ghrelin and its functional receptor, the type 1a growth hormone secretagogue receptor, in the cyclic human ovary. J Clin Endocrinol Metab. 2003;88:879–87.
13. Tortorella C, Macchi C, Spinazzi R, Malendowicz LK, Trejter M, Nussdorfer GG. Ghrelin, an endogenous ligand for the growth hormone-secretagogue receptor, is expressed in the human adrenal cortex. Int J Mol Med. 2003;12:213–7.
14. Ghelardoni S, Carnicelli V, Frascarelli S, Ronca-Testoni S, Zucchi R. Ghrelin tissue distribution: comparison between gene and protein expression. J Endocrinol Invest. 2006;29:115–21.
15. Korbonits M, Kojima M, Kangawa K, Grossman AB. Presence of ghrelin in normal and adenomatous human pituitary. Endocrine. 2001;14:101–4.
16. Cowley MA, Smith RG, Diano S, et al. The distribution and mechanism of action of ghrelin in the CNS demonstrates a novel hypothalamic circuit regulating energy homeostasis. Neuron. 2003;37:649–61.
17. Kojima M, Kangawa K. Ghrelin: structure and function. Physiol Rev. 2005;85:495–522.
18. Soares JB, Leite-Moreira AF. Ghrelin, des-acyl ghrelin and obestatin: three pieces of the same puzzle. Peptides. 2008;29:1255–70.
19. Smith RG, Van der Ploeg LH, Howard AD, et al. Peptidomimetic regulation of growth hormone secretion. Endocr Rev. 1997;18:621–45.
20. Momany FA, Bowers CY, Reynolds GA, Chang D, Hong A, Newlander K. Design, synthesis, and biological activity of peptides which release growth hormone in vitro. Endocrinology. 1981;108:31–9.

21. Momany FA, Bowers CY, Reynolds GA, Hong A, Newlander K. Conformational energy studies and in vitro and in vivo activity data on growth hormone-releasing peptides. Endocrinology. 1984;114:1531–6.
22. Bowers CY. GH releasing peptides – structure and kinetics. J Pediatr Endocrinol. 1993;6:21–31.
23. Bowers CY. Growth hormone-releasing peptide (GHRP). Cell Mol Life Sci. 1998;54:1316–29.
24. Camanni F, Ghigo E, Arvat E. Growth hormone-releasing peptides and their analogs. Front Neuroendocrinol. 1998;19:47–72.
25. Ghigo E, Arvat E, Giordano R, Broglio F, et al. Biologic activities of growth hormone secretagogues in humans. Endocrine. 2001;14:87–93.
26. Bowers CY, Sartor AO, Reynolds GA, Badger TM. On the actions of the growth hormone-releasing hexapeptide, GHRP. Endocrinology. 1991;128:2027–35.
27. Guan X-M, Yu H, Palyha OC, et al. Distribution of mRNA encoding the growth hormone secretagogue receptor in brain and peripheral tissues. Mol Brain Res. 1997;48:23–9.
28. Zigman JM, Jones JE, Lee CE, Saper CB, Elmquist JK. Expression of ghrelin receptor mRNA in the rat and mouse brain. J Comp Neurol. 2006;494:528–48.
29. Willensen MG, Kristensen P, Romer J. Co-localization of growth hormone secretagogue receptor and NPY mRNA in the arcuate nucleus of the rat. Neuroendocrinology. 1999;70:306–16.
30. Mondal MS, Date Y, Yamaguchi H, et al. Identification of ghrelin and its receptor in neurons of the rat arcuate nucleus. Regul Pept. 2005;126:55–9.
31. Papotti M, Ghè C, Cassoni P, et al. Growth hormone secretagogue binding sites in peripheral human tissue. J Clin Endocrinol Metab. 2000;85:3803–7.
32. Bodosi B, Gardi J, Hajdu I, Szentirmai E, Obal Jr F, Krueger JM. Rhythms of ghrelin, leptin and sleep in rats: effects of the normal diurnal cycle, restricted feeding, and sleep deprivation. Am J Physiol Regul Integr Comp Physiol. 2004;287:R1071–9.
33. Tolle V, Bassant MH, Zizzari P, et al. Ultradian rhythmicity of ghrelin secretion in relation with GH, feeding behavior, and sleep-wake patterns in rats. Endocrinology. 2002;143:1353–61.
34. Liu YL, Yakar S, Otero-Corchon V, Low MJ, Liu J-L. Ghrelin gene expression is age-dependent and influenced by gender and the level of circulating IGF-1. Mol Cell Endocrinol. 2002;189:97–103.
35. Kim M-S, Yoon C-Y, Park K-H, et al. Changes in ghrelin and ghrelin receptor expression according to feeding status. Neuroreport. 2003;14:1317–20.
36. Cummings DE, Purnell JQ, Frayo RS, Schmidova K, Wisse BE, Weigle DS. A preprandial rise in plasma ghrelin levels suggests a role in meal initiation in humans. Diabetes. 2001;50:1714–9.
37. Koutkia P, Canavan B, Breu J, Johnson ML, Grinspoon SK. Nocturnal ghrelin pulsatility and response to growth hormone secretagogues in healthy men. Am J Physiol Endocrinol Metab. 2004;287:E506–12.
38. Barkan AL, Dimaraki EV, Jessup SK, Symons KV, Ermolenko M, Jaffe CA. Ghrelin secretion in humans is sexually dimorphic, suppressed by somatostatin, and not affected by the ambient growth hormone levels. J Clin Endocrinol Metab. 2003;88:2180–4.
39. Tschop M, Wawarta R, Riepl RL, et al. Post-prandial decrease of circulating human ghrelin levels. J Endocrinol Invest. 2001;24:RC19–21.
40. Ariyasu H, Takaya K, Tagami T, et al. Stomach is a major source of circulating ghrelin, and feeding state determines plasma ghrelin-like immunoreactivity levels in humans. J Clin Endocrinol Metab. 2001;86:4753–8.
41. Klok MD, Jakobsdottir S, Drent ML. The role of leptin and ghrelin in the regulation of food intake and body weight in humans: a review. Obes Rev. 2007;8:21–34.
42. Cassoni P, Papotti M, Ghè C, et al. Identification, characterization, and biological activity of specific receptors for natural (ghrelin) and synthetic growth hormone secretagogues and analogs in human breast carcinomas and cell lines. J Clin Endocrinol Metab. 2001;86:1738–45.
43. Baldanzi G, Filigheddu N, Cutrupi S, et al. Ghrelin and des-acyl ghrelin inhibit cell death in cardiomyocytes and endothelial cells through ERK1/2 and PI 3-kinase/AKT. J Cell Biol. 2002;159:1029–37.

44. Muccioli G, Pons N, Ghè C, Catapano F, Granata R, Ghigo E. Ghrelin and des-acyl ghrelin both inhibit isoproterenol-induced lipolysis in rat adipocytes via a non-type 1a growth hormone secretagogue receptor. Eur J Pharmacol. 2004;498:27–35.
45. Filigheddu N, Gnocchi VF, Coscia M, et al. Ghrelin and des-acyl ghrelin promote differentiation and fusion of C_2C_{12} skeletal muscle cells. Mol Biol Cell. 2007;18:986–94.
46. Ghigo E, Broglio F, Arvat E, Maccario M, Papotti M, Muccioli G. Ghrelin: more than a natural GH secretagogue and/or an orexigenic factor. Clin Endocrinol. 2005;62:1–17.
47. Mackelvie KJ, Meneilly GS, Elahi D, Wong AC, Barr SI, Chanoine JP. Regulation of appetite in lean and obese adolescents after exercise: role of acylated and desacyl ghrelin. J Clin Endocrinol Metab. 2007;92:648–54.
48. Giovambattista A, Gaillard RC, Spinedi E. Ghrelin gene-related peptides modulate rat white adiposity. Vitam Horm. 2008;77:171–205.
49. Zhang JV, Ren PG, Avsian-Kretchmer O, Luo CW, Rauch R, Klein C, et al. Obestatin, a peptide encoded by the ghrelin gene, opposes ghrelin's effects on food intake. Science. 2005;310:996–9.
50. Zhang JV, Jahr H, Luo CW, et al. Obestatin induction of early-response gene expression in gastrointestinal and adipose tissues and the mediatory role of G protein-coupled receptor, GPR39. Mol Endocrinol. 2008;22:1464–75.
51. Tang SQ, Jiang QY, Zhang YL, et al. Obestatin: its physicochemical characteristics and physiological functions. Peptides. 2008;29:639–45.
52. Granata R, Settanni F, Gallo D, et al. Obestatin promotes survival of pancreatic beta-cells and human islets and induces expression of genes involved in the regulation of beta-cell mass and function. Diabetes. 2008;57:967–79.
53. Volante M, Rosas R, Ceppi P, et al. Obestatin in human neuroendocrine tissues and tumours: expression and effect on tumour growth. J Pathol. 2009;218:458–66.
54. Broglio F, Prodam F, Riganti F, Muccioli G, Ghigo E. Ghrelin: from somatotrope secretion to new perspectives in the regulation of peripheral metabolic functions. Front Horm Res. 2006;35:102–14.
55. Arvat E, Di Vito L, Broglio F, et al. Preliminary evidence that ghrelin, the natural GH secretagogue (GHS)-receptor ligand, strongly stimulates GH secretion in humans. J Endocrinol Invest. 2000;23:493–5.
56. Goth MI, Lyons CE, Canny BJ, Thorner MO. Pituitary adenylate cyclase activating polypeptide, growth hormone (GH)-releasing peptide and GH-releasing hormone stimulate GH release through distinct pituitary receptors. Endocrinology. 1992;130:939–44.
57. Smith RG, Cheng K, Schoen WR, et al. A nonpeptidyl growth hormone secretagogue. Science. 1993;260:1640–3.
58. Clark RG, Carlsson MS, Trojnar J, Robinson IC. The effects of a growth hormone-releasing peptide and growth hormone-releasing factor in conscious and anaesthetized rats. J Neuroendocrinol. 1989;1:249–55.
59. Drisko JE, Faidley TD, Zhang D, et al. Administration of a nonpeptidyl growth hormone secretagogue, L-163, 255, changes somatostatin pattern, but has no effect on patterns of growth hormone-releasing factor in the hypophyseal-portal circulation of the conscious pig. Proc Soc Exp Biol Med. 1999;222:70–7.
60. Korbonits M, Little JA, Forsling ML, et al. The effect of growth hormone secretagogues and neuropeptide Y on hypothalamic hormone release from acute rat hypothalamic explants. J Neuroendocrinol. 1999;11:521–8.
61. Tannenbaum GS, Epelbaum J, Bowers CY. Interrelationship between the novel peptide ghrelin and somatostatin/growth hormone-releasing hormone in regulation of pulsatile growth hormone secretion. Endocrinology. 2003;144:967–74.
62. Malozowski S, Hao EH, Ren SG, et al. Growth hormone (GH) responses to the hexapeptide GH-releasing peptide and GH-releasing hormone (GHRH) in the cynomolgus macaque: evidence for non-GHRH-mediated responses. J Clin Endocrinol Metab. 1991;73:314–7.
63. Sun Y, Ahmed S, Smith RG. Deletion of ghrelin impairs neither growth nor appetite. Mol Cell Biol. 2003;23:7973–81.
64. Wortley KE, Anderson KD, Garcia K, et al. Genetic deletion of ghrelin does not decrease food intake but influences metabolic fuel preference. Proc Natl Acad Sci USA. 2004;101:8227–32.

65. Zizzari P, Halem H, Taylor J, et al. Endogenous ghrelin regulates episodic growth hormone (GH) secretion by amplifying GH pulse amplitude: evidence from antagonism of the GH secretagogue-R1a receptor. Endocrinology. 2005;146:3836–42.
66. Halem HA, Taylor JE, Dong JZ, et al. Novel analogs of ghrelin: physiological and clinical implications. Eur J Endocrinol. 2004;151:S071–205.
67. Iwakura H, Ariyasu H, Li Y, et al. A mouse model of ghrelinoma exhibited activated GH-IGF-I axis and glucose intolerance. Am J Physiol Endocrinol Metab. 2009;297:E802–11.
68. Arvat E, Camanni F, Ghigo E. Age-related growth hormone releasing activity of growth hormone secretagogues in humans. Acta Paediatr Suppl. 1997;423:92–6.
69. Loche S, Colao A, Cappa M, et al. The growth hormone response to hexarelin in children: reproducibility and effect of sex steroids. J Clin Endocrinol Metab. 1997;82:861–4.
70. Arvat E, Ceda GP, Di Vito L, et al. Age-related variations in the neuroendocrine control, more than impaired receptor sensitivity, cause the reduction in the GH-releasing activity of GHRPs in human aging. Pituitary. 1998;1:51–8.
71. Arvat E, Ramunni J, Bellone J, et al. The GH, prolactin, ACTH and cortisol responses to hexarelin, a synthetic hexapeptide, undergo different age-related variations. Eur J Endocrinol. 1997;137:635–42.
72. Anderson SM, Wideman L, Patrie JT, Weltman A, Bowers CY, Veldhuis JD. E2 supplementation selectively relieves GH's autonegative feedback on GH-releasing peptide-2-stimulated GH secretion. J Clin Endocrinol Metab. 2001;86:5904–11.
73. Anderson SM, Shah N, Evans WS, Patrie JT, Bowers CY, Veldhuis JD. Short-term estradiol supplementation augments growth hormone (GH) secretory responsiveness to dose-varying GH-releasing peptide infusions in healthy postmenopausal women. J Clin Endocrinol Metab. 2001;86:551–60.
74. Giustina A, Veldhuis JD. Pathophysiology of the neuroregulation of growth hormone secretion in experimental animals and the human. Endocr Rev. 1998;19:717–97.
75. Ghigo E, Arvat E, Aimaretti G, Broglio F, Giordano R, Camanni F. Diagnostic and therapeutic uses of growth hormone-releasing substances in adult and elderly subjects. Baillières Clin Endocrinol Metab. 1998;12:341–58.
76. Veldhuis J, Reynolds G, Iranmanesh A, Bowers C. Twenty-four hour continuous ghrelin infusion augments physiologically pulsatile, nycthemeral, and entropic (feedback-regulated) modes of growth hormone secretion. J Clin Endocrinol Metab. 2008;93:3597–603.
77. Rubinfeld H, Hadani M, Taylor JE, et al. Novel ghrelin analogs with improved affinity for the GH secretagogue receptor stimulate GH and prolactin release from human pituitary cells. Eur J Endocrinol. 2004;151:787–95.
78. Bowers CY. On a peptidomimetic growth hormone-releasing peptide. J Clin Endocrinol Metab. 1994;79:940–2.
79. Arvat E, Maccario M, Di Vito L, et al. Endocrine activities of ghrelin, a natural growth hormone secretagogue (GHS), in humans: comparison and interactions with hexarelin, a nonnatural peptidyl GHS, and GH-releasing hormone. J Clin Endocrinol Metab. 2001;86:1169–74.
80. Broglio F, Benso A, Castiglioni C, et al. The endocrine response to ghrelin as a function of gender in humans in young and elderly subjects. J Clin Endocrinol Metab. 2003;88:1537–42.
81. Tena-Sempere M, Aguilar E, Fernandez-Fernandez R, Pinilla L. Ghrelin inhibits prolactin secretion in prepubertal rats. Neuroendocrinology. 2004;79:133–41.
82. Jiang H, Betancourt L, Smith RG. Ghrelin amplifies dopamine signaling by cross talk involving formation of growth hormone secretagogue receptor/dopamine receptor subtype 1 heterodimers. Mol Endocrinol. 2006;20:1772–85.
83. Abizaid A, Liu ZW, Andrews ZB, et al. Ghrelin modulates the activity and synaptic input organization of midbrain dopamine neurons while promoting appetite. J Clin Invest. 2006;116:3229–39.
84. Jerlhag E, Egecioglu E, Dickson SL, Andersson M, Svensson L, Engel JA. Ghrelin stimulates locomotor activity and accumbal dopamine-overflow via central cholinergic systems in mice: implications for its involvement in brain reward. Addict Biol. 2006;11:45–54.
85. Quarta D, Di Francesco C, Melotto S, Mangiarini L, Heidbreder C, Hedou G. Systemic administration of ghrelin increases extracellular dopamine in the shell but not the core subdivision of the nucleus accumbens. Neurochem Int. 2009;54:89–94.

86. Jiang H, Li LJ, Wang J, Xie JX. Ghrelin antagonizes MPTP-induced neurotoxicity to the dopaminergic neurons in mouse substantia nigra. Exp Neurol. 2008;212:532–7.
87. Messini CI, Dafopoulos K, Chalvatzas N, Georgoulias P, Anifandis G, Messinis IE. Blockage of ghrelin-induced prolactin secretion in women by bromocriptine. Fertil Steril. 2010;94:1478–81.
88. Messini CI, Dafopoulos K, Chalvatzas N, Georgoulias P, Anifandis G, Messinis IE. Effect of ghrelin and thyrotropin-releasing hormone on prolactin secretion in normal women. Horm Metab Res. 2010;42:204–8.
89. Clark RG, Thomas GB, Mortensen DL, et al. Growth hormone secretagogues stimulate the hypothalamic-pituitary-adrenal axis and are diabetogenic in the Zucker diabetic fatty rat. Endocrinology. 1997;138:4316–23.
90. Thomas GB, Fairhall KM, Robinson IC. Activation of the hypothalamo-pituitary-adrenal axis by the growth hormone (GH) secretagogue, GH-releasing peptide-6, in rats. Endocrinology. 1997;138:1585–91.
91. Takaya K, Ariyasu H, Kanamoto N, et al. Ghrelin strongly stimulates growth hormone release in humans. J Clin Endocrinol Metab. 2000;85:4908–11.
92. Ghigo E, Arvat E, Camanni F. Growth hormone secretagogues as corticotrophin-releasing factors. Growth Horm IGF Res. 1998;8(Suppl B):145–8.
93. Bowers CY. Unnatural growth hormone-releasing peptide begets natural ghrelin. J Clin Endocrinol Metab. 2001;86:1464–9.
94. Shimon I, Yan X, Melmed S. Human fetal pituitary expresses functional growth hormone-releasing peptide receptors. J Clin Endocrinol Metab. 1998;83:174–8.
95. Ishizaki S, Murase T, Sugimura Y, et al. Role of ghrelin in the regulation of vasopressin release in conscious rats. Endocrinology. 2002;143:1589–93.
96. Dickson SL, Luckman SM. Induction of c-fos messenger ribonucleic acid in neuropeptide Y and growth hormone (GH)-releasing factor neurons in the rat arcuate nucleus following systemic injection of the GH secretagogue, GH-releasing peptide-6. Endocrinology. 1997;138:771–7.
97. Arvat E, Maccagno B, Ramunni J, et al. Effects of dexamethasone and alprazolam, a benzodiazepine, on the stimulatory effect of hexarelin, a synthetic GHRP, on ACTH, cortisol and GH secretion in humans. Neuroendocrinology. 1998;67:310–6.
98. Arvat E, Giordano R, Ramunni J, et al. Adrenocorticotropin and cortisol hyperresponsiveness to hexarelin in patients with Cushing's disease bearing a pituitary microadenoma, but not in those with macroadenoma. J Clin Endocrinol Metab. 1998;83:4207–11.
99. Barlier A, Zamora AJ, Grino M, et al. Expression of functional growth hormone secretagogue receptors in human pituitary adenomas: polymerase chain reaction, triple in-situ hybridization and cell culture studies. J Neuroendocrinol. 1999;11:491–502.
100. Leal-Cerro A, Torres E, Soto A, et al. Ghrelin is no longer able to stimulate growth hormone secretion in patients with Cushing's syndrome but instead induces exaggerated corticotropin and cortisol responses. Neuroendocrinology. 2002;76:390–6.
101. Martínez-Fuentes AJ, Moreno-Fernández J, Vázquez-Martínez R, et al. Ghrelin is produced by and directly activates corticotrope cells from adrenocorticotropin-secreting adenomas. J Clin Endocrinol Metab. 2006;91:2225–31.
102. van der Toorn FM, Janssen JA, De Herder WW, Broglio F, Ghigo E, Van Der Lely AJ. Central ghrelin production does not substantially contribute to systemic ghrelin concentrations; a study in two subjects with active acromegaly. Eur J Endocrinol. 2002;147:195–9.
103. de Keyzer Y, Lenne F, Bertagna X. Widespread transcription of the growth hormone-releasing peptide receptor gene in neuroendocrine human tumors. Eur J Endocrinol. 1997;137:715–8.
104. Lucidi P, Murdolo G, Di Loreto C, et al. Metabolic and endocrine effects of physiological increments in plasma ghrelin concentrations. Nutr Metab Cardiovasc Dis. 2005;15:410–7.
105. Barreiro ML, Tena-Sempere M. Ghrelin and reproduction: a novel signal linking energy status and fertility? Mol Cell Endocrinol. 2004;226:1–9.
106. Tena-Sempere M. Ghrelin: novel regulator of gonadal function. J Endocrinol Invest. 2005;28:26–9.

107. Tena-Sempere M, Barreiro ML, Gonzalez LC, et al. Novel expression and functional role of ghrelin in rat testis. Endocrinology. 2002;143:717–25.
108. Caminos JE, Tena-Sempere M, Gaytán F, et al. Expression of ghrelin in the cyclic and pregnant rat ovary. Endocrinology. 2003;144:1594–602.
109. Kawamura K, Sato N, Fukuda J, et al. Ghrelin inhibits the development of mouse preimplantation embryos in vitro. Endocrinology. 2003;144:2623–33.
110. Gaytan F, Barreiro ML, Caminos JE, et al. Expression of ghrelin and its functional receptor, the type 1a growth hormone secretagogue receptor, in normal human testis and testicular tumors. J Clin Endocrinol Metab. 2004;89:400–9.
111. Furuta M, Funabashi T, Rimura F. Intracerebroventricular administration of ghrelin rapidly suppresses pulsatile luteinizing hormone secretion in ovariectomized rats. Biochem Biophys Res Commun. 2001;288:780–5.
112. Fernández-Fernández R, Tena-Sempere M, Aguilar E, Pinilla L. Ghrelin effects on gonadotropin secretion in male and female rats. Neurosci Lett. 2004;362:103–7.
113. Vulliémoz NR, Xiao E, Xia-Zhang L, Germond M, Rivier J, Ferin M. Decrease in luteinizing hormone pulse frequency during a five-hour peripheral ghrelin infusion in the ovariectomized rhesus monkey. J Clin Endocrinol Metab. 2004;89:5718–23.
114. Fernández-Fernández R, Tena-Sempere M, Navarro VM, et al. Effects of ghrelin upon gonadotropin-releasing hormone and gonadotropin secretion in adult female rats: in vivo and in vitro studies. Neuroendocrinology. 2005;82:245–55.
115. Iqbal J, Kurose Y, Canny B, Clarke IJ. Effects of central infusion of ghrelin on food intake and plasma levels of growth hormone, luteinizing hormone, prolactin, and cortisol secretion in sheep. Endocrinology. 2006;147:510–9.
116. Lanfranco F, Bonelli L, Baldi M, Me E, Broglio F, Ghigo E. Acylated ghrelin inhibits spontaneous LH pulsatility and responsiveness to naloxone, but not that to GnRH in young men: evidence for a central inhibitory action of ghrelin on the gonadal axis. J Clin Endocrinol Metab. 2008;93:3633–9.
117. Kluge M, Schüssler P, Uhr M, Yassouridis A, Steiger A. Ghrelin suppresses secretion of luteinizing hormone in humans. J Clin Endocrinol Metab. 2007;92:3202–5.
118. Vulliémoz NR, Xiao E, Xia-Zhang L, Wardlaw SL, Ferin M. Central infusion of agouti-related peptide suppresses pulsatile luteinizing hormone release in the ovariectomized rhesus monkey. Endocrinology. 2005;146:784–9.
119. Martini AC, Fernández-Fernández R, Tovar S, et al. Comparative analysis of the effects of ghrelin and unacylated ghrelin on luteinizing hormone secretion in male rats. Endocrinology. 2006;147:2374–82.
120. Forbes S, Li XF, Kinsey-Jones J, O'Byrne K. Effects of ghrelin on *Kisspeptin* mRNA in the hypothalamic medial preoptic area and pulsatile luteinising hormone secretion in the female rat. Neurosci Lett. 2009;460:143–7.
121. Vulliémoz NR, Xiao E, Xia-Zhang L, Rivier J, Ferin M, Astressin B. a nonselective corticotropin-releasing hormone receptor antagonist, prevents the inhibitory effect of ghrelin on luteinizing hormone pulse frequency in the ovariectomized rhesus monkey. Endocrinology. 2008;149:869–74.
122. Kinsey-Jones JS, Li XF, Knox AMI, et al. Down-regulation of hypothalamic kisspeptin and its receptor, kiss1r, mRNA expression is associated with stress-induced suppression of luteinising hormone secretion in the female rat. J Neuroendocrinol. 2009;21:20–9.
123. Vanhorebeek I, Langouche L, Van den Berghe G. Endocrine aspects of acute and prolonged critical illness. Nat Clin Pract Endocrinol Metab. 2006;2:20–31.
124. Broglio F, Gianotti L, Destefanis S, et al. The endocrine response to acute ghrelin administration is blunted in patients with anorexia nervosa, a ghrelin hypersecretory state. Clin Endocrinol. 2004;60:592–9.
125. Shimizu Y, Nagaya N, Isobe T, et al. Increased plasma ghrelin level in lung cancer cachexia. Clin Cancer Res. 2003;9:774–8.

Chapter 3
Growth Hormone Pulsatility and its Impact on Growth and Metabolism in Humans

Antonio Ribeiro-Oliveira Jr. and Ariel L. Barkan

Abstract Classical endocrinology was based on the quantitative features of hormone secretion, i.e., its deficiency or excess. Recent studies have shown that qualitative features of hormone delivery to the target cells may have an independent effect on tissue responses. Whether growth hormone (GH) may share a similar mechanism, through its continuous or pulsatile release, is a burgeoning field of recent endocrine research. This chapter provides an overview of the different regulators of GH signaling and the impact of these signals upon GH pulsatility, concentrating primarily on human studies, outlining the roles of total GH output, GH pulses, and interpulse levels in determining generation of IGF-1 (i.e., growth) and metabolic effects (primarily, lipolysis) in health and disease. These data suggest that it is not only the gross quantity of GH output, but also the pattern of presentation of GH to the peripheral tissues that plays an important role in determining its biological properties. The understanding of the kinetic properties of GH secretion and their potential impact on growth and metabolism may alter our understanding of GH physiology and action, and potentially devise novel and superior strategies to optimize the effects of exogenously administered GH tailored to a specific therapeutic goal.

Keywords GHRH • GH pulsatility • Somatostatin • Ghrelin and analogs • GH and IGF-1 feedback regulation

A.L. Barkan (✉)
Division of Metabolism, Endocrinology and Diabetes, Department of Neurosurgery, University of Michigan, 24 Frank Lloyd Wright Drive, G-1500, Ann Arbor, MI 48106, USA
e-mail: abarkan@umich.edu

K. Ho (ed.), *Growth Hormone Related Diseases and Therapy: A Molecular and Physiological Perspective for the Clinician*, Contemporary Endocrinology,
DOI 10.1007/978-1-60761-317-6_3, © Springer Science+Business Media, LLC 2011

Introduction

Classical endocrinology was based on the quantitative features of hormone secretion, i.e., its deficiency or excess. Pioneering studies of Belchetz et al. [1] have revolutionized this simple concept by demonstrating that qualitative features of gonadotropin-releasing hormone (GnRH) delivery to the pituitary, i.e., pulsatile vs. continuous, had opposite effects on LH/FSH secretion, by either stimulating or inhibiting their output. This was rapidly translated into a development of pulsatile GnRH delivery for the induction of puberty and ovulation, and the use of a constant GnRH signal (long-acting GnRH analogs) for the induction of hypogonadotropic hypogonadism to treat prostate and breast cancer or to arrest puberty. Similarly, intermittent and continuous PTH administration have disparate effects on bone accretion [2] leading to the use of daily PTH injections as treatment for osteoporosis.

These developments radically altered the view of hormone action, by introducing a concept of kinetic properties of hormone delivery to the target tissues as an important component of hormone's action. Whether other hormonal systems may include a similar mechanism is a burgeoning field of current endocrine research.

Growth hormone (GH) secretion is pulsatile in nature in all species. Similar to other hormonal systems, the somatotropic axis consists of a trophic hormone (pituitary GH) and a target hormone (diffusely produced IGF-1). The transmitted GH message depends on the periodic pattern of GH release in a tissue-specific manner. For example, in rats, only pulsatile GH can normalize muscle and cartilage insulin-like growth factor (IGF)-1 mRNA levels [3], and only the continuous component of GH's secretory profile induces hepatic mRNA for certain cytochrome P-450 enzymes [4, 5]. Therefore, the question of what regulates pulsatile GH secretion pattern and its potential importance in human physiology are the issues of considerable practical importance for designing different GH therapies for a variety of human diseases.

The primary control of GH secretion is exerted by a dual interplay between hypothalamic growth hormone-releasing hormone (GHRH) and somatotropin release-inhibiting factor (somatostatin [SRIF]). The end product of GH's action, IGF-1, exerts a negative feedback on GH secretion. In addition, somatotrope function is modulated by other hypothalamic and peripheral factors (e.g., ghrelin, leptin) which together enable a balanced integrated feedforward and feedback signaling, determining processes in which GH plays a relevant regulatory role, including its unique functions in growth and metabolism. The amount of GH secreted and the pattern of its release are also subjected to the nutritional state, to nutrients themselves, and to the prevailing gonadal steroid milieu. All these factors interact with each other in a precise and coordinated manner and the interplay between them is necessarily complex. Overall, GH pulsatility may determine important metabolic actions in humans, ranging from physiological actions to those observed in human disease conditions.

This chapter provides an introductory overview of the different regulators of GH signaling and the impact of these signals upon GH secretion, concentrating primarily on human studies, to highlight the impact of the components of GH secretory pattern upon fuel mobilization in different nutritional states, as well as in certain disease conditions.

Growth Hormone-Releasing Hormone

The stimulatory hypothalamic influence on GH secretion was first suggested by Reichlin [6], who first demonstrated that hypothalamic destruction abolished somatic growth in young rats. Further studies from Frohman and coworkers [7, 8] showed that destruction of the ventromedial hypothalamus reduced pituitary and plasma GH, whereas electrical stimulation of the ventromedial and arcuate nuclei acutely increased GH release. The long search for the putative GH-releasing hormone (GHRH) culminated in the isolation of this neurohormone from two malignant extrapituitary tumors producing acromegaly in humans [9, 10] and the subsequent identification of the same peptide in hypothalamus [11].

The importance of GHRH pulses as generators of pulsatile GH secretion has been directly studied in vivo in several animal species. In the rat, both immunoneutralization of GHRH [12, 13] and abolition of GHRH secretion [14, 15] eliminate GH pulses [12–16]. Interestingly, the same approach to female rats not only abolished the pulses, but also decreased the tonic, interpulse GH levels [17] and suggested sexual dimorphism of GH regulation. Direct sampling of rat [12] or sheep [18] hypophysial-portal blood has found GHRH peaks to be largely concordant with GH pulses, validating the analysis of GH pulsatility as a method for the study of endogenous GHRH secretion in these species. Indeed, only a pulsatile mode of GHRH administration increased growth in normal and GHRH-deficient rats [19] and upregulated GH-mRNA and protein content in the pituitaries of young female rats [20]. However, this mechanism seems to be different in humans, as continuous GHRH infusion augmented pulsatile GH release in normal and GHRH-deficient humans [21].

Studies utilizing GHRH antagonist (N-Acl, D-Tyr2) [22] either as a bolus night dose [23] or as a continuous 24-h intravenous infusion [24] have shown severe impairment in pulsatile GH release and proportionately suppressed acute GH responses to bolus doses of exogenous GHRH. These findings are in accord to what has been described for human dwarfs with inactivating mutations of GHRH receptor [25].

Multiple human studies have assessed GH pulsatility in various physiological conditions, such as puberty [26], aging [27], fasting [28], and the menstrual cycle [29], as well as in human illnesses such as growth delay [30], acromegaly [31, 32], diabetes [33], and obesity [34]. Indeed, the general assumption that GH pulses reflect periodic discharges of hypothalamic GHRH in humans was confirmed by studies where GHRH antagonist has been shown to inhibit GH pulses [22, 23, 35]. This was the case, for example, in studies showing that somatopause shifts the dose-inhibition curve of GH output to the left in elderly men, corroborating the assumption that somatopause is a GHRH-deficient state. Furthermore, other studies utilizing GHRH antagonist have demonstrated the importance of GHRH to the genesis of acute GH responses to several provocative stimuli commonly used in clinical practice, including clonidine, L-DOPA, arginine, pyridostigmine, GHRP-6, and hypoglycemia [35, 36].

Another point of interest is the sexual dimorphism of GH secretion and its relationship to GHRH. In men, most GH is secreted at the early hours of night, whereas daytime GH secretion is relatively hypopulsatile and has low baseline component [37]. In contrast, young women have similar total 24-h GH output but high basal GH output during the day and relatively attenuated nocturnal GH release [37]. Unlike males, women have higher sensitivity to GHRH antagonist during the night, but also suppress daytime basal GH levels [24]. This is in agreement with the data in male and female rats [12, 17] and indicates higher GHRH output at night in men and the involvement of endogenous GHRH in the maintenance of basal GH secretion in women.

Endogenous GHRH is therefore the main regulator of GH pulse generation in humans, and it participates in the generation of GH responses to a variety of pharmacological stimuli. The age-related decline in humans is another example of GHRH-mediated phenomenon, as is the sexual dimorphism of the structure of pulsatile GH secretion during both daytime and night time.

Somatostatin

SRIF, a 14-amino acid peptide, is the main negative regulator of GH secretion. Five subtypes of SRIF receptors have been identified, of which types 2 and 5 are the main mediators of the suppressive GH effect. SRIF antagonizes the mitogenic effect of GHRH on somatotrophs, but does not inhibit GH synthesis [38]. It suppresses spontaneous GH release and GH responses to all stimuli tested so far. SRIF has a very short circulating half-life (approximately 3 min) and termination of its infusion elicits modest rebound GH rise and significantly increases GH response to GHRH. This is likely caused by the accumulation of the readily releasable GH in the somatotrophs during SRIF exposure. A continuous infusion of GHRH almost completely desensitizes pituitary responses to GHRH in humans, but actually augments pulsatile GH release [21], and the rebound GH release post-SRIF withdrawal does not require GHRH [35].

The hypothesis is, therefore, that the periodic, pulsatile GH release is the result of a coordinated SRIF and GHRH secretion. However, direct measurements of SRIF in the pituitary-portal circulation in conscious sheep failed to disclose any relations between SRIF and GH pulses [39].

In the rat model, administration of SRIF antibody does not alter pulsatile GH release, but elevates the interpulse GH levels [12]. In humans, continuous infusion of octreotide, a type 2 SRIF agonist, created circulating octreotide levels in excess of 1–2 ng/mL [40, 41], at least 50 times higher than the pituitary-portal SRIF concentrations in rats [12] and sheep [39]. In this model, GH pulse occurrence remained unmodified, but both the interpulse GH levels and the amplitude of GH pulses were powerfully suppressed. Therefore, it seems that the role of SRIF in humans may be limited to the adjustment of the magnitude of basal and pulsatile GH release, but not to generation of GH pulsatility. Indeed, a recent study in women with low-somatostatin milieu has shown that age and estrogen availability are predictive of GH pulsatility [42].

Interestingly, experiments utilizing SRIF receptor type-2 knock-out mice [43] or those with constitutive hypersecretion of SRIF [44] showed that these grew at a normal rate and passive immunization to SRIF in sheep also did not increase somatic growth [45]. Thus, further studies are necessary to address better the exact role of SRIF in human physiology and pathology.

Ghrelin and Analogs

Synthetic analogs of Ghrelin (GHS) had been introduced as early as 25 years ago and were instrumental for the discoveries of ghrelin receptor and, subsequently, ghrelin itself [46]. Indeed, most of the information regarding the GH-promoting effects of ghrelin has been obtained with these synthetic peptides.

Ghrelin, a 28-amino acid hormone of primarily gastric origin, which is acylated posttranslation, is the endogenous ligand for the GH secretagogue receptor. It circulates in two forms: deoctanoylated and octanoylated, with only the latter possessing the GH-releasing properties. Although of primarily gastric origin, ghrelin mRNA had been found in the hypothalamus [47] where it exerted its GH-releasing properties. Acute administration of GHS produces an immediate and massive release of GH [48], although the coadministration of GHS and GHRH results in a powerful GH increase that is even greater than the effect of either peptide administered alone [48]. The GHS synergism with GHRH can be explained by its possible functional SRIF antagonism [49]. Indeed, although GHS act directly at the pituitary level as suggested by a recent study [50], their action is much more powerful when applied to combined hypothalamic-pituitary segments in vitro or to intact animals [48], suggesting that the presence of hypothalamic GHRH is essential for their full action. Furthermore, pretreatment of humans with GHRH receptor antagonist attenuates their responsiveness to GHS [36], and continuous infusion of GHS amplifies the magnitude of GH pulses and GH responsiveness to GHRH [51].

However, most studies aimed at the elucidation of the potential role of endogenous ghrelin failed to find a significant role for this peptide as a regulator of spontaneous GH secretion. Whereas good concordance between octanoylated ghrelin and GH pulses was seen in fed men [52], fasting resulted in a rise in GH pulse amplitude despite virtual abolition of octanoylated ghrelin secretion [52, 53]. Likewise, ghrelin-producing tumors were not able to augment GH secretion in humans [54, 55]. In addition, knock-out animals for ghrelin or ghrelin receptor genes did not show any attenuation of growth [56–59].

Although pulsatile GH secretion and growth continue in the absence of ghrelin, a recent study [60] showed that a GHS-R missense mutation segregated with short stature in two independent families. This mutation, which results in decreased cell-surface expression of the receptor, selectively impairs the constitutive activity of the GHS-R, while preserving its ability to respond to ghrelin [60]. Further studies are therefore necessary to address better the role of GHS in growth. In the meanwhile, vit seems that neither circulating nor hypothalamic ghrelin seems to be involved in meaningful physiologic regulation of GH secretion in the majority of situations.

GH and IGF-1 Feedback Regulation

GH secretion is a target of multiple negative feedback loops at multiple levels. First, there are ultrashort feedback loops in which GHRH inhibits its own secretion [61, 62] and SRIF suppresses its own neuronal release [63]. Second, there are short feedback loops in which SRIF secretion is directly stimulated by GHRH [64] and GHRH secretion is inhibited by SRIF [65]. Pituitary GH inhibits hypothalamic GHRH secretion [66, 67] and stimulates SRIF release from hypothalamic neurons [68]. It is thought that a hypothalamic stimulus altering GH secretion automatically extinguishes its own effect. This likely participates in the creation of pulsatile GHRH secretion and SRIF secretion patterns and, ultimately, in the maintenance of pulsatile GH release from the pituitary.

However, these mechanisms are likely to be operative on a short-term basis, while the long-term negative feedback of GH secretion is accomplished mainly through a classical "trophic-target" loop involving the ultimate product of GH action (i.e., IGF-1). Interestingly, although both liver-derived circulating IGF-1 and local bone-derived IGF-1 can redundantly replace each other in the maintenance of normal longitudinal bone growth, locally derived IGF-1 cannot replace liver-derived IGF-1 for the regulation of several parameters, including GH secretion [69].

The regulation of IGF-1 production by GH is mediated by signaling through the Janus kinase (JAK)-2 pathways, via the phosphorylation of the transcriptional factor, signal transducer, and activator of transcription (STAT)-5b [70]. Nevertheless, the exact site of IGF-1 negative feedback on GH secretory mechanisms is still uncertain.

A continuous infusion of IGF-1 in young men and women increased plasma IGF-1 concentrations three to four times above the upper limit of the normal range [37, 71]. This reliably suppressed plasma GH concentrations by approximately 50–80% in both sexes, at the expense of grossly diminished GH pulse amplitude. Administration of exogenous GHRH to the same individuals, however, showed a sexually dimorphic effect. In men, IGF-1 infusion grossly suppressed plasma GH response to GHRH, suggesting that the negative feedback of IGF-1 was expressed either directly at the pituitary level or at the hypothalamic level by stimulating SRIF secretion (Fig. 3.1). On the other hand, administration of exogenous IGF-1 to women was completely ineffective in suppressing their GH responses to GHRH. This suggests that, in women, elevated IGF-1 suppresses GH secretion by selective suppression of the hypothalamic GHRH output.

Studies have successfully analyzed the reverse model in order to evaluate consequence of a decline in IGF-1 upon GH secretion. A rapid increase in GH secretion occurs after termination of IGF-1 infusion [72] corresponding to a decline of free IGF-1 in the peripheral circulation. Fasting lowers free and total IGF-1 concentrations in the peripheral circulation [72, 73]. Indeed, free IGF-1 declines rapidly, while total IGF-1 takes up to 5 days to fall, due to the prolonged half-life of the bound hormone. As opposed to IGF-1 decline, plasma GH concentrations increase 2–3-fold and reach a plateau after approximately 3 days of fasting [73]. Interestingly, this fasting-mediated increase in GH secretion is accomplished almost exclusively at the expense

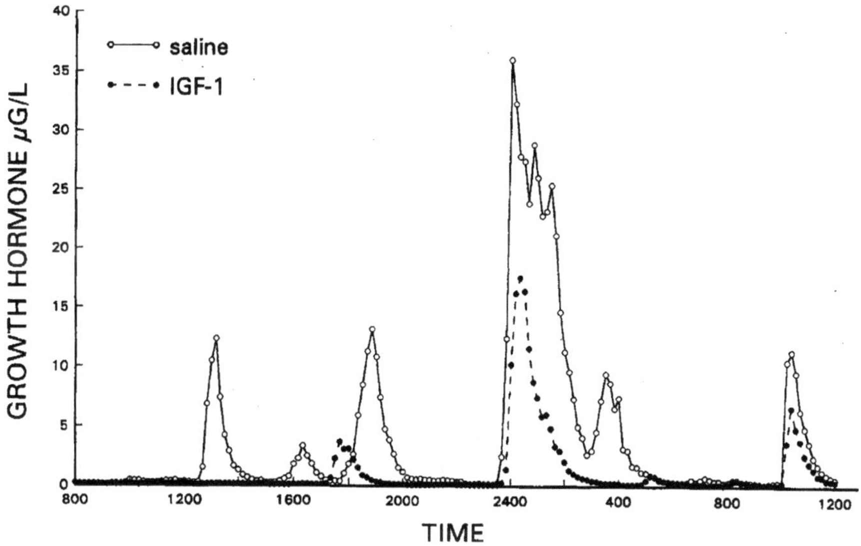

Fig. 3.1 Suppression of GH secretion by continuous insulin-like growth factor (IGF)-1 infusion in a young man

of augmented GH pulse amplitude, whereas GH interpulse and GH pulse frequency minimally contributed to this effect.

Fasting in radiation-induced hypothalamic GH-deficient patients augments GH secretion to a lesser degree when compared to healthy subjects [74], suggesting that an intact hypothalamus is necessary for an intact IGF-1 feedback in these patients.

Changes in GH Secretion During Physiological Situations (Sleep, Gonadal Steroids, and Nutrition)

There are several physiological and pathophysiological conditions associated with changes in GH secretion. One of the hallmarks of pulsatile GH architecture in humans is the sleep-associated increase of GH secretion. It usually occurs around midnight and the GH levels at that time are at their highest during the 24-h period. Partially, this phenomenon is time-entrained and partially related to sleep itself [75, 76]. The maximum GH levels occur within minutes of the onset of slow wave sleep [77]. Interestingly, there is marked sexual dimorphism of the nocturnal GH increase in humans, constituting only a fraction of the total daily GH release in women, but otherwise the bulk of GH output in men [51]. The neuroendocrine mechanisms of nocturnal GH increase as well as its gender disparity are not well understood, although GHRH is supposed to play a major role [78]. Changes in ghrelin secretion do not seem to play a role [52, 53]. While continuous infusion of octreotide in pharmacological concentration inhibited the amplitude but not the occurrence of the nocturnal GH rise [40, 41], pituitary sensitivity to GHRH boluses was clearly augmented at the time of the expected nocturnal GH surge [79], implying changed pituitary sensitivity to GHRH at that time. Further studies are necessary to better clarify the involvement of hypothalamic SRIF in the generation of the nocturnal GH pulses.

Activation of the gonadal system during puberty is accompanied by increased GH and IGF-1 concentrations [80, 81]. Conversely, attenuation of gonadal activity with aging is temporarily accompanied by a decline in both GH and IGF-1 concentrations [22]. The relationship between gonadal steroids and somatotropic axis has been a matter of discussion, although at least a temporary link between these two systems is already well established in puberty and aging.

The sexual dimorphism of GH secretion suggests that androgens and estrogens may play different roles in the regulation of somatotrope axis [13]. Data coming from rat studies are conflicting regarding gender differences for GHRH and SRIF content and release, and it does not seem that these parameters are regulated by androgens or estrogens. Conversely, there is a major sexual dimorphism in pituitary GH synthesis. Male rats have 2–4-fold higher levels of GH-mRNA and GH protein and higher sensitivity to exogenous GHRH. Furthermore, administration of androgens, but not estrogens, increased GH responses to GHRH in castrated animals. Therefore, these data suggest that high GH pulses in male rats may be affected by the sexually dimorphic effects of gonadal steroids at the pituitary level.

In humans, gonadal-somatotropic relations are often diametrically opposed to those found in animal models. Women secrete more GH in response to GHRH than men, and pulsatile GH secretion in women has been shown to be influenced by age and estrogen status [82]. Interestingly, studies with androgen or estrogen receptor blockers have shown that puberty-related increase of GH secretion is an estrogen, but not an androgen-related phenomenon [83, 84]. Likewise, only aromatizable (testosterone), but not nonaromatizable (dihydrotestosterone or oxandrolone), androgens increase GH pulse amplitude in boys [85]. This effect seems to be regulated at the pituitary level, as increasing plasma testosterone to the "young adult" levels in prepubertal boys increased both GH pulse amplitude and plasma IGF-1 concentrations without altering set point for the inhibition of GH secretion by GHRH receptor antagonist [86]. Interestingly, when the same testosterone levels were reproduced in elderly men, plasma GH actually fell twofold, whereas their plasma IGF-1 minimally increased [87].

The gonadal-somatotropic interactions are further complicated by the direct effects of gonadal steroids at the level of hepatic IGF-1 synthesis and the resultant alterations in the negative feedback effects of IGF-1 on GH. Castrated female rats given physiologic male concentrations of dihydrotestosterone exhibited increased hepatic IGF-1 mRNA concentrations, plasma IGF-1 levels, and somatic growth [88]. However, pituitary GH content in these animals was grossly increased, but plasma GH levels were low. In contrast, estradiol-treated animals exhibited virtually abolished somatic growth, high plasma GH concentrations, low pituitary GH content, and low hepatic IGF-1 mRNA and plasma IGF-1 levels [88]. Taken together, these data suggest that androgens increase hepatic IGF-1 synthesis, which subsequently suppresses pituitary GH release. Estrogen, in contrast, suppresses hepatic IGF-1 synthesis and the lower circulating IGF-1 level removes the negative feedback at the pituitary level. These mechanisms seem to be also operative in humans, as evidenced by the inhibitory action of estrogen on IGF-1 synthesis [89–91], probably as a result of interference with GH signaling mechanisms [92].

The nutritional influence upon GH secretion and its pulsatility has also been a matter of active interest. It has long been demonstrated that hyperglycemia causes transient GH suppression for 1–3 h, followed by GH rise at 3–5 h after oral glucose administration [93, 94]. The GH responses to GHRH or GHS given after glucose load are attenuated in healthy volunteers [95, 96], suggesting that SRIF discharge may be the cause of the rapid inhibitory effect of glucose on GH release. When SRIF release declines, endogenous GHRH secretion is activated reciprocally, and available pituitary stores of GH are released, leading to the rebound increase in serum GH levels [97, 98]. Conversely, hypoglycemia leads to acute GH secretion, and it requires intact GHRH signaling [35].

Elevated free fatty acids are a strong inhibitor of GH release in normal humans [98]. The response of GH to GHRH is markedly decreased during lipid/heparin infusion, suggesting significant elevation of SRIF tone [98]. Indeed, the administration of an inhibitor of lipolysis, Acipimox, concomitantly increases GH secretion [98, 99] and the GH secretory response to GHRH in normal subjects [100]. Furthermore, GH levels are often suppressed in conditions associated with chronic elevations of lipolysis, such as obesity [101]. The degree of GH attenuation correlates with the amount of total and visceral fat [102]. Indeed, suppression of free fatty acids in obesity normalizes GH levels [99, 103]. In obesity, however, free IGF-1 levels are elevated, and this may suppress pituitary GH secretion directly [104], as suggested in humans and sheep [105, 106]. The response of somatotrophs to GHRH and GHS stimulations is reduced [107, 108] in obese people, suggesting increased SRIF tone or a direct pituitary effect of free fatty acids. Contrary to the state of energy excess, fasting powerfully increases GH secretion, despite high free fatty acid levels, most likely as a result of the decrease in IGF-1 levels [109, 110].

GH Secretory Pattern and its Actions

It has been estimated that GH is secreted in humans with a periodicity of ~110 min [111], a pulse duration of ~60 min [112], and an average secretion rate of 0.18 and 0.27 μmol kg^{-1} min^{-1} in adult men and women, respectively [111], with a serum half-life on the order of minutes [37, 112]. Although there is usually a dominant rhythmicity to the daily GH pulses overall, the frequency and amplitude of these pulses are quite variable between individuals [113]. In between these pulses, GH levels are low and relatively stable. This tonic "interpulse" or "basal" GH secretion accounts for a relatively small portion of daily GH output.

It has been known that the mode of GH delivery in rats regulates hormone action and the importance of GH pulsatility has been demonstrated in rats [3, 19, 114]. For example, it has been shown that pulsatile GH is more effective than continuous GH infusion in stimulating somatic growth [3, 19] and muscle growth plate IGF-1 mRNA [114] or to determine sex difference in predisposition to thrombosis [115]. In contrast, continuous GH component appears to be predominant in inducing hepatic IGF-1 mRNA [3] and CYP 3A4 enzymes [5, 116, 117] and GH receptors [118]. However, the importance of GH secretion pattern in humans has been a matter of debate. This issue has begun to be addressed only recently.

Previous studies in humans attempting to address the importance of GH secretion pattern in growth and metabolism have failed to show the importance of GH pulsatility on metabolism [119, 120] due to the lack of an appropriate model. Subsequent studies, however, utilizing a methodology which mimicked GH pulsatility through bolus GH infusions have provided compelling evidence for the importance of GH pulsatility in the regulation of a variety of GH targets [121]. For example, pulsatile GH infusion increased parameters of bone formation and resorption more than continuous GH infusion [121]. Furthermore, sexual dimorphism of GH secretion pattern seems to be relevant to the regulation of GH-stimulated P450 liver enzymes [121].

The mechanisms underlying the GH secretion patterns and its effects on human growth and metabolism are poorly understood. Pulsatile or continuous GH patterns may differentially alter GH-mediated control of the signal transducers and transcriptional activators STAT5a and STAT5b, which are ultimately central to several GH effects [122–125].

GH has been found to participate in the regulation of lipolysis, proteolysis, and hepatic glucose production [126], but the specific impact of GH and its pulsatility on fuel mobilization in normal healthy humans has only recently been investigated [127, 128]. GH is recognized as a lipolytic regulator, but fasting-associated increase in lipolysis is commonly attributed to suppression of insulin secretion. Nevertheless, the majority of studies available so far on the role of GH on energy metabolism in humans were performed in GH-deficient patients who were given chronic exogenous GH in a nonphysiological manner. This model, however, is obviously not ideal as compared to the model of normal healthy subjects who are made temporarily and selectively GH-deficient through administration of GHRH receptor antagonist (Fig. 3.2).

The administration of GHRH receptor antagonist to normal humans during physiological overnight fasting has shown no changes in lipolysis, proteolysis, and hepatic glucose production. However, when fasting was prolonged for up to 2 days, the administration of GHRH antagonist prevented the fasting-induced rise in lipolysis despite suppressed insulin secretion. These findings were also coupled to decreased hepatic glucose production and no significant changes in proteolysis [128] (Fig. 3.3). Thus, although GH does not seem to play a critical role in daily feed/fast cycles, it may be an important fuel metabolic regulator when insulin is suppressed, as suggested over 40 years ago [129], constituting a major indispensable mechanism for the starvation-induced increase in the rate of lipolysis. Indeed, insulin suppression alone was not able to accelerate lipolysis when GH secretion was low. These data

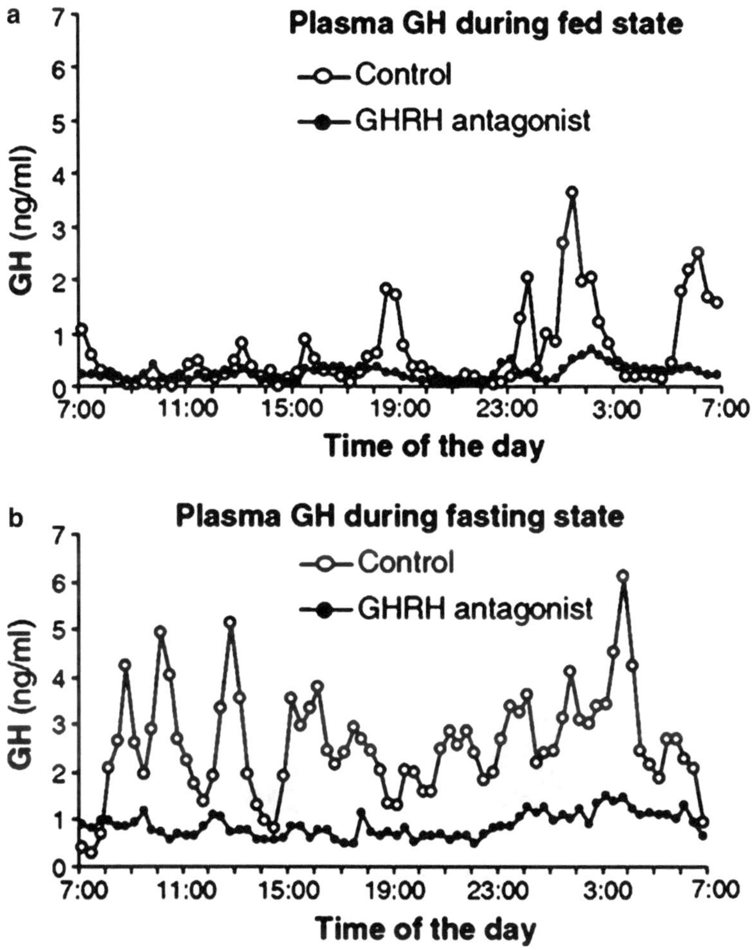

Fig. 3.2 Effects of GHRH-A on 24-h serum GH levels in the fed state (**a**) and during the last 24-h of a 2-day fast (**b**). Only the mean data are shown, and the SE bars are omitted for clarity (reproduced by permission of Sakharova et al. [128])

were completely confirmed in another model, when GH-receptor antagonist was given to fasting humans [130]. Furthermore, since GHRH antagonist abolished GH pulses but otherwise kept a "basal" secretion during fasting state, it could be hypothesized that GH pulsatility per se might have been an underlying mechanism regulating human lipolysis [128].

According to this, other studies found minor activation of lipolysis after continuous GH infusion, but far more robust lipolytic activation with pulsatile GH delivery in lean humans [127].

The questions which then come up are: what is the role of total GH output, GH pulses, and interpulse levels in humans? Which of these are most important to IGF-1

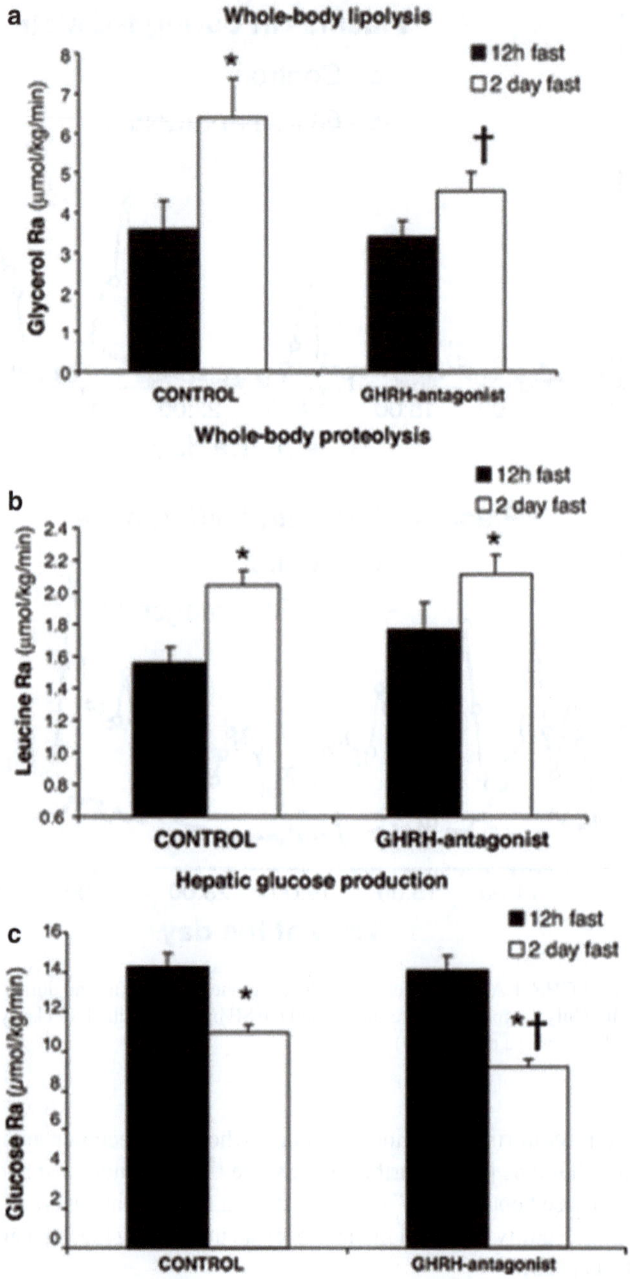

Fig. 3.3 Effects of GHRH-A on rate of whole-body lipolysis (**a**) glycerol Ra, rate of proteolysis (**b**) leucine Ra, and HGP rate (i.e., glucose Ra) measured after an overnight (12-h) fast and again after 2 days of fasting (**c**). *Asterisks* significantly different from 12-h fast ($P < 0.05$); *plus* significantly different from 2-day control (saline) fast value ($P < 0.05$) (reproduced by permission of Sakharova et al. [128])

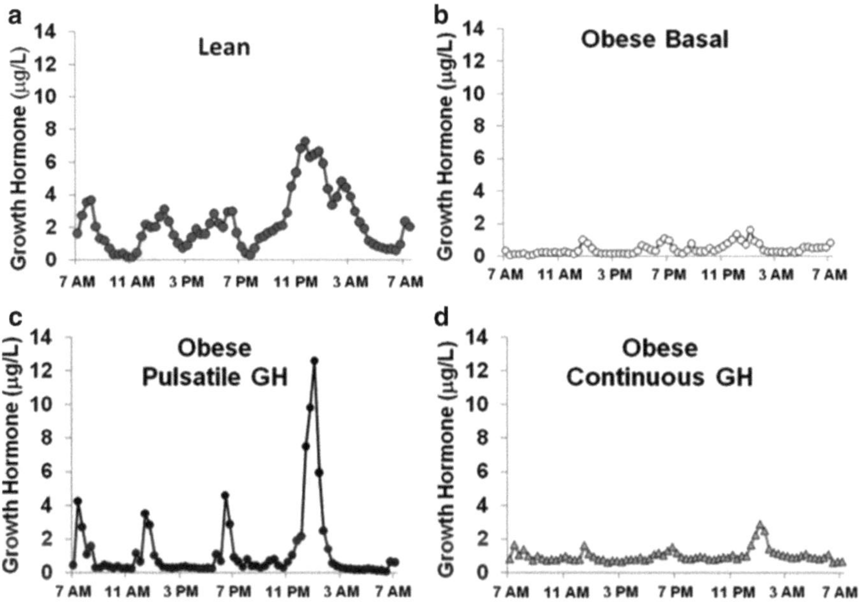

Fig. 3.4 Mean 24-h GH profile for lean subjects under basal conditions (**a**), obese subjects under basal/control conditions (**b**), obese subjects pulsatile GH infusion (**c**), and obese subjects during continuous GH infusion (**d**). Relatively minor variability in timing of the GH pulses among the lean subjects distorts the GH profile to make the peak shapes more broad and less defined. SE bars have been removed for clarity (reproduced by permission of Surya et al. [131])

(i.e., growth) and metabolic effects? Do they play a role in pathological conditions? While definitive answers are still not finally available, some insights coming out from novel human studies have begun to shed light into this field.

A recent study [131] aimed to determine the metabolic roles of GH pulses vs. interpulse GH concentrations in human obesity, a well-documented GH-impoverished state. To this purpose, GH was administered to obese patients in either pulsatile or continuous infusion and compared to both baseline and to lean-control subjects (Fig. 3.4).

The analysis of plasma GH profiles, plasma IGF-1 and skeletal muscle IGF-1mRNA, whole-body lipolytic rate, and glucose metabolism indicates that GH differentially affects various physiological processes in a tissue- and pattern-specific mode in humans. While continuous GH administration was more effective than the pulsatile GH pattern at increasing plasma IGF-1 levels and IGF-1 mRNA abundance, only the pulsatile GH secretion augments the lipolytic rate in obese subjects (Figs. 3.5 and 3.6).

On the other hand, insulin sensitivity was equally affected by either pulsatile or continuous GH administration (Fig. 3.7). Furthermore, a higher lipolytic rate in response to the pulsatile GH infusion could be observed in male subjects, providing further evidence for a sex-related difference in this GH-related parameter in humans.

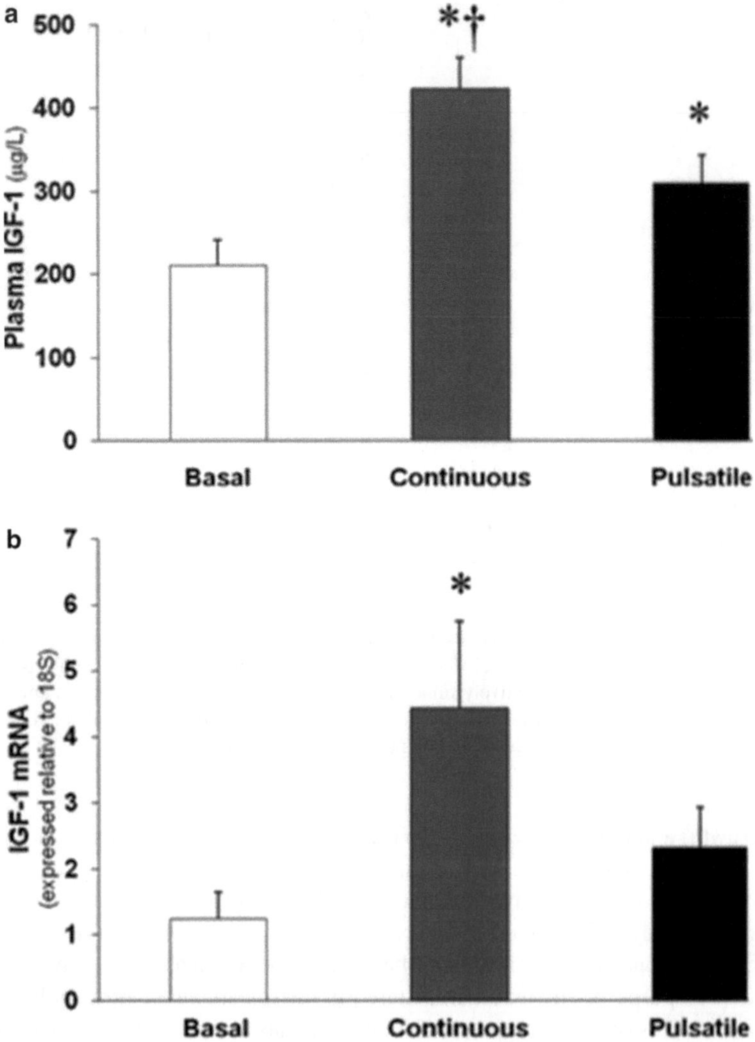

Fig. 3.5 Plasma IGF-1 concentration (**a**) and IGF-1 mRNA abundance (**b**) in skeletal muscle during basal/control conditions, continuous GH infusion, and pulsatile GH infusion. *Asterisks* significantly different than basal conditions, $P<0.05$ (reproduced by permission of Surya et al. [131])

These data show that the manner of GH presentation to the peripheral tissues is an important parameter of GH action and that it is at least in part independent of the average daily GH milieu. The continuous administration (and by inference, the tonic or interpulse component of GH secretion) is largely responsible for increasing plasma IGF-1 concentrations, likely by augmenting hepatic IGF-1 production [132]. Similarly, it is also predominant in increasing muscle IGF-1 mRNA concentrations;

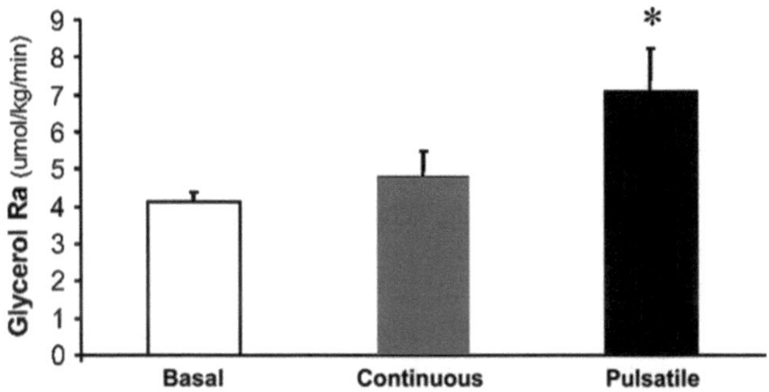

Fig. 3.6 Glycerol rate of appearance in plasma (Ra), an index of whole-body lipolytic rate, during basal/control conditions, continuous GH infusion, and pulsatile GH infusion. *Asterisk* significantly different than basal conditions, $P<0.05$ (reproduced by permission of Surya et al. [131])

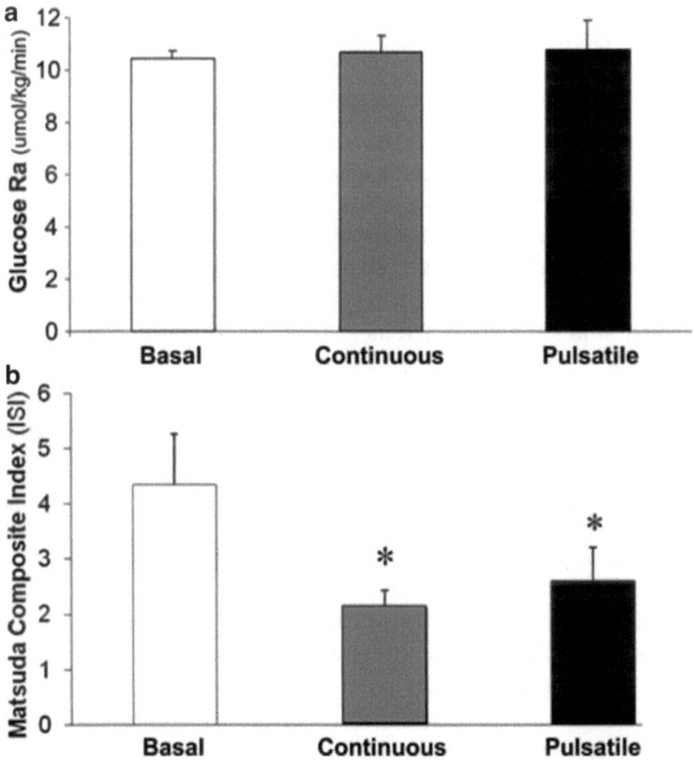

Fig. 3.7 Glucose Ra, an index of hepatic glucose production (**a**) and Matsuda Composite index, as a marker for insulin sensitivity (**b**) during basal/control conditions, continuous GH infusion, and pulsatile GH infusion. *Asterisks* significantly different than basal conditions, $P<0.05$ (reproduced by permission of Surya et al. [131])

as opposed to the data from rodents [3]. We may thus speculate that strategies to improve growth might use GH formulations that maintain elevated steady-state GH levels over long periods of time in order to improve growth-promoting efficacy without requiring supraphysiological GH doses. Indeed, a recent study in dwarf rats [133] subjected to different GH treatment protocols corroborates the hypothesis that growth may be improved by changing scheduled GH treatments.

In contrast, only pulsatile GH administration (and by inference, the pulsatile component of GH secretion) augments adipose tissue lipolysis. Concerning insulin sensitivity, however, it is the total GH output, rather than the mode of administration to humans, which is responsible for the worsening of glucose tolerance. It thus seems that the decline in GH pulsatility occurring in obesity may be protective against lipolysis and, consequently, against further worsening of insulin sensitivity. These data may also explain why men and women have similar IGF-1 and mean daily GH concentrations despite the well-known effect of the estrogen on the somatotropic system (lower IGF-1 and, consequently, higher GH output). Since the inherent pattern of GH secretion in women is characterized by higher basal GH levels, it may maintain normal plasma IGF-1 concentrations and GH output even in the presence of inhibitory estrogen action on IGF-1 synthesis.

It would be interesting to know if these data would also apply to pathological human conditions such as the state of GH excess in acromegaly as compared to healthy controls. To this end, patients and controls have been studied, through frequent 24-h sampling, to ascertain the relative roles of total GH output (24 h mean), GH pulses, and interpulse GH levels in determining plasma IGF-1 levels. Total GH output correlated with serum IGF-1 levels in normal and acromegalic patients, but it was due exclusively to the contribution of valley/nadir GH concentrations for both genders. Conversely, GH pulse mass did not correlate with IGF-1 [134].

These data suggest that the nadir levels of GH are the determining factor of IGF-1 production in normal and acromegalic adults. It may explain, therefore, the mechanism(s) whereby acromegalic patients with low or normal mean 24 h GH levels can generate elevated IGF-1 and also why GH-deficient patients may sometimes maintain normal IGF-1 [135–137]. Lastly, it reinforces the importance of GH kinetics delivery to the GH effects on peripheral tissues, besides raising potential applications for exogenous GH replacement strategies.

The understanding of the kinetic properties of GH secretion and their potential impact on growth and metabolism is still in its infancy. However, enough preliminary evidence has been collected to hypothesize that it is not only the gross quantity of GH output, but also the pattern of presentation of GH to the peripheral tissues that plays an important role in determining its biological properties. Studies in this direction may alter our understanding of GH physiology and action and potentially devise novel and superior strategies to optimize the effects of exogenously administered GH tailored to a specific therapeutic goal.

Acknowledgments AROJr was supported by Fulbright/CAPES institutions. ALB was supported by the R01-DK07955 from the NIH.

References

1. Belchetz PE, Plant TM, Nakai Y, Keogh EJ, Knobil E. Hypophysial responses to continuous and intermittent delivery of hypothalamic gonadotropin-releasing hormone. Science. 1978;202:631–3.
2. Podbesek R, Edouard C, Meunier PJ, Parsons JA, Reeve J, Stevenson RW, et al. Effects of two treatment regimes with synthetic human parathyroid hormone fragment on bone formation and the tissue balance of trabecular bone in greyhounds. Endocrinology. 1983;112:1000–6.
3. Isgaard J, Carlsson L, Isaksson OG, Jansson JO. Pulsatile intravenous growth hormone (GH) infusion to hypophysectomized rats increases insulin-like growth factor I messenger ribonucleic acid in skeletal tissues more effectively than continuous GH infusion. Endocrinology. 1988;123:2605–10.
4. Cheung C, Yu AM, Chen CS, Krausz KW, Byrd LG, Feigenbaum L, et al. Growth hormone determines sexual dimorphism of hepatic cytochrome P450 3A4 expression in transgenic mice. J Pharmacol Exp Ther. 2006;316:1328–34.
5. Waxman DJ, Holloway MG. Sex differences in the expression of hepatic drug metabolizing enzymes. Mol Pharmacol. 2009;76:215–28.
6. Reichlin S. Growth and the hypothalamus. Endocrinology. 1960;67:760–3.
7. Frohman LA, Bernardis LL, Kant KJ. Hypothalamic stimulation of growth hormone secretion. Science. 1968;162:580–2.
8. Frohman LA, Bernardis LL. Growth hormone and insulin levels in weanling rats with ventromedial hypothalamic lesions. Endocrinology. 1968;82:1125–32.
9. Thorner MO, Perryman RL, Cronin MJ, Rogol AD, Draznin M, Johanson A, et al. Somatotroph hyperplasia: successful treatment of acromegaly by removal of a pancreatic islet tumor secreting growth hormone-releasing factor. J Clin Invest. 1982;70:967–77.
10. Sassolas G, Chayvialle JA, Partensky C, Berger G, Trouillas J, Berger F, et al. Acromegaly, clinical expression of the production of growth hormone releasing factor in pancreatic tumors. Ann Endocrinol (Paris). 1983;44:347–54.
11. Ling N, Esch F, Böhlen P, Brazeau P, Wehrenberg WB, Guillemin R. Isolation, primary structure, and synthesis of human hypothalamic somatocrinin: growth hormone-releasing factor. Proc Natl Acad Sci U S A. 1984;81:4302–6.
12. Plotsky PM, Vale W. Patterns of growth hormone-releasing factor and somatostatin secretion into the hypophysial-portal circulation of the rat. Science. 1985;230:461–3.
13. Wehrenberg WB, Brazeau P, Luben R, Böhlen P, Guillemin R. Inhibition of the pulsatile secretion of growth hormone by monoclonal antibodies to the hypothalamic growth hormone releasing factor (GRF). Endocrinology. 1982;111:2147–8.
14. Tannenbaum GS, Eikelboom R, Ling N. Human pancreas GH-releasing factor analog restores high-amplitude GH pulses in CNS lesion-induced GH deficiency. Endocrinology. 1983;113:1173–5.
15. Millard WJ, Martin Jr JB, Audet J, Sagar SM, Martin JB. Evidence that reduced growth hormone secretion observed in monosodium glutamate-treated rats is the result of a deficiency in growth hormone-releasing factor. Endocrinology. 1982;110:540–50.
16. Ferland L, Labrie F, Jobin M, Arimura A, Schally AV. Physiological role of somatostatin in the control of growth hormone and thyrotropin secretion. Biochem Biophys Res Commun. 1976;68:149–56.
17. Ono M, Miki N, Demura H. Effect of antiserum to rat growth hormone (GH)-releasing factor on physiological GH secretion in the female rat. Endocrinology. 1991;129:1791–6.
18. Frohman LA, Downs TR, Clarke IJ, Thomas GB. Measurement of growth hormone-releasing hormone and somatostatin in hypothalamic-portal plasma of unanesthetized sheep. J Clin Invest. 1990;86:17–24.
19. Clark RG, Robinson IC. Growth induced by pulsatile infusion of an amidated fragment of human growth hormone releasing factor in normal and GHRF-deficient rats. Nature. 1985;314:281–3.

20. Borski RJ, Tsai W, Demott-Friberg R, Barkan AL. Induction of growth hormone (GH) mRNA by pulsatile GH-releasing hormone in rats is pattern specific. Am J Physiol Endocrinol Metab. 2000;278:885–91.
21. Vance ML, Kaiser DL, Martha Jr PM, Furlanetto R, Rivier J, Vale W, et al. Lack of in vivo somatotroph desensitization or depletion after 14 days of continuous growth hormone (GH)-releasing hormone administration in normal men and a GH-deficient boy. J Clin Endocrinol Metab. 1989;68:22–8.
22. Russell-Aulet M, Dimaraki EV, Jaffe CA, DeMott-Friberg R, Barkan AL. Aging-related growth hormone (GH) decrease is a selective hypothalamic GH-releasing hormone pulse amplitude mediated phenomenon. J Gerontol A Biol Sci Med Sci. 2001;56:124–9.
23. Jaffe CA, Friberg RD, Barkan AL. Suppression of growth hormone (GH) secretion by a selective GH-releasing hormone (GHRH) antagonist. Direct evidence for involvement of endogenous GHRH in the generation of GH pulses. J Clin Invest. 1993;92:695–701.
24. Jessup SK, Dimaraki EV, Symons KV, Barkan AL. Sexual dimorphism of growth hormone (GH) regulation in humans: endogenous GH-releasing hormone maintains basal GH in women but not in men. J Clin Endocrinol Metab. 2003;88:4776–80.
25. Maheshwari HG, Pezzoli SS, Rahim A, Shalet SM, Thorner MO, Baumann G. Pulsatile growth hormone secretion persists in genetic growth hormone-releasing hormone resistance. Am J Physiol Endocrinol Metab. 2002;282:943–51.
26. Martha Jr PM, Rogol AD, Veldhuis JD, Kerrigan JR, Goodman DW, Blizzard RM. Alterations in the pulsatile properties of circulating growth hormone concentrations during puberty in boys. J Clin Endocrinol Metab. 1989;69:563–70.
27. Ho KY, Evans WS, Blizzard RM, Veldhuis JD, Merriam GR, Samojlik E, et al. Effects of sex and age on the 24-hour profile of growth hormone secretion in man: importance of endogenous estradiol concentrations. J Clin Endocrinol Metab. 1987;64:51–8.
28. Hartman ML, Veldhuis JD, Johnson ML, Lee MM, Alberti KG, Samojlik E, et al. Augmented growth hormone (GH) secretory burst frequency and amplitude mediate enhanced GH secretion during a two-day fast in normal men. J Clin Endocrinol Metab. 1992;74:757–65.
29. Faria AC, Bekenstein LW, Booth Jr RA, Vaccaro VA, Asplin CM, Veldhuis JD, et al. Pulsatile growth hormone release in normal women during the menstrual cycle. Clin Endocrinol (Oxf). 1992;36(6):591–6.
30. Zadik Z, Chalew SA, Kowarski A. The definition of a spontaneous growth hormone (GH) peak: studies in normally growing and GH-deficient children. J Clin Endocrinol Metab. 1992;74:801–5.
31. Barkan AL, Stred SE, Reno K, Markovs M, Hopwood NJ, Kelch RP, et al. Increased growth hormone pulse frequency in acromegaly. J Clin Endocrinol Metab. 1989;69:1225–33.
32. Hartman ML, Veldhuis JD, Vance ML, Faria AC, Furlanetto RW, Thorner MO. Somatotropin pulse frequency and basal concentrations are increased in acromegaly and are reduced by successful therapy. J Clin Endocrinol Metab. 1990;70:1375–84.
33. Asplin CM, Faria AC, Carlsen EC, Vaccaro VA, Barr RE, Iranmanesh A, et al. Alterations in the pulsatile mode of growth hormone release in men and women with insulin-dependent diabetes mellitus. J Clin Endocrinol Metab. 1989;69:239–45.
34. Veldhuis JD, Iranmanesh A, Ho KK, Waters MJ, Johnson ML, Lizarralde G. Dual defects in pulsatile growth hormone secretion and clearance subserve the hyposomatotropism of obesity in man. J Clin Endocrinol Metab. 1991;72:51–9.
35. Jaffe CA, DeMott-Friberg R, Barkan AL. Endogenous growth hormone (GH)-releasing hormone is required for GH responses to pharmacological stimuli. J Clin Invest. 1996;97:934–40.
36. Pandya N, DeMott-Friberg R, Bowers CY, Barkan AL, Jaffe CA. Growth hormone (GH)-releasing peptide-6 requires endogenous hypothalamic GH-releasing hormone for maximal GH stimulation. J Clin Endocrinol Metab. 1998;83:1186–9.
37. Jaffe CA, Ocampo-Lim B, Guo W, Krueger K, Sugahara I, DeMott-Friberg R, et al. Regulatory mechanisms of growth hormone secretion are sexually dimorphic. J Clin Invest. 1998;102:153–64.

38. Billestrup N, Swanson LW, Vale W. Growth hormone-releasing factor stimulates proliferation of somatotrophs in vitro. Proc Natl Acad Sci U S A. 1986;83:6854–7.
39. Cataldi M, Magnan E, Guillaume V, Dutour A, Conte-Devolx B, Lombardi G, et al. Relationship between hypophyseal portal GH-RH and somatostatin and peripheral GH levels in the conscious sheep. J Endocrinol Invest. 1994;17:717–22.
40. Dimaraki EV, Jaffe CA, Demott-Friberg R, Russell-Aulet M, Bowers CY, Marbach P, et al. Generation of growth hormone pulsatility in women: evidence against somatostatin withdrawal as pulse initiator. Am J Physiol Endocrinol Metab. 2001;280:489–95.
41. Dimaraki EV, Jaffe CA, Bowers CY, Marbach P, Barkan AL. Pulsatile and nocturnal growth hormone secretions in men do not require periodic declines of somatostatin. Am J Physiol Endocrinol Metab. 2003;285:163–70.
42. Veldhuis JD, Hudson SA, Bailey JN, Erickson D. Regulation of basal, pulsatile, and entropic (patterned) modes of GH secretion in a putatively low-somatostatin milieu in women. Am J Physiol Endocrinol Metab. 2009;297:483–9.
43. Zheng H, Bailey A, Jiang MH, Honda K, Chen HY, Trumbauer ME, et al. Somatostatin receptor subtype 2 knockout mice are refractory to growth hormone-negative feedback on arcuate neurons. Mol Endocrinol. 1997;11:1709–17.
44. Low MJ, Hammer RE, Goodman RH, Habener JF, Palmiter RD, Brinster RL. Tissue-specific posttranslational processing of pre-prosomatostatin encoded by a metallothionein-somatostatin fusion gene in transgenic mice. Cell. 1985;41(1):211–9.
45. Bass JJ, Gluckman PD, Fairclough RJ, Peterson AJ, Davis SR, Carter WD. Effect of nutrition and immunization against somatostatin on growth and insulin-like growth factors in sheep. J Endocrinol. 1987;112:27–31.
46. Smith RG, Jiang H, Sun Y. Developments in ghrelin biology and potential clinical relevance. Trends Endocrinol Metab. 2005;16:436–42.
47. Mozid AM, Tringali G, Forsling ML, Hendricks MS, Ajodha S, Edwards R, et al. Ghrelin is released from rat hypothalamic explants and stimulates corticotrophin-releasing hormone and arginine-vasopressin. Horm Metab Res. 2003;35:455–9.
48. Bowers CY, Sartor AO, Reynolds GA, Badger TM. On the actions of the growth hormone-releasing hexapeptide, GHRP. Endocrinology. 1991;128:2027–35.
49. Herrington J, Hille B. Growth hormone-releasing hexapeptide elevates intracellular calcium in rat somatotropes by two mechanisms. Endocrinology. 1994;135:1100–8.
50. Kineman RD, Luque RM. Evidence that ghrelin is as potent growth hormone (GH) in releasing GH from primary pituitary cell cultures of a nonhuman primate (Papio anubis), acting through intracellular signaling pathways distinct from GHRH. Endocrinology. 2007;148: 4440–9.
51. Jaffe CA, Ho PJ, Demott-Friberg R, Bowers CY, Barkan AL. Effects of a prolonged growth hormone (GH)-releasing peptide infusion on pulsatile GH secretion in normal men. J Clin Endocrinol Metab. 1993;77:1641–7.
52. Nass R, Farhy LS, Liu J, Prudom CE, Johnson ML, Veldhuis P, et al. Evidence for acyl-ghrelin modulation of growth hormone release in the fed state. J Clin Endocrinol Metab. 2008;93:1988–94.
53. Avram AM, Jaffe CA, Symons KV, Barkan AL. Endogenous circulating ghrelin does not mediate growth hormone rhythmicity or response to fasting. J Clin Endocrinol Metab. 2005;90:2982–7.
54. Corbetta S, Peracchi M, Cappiello V, Lania A, Lauri E, Vago L, et al. Circulating ghrelin levels in patients with pancreatic and gastrointestinal neuroendocrine tumors: identification of one pancreatic ghrelinoma. J Clin Endocrinol Metab. 2003;88:3117–20.
55. Tsolakis AV, Portela-Gomes GM, Stridsberg M, Grimelius L, Sundin A, Eriksson BK, et al. Malignant gastric ghrelinoma with hyperghrelinemia. J Clin Endocrinol Metab. 2004;89: 3739–44.
56. Sun Y, Ahmed S, Smith RG. Deletion of Ghrelin impairs neither nor appetite. Mol Cell Biol. 2003;23:7973–81.

57. Wortley KE, Anderson KD, Garcia K, Murray JD, Malinova L, Liu R, et al. Genetic deletion of ghrelin does not decrease food intake but influences metabolic fuel preference. Proc Natl Acad Sci U S A. 2004;101:8227–32.
58. Wortley KE, del Rincon JP, Murray JD, Garcia K, Iida K, Thorner MO, et al. Absence of ghrelin protects against early-onset obesity. J Clin Invest. 2005;115:3573–8.
59. Zigman JM, Nakano Y, Coppari R, Balthasar N, Marcus JN, Lee CE, et al. Mice lacking ghrelin receptors resist the development of diet-induced obesity. J Clin Invest. 2005;115:3564–72.
60. Pantel J, Legendre M, Cabrol S, Hilal L, Hajaji Y, Morisset S, et al. Loss of constitutive activity of the growth hormone secretagogue receptor in familial short stature. J Clin Invest. 2006;116:760–8.
61. Lumpkin MD, McDonald JK. Blockade of growth hormone-releasing factor (GRF) activity in the pituitary and hypothalamus of the conscious rat with a peptidic GRF antagonist. Endocrinology. 1989;124:1522–31.
62. Lumpkin MD, Mulroney SE, Haramati A. Inhibition of pulsatile growth hormone (GH) secretion and somatic growth in immature rats with a synthetic GH-releasing factor antagonist. Endocrinology. 1989;124:1154–9.
63. Peterfreund RA, Vale WW. Somatostatin analogs inhibit somatostatin secretion from cultured hypothalamus cells. Neuroendocrinology. 1984;39:397–402.
64. Mitsugi N, Arita J, Kimura F. Effects of intracerebroventricular administration of growth hormone-releasing factor and corticotropin-releasing factor on somatostatin secretion into rat hypophysial portal blood. Neuroendocrinology. 1990;51:93–6.
65. Yamauchi N, Shibasaki T, Ling N, Demura H. In vitro release of growth hormone-releasing factor (GRF) from the hypothalamus: somatostatin inhibits GRF release. Regul Pept. 1991;33:71–8.
66. Frohman MA, Downs TR, Chomczynski P, Frohman LA. Cloning and characterization of mouse growth hormone-releasing hormone (GRH) complementary DNA: increased GRH messenger RNA levels in the growth hormone-deficient lit/lit mouse. Mol Endocrinol. 1989;3:1529–36.
67. Chomczynski P, Downs TR, Frohman LA. Feedback regulation of growth hormone (GH)-releasing hormone gene expression by GH in rat hypothalamus. Mol Endocrinol. 1988;2:236–41.
68. Chihara K, Minamitani N, Kaji H, Arimura A, Fujita T. Intraventricularly injected growth hormone stimulates somatostatin release into rat hypophysial portal blood. Endocrinology. 1981;109:2279–81.
69. Ohlsson C, Mohan S, Sjögren K, Tivesten A, Isgaard J, Isaksson O, et al. The role of liver-derived insulin-like growth factor-I. Endocr Rev. 2009;30:494–535.
70. Rosenfeld RG, Hwa V. The growth hormone cascade and its role in mammalian growth. Horm Res. 2009;71 Suppl 2:36–40.
71. Bermann M, Jaffe CA, Tsai W, DeMott-Friberg R, Barkan AL. Negative feedback regulation of pulsatile growth hormone secretion by insulin-like growth factor I. Involvement of hypothalamic somatostatin. J Clin Invest. 1994;94:138–45.
72. Chapman IM, Hartman ML, Pieper KS, Skiles EH, Pezzoli SS, Hintz RL, et al. Recovery of growth hormone release from suppression by exogenous insulin-like growth factor I (IGF-I): evidence for a suppressive action of free rather than bound IGF-I. J Clin Endocrinol Metab. 1998;83:2836–42.
73. Ho PJ, Friberg RD, Barkan AL. Regulation of pulsatile growth hormone secretion by fasting in normal subjects and patients with acromegaly. J Clin Endocrinol Metab. 1992;75:812–9.
74. Darzy KH, Murray RD, Gleeson HK, Pezzoli SS, Thorner MO, Shalet SM. The impact of short-term fasting on the dynamics of 24-hour growth hormone (GH) secretion in patients with severe radiation-induced GH deficiency. J Clin Endocrinol Metab. 2006;91:987–94.
75. Obal Jr F, Krueger JM. GHRH and sleep. Sleep Med Rev. 2004;8:367–77.
76. Van Cauter E, Plat L, Copinschi G. Interrelations between sleep and the somatotropic axis. Sleep. 1998;21:553–66.

77. Holl RW, Hartman ML, Veldhuis JD, Taylor WM, Thorner MO. Thirty-second sampling of plasma growth hormone in man: correlation with sleep stages. J Clin Endocrinol Metab. 1991;72:854–61.
78. Jessup SK, Malow BA, Symons KV, Barkan AL. Blockade of endogenous growth hormone-releasing hormone receptors dissociates nocturnal growth hormone secretion and slow-wave sleep. Eur J Endocrinol. 2004;151:561–6.
79. Jaffe CA, Turgeon DK, Friberg RD, Watkins PB, Barkan AL. Nocturnal augmentation of growth hormone (GH) secretion is preserved during repetitive bolus administration of GH-releasing hormone: potential involvement of endogenous somatostatin – a clinical research center study. J Clin Endocrinol Metab. 1995;80:3321–6.
80. Veldhuis JD, Roemmich JN, Richmond EJ, Bowers CY. Somatotropic and gonadotropic axes linkages in infancy, childhood, and the puberty-adult transition. Endocr Rev. 2006;27: 101–40.
81. Giordano R, Lanfranco F, Bo M, Pellegrino M, Picu A, Baldi M, et al. Somatopause reflects age-related changes in the neural control of GH/IGF-I axis. J Endocrinol Invest. 2005;28:94–8.
82. Veldhuis JD, Hudson SB, Erickson D, Bailey JN, Reynolds GA, Bowers CY. Relative effects of estrogen, age, and visceral fat on pulsatile growth hormone secretion in healthy women. Am J Physiol Endocrinol Metab. 2009;297:367–74.
83. Metzger DL, Kerrigan JR. Estrogen receptor blockade with tamoxifen diminishes growth hormone secretion in boys: evidence for a stimulatory role of endogenous estrogens during male adolescence. J Clin Endocrinol Metab. 1994;79:513–8.
84. Metzger DL, Kerrigan JR. Androgen receptor blockade with flutamide enhances growth hormone secretion in late pubertal males: evidence for independent actions of estrogen and androgen. J Clin Endocrinol Metab. 1993;76:1147–52.
85. Veldhuis JD, Metzger DL, Martha Jr PM, Mauras N, Kerrigan JR, Keenan B, et al. Estrogen and testosterone, but not a nonaromatizable androgen, direct network integration of the hypothalamo-somatotrope (growth hormone)-insulin-like growth factor I axis in the human: evidence from pubertal pathophysiology and sex-steroid hormone replacement. J Clin Endocrinol Metab. 1997;82:3414–20.
86. Racine MS, Symons KV, Foster CM, Barkan AL. Augmentation of growth hormone secretion after testosterone treatment in boys with constitutional delay of growth and adolescence: evidence against an increase in hypothalamic secretion of growth hormone-releasing hormone. J Clin Endocrinol Metab. 2004;89:3326–31.
87. Orrego JJ, Dimaraki E, Symons K, Barkan AL. Physiological testosterone replenishment in healthy elderly men does not normalize pituitary growth hormone output: evidence against the connection between senile hypogonadism and somatopause. J Clin Endocrinol Metab. 2004;89:3255–60.
88. Borski RJ, Tsai W, DeMott-Friberg R, Barkan AL. Regulation of somatic growth and the somatotropic axis by gonadal steroids: primary effect on insulin-like growth factor I gene expression and secretion. Endocrinology. 1996;137:3253–9.
89. Dimaraki EV, Symons KV, Barkan AL. Raloxifene decreases serum IGF-I in male patients with active acromegaly. Eur J Endocrinol. 2004;150(4):481–7.
90. Cozzi R, Barausse M, Lodrini S, Lasio G, Attanasio R. Estroprogestinic pill normalizes IGF-I levels in acromegalic women. J Endocrinol Invest. 2003;26:347–52.
91. Leung KC, Johannsson G, Leong GM, Ho KK. Estrogen regulation of growth hormone action. Endocr Rev. 2004;25:693–721.
92. Leung KC, Doyle N, Ballesteros M, Sjogren K, Watts CK, Low TH, et al. Estrogen inhibits GH signaling by suppressing GH-induced JAK2 phosphorylation, an effect mediated by SOCS-2. Proc Natl Acad Sci U S A. 2003;100:1016–21.
93. Roth J, Glick SM, Yalow RS, Berson SA. Secretion of human growth hormone: physiologic and experimental modification. Metabolism. 1963;12:577–9.
94. Yalow RS, Goldsmith SJ, Berson SA. Influence of physiologic fluctuations in plasma growth hormone on glucose tolerance. Diabetes. 1969;18(6):402–8.

95. Masuda A, Shibasaki T, Nakahara M, Imaki T, Kiyosawa Y, Jibiki K, et al. The effect of glucose on growth hormone (GH)-releasing hormone-mediated GH secretion in man. J Clin Endocrinol Metab. 1985;60:523–6.
96. Broglio F, Benso A, Gottero C, Prodam F, Grottoli S, Tassone F, et al. Effects of glucose, free fatty acids or arginine load on the GH-releasing activity of ghrelin in humans. Clin Endocrinol (Oxf). 2002;57:265–71.
97. Valcavi R, Zini M, Davoli S, Portioli I. The late growth hormone rise induced by oral glucose is enhanced by cholinergic stimulation with pyridostigmine in normal subjects. Clin Endocrinol (Oxf). 1992;37:360–4.
98. Imaki T, Shibasaki T, Shizume K, Masuda A, Hotta M, Kiyosawa Y, et al. The effect of free fatty acids on growth hormone (GH)-releasing hormone-mediated GH secretion in man. J Clin Endocrinol Metab. 1985;60:290–3.
99. Cordido F, Peino R, Peñalva A, Alvarez CV, Casanueva FF, Dieguez C. Impaired growth hormone secretion in obese subjects is partially reversed by acipimox-mediated plasma free fatty acid depression. J Clin Endocrinol Metab. 1996;81:914–8.
100. Pontiroli AE, Lanzi R, Monti LD, Pozza G. Effect of acipimox, a lipid lowering drug, on growth hormone (GH) response to GH-releasing hormone in normal subjects. J Endocrinol Invest. 1990;13:539–42.
101. Maccario M, Grottoli S, Procopio M, Oleandri SE, Rossetto R, Gauna C, et al. The GH/IGF-I axis in obesity: influence of neuro-endocrine and metabolic factors. Int J Obes Relat Metab Disord. 2000;24:96–9.
102. Clasey JL, Weltman A, Patrie J, Weltman JY, Pezzoli S, Bouchard C, et al. Abdominal visceral fat and fasting insulin are important predictors of 24-hour GH release independent of age, gender, and other physiological factors. J Clin Endocrinol Metab. 2001;86:3845–52.
103. Kok P, Buijs MM, Kok SW, Van Ierssel IH, Frölich M, Roelfsema F, et al. Acipimox enhances spontaneous growth hormone secretion in obese women. Am J Physiol Regul Integr Comp Physiol. 2004;286:693–8.
104. Sandhu MS, Gibson JM, Heald AH, Dunger DB, Wareham NJ. Association between insulin-like growth factor-I: insulin-like growth factor-binding protein-1 ratio and metabolic and anthropometric factors in men and women. Cancer Epidemiol Biomarkers Prev. 2004;13:166–70.
105. Hartman ML, Clayton PE, Johnson ML, Celniker A, Perlman AJ, Alberti KG, et al. A low dose euglycemic infusion of recombinant human insulin-like growth factor I rapidly suppresses fasting-enhanced pulsatile growth hormone secretion in humans. J Clin Invest. 1993;91:2453–62.
106. Fletcher TP, Thomas GB, Dunshea FR, Moore LG, Clarke IJ. IGF feedback effects on growth hormone secretion in ewes: evidence for action at the pituitary but not the hypothalamic level. J Endocrinol. 1995;144:323–31.
107. Ghigo E, Procopio M, Boffano GM, Arvat E, Valente F, Maccario M, et al. Arginine potentiates but does not restore the blunted growth hormone response to growth hormone-releasing hormone in obesity. Metabolism. 1992;41:560–3.
108. Alvarez-Castro P, Isidro ML, Garcia-Buela J, Leal-Cerro A, Broglio F, Tassone F, et al. Marked GH secretion after ghrelin alone or combined with GH-releasing hormone (GHRH) in obese patients. Clin Endocrinol (Oxf). 2004;61:250–5.
109. Koutkia P, Schurgin S, Berry J, Breu J, Lee BS, Klibanski A, et al. Reciprocal changes in endogenous ghrelin and growth hormone during fasting in healthy women. Am J Physiol Endocrinol Metab. 2005;289:814–22.
110. Katz LE, DeLeón DD, Zhao H, Jawad AF. Free and total insulin-like growth factor (IGF)-I levels decline during fasting: relationships with insulin and IGF-binding protein-1. J Clin Endocrinol Metab. 2002;87(6):2978–83.
111. Winer LM, Shaw MA, Baumann G. Basal plasma growth hormone levels in man: new evidence for rhythmicity of growth hormone secretion. J Clin Endocrinol Metab. 1990;70(6):1678–86.

112. Ulloa-Aguirre A, Blizzard RM, Garcia-Rubi E, Rogol AD, Link K, Christie CM, et al. Testosterone and oxandrolone, a nonaromatizable androgen, specifically amplify the mass and rate of growth hormone (GH) secreted per burst without altering GH secretory burst duration or frequency or the GH half-life. J Clin Endocrinol Metab. 1990;71:846–54.
113. Surya S, Symons K, Rothman E, Barkan AL. Complex rhythmicity of growth hormone secretion in humans. Pituitary. 2006;9:121–5.
114. Isgaard J, Möller C, Isaksson OG, Nilsson A, Mathews LS, Norstedt G. Regulation of insulin-like growth factor messenger ribonucleic acid in rat growth plate by growth hormone. Endocrinology. 1988;122:1515–20.
115. Wong JH, Dukes J, Levy RE, Sos B, Mason SE, Fong TS, et al. Sex differences in thrombosis in mice are mediated by sex-specific growth hormone secretion patterns. J Clin Invest. 2008;118:2969–78.
116. Waxman DJ, O'Connor C. Growth hormone regulation of sex-dependent liver gene expression. Mol Endocrinol. 2006;20:2613–29.
117. Choi HK, Waxman DJ. Pulsatility of growth hormone (GH) signalling in liver cells: role of the JAK-STAT5b pathway in GH action. Growth Horm IGF Res. 2000;10:S1–8.
118. Bick T, Hochberg Z, Amit T, Isaksson OG, Jansson JO. Roles of pulsatility and continuity of growth hormone (GH) administration in the regulation of hepatic GH-receptors, and circulating GH-binding protein and insulin-like growth factor-I. Endocrinology. 1992;131:423–9.
119. Jørgensen JO, Møller N, Lauritzen T, Christiansen JS. Pulsatile versus continuous intravenous administration of growth hormone (GH) in GH-deficient patients: effects on circulating insulin-like growth factor-I and metabolic indices. J Clin Endocrinol Metab. 1990;70:1616–23.
120. Laursen T, Gravholt CH, Heickendorff L, Drustrup J, Kappelgaard AM, Jørgensen JO, et al. Long-term effects of continuous subcutaneous infusion versus daily subcutaneous injections of growth hormone (GH) on the insulin-like growth factor system, insulin sensitivity, body composition, and bone and lipoprotein metabolism in GH-deficient adults. J Clin Endocrinol Metab. 2001;86:1222–8.
121. Jaffe CA, Turgeon DK, Lown K, Demott-Friberg R, Watkins PB. Growth hormone secretion pattern is an independent regulator of growth hormone actions in humans. Am J Physiol Endocrinol Metab. 2002;283:1008–15.
122. Choi HK, Waxman DJ. Growth hormone, but not prolactin, maintains, low-level activation of STAT5a and STAT5b in female rat liver. Endocrinology. 1999;140:5126–35.
123. Davey HW, Park SH, Grattan DR, McLachlan MJ, Waxman DJ. STAT5b-deficient mice are growth hormone pulse-resistant. Role of STAT5b in sex-specific liver p450 expression. J Biol Chem. 1999;274:35331–6.
124. Park SH, Liu X, Hennighausen L, Davey HW, Waxman DJ. Distinctive roles of STAT5a and STAT5b in sexual dimorphism of hepatic P450 gene expression. Impact of STAT5a gene disruption. J Biol Chem. 1999;274:7421–30.
125. Udy GB, Towers RP, Snell RG, Wilkins RJ, Park SH, Ram PA, et al. Requirement of STAT5b for sexual dimorphism of body growth rates and liver gene expression. Proc Natl Acad Sci U S A. 1997;94:7239–44.
126. Nørrelund H. Consequences of growth hormone deficiency for intermediary metabolism and effects of replacement. Front Horm Res. 2005;33:103–20.
127. Cersosimo E, Danou F, Persson M, Miles JM. Effects of pulsatile delivery of basal growth hormone on lipolysis in humans. Am J Physiol. 1996;271:123–6.
128. Sakharova AA, Horowitz JF, Surya S, Goldenberg N, Harber MP, Symons K, et al. Role of growth hormone in regulating lipolysis, proteolysis, and hepatic glucose production during fasting. J Clin Endocrinol Metab. 2008;93:2755–9.
129. Rabinowitz D, Zierler KL. A metabolic regulating device based on the actions of human growth hormone and of insulin, singly and together, on the human forearm. Nature. 1963;199:913–5.

130. Moller L, Norrelund H, Jessen N, Flyvbjerg A, Pedersen SB, Gaylinn BD, et al. Impact of growth hormone receptor blockade on substrate metabolism during fasting in healthy subjects. J Clin Endocrinol Metab. 2009;94:4524–32.
131. Surya S, Horowitz JF, Goldenberg N, Sakharova A, Harber M, Cornford AS, et al. The pattern of growth hormone delivery to peripheral tissues determines insulin-like growth factor-1 and lipolytic responses in obese subjects. J Clin Endocrinol Metab. 2009;94:2828–34.
132. Yakar S, Liu JL, Stannard B, Butler A, Accili D, Sauer B, et al. Normal growth and development in the absence of hepatic insulin-like growth factor I. Proc Natl Acad Sci U S A. 1999;96:7324–9.
133. Westwood M, Maqsood AR, Solomon M, Whatmore AJ, Davis JR, Baxter RC, Gevers EF, Robinson IC, Clayton PE. The effect of different patterns of growth hormone administration on the IGF axis, somatic and skeletal growth of the dwarf rat. Am J Physiol Endocrinol Metab. 2009;27. doi:10.1152/ajpendo.00234.2009.
134. Faje AT, Barkan AL. Basal, but not pulsatile, GH secretion determines the ambient circulating levels of IGF-1. J Clin Endocrinol Metab. 2010;95:2486–91.
135. Dimaraki EV, Jaffe CA, DeMott-Friberg R, Chandler WF, Barkan AL. Acromegaly with apparently normal GH secretion: implications for diagnosis and follow-up. J Clin Endocrinol Metab. 2002;87:3537–42.
136. Peacey SR, Toogood AA, Veldhuis JD, Thorner MO, Shalet SM. The relationship between 24-hour growth hormone secretion and insulin-like growth factor I in patients with successfully treated acromegaly: impact of surgery or radiotherapy. J Clin Endocrinol Metab. 2001;86:259–66.
137. Reutens AT, Veldhuis JD, Hoffman DM, Leung KC, Ho KK. A highly sensitive growth hormone (GH) enzyme-linked immunosorbent assay uncovers increased contribution of a tonic mode of GH secretion in adults with organic GH deficiency. J Clin Endocrinol Metab. 1996;81:1591–7.

Chapter 4
Metabolic Actions of Growth Hormone

Morton G. Burt

Abstract Growth hormone (GH) and insulin-like growth factor-1 (IGF-1) exert profound effects on energy, lipid, protein, and carbohydrate metabolism. GH increases energy expenditure, lipolysis, and lipid oxidation, actions that account for the reduction in fat mass during GH replacement. The increase in lean body and skeletal muscle mass induced by GH is mediated by an increase in protein synthesis and a reduction in irreversible oxidative loss of amino acids. While GH excess causes insulin resistance, the effect of lower doses of GH on carbohydrate metabolism is variable and dependent on changes in lipolysis, abdominal adiposity, and IGF-1.

IGF-1 augments some of the metabolic effects of GH, but counteracts others. Like GH, IGF-1 increases protein mass by reducing protein oxidation and increasing protein synthesis. However, IGF-1 has an insulin-like effect on carbohydrate metabolism that partially offsets the effect of GH excess. The effect of IGF-1 on lipid metabolism is likely to be indirectly mediated by changes in GH and insulin secretion.

Finally, there is not a single effect of GH and IGF-1 on lipid, protein, and carbohydrate metabolism. Rather, there is a spectrum of effects determined by dose, duration of use, and underlying disease state. Integration of many variables is needed to explain the metabolic actions of GH and IGF-1 in acromegaly and GH deficiency and the effect of GH replacement in the latter.

Keywords Resting energy expenditure • Lipolysis • Lipid oxidation • Protein metabolism • Insulin sensitivity

---

M.G. Burt (✉)
Southern Adelaide Diabetes and Endocrine Services, Repatriation General Hospital and Flinders University, Daws Road, Adelaide, SA 5041, Australia
e-mail: morton.burt@health.sa.gov.au

K. Ho (ed.), *Growth Hormone Related Diseases and Therapy: A Molecular and Physiological Perspective for the Clinician*, Contemporary Endocrinology,
DOI 10.1007/978-1-60761-317-6_4, © Springer Science+Business Media, LLC 2011

Introduction

Growth hormone (GH) was named because of its potent action to stimulate longitudinal growth in children with GH deficiency, for which it has been used therapeutically for over 50 years [1]. However, early studies of GH action demonstrated it also exerted important effects on lipid, protein, and carbohydrate metabolism [2–4]. It was not until 1985 that the availability of recombinant GH provided sufficient quantities of GH to explore its metabolic action in detail. The first randomized controlled studies in adults with GH deficiency, published in 1989, demonstrated that GH is an important metabolic hormone in adults [5, 6]. The subsequent two decades saw an extensive body of literature demonstrating the metabolic actions of GH. Changes in energy, lipid, protein, and carbohydrate metabolism during GH deficiency and excess affect body composition, plasma glucose and lipid concentrations and probably risk of cardiovascular disease. Many of the metabolic abnormalities found in GH deficiency are reversed by GH replacement, providing the rationale for GH therapy in adults.

In the original somatomedin hypothesis, GH exerted its effect following stimulation of circulating insulin-like growth factor-1 (IGF-1) [7]. The situation was subsequently shown to be more complex, with IGF-1 also having autocrine and paracrine effects and GH having IGF-1 independent actions [8]. This complexity is evident when considering the metabolic action of GH. While some of the metabolic actions of IGF-1 are similar to GH, others are the opposite. When both GH and IGF-1 are perturbed, the metabolic effects of GH usually dominate.

This chapter reviews studies investigating the metabolic actions of GH and IGF-1 in humans, with particular focus on the metabolic changes in acromegaly and GH deficiency and the effect of GH replacement in the latter. The changes in energy/lipid and protein metabolism induced by GH and IGF-1 are related to changes in fat and protein mass, respectively. The effects of GH and IGF-1 on carbohydrate metabolism are then reviewed. These studies demonstrate a complex range of metabolic actions that are influenced by GH dose, duration of GH deficiency/excess, and underlying disease state.

Energy and Lipid Metabolism

Introduction

The effect of GH on total and visceral abdominal fat is important because of their strong association with insulin resistance and cardiovascular disease. Changes in energy and lipid metabolism exert a major effect on fat mass. Energy metabolism refers to the generation of energy from the oxidative metabolism of body fuels. The body stores energy in the form of adipose tissue when the dietary intake of energy exceeds energy expenditure. As such, an increase in energy metabolism reduces the amount of calories that are stored as fat and vice versa. Lipid metabolism includes the breakdown of triglycerides prior to uptake into adipose and muscle cells, lipolysis of triglycerides in adipose tissue, and oxidation of free fatty acids (FFA) in hepatic and muscle cells. Thus, changes in lipid metabolism influence both the size of fat depots and circulating lipid components.

GH/IGF-1 and Fat Mass

GH has a major effect on fat mass. Total fat mass is approximately 20–40% greater in adults with GH deficiency than age-matched controls [9, 10], with visceral fat increased to a greater extent than other fat depots [11]. Randomized controlled studies of GH replacement in subjects with GH deficiency demonstrate a reduction in fat mass by an average of 3 kg [12]. Typically, fat mass reduces over the first 12 months of GH replacement and then plateaus [13], with the greatest reduction in fat mass occurring in visceral adipose tissue [11, 14].

Changes in fat mass with GH excess are the opposite of those in GH deficiency. In acromegaly, total fat mass is about 15% lower than that in healthy subjects [15]. Visceral abdominal adipose tissue is reduced more than subcutaneous abdominal adipose tissue, while intramuscular adipose tissue is greater in acromegaly [16]. Treatment of acromegaly that reduces GH secretion increases fat mass [15].

The effect of IGF-1 per se on fat mass has been assessed in small studies of patients with GH deficiency or resistance. During the first few months after administration of IGF-1, there is a reduction in fat mass [17–19]. However, during longer-term therapy with IGF-1, waist-hip ratio and fat-fold thickness increase [20]. A small study that characterized the long-term effect of IGF-1 on fat mass reported it is increased to 2–3 times predicted [21]. These data suggest that IGF-1 acutely reduces fat mass, but that this is not sustained during longer-term treatment and fat mass markedly increases.

GH/IGF-1 and Energy Metabolism

GH-induced changes in energy metabolism contribute to its effect on fat mass. Resting energy expenditure (REE) is higher in patients with acromegaly than in healthy controls [15, 22, 23] and is reduced by successful treatment [23, 24]. Similarly, REE is increased by acute GH administration to healthy adults [25] and adults with GH deficiency [26]. Furthermore, REE is reduced by acute GH receptor blockade in normal subjects [27].

The rate of REE in GH deficiency is less clear. Early studies reported that REE was not significantly different in subjects with GH deficiency and normal controls [9, 26, 28]. However, the sample size in the above studies was small and they may have been underpowered to detect a significant change in REE. In two of the above studies, REE was lower in subjects with GH deficiency, although the differences were not statistically significant [26, 28]. A larger study recently reported that REE was significantly lower in subjects with GH deficiency by 9% [10], consistent with the nonsignificant reduction in smaller studies [26, 28], the effect of GH administration on REE, and the findings in acromegaly.

The mechanisms by which GH stimulates REE have been partially elucidated. REE is strongly correlated to lean body mass (LBM), so any change in LBM will result in a parallel change in REE. However, GH also directly stimulates REE, as the GH-induced increase in REE persists after correction for the increase in LBM [26]. GH increases skeletal muscle uncoupling protein-2 [29] and stimulates conversion of peripheral thyroxine to tri-iodothyronine [30], which are likely to contribute to an increase in REE. However, the extent to which these mechanisms explain changes in REE has not been determined.

The effect of IGF-1 on REE is similar to GH. When administered to healthy [31] and GH-deficient adults [32], IGF-1 increases REE. The increase in REE with combined GH and IGF-1 therapy is greater than that induced by GH alone [32]. It is not clear whether the effect of GH on REE is mediated by IGF-1 or whether GH and IGF-1 increase REE independently.

GH and Lipolysis

GH exerts a major effect on lipid metabolism. A single pulse of GH in normal adults stimulates lipolysis and ketogenesis with FFA concentration peaking 2–3 h after GH administration [33]. Consistent with these findings, acute GH receptor antagonism reduces FFA and ketogenesis [27] and GH-releasing hormone antagonism inhibits the characteristic increase in lipolysis during fasting [34]. Lipolysis and ketogenesis increase in a dose-dependent manner across the physiologic range of GH concentrations, with a plateau above this range [35]. GH stimulates lipolysis in both femoral and abdominal subcutaneous adipose tissue [36]. No studies have examined the lipolytic effect of GH in visceral adipose tissue, but as visceral fat mass is predominantly reduced during GH treatment, this adipose depot is likely to be GH-responsive [37].

GH-induced stimulation of lipolysis is likely to play an important role in normal physiology. The time between exogenous GH administration and FFA peak matches the interval between the nocturnal peak in GH and the early morning rise in FFA [38]. Furthermore, suppression of the nocturnal rise in GH with octreotide reduces subcutaneous abdominal tissue lipolysis [39]. These findings suggest that GH is a major regulator of the circadian changes in lipid release.

Few studies have investigated whether GH-induced changes in lipolysis are sustained during long-term therapy. In healthy elderly subjects, subcutaneous abdominal tissue lipolysis was not stimulated after 12 weeks of GH [40]. In contrast, during a hyperinsulinemic clamp, FFAs and some lipid intermediates were increased in acromegaly [23]. As the reduction in fat mass plateaus during long-term GH replacement, it is possible that GH-induced changes in lipolysis may be attenuated over time.

GH induces changes in key enzymes that regulate the uptake and release of FFAs in adipose tissue. The lipolytic effect of GH is predominantly mediated by hormone-sensitive lipase, the enzyme responsible for breakdown of triglycerides into FFA in adipose tissue, as inhibition of hormone-sensitive lipase by acipimox suppresses GH-induced lipolysis [41, 42]. GH also inhibits lipoprotein lipase, the major enzyme responsible for breakdown of triglycerides into FFAs prior to uptake by adipocytes [43]. Therefore, GH reduces uptake and increases output of FFA in adipose tissue.

GH and Lipid Oxidation

Studies using indirect calorimetry demonstrate that GH acutely stimulates lipid oxidation in healthy adults [44, 45]. Short-term administration of a GH receptor antagonist to healthy adults results in the opposite effect; lipid oxidation is reduced [27]. The rate of lipid oxidation is increased after 2–12 weeks of GH administration in healthy adults [25, 40] and subjects with GH deficiency [26].

It is less certain whether acute GH-induced changes in lipid oxidation are sustained. Basal [10] and exercise-induced [46] lipid oxidation appear lower in subjects with GH deficiency, but not all studies demonstrate a significant difference from controls [9]. Some, but not all, studies demonstrate an ongoing effect on lipid oxidation during long-term GH therapy. Withdrawal of GH therapy in GH-deficient adults was associated with a sustained reduction in lipid oxidation at 12 months [47] and lipid oxidation during exercise was increased after 12 months of GH replacement [46]. In contrast, other data suggest that the initial increase in lipid oxidation after 6 weeks of GH replacement is attenuated after 6 months [26].

Similarly, the rate of lipid oxidation in acromegaly is not clearly established. O'Sullivan et al. reported that basal lipid oxidation was not significantly different from healthy controls in acromegaly and that postprandial lipid oxidation was paradoxically reduced [22]. In contrast, another study reported lipid oxidation was increased in acromegaly and reduced by successful treatment [23]. Measurement of lipid oxidation using indirect calorimetry has poor day-to-day reproducibility, which may contribute to the discordant results in these studies with a relatively small sample size.

The site of increased lipid oxidation is uncertain. Skeletal muscle is the major site of lipid oxidation, but one study reported that genes responsible for lipid oxidation are downregulated in skeletal muscle during acute GH replacement [48]. Further work is needed to clarify the relative contributions of skeletal muscle and liver to GH-induced changes in lipid oxidation.

IGF-1 and Lipolysis

The increase in fat mass with long-term IGF-1 administration is probably predominantly secondary to stimulation of differentiation of adipocytes, which has been well characterized in vitro [49]. As mature adipocytes have few functional type 1

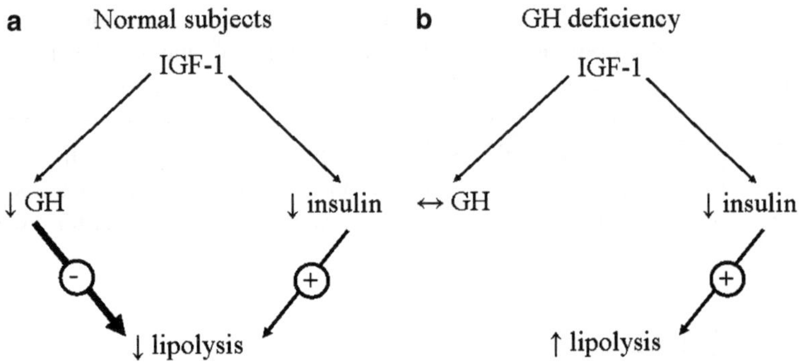

Fig. 4.1 Proposed mechanisms for the contrasting effects of IGF-1 on lipolysis in normal subjects (**a**) and subjects with GH deficiency (**b**). See text for details

IGF-1 receptors, IGF-1 is unlikely to exert a direct effect on lipolysis in adipocytes [50–52]. Despite these observations, short-term administration of IGF-1 in healthy adults decreases FFAs, beta-hydroxybutyrate, and whole-body palmitate flux, reflective of decreased lipolysis and ketogenesis [53–55]. In contrast, IGF-1 increases lipolysis in subjects with GH deficiency [32] or insensitivity [19]. These variable changes in lipolysis are probably indirectly mediated by IGF-1-induced changes in other hormones that affect lipolysis.

Differential effects of IGF-1 on GH and insulin secretion provide an explanation for the discordant changes in lipolysis in normal and GH-deficient subjects. In healthy subjects, IGF-1 administration reduces both GH, which stimulates, and insulin, which inhibits lipolysis. The reduction in lipolysis in healthy subjects administered IGF-1 is likely to occur because of a reduction in GH-induced lipolysis, which exerts a greater effect than insulin (Fig. 4.1a). In contrast, in subjects with GH deficiency, GH is already low and unlikely to fall further during IGF-1 administration. As GH does not change, the main effect on lipolysis with IGF-1 is secondary to a reduction in insulin-induced inhibition of lipolysis, resulting in an increase in lipolysis (Fig. 4.1b). In a study of subjects with type 1 diabetes, where plasma insulin was clamped and GH secretion inhibited by octreotide, administration of IGF-1 did not affect lipolysis [56].

IGF-1 and Lipid Oxidation

Similar to its effect on lipolysis, IGF-1 produces variable changes in lipid oxidation depending on GH status. Acute IGF-1 decreases lipid oxidation in healthy adults [57]. In contrast, lipid oxidation was either unaffected [18] or increased [32] by IGF-1 administration in subjects with GH deficiency and was increased in adults with GH receptor deficiency [19]. As changes in lipid oxidation usually parallel changes in lipolysis [41, 42, 58], variable effects on lipolysis and FFA availability could underlie the discordant effects of IGF-1 in normal and GH-deficient or resistant subjects.

Lipids

Changes in lipid metabolism may directly contribute to the lipid abnormalities associated with changes in GH status. As would be expected in a setting of sustained stimulation of lipolysis, hypertriglyceridemia is the major lipid abnormality in acromegaly and is corrected by biochemical control of GH excess [23, 59, 60]. Subjects with GH deficiency have increased total and LDL-cholesterol and reduced HDL-cholesterol, but no change in triglycerides [61]. In a large meta-analysis, GH replacement in subjects with GH deficiency significantly reduced total and LDL-cholesterol, but did not affect HDL-cholesterol or triglycerides [12]. It is likely that indirect effects of GH deficiency on body composition and carbohydrate metabolism contribute to changes in plasma lipids in GH deficiency and during replacement.

Summary

GH acutely induces changes in energy and lipid metabolism which are consistent with an acute reduction in fat mass. GH stimulates REE, lipolysis (mediated by hormone-sensitive lipase), and lipid oxidation. Furthermore, it reduces uptake of triglyceride into adipose depots by inhibiting lipoprotein lipase. It is less clear whether the acute GH-induced increase in lipolysis and lipid oxidation is sustained long-term. Attenuation of the effect of GH on lipid metabolism may explain the plateau in fat mass during long-term GH replacement after an initial decrease.

Stimulation of REE, lipolysis, and lipid oxidation by IGF-1 in GH deficiency and GH resistance is consistent with an early reduction in fat mass. However, these mechanisms do not explain the increase in fat mass during long-term IGF-1 administration. Stimulation of adipocyte differentiation by IGF-1 is a likely mechanism that counteracts the metabolic action of IGF-1 on fat mass during long-term treatment.

Protein Metabolism

Introduction

Studies of body composition and protein metabolism produce complementary information on protein mass. Measurement of LBM by dual-energy X-ray absorptiometry (DXA) provides an estimate of whole-body protein mass. As GH is an antinatiuretic hormone, some studies have combined DXA with an estimate of extracellular water (ECW), which is subtracted from LBM to estimate body cell mass. This provides a better estimate of protein mass during changes in GH status, as GH-induced changes in ECW are accounted for. There is also great interest in the effect of GH on skeletal muscle mass, which can be quantified using computed tomography or magnetic resonance imaging.

Protein metabolism is also assessed at the "whole body" level and in skeletal muscle. Whole-body protein turnover studies employ intravenously infused isotopically labeled amino acids as tracers to quantify net rates in all tissues of protein breakdown, oxidation, and synthesis, the key components of protein metabolism. However, they provide limited information on skeletal muscle protein metabolism, which turns over relatively slowly and only accounts for a proportion of whole-body protein turnover [62]. Therefore, stable isotopic techniques have been developed that estimate skeletal muscle protein metabolism, including measurement of amino acid incorporation into muscle reflecting muscle protein synthesis, measurement of limb arteriovenous amino acid differences to quantify muscle breakdown and synthesis, and techniques that combine these two approaches [63].

GH/IGF-1 and Protein Mass

Subjects with GH deficiency have less LBM than healthy controls, which includes both a reduction in ECW and body cell mass [9]. Similarly, skeletal muscle volume and strength are lower in subjects with GH deficiency [64]. In a meta-analysis of randomized controlled studies, GH significantly increased LBM in subjects with GH deficiency [12]. The increase in LBM includes both an increase in ECW and body cell mass [14] and occurs within the first 12 months of GH therapy with a plateau thereafter [65, 66]. Skeletal muscle mass and strength are also slightly but significantly increased by GH in subjects with GH deficiency [64, 67].

It is less certain whether elevating GH into the supraphysiologic range increases protein mass. In acromegaly, LBM is greater than in matched controls [68], but the increase is predominantly secondary to an increase in ECW [15]. There is no significant difference in skeletal muscle mass between subjects with acromegaly and matched controls [68]. Similarly, administration of supraphysiologic GH to healthy adults increases LBM, but not skeletal muscle mass and strength [69–72]. These findings suggest that either the increase in LBM with supraphysiologic GH is an increase in ECW or an increase of protein in tissues other than skeletal muscle.

Short-term studies report that IGF-1 also increases LBM in subjects with GH insufficiency [19, 73] and deficiency [18]. The long-term effect of IGF-1 on protein mass and its effect on skeletal muscle have not been studied.

GH and Protein Metabolism

In healthy subjects after an overnight fast, approximately 15–20% of amino acids are oxidized and 80–85% reincorporated into protein (Fig. 4.2a). In subjects with GH deficiency, administration of GH for 1 week to 2 months increases postabsorptive whole-body protein synthesis [18, 74, 75]. A similar increase has been reported in GH-sufficient subjects administered supraphysiologic GH [76, 77] and postprandially in GH-deficient subjects [78]. Some studies, particularly using higher GH doses, report that whole-body protein breakdown is also increased [18, 74]. However, as stimulation of protein synthesis exceeds breakdown, irreversible oxidative loss of amino acids is reduced (Fig. 4.2b). The reduction in protein oxidation is predictive of the increase in LBM during GH replacement [79].

The effect of a chronic change in GH status on whole-body protein metabolism differs from that during acute GH administration. Whole-body protein breakdown and synthesis are proportionately reduced in subjects with chronic GH deficiency [80, 81] and increased in acromegaly [82], after correction for differences in body composition (Fig. 4.2c, d). Therefore, in both settings, LBM-corrected protein oxidation is not significantly different from controls [80–82], which is not consistent with an ongoing change in protein mass. Hoffman et al. hypothesized that normalization of protein oxidation during chronic GH deficiency represented a metabolic adaptation that stabilized protein mass at a new, but reduced, steady state [81]. Evidence to support this concept has emerged from studies of GH replacement that report the acute reduction in protein oxidation is attenuated during longer-term treatment [79, 83]. This mechanism would explain the plateau in LBM during GH replacement.

The action of GH in skeletal muscle is less well established, but most data suggest that GH increases muscle mass by increasing protein synthesis. An early study reported that direct infusion of GH into the brachial artery increased skeletal muscle protein synthesis, but not breakdown [84]. Studies that have examined the acute effect of systemic GH administration in healthy adults have reported discordant results, with both an increase in skeletal muscle protein synthesis [85] and no change [86] reported. Very high GH doses (~2 mg/day) increased skeletal muscle protein synthesis in malnourished patients with chronic renal failure [87] and elderly women [88]. Skeletal muscle protein metabolism in acromegaly and GH deficiency and the long-term effect of GH replacement on skeletal muscle protein metabolism have not been characterized.

The mechanisms underlying the protein anabolic action of GH remain to be fully elucidated. While changes in availability of glucose and amino acids may play a role, most investigation has focused on the hypothesis that GH-induced lipolysis produces FFA substrate for oxidation and spares protein, thereby indirectly exerting an anabolic effect. There is an inverse relationship between FFA and GH secretion [89, 90] and action in skeletal muscle [91], suggesting that GH and FFA are physiologically related via a negative feedback mechanism. In healthy subjects, infusion of beta-hydroxybutyrate induces an anabolic response

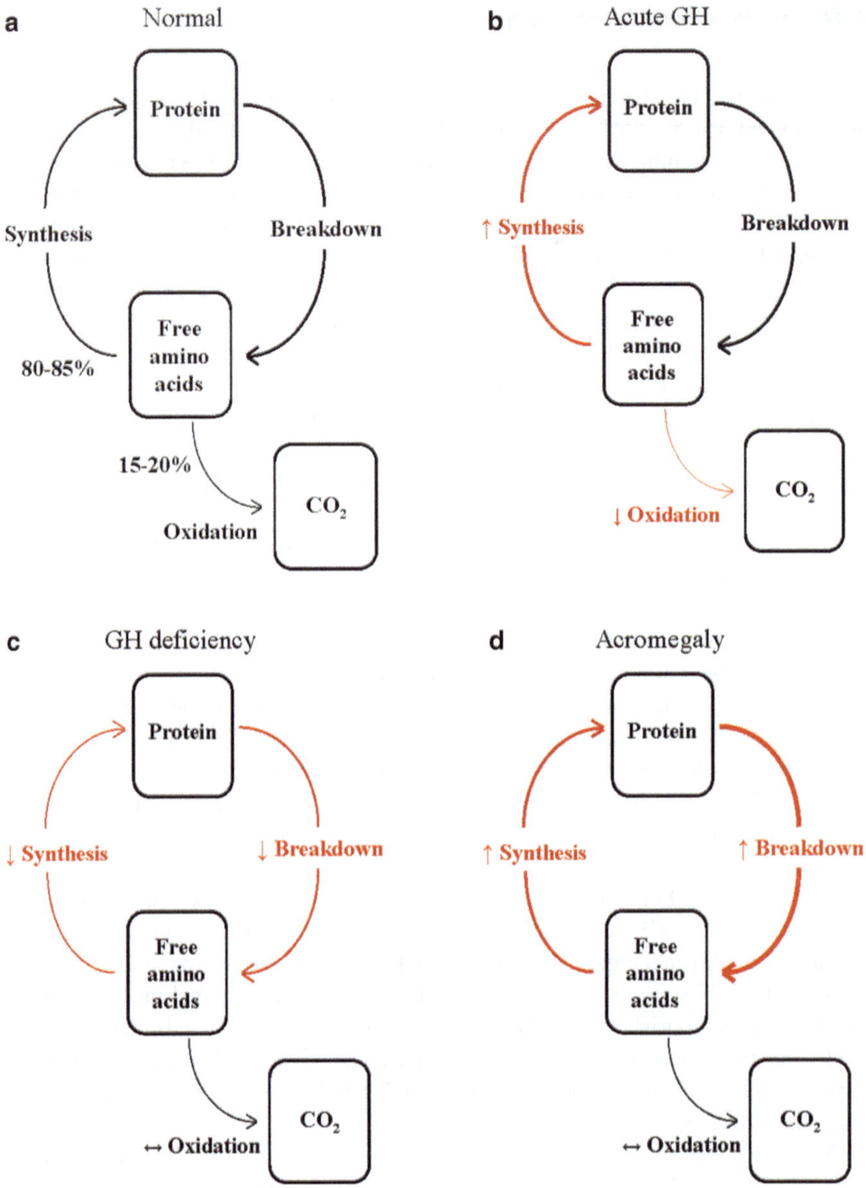

Fig. 4.2 Models of whole-body protein metabolism in normal subjects (**a**), subjects receiving acute GH replacement (**b**), chronic GH deficiency (**c**), and acromegaly (**d**)

in skeletal muscle [92], while inhibition of lipolysis by acipimox has the opposite effect [93]. However, in subjects with GH deficiency receiving GH replacement, inhibition of lipolysis stimulated protein breakdown and synthesis equally, resulting in no effect on net protein balance [94]. Thus, although changes in lipolysis

Table 4.1 Effect of changes in GH status on whole body and skeletal muscle protein metabolism and mass

	Whole body		Skeletal muscle	
	Oxidation	LBM	Synthesis	Mass
Acute GH replacement	↓	↑	↔/↑	↑
Chronic GH replacement	↔	↔	?	?
GH deficiency	↔	↓ (? plateau)	?	↓
Acromegaly	↔	↑ (? plateau)	?	↔

LBM lean body mass; ↑ increased; ↔ unchanged; ↓ reduced; ? unknown

and FFA concentration clearly affect protein metabolism, it is not clear to what extent increased FFA flux contributes to the anabolic action of GH.

IGF-1 and Protein Metabolism

Early studies using high doses of IGF-1 reported a reduction in whole-body protein breakdown and oxidation with no effect on synthesis, suggesting an insulin-like effect on protein metabolism [53, 95]. However, the physiologic relevance of these studies was questioned because amino acid concentrations were reduced to below the normal range [96]. When amino acid are "clamped" at physiologic concentrations, the effect of IGF-1 was similar to GH and not insulin; whole-body protein synthesis is increased with no effect on breakdown [96]. Subsequent studies of usual IGF-1 replacement doses in subjects with GH deficiency [18] and GH resistance [19] also report a reduction in whole-body protein oxidation, increase in protein synthesis, and no significant change in protein breakdown. Furthermore, studies of skeletal muscle protein metabolism reveal that the predominant effect of IGF-1 is to increase protein synthesis [97, 98]. In summary, administration of standard therapeutic doses of IGF-1 reduces protein oxidation and increases protein synthesis. At higher IGF-1 doses, protein breakdown is reduced if amino acid supply is not maintained.

Summary

Changes in protein metabolism and protein mass induced by GH are closely related (Table 4.1). Acute GH replacement in subjects with GH deficiency reduces whole-body protein oxidation and increases skeletal muscle protein synthesis, leading to increased lean body and skeletal muscle mass. During long-term GH replacement, the reduction in protein oxidation is attenuated, which is likely to represent a metabolic adaptation that stabilizes protein mass at a new steady state. A similar mechanism is likely to occur in acromegaly and chronic GH deficiency, with a normal rate of protein oxidation and a stable increased and reduced LBM, respectively. Finally, IGF-1 exerts a similar effect to GH on protein metabolism and mass, and at least in part, mediates its anabolic action.

Carbohydrate Metabolism

Introduction

Seventy years ago, Houssay reported that hypophysectomy improved the hyperglycemia of experimental diabetes in dogs, suggesting that the pituitary gland secreted a hormone that opposed insulin action [2]. It is now recognized that these changes in insulin sensitivity are mediated by GH. When acutely administered in supraphysiologic doses, GH antagonizes the action of insulin, resulting in an increase in hepatic glucose production and a reduction in carbohydrate oxidation and hepatic and peripheral insulin sensitivity [99, 100]. IGF-1 exerts the opposite effect on carbohydrate metabolism to GH and improves insulin sensitivity [32]. The opposing effects of GH and IGF-1 cause complex changes in carbohydrate metabolism which are influenced by the dose of GH administered and duration of therapy.

Randle et al. first proposed that high FFA concentrations induce insulin resistance [101]. As GH both stimulates lipolysis and causes insulin resistance, it was hypothesized that the two metabolic actions were causally linked. This is strongly supported by the observation that inhibition of lipolysis by acipimox increases insulin-stimulated glucose uptake and glucose oxidation in GH-deficient subjects receiving acute [102] and chronic [41, 42] GH replacement. Coadministration of acipimox does not reduce hepatic glucose production, suggesting that hepatic glucose output is directly mediated by GH, while the reduction in peripheral insulin sensitivity is secondary to stimulation of lipolysis [103].

Acromegaly

Acromegaly represents a model of chronic GH and IGF-1 excess. In this setting, the changes in carbohydrate metabolism parallel those reported during acute GH administration; hepatic glucose production is increased and glucose oxidation and peripheral glucose uptake decreased [23, 104]. As such, the prevalence of diabetes is high in acromegaly, ranging from 19 to 56% in different series [105]. Insulin resistance in acromegaly develops despite an increase in circulating IGF-1, which acts independently of GH to counterbalance GH-induced insulin resistance [106]. However, the effect of GH clearly dominates.

The different therapeutic options for acromegaly exert differing effects on carbohydrate metabolism. Successful surgical treatment reverses the adverse effects of GH on carbohydrate metabolism [23, 104]. The effect of somatostatin analogs is more complex as they reduce insulin as well as GH secretion, which would be expected to increase postprandial glucose concentration and improve insulin sensitivity, respectively. In general, improvement in glycemia occurs in those with impaired glucose tolerance, whereas a mild deterioration may occur in those with normal glucose tolerance [107]. Glycemia predominantly improves in patients who achieve biochemical control of GH secretion [108]. Pegvisomant reduces hepatic glucose production and improves peripheral insulin sensitivity [24, 109] and results in favorable change in glucose homeostasis compared to somatostatin analogs [110, 111].

GH Deficiency and Replacement

Untreated children with GH deficiency are prone to hypoglycemia. Thus, it was initially considered paradoxical that insulin resistance is increased in adults with GH deficiency. However, insulin resistance probably develops because of increased abdominal adiposity. Patients with GH deficiency have reduced peripheral insulin-stimulated glucose storage and muscle glycogen synthase activity, but no difference has been noted in hepatic glucose production from controls [112]. As the rate of carbohydrate oxidation is greater in subjects with GH deficiency [10], the reduction in peripheral insulin sensitivity is likely to arise from a reduction in nonoxidative glucose disposal.

A meta-analysis of major randomized studies of GH replacement in subjects with GH deficiency reported that fasting glucose increased slightly but significantly by a mean of 0.2 mmol/L and fasting insulin increased by 8.7 pmol/L [12]. However, while short-term studies of GH replacement consistently show a reduction in insulin sensitivity, the results of longer-term studies are more variable, with some studies reporting an improvement in insulin sensitivity and others persistent insulin resistance despite favorable changes in body composition [113]. Variability in GH dose and changes in body composition are likely to underlie the heterogeneity in these findings.

Recent studies investigating the effect of very low GH doses on insulin sensitivity have provided new insight into the effect of GH on carbohydrate metabolism. When administered for 7–14 days, low-dose GH improves insulin sensitivity in normal [114, 115] and GH-deficient [116] subjects, demonstrating an improvement in insulin sensitivity that is not mediated by body compositional change. Low-dose GH did not increase FFA and the improvement in insulin sensitivity was correlated to the increase in free IGF-1 [115]. A longer-term study demonstrated that insulin sensitivity was improved after 12 months low-dose GH (0.1 mg/day) in subjects with GH deficiency, despite no significant change in truncal fat mass [117].

The above findings demonstrate that there is not a single action of GH on carbohydrate metabolism. The effect of GH is influenced by GH dose, duration of GH use, and changes in circulating IGF-1 and abdominal fat. Low-dose GH improves insulin sensitivity as it increases IGF-1 without stimulating lipolysis (Fig. 4.3a). Higher GH doses also acutely increase IGF-1, but the benefit is offset by lipolysis-induced insulin resistance (Fig. 4.3b). During long-term GH replacement, there is a complex interplay between increased IGF-1, reduced abdominal fat, and possibly an attenuation of GH-induced lipolysis. The relative effects of these three mechanisms will determine whether insulin sensitivity is increased, unchanged, or reduced (Fig. 4.3c).

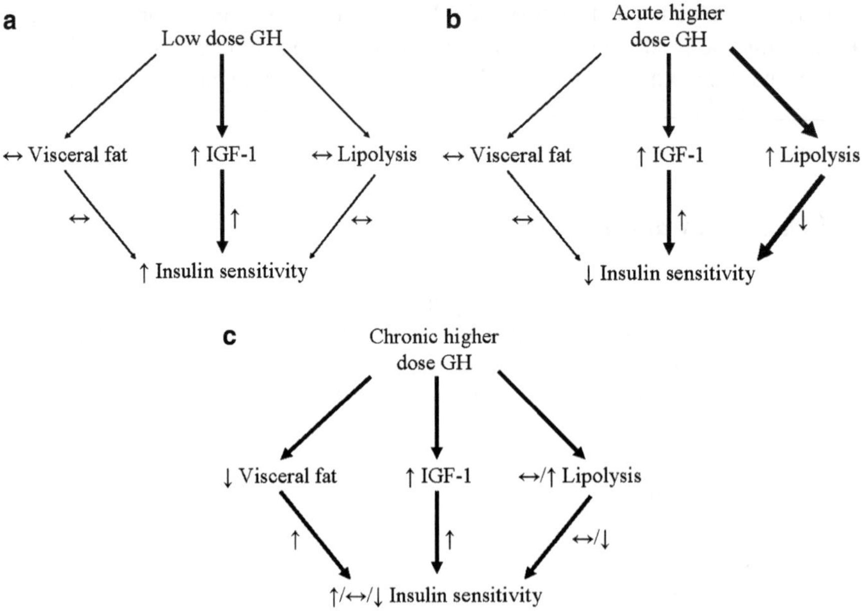

Fig. 4.3 Models of the effect of low-dose GH (**a**), acute higher dose GH (**b**), and long-term higher dose GH (**c**) on insulin sensitivity

IGF-1 and Carbohydrate Metabolism

IGF-1 and the IGF-1 receptor bear close structural homology to proinsulin/insulin and the insulin receptor, respectively [118]. When acutely administered to healthy humans [119] and patients with GH insensitivity [120], IGF-1 causes hypoglycemia. These effects are mediated by "insulin-like" effects on carbohydrate metabolism; IGF-1 inhibits hepatic glucose production and increases oxidative and nonoxidative glucose disposal [53, 54]. When administered at doses that have a comparable effect on glucose uptake, the reduction in hepatic glucose production is relatively lesser and increase in muscle glucose uptake greater than insulin [53], reflecting differences in IGF-1 and insulin receptor distribution [121]. Molar concentrations of IGF-1 approximately 10–20-fold higher than insulin are required to achieve an equivalent increase in peripheral glucose uptake [53, 54, 119]. However, this reflects the effect of the IGF-binding protein system to limit the bioavailability of IGF-1 rather than a major difference in potency.

There has been interest in the potential role of IGF-1 as treatment for type 1 diabetes. Portal insulin delivery to the liver is necessary for hepatic IGF-1 production (Fig. 4.4a). Therefore, systemic insulin administration in type 1 diabetes does not achieve concentrations sufficient to optimize IGF-1 production and results in relative IGF-1 deficiency. The loss of negative feedback from IGF 1 stimulates GH

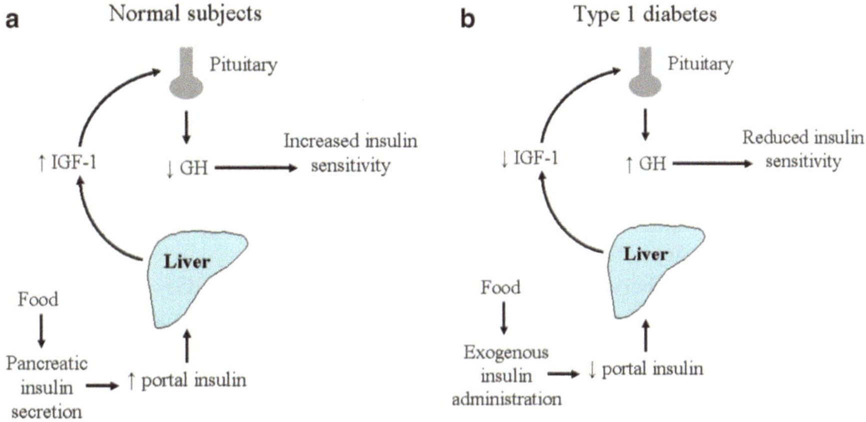

Fig. 4.4 The relationship between insulin and the GH/IGF-1 axis in normal subjects (**a**) and subjects with type 1 diabetes (**b**) (adapted from ref. [113] with permission)

secretion, resulting in increased lipolysis and insulin resistance (Fig. 4.4b). Short-term studies have demonstrated that IGF-1, alone or in combination with IGF-binding protein-3, improves insulin sensitivity in type 1 diabetes [122, 123]. An improvement in glycemic control has also been reported in patients with type 2 diabetes [124]. The long-term cost-benefit of IGF-1 therapy in diabetes remains to be clarified.

The mechanism by which IGF-1 improves insulin sensitivity in type 1 diabetes has been explored. IGF-1 could act directly through its receptor to improve insulin sensitivity or the effect could be secondary to a reduction in GH secretion and lipolysis. In subjects with type 1 diabetes in whom GH secretion was suppressed by octreotide, IGF-1 reduced hepatic glucose production and increased peripheral insulin sensitivity [56]. These findings suggest that the improvement in insulin sensitivity with IGF-1 is, at least in part, independent of changes in GH secretion and action.

Summary

High-dose exogenous and endogenous GH causes hepatic and peripheral insulin resistance. This occurs despite an increase in IGF-1 which has an insulin-like action and improves insulin sensitivity. The deterioration in insulin sensitivity induced by GH is a combination of a direct effect on hepatic insulin sensitivity and an indirect effect on peripheral insulin sensitivity secondary to GH-induced lipolysis. However, low doses of GH improve insulin sensitivity as they increase IGF-1 without stimulating lipolysis. The change in insulin sensitivity during long-term GH is dependent on relative changes in lipolysis, IGF-1, and abdominal adiposity.

Conclusion

GH and IGF-1 are major regulators of a range of metabolic processes in humans (Fig. 4.5). GH directly stimulates REE, lipolysis, and lipid oxidation, which mediates the effect of GH on fat mass. GH and IGF-1 both induce changes in protein metabolism associated with an increase in protein mass. Changes in insulin sensitivity with GH are complex and mediated by a direct effect of GH and indirectly by changes in lipolysis, IGF-1, and abdominal adiposity.

The relationship between the metabolic action of GH and IGF-1 is complex. While IGF-1 exerts a similar effect to GH on protein metabolism and may mediate its anabolic action, IGF-1 opposes the effect of GH excess on carbohydrate metabolism. The effect of IGF-1 on lipid metabolism is indirect and secondary to changes in GH and insulin secretion. Kaplan and Cohen have proposed an augmentative/counteractive hypothesis of the GH/IGF-1 axis to explain this complexity, hypothesizing that this system has developed to enhance the anabolic action of GH while minimizing its deleterious effects [125].

Finally, the metabolic actions of GH and IGF-1 vary depending on GH dose, disease state, and duration of GH deficiency or excess. While high GH doses cause insulin resistance, lower GH doses improve insulin sensitivity. Administration of IGF-1 inhibits and stimulates lipolysis in normal subjects and subjects with GH deficiency, respectively. GH acutely reduces protein oxidation, but this effect is attenuated during long-term use. Therefore, there is not a single metabolic action of GH and IGF-1 on lipid, protein, and carbohydrate metabolism. Integration of many variables is needed to explain the metabolic actions of GH and IGF-1 in acromegaly and GH deficiency and the effect of GH replacement in the latter.

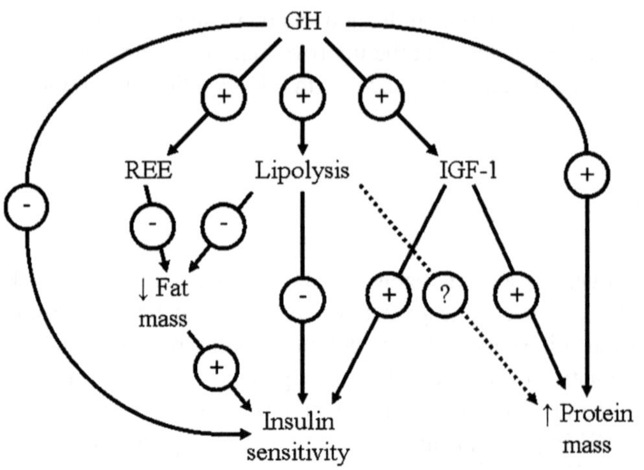

Fig. 4.5 Schematic representation of the integrated metabolic effects of GH and IGF

References

1. Soyka LF, Ziskind A, Crawford JD. Treatment of short stature in children and adolescents with human pituitary growth hormone (Raben). N Engl J Med. 1964;271:754–64.
2. Houssay BA. The hypophysis and metabolism. N Engl J Med. 1936;214(20):961–86.
3. Raben MS. Growth hormone. 1. Physiologic aspects. N Engl J Med. 1962;266:31–5.
4. Raben MS. Growth hormone. 2. Clinical use of human growth hormone. N Engl J Med. 1962;266:82–6.
5. Salomon F, Cuneo RC, Hesp R, Sonksen PH. The effects of treatment with recombinant human growth hormone on body composition and metabolism in adults with growth hormone deficiency. N Engl J Med. 1989;321(26):1797–803.
6. Jorgensen JO, Pedersen SA, Thuesen L, et al. Beneficial effects of growth hormone treatment in GH-deficient adults. Lancet. 1989;1(8649):1221–5.
7. Daughaday WH, Hall K, Raben MS, Salmon Jr WD, van den Brande JL, van Wyk JJ. Somatomedin: proposed designation for sulphation factor. Nature. 1972;235(5333):107.
8. Le Roith D, Bondy C, Yakar S, Liu JL, Butler A. The somatomedin hypothesis: 2001. Endocr Rev. 2001;22(1):53–74.
9. Hoffman DM, O'Sullivan AJ, Freund J, Ho KK. Adults with growth hormone deficiency have abnormal body composition but normal energy metabolism. J Clin Endocrinol Metab. 1995;80(1):72–7.
10. Burt MG, Gibney J, Ho KK. Characterization of the metabolic phenotypes of Cushing's syndrome and growth hormone deficiency: a study of body composition and energy metabolism. Clin Endocrinol (Oxf). 2006;64(4):436–43.
11. Snel YE, Doerga ME, Brummer RM, Zelissen PM, Koppeschaar HP. Magnetic resonance imaging-assessed adipose tissue and serum lipid and insulin concentrations in growth hormone-deficient adults. Effect of growth hormone replacement. Arterioscler Thromb Vasc Biol. 1995;15(10):1543–8.
12. Maison P, Griffin S, Nicoue-Beglah M, Haddad N, Balkau B, Chanson P. Impact of growth hormone (GH) treatment on cardiovascular risk factors in GH-deficient adults: a metaanalysis of blinded, randomized, placebo-controlled trials. J Clin Endocrinol Metab. 2004;89(5):2192–9.
13. Gotherstrom G, Bengtsson BA, Bosaeus I, Johannsson G, Svensson J. A 10-year, prospective study of the metabolic effects of growth hormone replacement in adults. J Clin Endocrinol Metab. 2007;92(4):1442–5.
14. Bengtsson BA, Eden S, Lonn L, et al. Treatment of adults with growth hormone (GH) deficiency with recombinant human GH. J Clin Endocrinol Metab. 1993;76(2):309–17.
15. O'Sullivan AJ, Kelly JJ, Hoffman DM, Freund J, Ho KK. Body composition and energy expenditure in acromegaly. J Clin Endocrinol Metab. 1994;78(2):381–6.
16. Freda PU, Shen W, Heymsfield SB, et al. Lower visceral and subcutaneous but higher intermuscular adipose tissue depots in patients with growth hormone and insulin-like growth factor I excess due to acromegaly. J Clin Endocrinol Metab. 2008;93(6):2334–43.
17. Laron Z, Anin S, Klipper-Aurbach Y, Klinger B. Effects of insulin-like growth factor on linear growth, head circumference, and body fat in patients with Laron-type dwarfism. Lancet. 1992;339(8804):1258–61.
18. Mauras N, O'Brien KO, Welch S, et al. Insulin-like growth factor I and growth hormone (GH) treatment in GH-deficient humans: differential effects on protein, glucose, lipid, and calcium metabolism. J Clin Endocrinol Metab. 2000;85(4):1686–94.
19. Mauras N, Martinez V, Rini A, Guevara-Aguirre J. Recombinant human insulin-like growth factor I has significant anabolic effects in adults with growth hormone receptor deficiency: studies on protein, glucose, and lipid metabolism. J Clin Endocrinol Metab. 2000;85(9):3036–42.
20. Ranke MB, Savage MO, Chatelain PG, Preece MA, Rosenfeld RG, Wilton P. Long-term treatment of growth hormone insensitivity syndrome with IGF-I. Results of the European Multicentre Study. The Working Group on Growth Hormone Insensitivity Syndromes. Horm Res. 1999;51(3):128–34.

21. Laron Z, Ginsberg S, Lilos P, Arbiv M, Vaisman N. Long-term IGF-I treatment of children with Laron syndrome increases adiposity. Growth Horm IGF Res. 2006;16(1):61–4.
22. O'Sullivan AJ, Kelly JJ, Hoffman DM, Baxter RC, Ho KK. Energy metabolism and substrate oxidation in acromegaly. J Clin Endocrinol Metab. 1995;80(2):486–91.
23. Moller N, Schmitz O, Joorgensen JO, et al. Basal- and insulin-stimulated substrate metabolism in patients with active acromegaly before and after adenomectomy. J Clin Endocrinol Metab. 1992;74(5):1012–9.
24. Lindberg-Larsen R, Moller N, Schmitz O, et al. The impact of pegvisomant treatment on substrate metabolism and insulin sensitivity in patients with acromegaly. J Clin Endocrinol Metab. 2007;92(5):1724–8.
25. Hansen M, Morthorst R, Larsson B, et al. Effects of 2 wk of GH administration on 24-h indirect calorimetry in young, healthy, lean men. Am J Physiol Endocrinol Metab. 2005;289(6):E1030–8.
26. Stenlof K, Sjostrom L, Lonn L, et al. Effects of recombinant human growth hormone on basal metabolic rate in adults with pituitary deficiency. Metabolism. 1995;44(1):67–74.
27. Moller L, Norrelund H, Jessen N, et al. Impact of growth hormone receptor blockade on substrate metabolism during fasting in healthy subjects. J Clin Endocrinol Metab. 2009;94(11):4524–32.
28. Chong PK, Jung RT, Scrimgeour CM, Rennie MJ, Paterson CR. Energy expenditure and body composition in growth hormone deficient adults on exogenous growth hormone. Clin Endocrinol (Oxf). 1994;40(1):103–10.
29. Pedersen SB, Kristensen K, Fisker S, Jorgensen JO, Christiansen JS, Richelsen B. Regulation of uncoupling protein-2 and -3 by growth hormone in skeletal muscle and adipose tissue in growth hormone-deficient adults. J Clin Endocrinol Metab. 1999;84(11):4073–8.
30. Jorgensen JO, Moller J, Laursen T, Orskov H, Christiansen JS, Weeke J. Growth hormone administration stimulates energy expenditure and extrathyroidal conversion of thyroxine to triiodothyronine in a dose-dependent manner and suppresses circadian thyrotrophin levels: studies in GH-deficient adults. Clin Endocrinol (Oxf). 1994;41(5):609–14.
31. Hussain MA, Schmitz O, Mengel A, et al. Insulin-like growth factor I stimulates lipid oxidation, reduces protein oxidation, and enhances insulin sensitivity in humans. J Clin Invest. 1993;92(5):2249–56.
32. Hussain MA, Schmitz O, Mengel A, et al. Comparison of the effects of growth hormone and insulin-like growth factor I on substrate oxidation and on insulin sensitivity in growth hormone-deficient humans. J Clin Invest. 1994;94(3):1126–33.
33. Moller N, Jorgensen JO, Schmitz O, et al. Effects of a growth hormone pulse on total and forearm substrate fluxes in humans. Am J Physiol. 1990;258(1 Pt 1):E86–91.
34. Sakharova AA, Horowitz JF, Surya S, et al. Role of growth hormone in regulating lipolysis, proteolysis, and hepatic glucose production during fasting. J Clin Endocrinol Metab. 2008;93(7):2755–9.
35. Hansen TK, Gravholt CH, Ørskov H, Rasmussen MH, Christiansen JS, Jorgensen JO. Dose dependency of the pharmacokinetics and acute lipolytic actions of growth hormone. J Clin Endocrinol Metab. 2002;87(10):4691–8.
36. Gravholt CH, Schmitz O, Simonsen L, Bulow J, Christiansen JS, Moller N. Effects of a physiological GH pulse on interstitial glycerol in abdominal and femoral adipose tissue. Am J Physiol. 1999;277(5 Pt 1):E848–54.
37. Moller N, Jorgensen JO. Effects of growth hormone on glucose, lipid, and protein metabolism in human subjects. Endocr Rev. 2009;30(2):152–77.
38. Rosenthal MJ, Woodside WF. Nocturnal regulation of free fatty acids in healthy young and elderly men. Metabolism. 1988;37(7):645–8.
39. Samra JS, Clark ML, Humphreys SM, et al. Suppression of the nocturnal rise in growth hormone reduces subsequent lipolysis in subcutaneous adipose tissue. Eur J Clin Invest. 1999;29(12):1045–52.
40. Lange KH, Lorentsen J, Isaksson F, et al. Endurance training and GH administration in elderly women: effects on abdominal adipose tissue lipolysis. Am J Physiol Endocrinol Metab. 2001;280(6):E886–97.

41. Nielsen S, Moller N, Christiansen JS, Jorgensen JO. Pharmacological antilipolysis restores insulin sensitivity during growth hormone exposure. Diabetes. 2001;50(10):2301–8.
42. Segerlantz M, Bramnert M, Manhem P, Laurila E, Groop LC. Inhibition of the rise in FFA by Acipimox partially prevents GH-induced insulin resistance in GH-deficient adults. J Clin Endocrinol Metab. 2001;86(12):5813–8.
43. Richelsen B, Pedersen SB, Kristensen K, et al. Regulation of lipoprotein lipase and hormone-sensitive lipase activity and gene expression in adipose and muscle tissue by growth hormone treatment during weight loss in obese patients. Metabolism. 2000;49(7):906–11.
44. Moller N, Jorgensen JO, Alberti KG, Flyvbjerg A, Schmitz O. Short-term effects of growth hormone on fuel oxidation and regional substrate metabolism in normal man. J Clin Endocrinol Metab. 1990;70(4):1179–86.
45. Vahl N, Moller N, Lauritzen T, Christiansen JS, Jorgensen JO. Metabolic effects and pharmacokinetics of a growth hormone pulse in healthy adults: relation to age, sex, and body composition. J Clin Endocrinol Metab. 1997;82(11):3612–8.
46. Brandou F, Aloulou I, Razimbaud A, Fedou C, Mercier J, Brun JF. Lower ability to oxidize lipids in adult patients with growth hormone (GH) deficiency: reversal under GH treatment. Clin Endocrinol (Oxf). 2006;65(4):423–8.
47. Norrelund H, Vahl N, Juul A, et al. Continuation of growth hormone (GH) therapy in GH-deficient patients during transition from childhood to adulthood: impact on insulin sensitivity and substrate metabolism. J Clin Endocrinol Metab. 2000;85(5):1912–7.
48. Sjogren K, Leung KC, Kaplan W, Gardiner-Garden M, Gibney J, Ho KK. Growth hormone regulation of metabolic gene expression in muscle: a microarray study in hypopituitary men. Am J Physiol Endocrinol Metab. 2007;293(1):E364–71.
49. Grohmann M, Sabin M, Holly J, Shield J, Crowne E, Stewart C. Characterization of differentiated subcutaneous and visceral adipose tissue from children: the influences of TNF-alpha and IGF-I. J Lipid Res. 2005;46(1):93–103.
50. DiGirolamo M, Eden S, Enberg G, et al. Specific binding of human growth hormone but not insulin-like growth factors by human adipocytes. FEBS Lett. 1986;205(1):15–9.
51. Bluher S, Kratzsch J, Kiess W. Insulin-like growth factor I, growth hormone and insulin in white adipose tissue. Best Pract Res Clin Endocrinol Metab. 2005;19(4):577–87.
52. Back K, Arnqvist HJ. Changes in insulin and IGF-I receptor expression during differentiation of human preadipocytes. Growth Horm IGF Res. 2009;19(2):101–11.
53. Laager R, Ninnis R, Keller U. Comparison of the effects of recombinant human insulin-like growth factor-I and insulin on glucose and leucine kinetics in humans. J Clin Invest. 1993;92(4):1903–9.
54. Boulware SD, Tamborlane WV, Rennert NJ, Gesundheit N, Sherwin RS. Comparison of the metabolic effects of recombinant human insulin-like growth factor-I and insulin. Dose-response relationships in healthy young and middle-aged adults. J Clin Invest. 1994;93(3):1131–9.
55. Laager R, Ninnis R, Keller U. Comparative effects of recombinant human insulin-like growth factor I and insulin on whole-body and forearm palmitate metabolism in man. Clin Sci (Lond). 1995;88(6):681–6.
56. Simpson HL, Jackson NC, Shojaee-Moradie F, et al. Insulin-like growth factor I has a direct effect on glucose and protein metabolism, but no effect on lipid metabolism in type 1 diabetes. J Clin Endocrinol Metab. 2004;89(1):425–32.
57. Mauras N, Martha Jr PM, Quarmby V, Haymond MW. rhIGF-I administration in humans: differential metabolic effects of bolus vs. continuous subcutaneous delivery. Am J Physiol. 1997;272(4 Pt 1):E628–33.
58. Fery F, Plat L, Baleriaux M, Balasse EO. Inhibition of lipolysis stimulates whole body glucose production and disposal in normal postabsorptive subjects. J Clin Endocrinol Metab. 1997; 82(3):825–30.
59. Ronchi CL, Varca V, Beck-Peccoz P, et al. Comparison between six-year therapy with long-acting somatostatin analogs and successful surgery in acromegaly: effects on cardiovascular risk factors. J Clin Endocrinol Metab. 2006;91(1):121–8.
60. Boero L, Manavela M, Gomez Rosso L, et al. Alterations in biomarkers of cardiovascular disease (CVD) in active acromegaly. Clin Endocrinol (Oxf). 2009;70(1):88–95.

61. Colao A. The GH-IGF-I axis and the cardiovascular system: clinical implications. Clin Endocrinol (Oxf). 2008;69(3):347–58.
62. Tessari P, Garibotto G, Inchiostro S, et al. Kidney, splanchnic, and leg protein turnover in humans. Insight from leucine and phenylalanine kinetics. J Clin Invest. 1996;98(6):1481–92.
63. Wagenmakers AJ. Tracers to investigate protein and amino acid metabolism in human subjects. Proc Nutr Soc. 1999;58(4):987–1000.
64. Janssen YJ, Doornbos J, Roelfsema F. Changes in muscle volume, strength, and bioenergetics during recombinant human growth hormone (GH) therapy in adults with GH deficiency. J Clin Endocrinol Metab. 1999;84(1):279–84.
65. Gotherstrom G, Svensson J, Koranyi J, et al. A prospective study of 5 years of GH replacement therapy in GH-deficient adults: sustained effects on body composition, bone mass, and metabolic indices. J Clin Endocrinol Metab. 2001;86(10):4657–65.
66. Chrisoulidou A, Beshyah SA, Rutherford O, et al. Effects of 7 years of growth hormone replacement therapy in hypopituitary adults. J Clin Endocrinol Metab. 2000;85(10):3762–9.
67. Cuneo RC, Salomon F, Wiles CM, Hesp R, Sonksen PH. Growth hormone treatment in growth hormone-deficient adults. I. Effects on muscle mass and strength. J Appl Physiol. 1991;70(2):688–94.
68. Freda PU, Shen W, Reyes-Vidal CM, et al. Skeletal muscle mass in acromegaly assessed by magnetic resonance imaging and dual-photon x-ray absorptiometry. J Clin Endocrinol Metab. 2009;94(8):2880–6.
69. Papadakis MA, Grady D, Black D, et al. Growth hormone replacement in healthy older men improves body composition but not functional ability. Ann Intern Med. 1996;124(8):708–16.
70. Blackman MR, Sorkin JD, Munzer T, et al. Growth hormone and sex steroid administration in healthy aged women and men: a randomized controlled trial. JAMA. 2002;288(18):2282–92.
71. Ehrnborg C, Ellegard L, Bosaeus I, Bengtsson BA, Rosen T. Supraphysiological growth hormone: less fat, more extracellular fluid but uncertain effects on muscles in healthy, active young adults. Clin Endocrinol (Oxf). 2005;62(4):449–57.
72. Giannoulis MG, Sonksen PH, Umpleby M, et al. The effects of growth hormone and/or testosterone in healthy elderly men: a randomized controlled trial. J Clin Endocrinol Metab. 2006;91(2):477–84.
73. Woods KA, Camacho-Hubner C, Bergman RN, Barter D, Clark AJ, Savage MO. Effects of insulin-like growth factor I (IGF-I) therapy on body composition and insulin resistance in IGF-I gene deletion. J Clin Endocrinol Metab. 2000;85(4):1407–11.
74. Russell-Jones DL, Weissberger AJ, Bowes SB, et al. The effects of growth hormone on protein metabolism in adult growth hormone deficient patients. Clin Endocrinol (Oxf). 1993;38(4):427–31.
75. Lucidi P, Lauteri M, Laureti S, et al. A dose-response study of growth hormone (GH) replacement on whole body protein and lipid kinetics in GH-deficient adults. J Clin Endocrinol Metab. 1998;83(2):353–7.
76. Horber FF, Haymond MW. Human growth hormone prevents the protein catabolic side effects of prednisone in humans. J Clin Invest. 1990;86(1):265–72.
77. Burt MG, Johannsson G, Umpleby AM, Chisholm DJ, Ho KK. Impact of growth hormone and dehydroepiandrosterone on protein metabolism in glucocorticoid-treated patients. J Clin Endocrinol Metab. 2008;93(3):688–95.
78. Russell-Jones DL, Bowes SB, Rees SE, et al. Effect of growth hormone treatment on postprandial protein metabolism in growth hormone-deficient adults. Am J Physiol. 1998;274(6 Pt 1):E1050–6.
79. Burt MG, Gibney J, Hoffman DM, Umpleby AM, Ho KK. Relationship between GH-induced metabolic changes and changes in body composition: a dose and time course study in GH-deficient adults. Growth Horm IGF Res. 2008;18(1):55–64.
80. Beshyah SA, Sharp PS, Gelding SV, Halliday D, Johnston DG. Whole-body leucine turnover in adults on conventional treatment for hypopituitarism. Acta Endocrinol (Copenh). 1993;129(2):158–64.

81. Hoffman DM, Pallasser R, Duncan M, Nguyen TV, Ho KK. How is whole body protein turnover perturbed in growth hormone-deficient adults? J Clin Endocrinol Metab. 1998;83(12):4344–9.
82. Gibney J, Wolthers T, Burt MG, Leung KC, Umpleby AM, Ho KK. Protein metabolism in acromegaly: differential effects of short- and long-term treatment. J Clin Endocrinol Metab. 2007;92(4):1479–84.
83. Shi J, Sekhar RV, Balasubramanyam A, et al. Short- and long-term effects of growth hormone (GH) replacement on protein metabolism in GH-deficient adults. J Clin Endocrinol Metab. 2003;88(12):5827–33.
84. Fryburg DA, Gelfand RA, Barrett EJ. Growth hormone acutely stimulates forearm muscle protein synthesis in normal humans. Am J Physiol. 1991;260(3 Pt 1):E499–504.
85. Fryburg DA, Barrett EJ. Growth hormone acutely stimulates skeletal muscle but not whole-body protein synthesis in humans. Metabolism. 1993;42(9):1223–7.
86. Copeland KC, Nair KS. Acute growth hormone effects on amino acid and lipid metabolism. J Clin Endocrinol Metab. 1994;78(5):1040–7.
87. Garibotto G, Barreca A, Russo R, et al. Effects of recombinant human growth hormone on muscle protein turnover in malnourished hemodialysis patients. J Clin Invest. 1997;99(1):97–105.
88. Butterfield GE, Thompson J, Rennie MJ, Marcus R, Hintz RL, Hoffman AR. Effect of rhGH and rhIGF-I treatment on protein utilization in elderly women. Am J Physiol. 1997;272(1 Pt 1):E94–9.
89. Peino R, Cordido F, Penalva A, Alvarez CV, Dieguez C, Casanueva FF. Acipimox-mediated plasma free fatty acid depression per se stimulates growth hormone (GH) secretion in normal subjects and potentiates the response to other GH-releasing stimuli. J Clin Endocrinol Metab. 1996;81(3):909–13.
90. Cordido F, Fernandez T, Martinez T, et al. Effect of acute pharmacological reduction of plasma free fatty acids on growth hormone (GH) releasing hormone-induced GH secretion in obese adults with and without hypopituitarism. J Clin Endocrinol Metab. 1998;83(12):4350–4.
91. Moller N, Gormsen LC, Schmitz O, Lund S, Jorgensen JO, Jessen N. Free fatty acids inhibit growth hormone/signal transducer and activator of transcription-5 signaling in human muscle: a potential feedback mechanism. J Clin Endocrinol Metab. 2009;94(6):2204–7.
92. Nair KS, Welle SL, Halliday D, Campbell RG. Effect of beta-hydroxybutyrate on whole-body leucine kinetics and fractional mixed skeletal muscle protein synthesis in humans. J Clin Invest. 1988;82(1):198–205.
93. Norrelund H, Nair KS, Nielsen S, et al. The decisive role of free fatty acids for protein conservation during fasting in humans with and without growth hormone. J Clin Endocrinol Metab. 2003;88(9):4371–8.
94. Nielsen S, Moller N, Pedersen SB, Christiansen JS, Jorgensen JO. The effect of long-term pharmacological antilipolysis on substrate metabolism in growth hormone (GH)-substituted GH-deficient adults. J Clin Endocrinol Metab. 2002;87(7):3274–8.
95. Turkalj I, Keller U, Ninnis R, Vosmeer S, Stauffacher W. Effect of increasing doses of recombinant human insulin-like growth factor-I on glucose, lipid, and leucine metabolism in man. J Clin Endocrinol Metab. 1992;75(5):1186–91.
96. Russell-Jones DL, Umpleby AM, Hennessy TR, et al. Use of a leucine clamp to demonstrate that IGF-I actively stimulates protein synthesis in normal humans. Am J Physiol. 1994;267(4 Pt 1):E591–8.
97. Fryburg DA. Insulin-like growth factor I exerts growth hormone- and insulin-like actions on human muscle protein metabolism. Am J Physiol. 1994;267(2 Pt 1):E331–6.
98. Fryburg DA, Jahn LA, Hill SA, Oliveras DM, Barrett EJ. Insulin and insulin-like growth factor-I enhance human skeletal muscle protein anabolism during hyperaminoacidemia by different mechanisms. J Clin Invest. 1995;96(4):1722–9.
99. Moller N, Butler PC, Antsiferov MA, Alberti KG. Effects of growth hormone on insulin sensitivity and forearm metabolism in normal man. Diabetologia. 1989;32(2):105–10.

100. Neely RD, Rooney DP, Bell PM, et al. Influence of growth hormone on glucose-glucose 6-phosphate cycle and insulin action in normal humans. Am J Physiol. 1992;263(5 Pt 1): E980–7.
101. Randle PJ, Garland PB, Hales CN, Newsholme EA. The glucose fatty-acid cycle. Its role in insulin sensitivity and the metabolic disturbances of diabetes mellitus. Lancet. 1963;1(7285): 785–9.
102. Segerlantz M, Bramnert M, Manhem P, Laurila E, Groop LC. Inhibition of lipolysis during acute GH exposure increases insulin sensitivity in previously untreated GH-deficient adults. Eur J Endocrinol. 2003;149(6):511–9.
103. Piatti PM, Monti LD, Caumo A, et al. Mediation of the hepatic effects of growth hormone by its lipolytic activity. J Clin Endocrinol Metab. 1999;84(5):1658–63.
104. Battezzati A, Benedini S, Fattorini A, et al. Insulin action on protein metabolism in acromegalic patients. Am J Physiol Endocrinol Metab. 2003;284(4):E823–9.
105. Colao A, Ferone D, Marzullo P, Lombardi G. Systemic complications of acromegaly: epidemiology, pathogenesis, and management. Endocr Rev. 2004;25(1):102–52.
106. O'Connell T, Clemmons DR. IGF-I/IGF-binding protein-3 combination improves insulin resistance by GH-dependent and independent mechanisms. J Clin Endocrinol Metab. 2002;87(9):4356–60.
107. Ho KY, Weissberger AJ, Marbach P, Lazarus L. Therapeutic efficacy of the somatostatin analog SMS 201-995 (octreotide) in acromegaly. Effects of dose and frequency and long-term safety. Ann Intern Med. 1990;112(3):173–81.
108. Colao A, Auriemma RS, Savastano S, et al. Glucose tolerance and somatostatin analog treatment in acromegaly: a 12-month study. J Clin Endocrinol Metab. 2009;94(8):2907–14.
109. Higham CE, Rowles S, Russell-Jones D, Umpleby AM, Trainer PJ. Pegvisomant improves insulin sensitivity and reduces overnight free fatty acid concentrations in patients with acromegaly. J Clin Endocrinol Metab. 2009;94(7):2459–63.
110. van der Lely AJ, Hutson RK, Trainer PJ, et al. Long-term treatment of acromegaly with pegvisomant, a growth hormone receptor antagonist. Lancet. 2001;358(9295):1754–9.
111. Barkan AL, Burman P, Clemmons DR, et al. Glucose homeostasis and safety in patients with acromegaly converted from long-acting octreotide to pegvisomant. J Clin Endocrinol Metab. 2005;90(10):5684–91.
112. Hew FL, Koschmann M, Christopher M, et al. Insulin resistance in growth hormone-deficient adults: defects in glucose utilization and glycogen synthase activity. J Clin Endocrinol Metab. 1996;81(2):555–64.
113. Yuen KC, Dunger DB. Therapeutic aspects of growth hormone and insulin-like growth factor-I treatment on visceral fat and insulin sensitivity in adults. Diabetes Obes Metab. 2007;9(1):11–22.
114. Yuen K, Ong K, Husbands S, et al. The effects of short-term administration of two low doses versus the standard GH replacement dose on insulin sensitivity and fasting glucose levels in young healthy adults. J Clin Endocrinol Metab. 2002;87(5):1989–95.
115. Yuen K, Frystyk J, Umpleby M, Fryklund L, Dunger D. Changes in free rather than total insulin-like growth factor-I enhance insulin sensitivity and suppress endogenous peak growth hormone (GH) release following short-term low-dose GH administration in young healthy adults. J Clin Endocrinol Metab. 2004;89(8):3956–64.
116. Yuen K, Cook D, Ong K, et al. The metabolic effects of short-term administration of physiological versus high doses of GH therapy in GH deficient adults. Clin Endocrinol (Oxf). 2002;57(3):333–41.
117. Yuen KC, Frystyk J, White DK, et al. Improvement in insulin sensitivity without concomitant changes in body composition and cardiovascular risk markers following fixed administration of a very low growth hormone (GH) dose in adults with severe GH deficiency. Clin Endocrinol (Oxf). 2005;63(4):428–36.
118. Jones JI, Clemmons DR. Insulin-like growth factors and their binding proteins: biological actions. Endocr Rev. 1995;16(1):3–34.

119. Guler HP, Zapf J, Froesch ER. Short-term metabolic effects of recombinant human insulin-like growth factor I in healthy adults. N Engl J Med. 1987;317(3):137–40.
120. Laron Z, Klinger B, Erster B, Anin S. Effect of acute administration of insulin-like growth factor I in patients with Laron-type dwarfism. Lancet. 1988;2(8621):1170–2.
121. Caro JF, Poulos J, Ittoop O, Pories WJ, Flickinger EG, Sinha MK. Insulin-like growth factor I binding in hepatocytes from human liver, human hepatoma, and normal, regenerating, and fetal rat liver. J Clin Invest. 1988;81(4):976–81.
122. Carroll PV, Christ ER, Umpleby AM, et al. IGF-I treatment in adults with type 1 diabetes: effects on glucose and protein metabolism in the fasting state and during a hyperinsulinemic-euglycemic amino acid clamp. Diabetes. 2000;49(5):789–96.
123. Saukkonen T, Amin R, Williams RM, et al. Dose-dependent effects of recombinant human insulin-like growth factor (IGF)-I/IGF binding protein-3 complex on overnight growth hormone secretion and insulin sensitivity in type 1 diabetes. J Clin Endocrinol Metab. 2004;89(9):4634–41.
124. Moses AC, Young SC, Morrow LA, O'Brien M, Clemmons DR. Recombinant human insulin-like growth factor I increases insulin sensitivity and improves glycemic control in type II diabetes. Diabetes. 1996;45(1):91–100.
125. Kaplan SA, Cohen P. The somatomedin hypothesis 2007: 50 years later. J Clin Endocrinol Metab. 2007;92(12):4529–35.

Part II
Genetics

Chapter 5
Molecular Genetics of Congenital Growth Hormone Deficiency

Christopher J. Romero, Elyse Pine-Twaddell, and Sally Radovick

Abstract Growth hormone deficiency (GHD) in patients is diagnosed as an isolated finding or in combination with other pituitary hormone deficiencies and classified as combined pituitary hormone deficiency (CPHD). Although any type of neurological insult such as trauma, radiation, or surgery place the pituitary at risk for injury and potential destruction of somatotrophs, a subgroup of patients develop idiopathic GHD who do not have a clear etiology. Advancements in understanding cellular and organ development as well as the availability of sophisticated genetic techniques have provided a molecular basis for some cases of idiopathic GHD. Genetic screening for mutations in pituitary development factors has resulted in the identification of a series of abnormalities in patients with GHD, resulting in hormone deficiency. Unfortunately, the incidence of identified mutations in this group remains low, and quite often, there is a wide spectrum of clinical presentations, which makes correlations between genotypes and abnormal phenotypes difficult. Nevertheless, the characterization of patients with GHD and the ongoing research efforts to decipher the complexities of pituitary development provide insight into the molecular basis of hypopituitarism. This chapter reviews the genetic defects that have been characterized in the hypothalamic-pituitary axis resulting in a deficiency of GH. Furthermore, it emphasizes the importance of continuing research efforts, which will offer physicians the means of determining a genetic etiology for patients with GHD and researchers the tools to develop definitive therapeutic options.

Keywords Growth hormone deficiency • Pituitary • Transcription factors • Hypopituitarism • GH1 gene

S. Radovick (✉)
Department of Pediatrics, Division of Endocrinology, Johns Hopkins University
School of Medicine, CMSC 4-106, 600 North Wolfe Street, Baltimore, MD 21287, USA
e-mail: sradovick@jhmi.edu

K. Ho (ed.), *Growth Hormone Related Diseases and Therapy: A Molecular and Physiological Perspective for the Clinician*, Contemporary Endocrinology,
DOI 10.1007/978-1-60761-317-6_5,

Introduction

Growth hormone (GH) is a 22 kd protein produced by the somatotrophs of the anterior pituitary [1, 2]. GH secretion follows a pulsatile pattern, which is reflective of the antagonistic influences of both growth hormone-releasing hormone (GHRH) and somatostatin (somatotropin release-inhibiting factor, (SRIF)). Growth hormone deficiency (GHD) is a condition that can occur as an isolated finding or in combination with deficiencies in other pituitary hormones (combined pituitary hormone deficiency [CPHD]) in addition to anatomic pituitary abnormalities. Although neurological insult such as trauma, radiation, and surgery may put the pituitary at risk for injury leading to destruction of somatotrophs, there is a subgroup of patients who develop GHD without a clear etiology, referred to as idiopathic. Several investigators have described familial forms of both GHD and CPHD [3, 4]. Among patients diagnosed with short stature secondary to GHD, as many as 3–30% have an affected first degree relative, therefore suggesting a genetic etiology for this condition [5–7]. The advances in understanding cell and organ development as well as the availability of sophisticated molecular techniques have helped provide a molecular basis for some of the "idiopathic" cases of GHD. Despite the genetic screening for mutations in genes that regulate pituitary development among cohorts of patients with CPHD and their families, the overall incidence of identified mutations remains low. Nevertheless, research efforts to characterize patients with idiopathic GHD biochemically and genetically have helped to delineate mechanisms of resulting pathology and provide us with correlations between abnormal genotypes and clinical phenotypes.

This chapter discusses the genetic defects that have been described and studied in the hypothalamic-pituitary axis resulting in the deficiency of GH. The structure of the growth hormone gene, GH1, has been well described and several reports have correlated mutations within the gene accounting for cases of isolated growth hormone deficiency (IGHD). In addition, abnormalities in the GHRH receptor (GHRHR) have been recognized as a basis for diminished or absent GH secretion. Despite appropriate somatotroph development, the complexities of GH posttranslational modifications and its ability to appropriately target and stimulate its receptor have also been a source in which genetic abnormalities prevent appropriate GH function. Finally, GH is often one of several pituitary hormones deficient in a patient with hypopituitarism and combined deficiencies have been shown in some cases to be a result of disrupted pituitary development. Several transcriptional signals and factors, which operate in a temporal and spatial pattern, have been identified and their function considered essential for the differentiation and function of specific pituitary cell types. This chapter will review the more commonly reported factors found to have mutations in patients with CPHD.

Isolated Growth Hormone Deficiency

The gene-encoding GH (GH1) is located on chromosome 17q23 within a cluster that includes two genes for placental lactogen (HPL), a pseudogene for HPL, and the GH2 gene that encodes placental GH (Fig. 5.1). *GH1* generates 20- and 22-kd

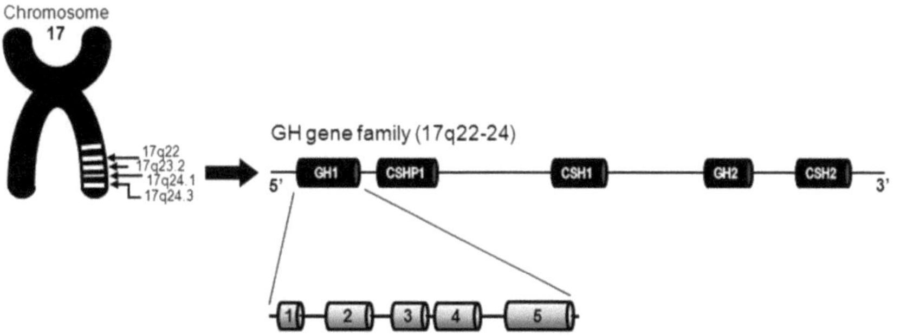

IGHD Type	Mutation	Inheritance	Phenotype
1A	gross/micro-deletions, nonsense mutations	AR	Severe short stature; GH replacement leads to Anti-GH Ab
1B	Splice site mutations (*GHRHR mutations)	AR	Severe short stature; good response to GH replacement; No Anti-GH Ab
2	Splice site mutations, intron/exon splice Enhancer, missense mutations	AD	Severe short stature; good response to GH replacement
3	Unclear	X-linked	Short stature and hypogammaglobulinemia.

Fig. 5.1 The GH gene family cluster on the long arm of chromosome 17 (17q11-14) consists of the growth hormone 1 gene (GH1), GH2, chorionic somatomammotropin pseudogene 1 (CSHP1), chorionic somatomammotropin 1 (CSH1), and CSH2. The GH1 gene is a 191-amino acid protein consisting of five exons and four introns. IGHD deficiency has been subdivided by the type of mutation, mode of inheritance, and phenotype (table below in figure). It should be noted that reported mutations in GHRHR have been classified under IGHD type 1B

proteins of approximately equal bioactivity. Although the majority of children with IGHD are considered idiopathic, the recognition of genetic defects suggests only up to 13% of these patients may harbor a GH1 gene mutation [8]. Four forms of IGHD have been described and its classification is based upon the clinical presentation, inheritance pattern, and GH secretion.

Isolated GHD Type 1 (GHD 1)

Isolated GHD type 1A occurs as a result of primarily large deletions along with microdeletions and single base pair substitutions of the GH1 gene, which ultimately prevent synthesis or secretion of the hormone. GHD 1A is considered the most severe form of IGHD as these patients present with no detectable serum GH and have an autosomal recessive inheritance pattern. Clinically, these patients present with decreased birth length, neonatal hypoglycemia, and severe postnatal growth retardation. Since GH is not produced in fetal life, these patients may be immunologically intolerant of GH and treatment with either pituitary-derived or recombinant DNA-derived GH leads to the development of anti-GH antibodies over time. As these patients become insensitive to GH replacement therapy and demonstrate a decreased clinical response, they become candidates for recombinant insulin-like growth factor 1 (IGF-1) therapy.

The less severe autosomal recessive form, GHD type IB, results from mutations or rearrangements of the *GH1* gene such as splice site mutations that lead to an aberrant GH molecule with reduced function. Clinically, there is greater phenotypic variability in these patients and treatment with GH replacement therapy does not lead to antibody production [9]. In addition to studying the GH1 gene, patients found to harbor mutations in the transmembrane and intracellular gene domains of the *GHRHR* gene have also been classified as IGHD type 1B [10]. The human *GHRHR* gene encodes a G-protein-coupled receptor with seven transmembrane domains [11, 12]. The little mouse (*lit/lit*), which demonstrates dwarfism and a decreased number of somatotrophs, has a recessively inherited missense mutation in the extracellular domain of the gene for *Ghrhr* [13]. The pituitary glands of these mice are found to be unresponsive to GHRH in vivo and in vitro [14]. Studies using this model suggest that GHRH is not essential for fetal somatotroph development; however, hypoplasia of somatotrophs has been illustrated postnatally [15, 16]. In humans, two cousins who presented clinically with the typical phenotype of severe GHD were found to have a nonsense mutation in the human *GHRHR* gene that introduced a stop codon at position 72 (E72X). This mutation resulted in a markedly truncated GHRHR protein that lacked the membrane-spanning regions and the G-protein-binding site. The patients demonstrated an undetectable serum GH level during provocative testing with GHRH and responded well to GH replacement therapy [17].

IGHD Type 2 (GHD2)

GHD2, which is classified as an autosomal dominant disorder, is considered the most common genetic form of IGHD. The majority of these cases result from mutations that inactivate the 5′ splice donor site of intron 3 of the *GH1* gene and result in a loss of exon 3 [18, 19]. Consequently, the resultant molecule cannot fold normally and leads to a toxic 17.5 kd GH-isoform that functions in a dominant negative manner to reduce the accumulation and secretion of wild-type GH. Others have described missense mutations in exon 4 or 5 with variable clinical presentation, but also potential reversibility of the impairment of intracellular GH storage and secretion after GH treatment [20]. In one multicenter study, the authors report that patients with IGHD2 have clinical variability in terms of not only onset, severity, and progression of GHD, but also can demonstrate late onset of ACTH or TSH deficiencies along with pituitary hypoplasia [9]. Furthermore, the assessment of *GH1* gene mutations in children with short stature revealed several heterozygous mutations [9]. Although these may not be the primary etiology, such abnormalities may contribute to the reported phenotypic variability. Recent in vitro research using cells to express the aberrant GH-isoform suggests that viral vectors which can integrate and "silence" the mRNA product of this isoform may show promise for the development of gene therapy to treat these patients [21].

IGHD Type 3

This type of IGHD is transmitted as an X-linked recessive trait and has been clinically characterized with both short stature and hypogammaglobulinemia [19, 22]. Although the immune deficiency syndrome in these patients is similar to X-linked agammaglobulinemia caused by mutations in the gene for Bruton's tyrosine kinase (BTK), the molecular defect for IGHD type 3 is not due to abnormalities of BTK [23]. A large Australian kindred demonstrating GHD along with variable pituitary hormonal deficiencies was found to have a duplication of the Xq25-Xq28 region suggesting a possible etiology [24]. In another study of patients with IGHD type 3, investigators suggest a possible etiology by reporting a mutation in the ets-family transcription factor myeloid elf-1-like factor (MEF), which activates a variety of promoters including those for granulocyte-macrophage colony-stimulating factor, interleukin-3, interleukin-8, and lysozyme [23]. Figure 5.1 summarizes in a table the mutation(s), inheritance pattern, and phenotype of the different types of IGHD.

Bioinactive GH

Serum GH exists in multiple molecular forms, reflecting alternative posttranscriptional or posttranslational processing of the mRNA or protein, respectively. Some of these forms are presumed to have defects in the amino acid sequences required for binding of GH to its receptor and different molecular forms of GH may have varying potencies for stimulating skeletal growth, although this remains to be rigorously proved. Short stature with normal GH immunoreactivity but reduced biopotency has been reported; these patients typically have high serum GH levels along with low concentrations of serum IGF-1 [25–27]. In addition, a marked response to exogenous GH administration is often reported. The molecular defects, however, have only been characterized in relatively few patients and often cases of suspected bioinactive GH have not been rigorously proven [28, 29].

One report describes a child with extreme short stature (−6.1 SDS), who was found to have a heterozygous single base substitution in the GH-1 gene that converted the amino acid at codon 77 from an arginine to a cysteine. This mutant GH demonstrated greater affinity than normal to GHBP and the GHR, but failed to stimulate tyrosine phosphorylation by itself and inhibited the action of normal GH [29]. Interestingly, the father had the same genetic abnormality, but did not express the mutant hormone.

In another patient with marked short stature (−3.6 SDS), a heterozygous single-base substitution (A→G) in exon 4 of the GH-1 gene which led to a glycine to arginine substitution was reported [28]. This mutation led to failure of appropriate molecular rotation of the dimerized receptor and subsequent diminished tyrosine phosphorylation and the GH-mediated intracellular cascade of events. Also, bioactivity of the mutant determined in a mouse B-cell lymphoma line was about 33% of immunoreactivity. The authors report that, clinically, administration of exogenous GH in this patient substantially increased her growth velocity (4.5–11.0 cm/year).

The complexity of GH's functional interaction with its receptor was demonstrated in a study that describes an Ile179Met substitution found in a short child. Receptor-binding studies showed no differences between wild type and mutant. Although the mutant's ability to activate STAT5 was normal, activation of ERK by the mutant was 50% than that observed by wild-type ligand [30]. Given that STAT5b is clearly the major (if not sole) GH-dependent mediator of IGF-I gene transcription, the mechanism of action of this mutation is not entirely clear. A screening of short children found six GH heterozygous variants with evidence of impairment of JAK/STAT activation, thus suggesting further studies are needed to determine not only the mechanism of interaction of GH with its receptor, but also to delineate genotype-phenotype correlations [31]. In a more recent report of a bioinactive GH, Besson and associates found a homozygous missense mutation (G705C) in a short child (−3.6 SDS) leading to absence of two disulfide bridges. Both GHR-binding and JAK2/STAT5-signaling activity were markedly reduced [25]. It should be recognized, however, that some patients may demonstrate decreased bioactivity when measured by sensitive in vitro assays, but not immunoreactivity. An absence of mutations may suggest that abnormal posttranslational modifications of GH or other peripheral mechanisms acting on the clearance or bioavailability of GH to target tissues may be responsible for the clinical phenotypes [32].

Combined Pituitary Hormone Deficiency

The five cell types of the anterior pituitary gland (AP) are produced in a multistep process that is dependent on several transcription and signaling factors in a temporal and spatial manner. These five cell types and their hormone product include: somatotrophs (GH), lactotrophs (prolactin [PRL]), thyrotrophs (thyroid-stimulating hormone [TSH]), corticotrophs (adrenocorticotrophic hormone [ACTH]), and gonadotrophs (lutenizing hormone [LH] and follicle-stimulating hormone [FSH]). Any disruption of this cascade adversely affects the development and/or survival of single or multiple cell types, thus ultimately leading to deficiencies in hormone products. Figure 5.2 illustrates a schematic of the cascade of transcription factors expressed during development. Modified from murine studies of pituitary development, the figure emphasizes the importance of timing when factors are expressed leading to differentiation of the specific pituitary cell types.

Early Developmental Factors

1. HESX1

   HESX1, a paired-like homeodomain gene, is expressed early in pituitary and forebrain development and acts as a promoter-specific transcriptional repressor

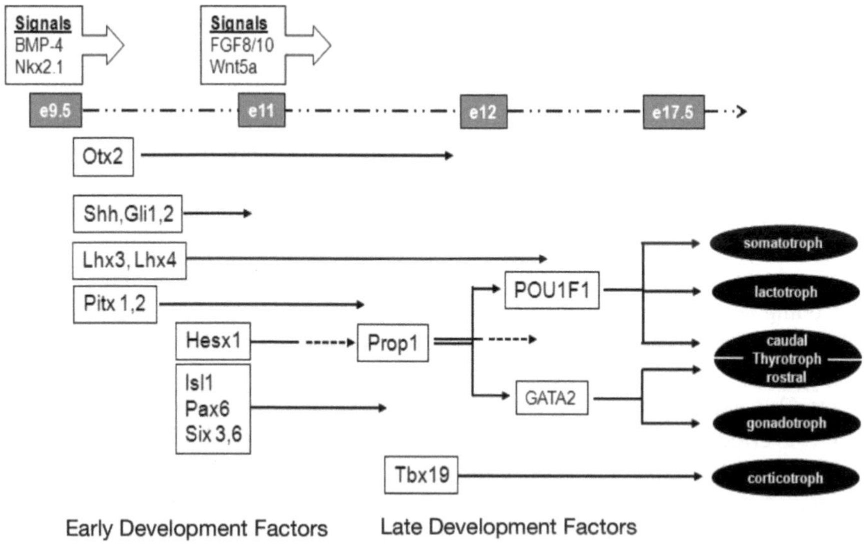

Fig. 5.2 Developmental factors and signals required for anterior pituitary development. This figure presents a modified overview of pituitary development adapted from previous embryological studies performed in murine species. The development of the mature pituitary gland initiates with the contact of the oral ectoderm with the neural ectoderm followed by a cascade of events consisting of both signaling molecules and transcription factors expressed in a specific temporal and spatial fashion. At approximately embryological day 9.5 (e9.5), expression of BMP-4 and Nkx2.1 along with sonic hedgehog (Shh) initiates the development of the primordial Rathke's pouch (RP). In addition, the expression of factors Gli 1,2, Lhx3 and Pitx 1,2 occurs for the development of progenitor pituitary cell types. These events are followed by the expression of Hesx1, Isl1, Pax6, and Six3, 6, which are also important for not only cellular development, but also proliferation and migration. The attenuation of Hesx1 (*dotted arrows* at approximately e12.5) is required for the expression of Prop1. By e12.5, RP has formed, and by e17.5, differentiation of specific pituitary cell types has been completed. The mature pituitary gland contains the differentiated cell types: somatotrophs, lactotrophs, thyrotrophs, gonadotrophs, and corticotrophs

[33, 34]. Hesx1 null mutant mice demonstrate forebrain and pituitary malformation along with abnormalities in the septum pellucidum and eye. This phenotype is similar to the human phenotype of septo-optic dysplasia (SOD) [34].

HESX1, one of the earliest known specific markers for the pituitary primordium, acts as a repressor and its downregulation is accompanied by a rise in Prophet of Pit1 (PROP1) and subsequent cell determination [35]. TLE1 and TLE3, members of the groucho-like gene family, as well as the nuclear corepressor (N-COR) act as corepressors of PROP1 activity [36–38].

Although mutations in HESX1 have been found in patients diagnosed with SOD and pituitary hypoplasia, there is still variability in phenotypes despite HESX1 mutations as some have reported patients without optic nerve involvement. Dattani et al. [34] identified two siblings with AP hypoplasia, an ectopic posterior lobe, corpus callosum agenesis, absent septum pellucidum, and optic nerve hypoplasia with a homozygous R160C mutation which caused loss of

DNA binding. A novel I26T mutation in exon 1 was reported in a patient with early GHD, FSH/LH deficiency, and evolving TSH and cortisol deficiency, along with pituitary structural abnormalities, but normal optic nerves [37]. Although mutations in HESX1 associated with pituitary disease appear to modulate the DNA-binding affinity of HESX1, this mutation appeared to impair transcriptional repression [33, 37].

Recently, approximately 850 patients have been studied for mutations in *HESX1* (300 with SOD, 410 with isolated pituitary dysfunction, optic nerve hypoplasia, or midline brain anomalies, and 126 patients with familial inheritance). Coding region mutations were found in merely 1% of this group, suggesting that mutations in *HESX1* are a rare cause of hypopituitarism and SOD [39].

2. OTX2

OTX2 is expressed early in anterior neuroectoderm development and is known for its critical role in retinal photoreceptor, rostral brain, and midbrain development [40–42]. Otx2 has a role in regulating Hesx1 and POU1F1 expression. A murine model showed mutant Otx2 causing anterior–posterior axis defects, which was rescued by expression of Dkk1, a Wnt antagonist, or following removal of one copy of the β-catenin gene [43].

Three novel heterozygous mutations in *OTX2* were reported in four patients, which include a frameshift and nonsense mutations along with a microdeletion of approximately a 2.9-Mb deletion involving OTX2 [44]. Two patients were diagnosed with isolated GHD, one patient with CPHD and two cases with normal pituitary function, yet all patients had ocular abnormalities. The authors demonstrated either reduced or lost transactivation of four promoters, GNRH1, HESX1, POU1F1, and IRBP, by these mutations as compared to the wild-type gene. This same group also reported a de novo heterozygous frameshift mutation, c.402insC, in a patient with IGHD and bilateral anophthalmia that led to a loss of the Otx2 transactivation region [45]. Furthermore, two unrelated patients diagnosed with CPHD and pituitary gland structural abnormalities, but without ocular pathology, were reported to have a heterozygous OTX2 mutation in exon 3 (N233S) [46]. Despite preserved binding to target genes, this mutant OTX2 was shown to act as a dominant negative inhibitor of HESX1 gene expression.

3. PITX2

PITX2 is a member of the bicoid-like homeobox transcription factor family, related to the *Otx2* gene, and is important in pituitary development [47]. PITX2 functions through activating the promoters of pituitary hormone target genes [48–50]. *Pitx2* null mice showed embryonic lethality with arrested pituitary and tooth organogenesis, while a hypomorphic allele model of *Pitx2* demonstrated pituitary hypoplasia and cellular differentiation defects in proportion to the reduced dosage of *Pitx2*, with gonadotrophs most severely affected, followed by somatotrophs and thyrotrophs [47, 51, 52]. Conversely, overexpression of *Pitx2* increases the gonadotroph population [53].

Clinical mutations of *PITX2* have been described in patients with Axenfeld–Rieger syndrome. Semina and associates described mutations in six out of ten families with autosomal dominant Rieger Syndrome, consisting of anomalies of the anterior chamber of the eye, dental hypoplasia, protuberant umbilicus, mental retardation, and pituitary anomalies; five of the six mutations were in the homeobox region and several show loss of DNA-binding capacity [54, 55].

4. SOX2

   Heterozygous *SOX2* mutations in humans have been associated with anophthalmia or microphthalmia and associated anomalies, such as esophageal atresia, male genital abnormalities, sensorineural hearing loss, hypoplasia of the corpus callosum, hypothalamic hamartoma, hippocampal malformation, learning difficulties, and pituitary dysfunction [56–65]. The pituitary dysfunction consists of GHD in some patients and gonadotropin deficiency in many with some males affected with genital abnormalities [59, 63, 66, 67]. A murine model with heterozygosity for a targeted disruption of Sox2 was found to have abnormal AP development with decreased GH, TSH, and LH [67]. Six of 235 screened patients were found to have mutations of SOX2; all had hypogonadotrophic hypogonadism and the males had abnormal genitalia [67].

5. SOX3

   A syndrome of mental retardation and hypopituitarism inherited in an X-linked fashion was described in multiple families. Males had AP and infundibular hypoplasia, an undescended posterior pituitary, and corpus callosum abnormalities. Duplications of the Xq26-27, which encompasses *SOX3,* an SRY-related HMG-Box gene, were found in several pedigrees [68]. GH deficiency was found in all affected males, with variable ACTH, TSH, and/or gonadotropin deficiencies. A duplication resulting in an expansion of 11 additional alanine residues was found in another family with males with X-linked hypopituitarism and mental retardation, while a family with seven additional alanine residues had complete panhypopituitarism and neurological structural abnormalities, but no mental retardation or craniofacial dysmorphism [69, 70]. Female carriers appear to be clinically unaffected, and men with 46 XX sex reversal, 46 XY gonadal dysgenesis, and idiopathic oligoazoospermia were not found to have any *SOX3* mutations [71, 72].

6. LHX3

   LHX3 is a LIM homeobox gene involved in the establishment and maintenance of the differentiated phenotype of pituitary cells [73]. In the LHX3(–/–) mutant mouse, Rathke's pouch development is arrested and regulation of cell differentiation into distinct pituitary lineages is affected [74].

   In humans, homozygous loss-of-function mutations in *LHX3* have been identified in patients with hypopituitarism including GH, TSH, PRL, LH, and FSH deficiencies, anterior pituitary defects, and cervical abnormalities with or without restricted neck rotation. ACTH deficiency has also been diagnosed in a subset of patients. The majority of patients have cervical abnormalities with limited rotation [75–79]. Three hundred and sixty-six patients with IGHD or CPHD were studied and only seven patients from four families were found to have LHX3 mutations, suggesting *LHX3* mutations are a rare cause of CPHD [77]. Most missense mutations identified in patients have decreased ability to activate transcription of the promoters of several potential *LHX3* target genes including *CGA*, *PRL*, *FSHB*, *TSHB*, and *POU1F1* [80, 81].

7. LHX4

   LHX4 is a LIM homeobox gene expressed in the developing brain, including the cortex, pituitary, and spinal cord [82]. The LHX4(−/−) mutant mouse demonstrates all five anterior pituitary cell lineages; however, there are reduced numbers leading to a hypoplastic lobe [74]. Further studies of Lhx4 in conjunction with Lhx3 demonstrate that Lhx4 is required for the proliferation of lineage precursors, while Lhx3 is necessary to establish the fate of pituitary precursor cells [74].

   In humans, there have been eight separate mutations reported of patients with heterozygous mutations in *LHX4*. All patients had GHD and hypoplastic AP glands; there was variable TSH, gonadotropin, and ACTH deficiency, and the majority had an undescended or ectopic posterior pituitary. There are also reports of a flat sella turcica and cerebellar defects [83–88]. Through functional studies, these heterozygous mutations have been shown to result in proteins that are unable to bind DNA and activate pituitary gene expression at potential promoter targets POU1F1, TSH-beta, FSH-beta, and CGA [83, 85, 87, 89].

8. GLI2/SHH

   Holoprosencephaly, due to abnormal midline development of the embryonic forebrain, may have hypothalamic insufficiency [90, 91]. Mutations in developmental proteins in the sonic hedgehog pathway (SHH) have been associated with development of holoprosencephaly, as SHH is critical in forebrain development [92]. Clinical features of holoprosencephaly include facial dysmorphism ranging from cyclopia to hypertelorism, absent nasal septum, midline defects such as cleft palate or lip, and a single central incisor. The GLI transcription factors, which are a part of the SHH pathway, have been associated with a clinical phenotype of holoprosencephaly. One report describes a GLI2 heterozygous loss-of-function mutation in patients diagnosed with holoprosencephaly, but also demonstrated pituitary hormone abnormalities [92, 93]. GH deficiency may be accompanied by other pituitary hormone insufficiencies. The incidence of GH deficiency is increased in cases of simple clefts of the lip and/or palate alone and children with cleft palates who grow abnormally require further evaluation [94, 95].

Late Developmental Factors

1. PROP1

   PROP1 is a paired-like homeodomain transcription factor with expression restricted to the anterior pituitary during development and is associated with CPHD [96]. *PROP1* acts as both a repressor in downregulating Hesx1 and as an activator of POU1F1 [36]. A proposed mechanism is through the Wnt/β-catenin signaling pathway, in which the binding of Prop-1 with β-catenin represses *Hesx1* expression and promotes *Pou1f1* development [97]. The expression of *PROP1* maintains the POU1F1 lineage and allows for differentiation of lactotrophs, somatotrophs, and thyrotrophs. It has been shown to be dependent on Notch signaling, which is thought to be important for late lineage differentiation [98]. The Ames dwarf mouse, which harbors a mutated Prop-1, demonstrates severe dwarfism, hypothyroidism, and infertility [99]. Recently, Ward and coworkers created a *Prop-1* bacterial artificial chromosome (BAC) transgene, which rescued the mutant mice with the *Prop-1* dwarf phenotype when crossed with wild-type mice [100].

There is considerable variation in clinical phenotypes of patients with *PROP1* mutations, even in patients bearing identical genotypes [101–103]. Hormone deficiency may be variable and dynamic, with deficiencies developing over time, for example, cortisol deficiency requiring hydrocortisone replacement has been described as early as 6 years and as late as the fifth decade [104, 105]. Additionally, hypogonadotrophic hypogonadism has been described in patients who have developed spontaneous puberty [103, 106]. The majority of patients develop GHD and TSH deficiency early in the course of disease. MRI findings include pituitary hypoplasia and pituitary enlargement, which, in some longitudinal studies, have been found to involute over time [107–112]. A GA repeat in exon 2 (295-CGA-GAG-AGT-303) has been reported to be a "hot spot" for mutations in *PROP1;* any combination of a GA or AG deletion in this repeat region results in a frameshift in the coding sequence and premature termination at codon 109 [106, 113]. Although *PROP1* mutations appear to be rare in sporadic cases, its prevalence is 29.5% in familial cases of CHPD as reported by Turton et al. [114, 115]. *PROP1* is considered the most commonly identified genetic cause of CPHD [106, 116, 117].

2. POU1F1 (PIT1)

   The *POU1F1* gene encodes POU1F1, a 290 amino acid protein, which contains two domains, the POU-specific and the POU-homeo, which are necessary for DNA binding and activation, the expression of somatotrophs, lactotrophs, and thyrotrophs, and is involved in the expression of *Ghrhr* [118–121]. The Snell dwarf mouse, which has a hypoplastic anterior pituitary gland, diminished length and GH, TSH, and PRL deficiency was found to have a missense mutation in the POU-homeodomain region (W261C), which generates a nonfunctional Pit1 protein [122]. POU1F1 regulates target genes by binding to response elements and recruitment of coactivator proteins, such as cAMP response element-binding protein (CREB)-binding protein (CBP) [123]. Regulation of POU1f1 also includes binding to the enhancer by Atbf1, which is a member of the giant, multiple-homeodomain and zinc finger family [124].

   Mutations in *POU1F1* in humans were described in 1992, by four different groups (Tatsumi, Radovick, Pfaeffle, Ohta), in patients with CPHD consisting of GHD, TSH, and PRL deficiency, and variable hypoplastic AP on MRI [119, 125–127]. At least 28 different mutations have been described, with 23 demonstrating autosomal recessive inheritance and five dominant inheritance [128]. The most common mutation is an R271W substitution affecting the POU-homeodomain, encoding a mutant protein that binds normally to DNA and acts as a dominant inhibitor of transcription [114, 125, 127, 129–135].

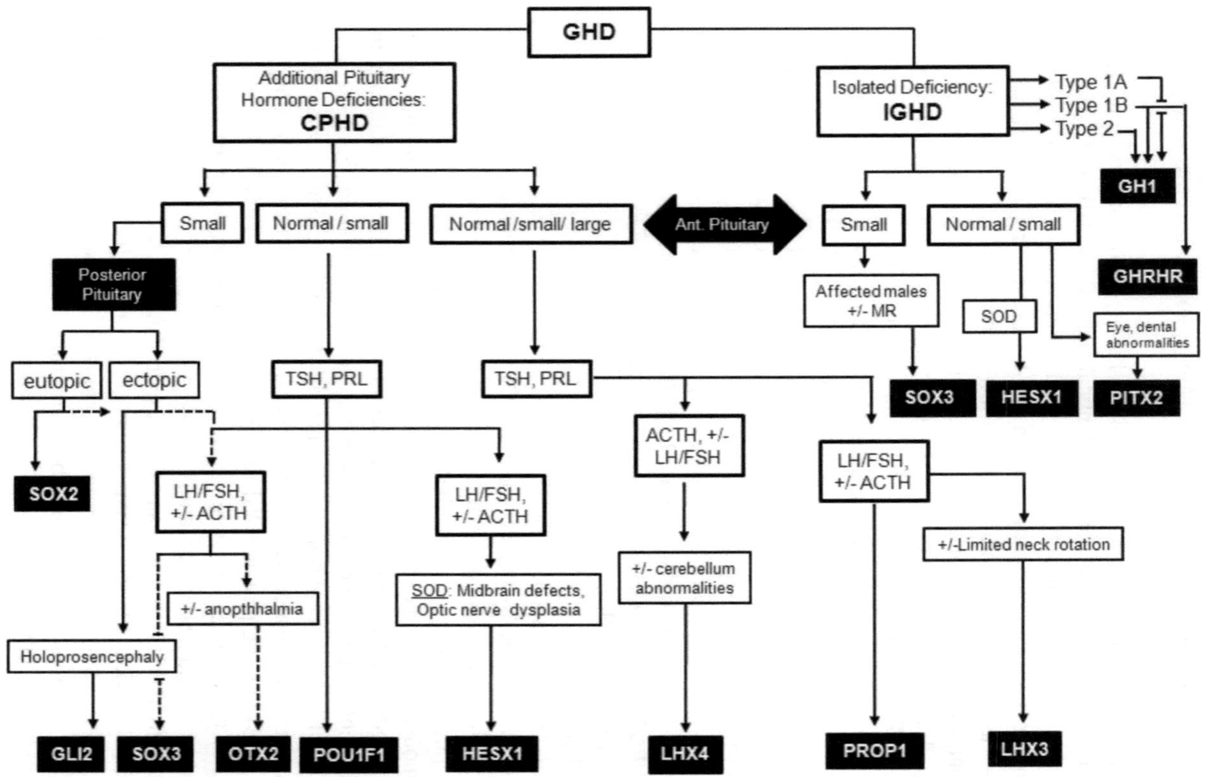

Fig. 5.3 Algorithm for molecular diagnosis of GHD. This figure delineates a systematic approach to diagnose a genetic etiology for GHD. The algorithm is divided into two main parts dependent on the clinical presentation of CPHD or IGHD. Both the anatomic appearance of the pituitary gland and clinical presentation will dictate the genetic evaluation. Pituitary genes are denoted by *black boxes*. The reader must keep in mind that pituitary deficiency(s) is variable and evolving

Conclusion

Advancements in molecular diagnostic techniques as well as the expanding knowledge of pituitary development have provided a genetic basis for hypopituitarism. Although the prevalence of mutations currently appears to be relatively low, this may simply reflect the complexities of normal pituitary development and the as-of-yet unknown factors. The availability of recombinant GH hormone has provided patients with GHD, a safe and effective therapy; however, it is not without adverse side effects and the schedule and method of delivery are not ideal. Gene transfer methodologies are currently being developed with the promise to provide novel treatment options. Recently, investigators have demonstrated the efficacy of delivery of plasmid DNA containing the genomic human GH sequence into the skeletal muscle of *lit/lit* mice, resulting in significant increases in GH levels within several days [136]. Understanding mechanisms of disease will allow for further therapeutic options in patients with hypopituitarism. For now, it remains vital that clinicians continue to appropriately characterize patients with GHD (Fig. 5.3) and utilize the available molecular techniques to evaluate gene mutations. Several centers, including ours, have established protocols to identify and screen patients with IGHD and CPHD in an effort to further delineate the correlations between genotype and phenotype. The importance and continued success of this translational research provides promise to decrease the morbidity and mortality faced by patients with pituitary disease.

References

1. Dixon JS, Li CH. The amino acid composition of human pituitary growth hormone. J Gen Physiol. 1962;45(4):176–8.
2. Li CH, Liu WK, Dixon JS. Human pituitary growth hormone. VI. Modified procedure of isolation and NH_2-terminal amino acid sequence. Arch Biochem Biophys Suppl. 1962;1:327–32.
3. Rimoin DL. Hereditary forms of growth hormone deficiency and resistance. Birth Defects Orig Artic Ser. 1976;12:15–29.
4. Rimoin DL. Genetic forms of pituitary dwarfism. Birth Defects Orig Artic Ser. 1971;7:12–20.
5. Perez Jurado LA, Argente J. Molecular basis of familial growth hormone deficiency. Horm Res. 1994;42:189–97.
6. Phillips III JA, Cogan JD. Genetic basis of endocrine disease. 6. Molecular basis of familial human growth hormone deficiency. J Clin Endocrinol Metab. 1994;78:11–6.
7. Parks JS, Pfaeffle RW, Brown MR, et al. Growth hormone deficiency. In: Weintraub BD, editor. Molecular endocrinology: basic concepts and clinical correlations. 1st ed. New York: Raven; 1995.
8. Wagner JK, Eble A, Hindmarsh PC, Mullis PE. Prevalence of human GH-1 gene alterations in patients with isolated growth hormone deficiency. Pediatr Res. 1998;43:105–10.
9. Mullis PE, Robinson IC, Salemi S, Eble A, Besson A, Vuissoz JM, et al. Isolated autosomal dominant growth hormone deficiency: an evolving pituitary deficit? A multicenter follow-up study. J Clin Endocrinol Metab. 2005;90:2089–96.
10. Salvatori R, Fan X, Phillips III JA, Espigares-Martin R, Martin De Lara I, Freeman KL, et al. Three new mutations in the gene for the growth hormone (gh)-releasing hormone receptor in familial isolated gh deficiency type ib. J Clin Endocrinol Metab. 2001;86:273–9.

11. Gaylinn BD. Molecular and cell biology of the growth hormone-releasing hormone receptor. Growth Horm IGF Res. 1999;9(Suppl A):37–44.
12. Gaylinn BD, Harrison JK, Zysk JR, Lyons CE, Lynch KR, Thorner MO. Molecular cloning and expression of a human anterior pituitary receptor for growth hormone-releasing hormone. Mol Endocrinol. 1993;7:77–84.
13. Gaylinn BD, Dealmeida VI, Lyons Jr CE, Wu KC, Mayo KE, Thorner MO. The mutant growth hormone-releasing hormone (GHRH) receptor of the little mouse does not bind GHRH. Endocrinology. 1999;140:5066–74.
14. Clark RG, Robinson IC. Effects of a fragment of human growth hormone-releasing factor in normal and 'Little' mice. J Endocrinol. 1985;106:1–5.
15. Godfrey P, Rahal JO, Beamer WG, Copeland NG, Jenkins NA, Mayo KE. GHRH receptor of little mice contains a missense mutation in the extracellular domain that disrupts receptor function. Nat Genet. 1993;4:227–32.
16. Lin SC, Lin CR, Gukovsky I, Lusis AJ, Sawchenko PE, Rosenfeld MG. Molecular basis of the little mouse phenotype and implications for cell type-specific growth. Nature. 1993;364:208–13.
17. Wajnrajch MP, Gertner JM, Harbison MD, Chua Jr SC, Leibel RL. Nonsense mutation in the human growth hormone-releasing hormone receptor causes growth failure analogous to the little (lit) mouse. Nat Genet. 1996;12:88–90.
18. Lopez-Bermejo A, Buckway CK, Rosenfeld RG. Genetic defects of the growth hormone-insulin-like growth factor axis. Trends Endocrinol Metab. 2000;11:39–49.
19. Procter AM, Phillips III JA, Cooper DN. The molecular genetics of growth hormone deficiency. Hum Genet. 1998;103:255–72.
20. Deladoey J, Stocker P, Mullis PE. Autosomal dominant GH deficiency due to an Arg183His GH-1 gene mutation: clinical and molecular evidence of impaired regulated GH secretion. J Clin Endocrinol Metab. 2001;86:3941–7.
21. Lochmatter D, Strom M, Eble A, Petkovic V, Fluck CE, Bidlingmaier M, et al. Isolated GH deficiency type II: knockdown of the harmful {Delta}3GH recovers wt-GH secretion in rat tumor pituitary cells. Endocrinology. 2010;151:4400–9.
22. Fleisher TA, White RM, Broder S, Nissley SP, Blaese RM, Mulvihill JJ, et al. X-linked hypogammaglobulinemia and isolated growth hormone deficiency. N Engl J Med. 1980;302:1429–34.
23. Stewart DM, Tian L, Notarangelo LD, Nelson DL. Update on X-linked hypogammaglobulinemia with isolated growth hormone deficiency. Curr Opin Allergy Clin Immunol. 2005;5:510–2.
24. Solomon NM, Nouri S, Warne GL, Lagerstrom-Fermer M, Forrest SM, Thomas PQ. Increased gene dosage at Xq26-q27 is associated with X-linked hypopituitarism. Genomics. 2002;79:553–9.
25. Besson A, Salemi S, Deladoey J, Vuissoz JM, Eble A, Bidlingmaier M, et al. Short stature caused by a biologically inactive mutant growth hormone (GH-C53S). J Clin Endocrinol Metab. 2005;90:2493–9.
26. Valenta LJ, Sigel MB, Lesniak MA, Elias AN, Lewis UJ, Friesen HG, et al. Pituitary dwarfism in a patient with circulating abnormal growth hormone polymers. N Engl J Med. 1985;312:214–7.
27. Kowarski AA, Schneider J, Ben-Galim E, Weldon VV, Daughaday WH. Growth failure with normal serum RIA-GH and low somatomedin activity: somatomedin restoration and growth acceleration after exogenous GH. J Clin Endocrinol Metab. 1978;47:461–4.
28. Takahashi Y, Shirono H, Arisaka O, Takahashi K, Yagi T, Koga J, et al. Biologically inactive growth hormone caused by an amino acid substitution. J Clin Invest. 1997;100:1159–65.
29. Takahashi Y, Kaji H, Okimura Y, Goji K, Abe H, Chihara K. Short stature caused by a mutant growth hormone with an antagonistic effect. Endocr J. 1996;43(Suppl):S27–32.
30. Lewis MD, Horan M, Millar DS, Newsway V, Easter TE, Fryklund L, et al. A novel dysfunctional growth hormone variant (Ile179Met) exhibits a decreased ability to activate the extracellular signal-regulated kinase pathway. J Clin Endocrinol Metab. 2004;89:1068–75.
31. Millar DS, Lewis MD, Horan M, Newsway V, Easter TE, Gregory JW, et al. Novel mutations of the growth hormone 1 (GH1) gene disclosed by modulation of the clinical selection criteria for individuals with short stature. Hum Mutat. 2003;21:424–40.

32. Radetti G, Bozzola M, Pagani S, Street ME, Ghizzoni L. Growth hormone immunoreactivity does not reflect bioactivity. Pediatr Res. 2000;48:619–22.
33. Brickman JM, Clements M, Tyrell R, McNay D, Woods K, Warner J, et al. Molecular effects of novel mutations in Hesx1/HESX1 associated with human pituitary disorders. Development. 2001;128:5189–99.
34. Dattani MT, Martinez-Barbera JP, Thomas PQ, Brickman JM, Gupta R, Martensson IL, et al. Mutations in the homeobox gene HESX1/Hesx1 associated with septo-optic dysplasia in human and mouse. Nat Genet. 1998;19:125–33.
35. Hermesz E, Mackem S, Mahon KA. Rpx: a novel anterior-restricted homeobox gene progressively activated in the prechordal plate, anterior neural plate and Rathke's pouch of the mouse embryo. Development. 1996;122:41–52.
36. Dasen JS, Rosenfeld MG. Signaling and transcriptional mechanisms in pituitary development. Annu Rev Neurosci. 2001;24:327–55.
37. Carvalho LR, Woods KS, Mendonca BB, Marcal N, Zamparini AL, Stifani S, et al. A homozygous mutation in HESX1 is associated with evolving hypopituitarism due to impaired repressor-corepressor interaction. J Clin Invest. 2003;112:1192–201.
38. Thomas P, Beddington R. Anterior primitive endoderm may be responsible for patterning the anterior neural plate in the mouse embryo. Curr Biol. 1996;6:1487–96.
39. McNay DE, Turton JP, Kelberman D, Woods KS, Brauner R, Papadimitriou A, et al. HESX1 mutations are an uncommon cause of septooptic dysplasia and hypopituitarism. J Clin Endocrinol Metab. 2007;92:691–7.
40. Henderson RH, Williamson KA, Kennedy JS, Webster AR, Holder GE, Robson AG, et al. A rare de novo nonsense mutation in OTX2 causes early onset retinal dystrophy and pituitary dysfunction. Mol Vis. 2009;15:2442–7.
41. Kurokawa D, Kiyonari H, Nakayama R, Kimura-Yoshida C, Matsuo I, Aizawa S. Regulation of Otx2 expression and its functions in mouse forebrain and midbrain. Development. 2004;131:3319–31.
42. Acampora D, Gulisano M, Simeone A. Genetic and molecular roles of Otx homeodomain proteins in head development. Gene. 2000;246:23–35.
43. Kimura-Yoshida C, Nakano H, Okamura D, Nakao K, Yonemura S, Belo JA, et al. Canonical Wnt signaling and its antagonist regulate anterior-posterior axis polarization by guiding cell migration in mouse visceral endoderm. Dev Cell. 2005;9:639–50.
44. Dateki S, Kosaka K, Hasegawa K, Tanaka H, Azuma N, Yokoya S, et al. Heterozygous orthodenticle homeobox 2 mutations are associated with variable pituitary phenotype. J Clin Endocrinol Metab. 2010;95:756–64.
45. Dateki S, Fukami M, Sato N, Muroya K, Adachi M, Ogata T. OTX2 mutation in a patient with anophthalmia, short stature, and partial growth hormone deficiency: functional studies using the IRBP, HESX1, and POU1F1 promoters. J Clin Endocrinol Metab. 2008;93: 3697–702.
46. Diaczok D, Romero C, Zunich J, Marshall I, Radovick S. A novel dominant negative mutation of OTX2 associated with combined pituitary hormone deficiency. J Clin Endocrinol Metab. 2008;93:4351–9.
47. Suh H, Gage PJ, Drouin J, Camper SA. Pitx2 is required at multiple stages of pituitary organogenesis: pituitary primordium formation and cell specification. Development. 2002;129:329–37.
48. Quentien MH, Manfroid I, Moncet D, Gunz G, Muller M, Grino M, et al. Pitx factors are involved in basal and hormone-regulated activity of the human prolactin promoter. J Biol Chem. 2002;277:44408–16.
49. Tremblay JJ, Goodyer CG, Drouin J. Transcriptional properties of Ptx1 and Ptx2 isoforms. Neuroendocrinology. 2000;71:277–86.
50. Tremblay JJ, Lanctot C, Drouin J. The pan-pituitary activator of transcription, Ptx1 (pituitary homeobox 1), acts in synergy with SF-1 and Pit1 and is an upstream regulator of the Lim-homeodomain gene Lim3/Lhx3. Mol Endocrinol. 1998;12:428–41.
51. Lin CR, Kioussi C, O'Connell S, Briata P, Szeto D, Liu F, et al. Pitx2 regulates lung asymmetry, cardiac positioning and pituitary and tooth morphogenesis. Nature. 1999;401:279–82.

52. Gage PJ, Suh H, Camper SA. Dosage requirement of Pitx2 for development of multiple organs. Development. 1999;126:4643–51.
53. Charles MA, Suh H, Hjalt TA, Drouin J, Camper SA, Gage PJ. PITX genes are required for cell survival and Lhx3 activation. Mol Endocrinol. 2005;19:1893–903.
54. Semina EV, Reiter R, Leysens NJ, Alward WL, Small KW, Datson NA, et al. Cloning and characterization of a novel bicoid-related homeobox transcription factor gene, RIEG, involved in Rieger syndrome. Nat Genet. 1996;14:392–9.
55. Amendt BA, Semina EV, Alward WL. Rieger syndrome: a clinical, molecular, and biochemical analysis. Cell Mol Life Sci. 2000;57:1652–66.
56. Bakrania P, Robinson DO, Bunyan DJ, Salt A, Martin A, Crolla JA, et al. SOX2 anophthalmia syndrome: 12 new cases demonstrating broader phenotype and high frequency of large gene deletions. Br J Ophthalmol. 2007;91:1471–6.
57. Chassaing N, Gilbert-Dussardier B, Nicot F, Fermeaux V, Encha-Razavi F, Fiorenza M, et al. Germinal mosaicism and familial recurrence of a SOX2 mutation with highly variable phenotypic expression extending from AEG syndrome to absence of ocular involvement. Am J Med Genet A. 2007;143:289–91.
58. Sisodiya SM, Ragge NK, Cavalleri GL, Hever A, Lorenz B, Schneider A, et al. Role of SOX2 mutations in human hippocampal malformations and epilepsy. Epilepsia. 2006;47:534–42.
59. Sato N, Kamachi Y, Kondoh H, Shima Y, Morohashi K, Horikawa R, et al. Hypogonadotropic hypogonadism in an adult female with a heterozygous hypomorphic mutation of SOX2. Eur J Endocrinol. 2007;156:167–71.
60. Williamson KA, Hever AM, Rainger J, Rogers RC, Magee A, Fiedler Z, et al. Mutations in SOX2 cause anophthalmia-esophageal-genital (AEG) syndrome. Hum Mol Genet. 2006;15:1413–22.
61. Ragge NK, Lorenz B, Schneider A, Bushby K, de Sanctis L, de Sanctis U, Salt A, Collin JR, Vivian AJ, Free SL, Thompson P, Williamson KA, Sisodiya SM, van Heyningen V, Fitzpatrick DR. SOX2 anophthalmia syndrome. Am J Med Genet A. 2005;135:1–7; discussion 8.
62. Hagstrom SA, Pauer GJ, Reid J, Simpson E, Crowe S, Maumenee IH, et al. SOX2 mutation causes anophthalmia, hearing loss, and brain anomalies. Am J Med Genet A. 2005;138A: 95–8.
63. Fantes J, Ragge NK, Lynch SA, McGill NI, Collin JR, Howard-Peebles PN, et al. Mutations in SOX2 cause anophthalmia. Nat Genet. 2003;33:461–3.
64. Zenteno JC, Gascon-Guzman G, Tovilla-Canales JL. Bilateral anophthalmia and brain malformations caused by a 20-bp deletion in the SOX2 gene. Clin Genet. 2005;68:564–6.
65. Guichet A, Triau S, Lepinard C, Esculapavit C, Biquard F, Descamps P, et al. Prenatal diagnosis of primary anophthalmia with a 3q27 interstitial deletion involving SOX2. Prenat Diagn. 2004;24:828–32.
66. Kelberman D, de Castro SC, Huang S, Crolla JA, Palmer R, Gregory JW, et al. SOX2 plays a critical role in the pituitary, forebrain, and eye during human embryonic development. J Clin Endocrinol Metab. 2008;93:1865–73.
67. Kelberman D, Rizzoti K, Avilion A, Bitner-Glindzicz M, Cianfarani S, Collins J, et al. Mutations within Sox2/SOX2 are associated with abnormalities in the hypothalamo-pituitary-gonadal axis in mice and humans. J Clin Invest. 2006;116:2442–55.
68. Solomon NM, Ross SA, Morgan T, Belsky JL, Hol FA, Karnes PS, et al. Array comparative genomic hybridisation analysis of boys with X linked hypopituitarism identifies a 3.9 Mb duplicated critical region at Xq27 containing SOX3. J Med Genet. 2004;41:669–78.
69. Laumonnier F, Ronce N, Hamel BC, Thomas P, Lespinasse J, Raynaud M, et al. Transcription factor SOX3 is involved in X-linked mental retardation with growth hormone deficiency. Am J Hum Genet. 2002;71:1450–5.
70. Woods KS, Cundall M, Turton J, Rizotti K, Mehta A, Palmer R, et al. Over- and underdosage of SOX3 is associated with infundibular hypoplasia and hypopituitarism. Am J Hum Genet. 2005;76:833–49.
71. Raverot G, Lejeune H, Kotlar T, Pugeat M, Jameson JL. X-linked sex-determining region Y box 3 (SOX3) gene mutations are uncommon in men with idiopathic oligoazoospermic infertility. J Clin Endocrinol Metab. 2004;89:4146–8.

72. Lim HN, Berkovitz GD, Hughes IA, Hawkins JR. Mutation analysis of subjects with 46, XX sex reversal and 46, XY gonadal dysgenesis does not support the involvement of SOX3 in testis determination. Hum Genet. 2000;107:650–2.
73. Zhadanov AB, Copeland NG, Gilbert DJ, Jenkins NA, Westphal H. Genomic structure and chromosomal localization of the mouse LIM/homeobox gene Lhx3. Genomics. 1995;27: 27–32.
74. Sheng HZ, Moriyama K, Yamashita T, Li H, Potter SS, Mahon KA, et al. Multistep control of pituitary organogenesis. Science. 1997;278:1809–12.
75. Kristrom B, Zdunek AM, Rydh A, Jonsson H, Sehlin P, Escher SA. A novel mutation in the LIM homeobox 3 gene is responsible for combined pituitary hormone deficiency, hearing impairment, and vertebral malformations. J Clin Endocrinol Metab. 2009;94:1154–61.
76. Rajab A, Kelberman D, de Castro SC, Biebermann H, Shaikh H, Pearce K, et al. Novel mutations in LHX3 are associated with hypopituitarism and sensorineural hearing loss. Hum Mol Genet. 2008;17:2150–9.
77. Pfaeffle RW, Savage JJ, Hunter CS, Palme C, Ahlmann M, Kumar P, et al. Four novel mutations of the LHX3 gene cause combined pituitary hormone deficiencies with or without limited neck rotation. J Clin Endocrinol Metab. 2007;92:1909–19.
78. Netchine I, Sobrier ML, Krude H, Schnabel D, Maghnie M, Marcos E, et al. Mutations in LHX3 result in a new syndrome revealed by combined pituitary hormone deficiency. Nat Genet. 2000;25:182–6.
79. Bhangoo AP, Hunter CS, Savage JJ, Anhalt H, Pavlakis S, Walvoord EC, et al. Clinical case seminar: a novel LHX3 mutation presenting as combined pituitary hormonal deficiency. J Clin Endocrinol Metab. 2006;91:747–53.
80. Sloop KW, Parker GE, Hanna KR, Wright HA, Rhodes SJ. LHX3 transcription factor mutations associated with combined pituitary hormone deficiency impair the activation of pituitary target genes. Gene. 2001;265:61–9.
81. Savage JJ, Hunter CS, Clark-Sturm SL, Jacob TM, Pfaeffle RW, Rhodes SJ. Mutations in the LHX3 gene cause dysregulation of pituitary and neural target genes that reflect patient phenotypes. Gene. 2007;400:44–51.
82. Mullen RD, Colvin SC, Hunter CS, Savage JJ, Walvoord EC, Bhangoo AP, et al. Roles of the LHX3 and LHX4 LIM-homeodomain factors in pituitary development. Mol Cell Endocrinol. 2007;265–266:190–5.
83. Tajima T, Yorifuji T, Ishizu K, Fujieda K. A novel mutation (V101A) of the LHX4 gene in a Japanese patient with combined pituitary hormone deficiency. Exp Clin Endocrinol Diabetes. 2010;118:405–9.
84. Dateki S, Fukami M, Uematsu A, Kaji M, Iso M, Ono M, et al. Mutation and gene copy number analyses of six pituitary transcription factor genes in 71 patients with combined pituitary hormone deficiency: identification of a single patient with LHX4 deletion. J Clin Endocrinol Metab. 2010;95:4043–7.
85. Pfaeffle RW, Hunter CS, Savage JJ, Duran-Prado M, Mullen RD, Neeb ZP, et al. Three novel missense mutations within the LHX4 gene are associated with variable pituitary hormone deficiencies. J Clin Endocrinol Metab. 2008;93:1062–71.
86. Tajima T, Hattori T, Nakajima T, Okuhara K, Tsubaki J, Fujieda K. A novel missense mutation (P366T) of the LHX4 gene causes severe combined pituitary hormone deficiency with pituitary hypoplasia, ectopic posterior lobe and a poorly developed sella turcica. Endocr J. 2007;54:637–41.
87. Castinetti F, Saveanu A, Reynaud R, Quentien MH, Buffin A, Brauner R, et al. A novel dysfunctional LHX4 mutation with high phenotypical variability in patients with hypopituitarism. J Clin Endocrinol Metab. 2008;93:2790–9.
88. Machinis K, Pantel J, Netchine I, Leger J, Camand OJ, Sobrier ML, et al. Syndromic short stature in patients with a germline mutation in the LIM homeobox LHX4. Am J Hum Genet. 2001;69:961–8.
89. Machinis K, Amselem S. Functional relationship between LHX4 and POU1F1 in light of the LHX4 mutation identified in patients with pituitary defects. J Clin Endocrinol Metab. 2005;90:5456–62.

90. Lieblich JM, Rosen SE, Guyda H, Reardan J, Schaaf M. The syndrome of basal encephalocele and hypothalamic-pituitary dysfunction. Ann Intern Med. 1978;89:910–6.
91. Hintz RL, Menking M, Sotos JF. Familial holoprosencephaly with endocrine dysgenesis. J Pediatr. 1968;72:81–7.
92. Roessler E, Du YZ, Mullor JL, Casas E, Allen WP, Gillessen-Kaesbach G, et al. Loss-of-function mutations in the human GLI2 gene are associated with pituitary anomalies and holoprosencephaly-like features. Proc Natl Acad Sci USA. 2003;100:13424–9.
93. Treier M, O'Connell S, Gleiberman A, Price J, Szeto DP, Burgess R, et al. Hedgehog signaling is required for pituitary gland development. Development. 2001;128:377–86.
94. Izenberg N, Rosenblum M, Parks JS. The endocrine spectrum of septo-optic dysplasia. Clin Pediatr (Phila). 1984;23:632–6.
95. Rudman D, Davis T, Priest JH, Patterson JH, Kutner MH, Heymsfield SB, et al. Prevalence of growth hormone deficiency in children with cleft lip or palate. J Pediatr. 1978;93:378–82.
96. Kelberman D, Dattani MT. Hypothalamic and pituitary development: novel insights into the aetiology. Eur J Endocrinol. 2007;157 Suppl 1:S3–14.
97. Olson LE, Tollkuhn J, Scafoglio C, Krones A, Zhang J, Ohgi KA, et al. Homeodomain-mediated beta-catenin-dependent switching events dictate cell-lineage determination. Cell. 2006;125:593–605.
98. Zhu X, Zhang J, Tollkuhn J, Ohsawa R, Bresnick EH, Guillemot F, et al. Sustained Notch signaling in progenitors is required for sequential emergence of distinct cell lineages during organogenesis. Genes Dev. 2006;20:2739–53.
99. Sornson MW, Wu W, Dasen JS, Flynn SE, Norman DJ, O'Connell SM, et al. Pituitary lineage determination by the Prophet of Pit-1 homeodomain factor defective in Ames dwarfism. Nature. 1996;384:327–33.
100. Ward RD, Davis SW, Cho M, Esposito C, Lyons RH, Cheng JF, et al. Comparative genomics reveals functional transcriptional control sequences in the Prop1 gene. Mamm Genome. 2007;18:521–37.
101. Vieira TC, da Silva MR, Abucham J. The natural history of the R120C PROP1 mutation reveals a wide phenotypic variability in two untreated adult brothers with combined pituitary hormone deficiency. Endocrine. 2006;30:365–9.
102. Lebl J, Vosahlo J, Pfaeffle RW, Stobbe H, Cerna J, Novotna D, et al. Auxological and endocrine phenotype in a population-based cohort of patients with PROP1 gene defects. Eur J Endocrinol. 2005;153:389–96.
103. Fluck C, Deladoey J, Rutishauser K, Eble A, Marti U, Wu W, et al. Phenotypic variability in familial combined pituitary hormone deficiency caused by a PROP1 gene mutation resulting in the substitution of Arg→Cys at codon 120 (R120C). J Clin Endocrinol Metab. 1998;83: 3727–34.
104. Bottner A, Keller E, Kratzsch J, Stobbe H, Weigel JF, Keller A, et al. PROP1 mutations cause progressive deterioration of anterior pituitary function including adrenal insufficiency: a longitudinal analysis. J Clin Endocrinol Metab. 2004;89:5256–65.
105. Pavel ME, Hensen J, Pfaffle R, Hahn EG, Dorr HG. Long-term follow-up of childhood-onset hypopituitarism in patients with the PROP-1 gene mutation. Horm Res. 2003;60:168–73.
106. Deladoey J, Fluck C, Buyukgebiz A, Kuhlmann BV, Eble A, Hindmarsh PC, et al. "Hot spot" in the PROP1 gene responsible for combined pituitary hormone deficiency. J Clin Endocrinol Metab. 1999;84:1645–50.
107. Voutetakis A, Argyropoulou M, Sertedaki A, Livadas S, Xekouki P, Maniati-Christidi M, et al. Pituitary magnetic resonance imaging in 15 patients with Prop1 gene mutations: pituitary enlargement may originate from the intermediate lobe. J Clin Endocrinol Metab. 2004;89:2200–6.
108. Pernasetti F, Toledo SP, Vasilyev VV, Hayashida CY, Cogan JD, Ferrari C, et al. Impaired adrenocorticotropin-adrenal axis in combined pituitary hormone deficiency caused by a two-base pair deletion (301-302delAG) in the prophet of Pit-1 gene. J Clin Endocrinol Metab. 2000;85:390–7.

109. Fofanova O, Takamura N, Kinoshita E, Vorontsov A, Vladimirova V, Dedov I, et al. MR imaging of the pituitary gland in children and young adults with congenital combined pituitary hormone deficiency associated with PROP1 mutations. AJR Am J Roentgenol. 2000;174: 555–9.
110. Riepe FG, Partsch CJ, Blankenstein O, Monig H, Pfaffle RW, Sippell WG. Longitudinal imaging reveals pituitary enlargement preceding hypoplasia in two brothers with combined pituitary hormone deficiency attributable to PROP1 mutation. J Clin Endocrinol Metab. 2001;86:4353–7.
111. Mendonca BB, Osorio MG, Latronico AC, Estefan V, Lo LS, Arnhold IJ. Longitudinal hormonal and pituitary imaging changes in two females with combined pituitary hormone deficiency due to deletion of A301, G302 in the PROP1 gene. J Clin Endocrinol Metab. 1999;84:942–5.
112. Rosenbloom AL, Almonte AS, Brown MR, Fisher DA, Baumbach L, Parks JS. Clinical and biochemical phenotype of familial anterior hypopituitarism from mutation of the PROP1 gene. J Clin Endocrinol Metab. 1999;84:50–7.
113. Vallette-Kasic S, Barlier A, Teinturier C, Diaz A, Manavela M, Berthezene F, et al. PROP1 gene screening in patients with multiple pituitary hormone deficiency reveals two sites of hypermutability and a high incidence of corticotroph deficiency. J Clin Endocrinol Metab. 2001;86:4529–35.
114. Turton JP, Mehta A, Raza J, Woods KS, Tiulpakov A, Cassar J, et al. Mutations within the transcription factor PROP1 are rare in a cohort of patients with sporadic combined pituitary hormone deficiency (CPHD). Clin Endocrinol (Oxf). 2005;63:10–8.
115. Rainbow LA, Rees SA, Shaikh MG, Shaw NJ, Cole T, Barrett TG, et al. Mutation analysis of POUF-1, PROP-1 and HESX-1 show low frequency of mutations in children with sporadic forms of combined pituitary hormone deficiency and septo-optic dysplasia. Clin Endocrinol (Oxf). 2005;62:163–8.
116. Vieira TC, Boldarine VT, Abucham J. Molecular analysis of PROP1, PIT1, HESX1, LHX3, and LHX4 shows high frequency of PROP1 mutations in patients with familial forms of combined pituitary hormone deficiency. Arq Bras Endocrinol Metabol. 2007;51:1097–103.
117. Cogan JD, Wu W, Phillips III JA, Arnhold IJ, Agapito A, Fofanova OV, et al. The PROP1 2-base pair deletion is a common cause of combined pituitary hormone deficiency. J Clin Endocrinol Metab. 1998;83:3346–9.
118. Andersen B, Rosenfeld MG. POU domain factors in the neuroendocrine system: lessons from developmental biology provide insights into human disease. Endocr Rev. 2001;22:2–35.
119. Tatsumi K, Miyai K, Notomi T, Kaibe K, Amino N, Mizuno Y, et al. Cretinism with combined hormone deficiency caused by a mutation in the PIT1 gene. Nat Genet. 1992;1:56–8.
120. Rhodes SJ, DiMattia GE, Rosenfeld MG. Transcriptional mechanisms in anterior pituitary cell differentiation. Curr Opin Genet Dev. 1994;4:709–17.
121. Rosenfeld MG. POU-domain transcription factors: pou-er-ful developmental regulators. Genes Dev. 1991;5:897–907.
122. Li S, Crenshaw III EB, Rawson EJ, Simmons DM, Swanson LW, Rosenfeld MG. Dwarf locus mutants lacking three pituitary cell types result from mutations in the POU-domain gene pit-1. Nature. 1990;347:528–33.
123. Xu L, Lavinsky RM, Dasen JS, Flynn SE, McInerney EM, Mullen TM, et al. Signal-specific co-activator domain requirements for Pit-1 activation. Nature. 1998;395:301–6.
124. Qi Y, Ranish JA, Zhu X, Krones A, Zhang J, Aebersold R, et al. Atbf1 is required for the Pit1 gene early activation. Proc Natl Acad Sci USA. 2008;105:2481–6.
125. Radovick S, Nations M, Du Y, Berg LA, Weintraub BD, Wondisford FE. A mutation in the POU-homeodomain of Pit-1 responsible for combined pituitary hormone deficiency. Science. 1992;257:1115–8.
126. Pfaffle RW, DiMattia GE, Parks JS, Brown MR, Wit JM, Jansen M, et al. Mutation of the POU-specific domain of Pit-1 and hypopituitarism without pituitary hypoplasia. Science. 1992;257:1118–21.

127. Ohta K, Nobukuni Y, Mitsubuchi H, Fujimoto S, Matsuo N, Inagaki H, et al. Mutations in the Pit-1 gene in children with combined pituitary hormone deficiency. Biochem Biophys Res Commun. 1992;189:851–5.
128. Kelberman D, Rizzoti K, Lovell-Badge R, Robinson IC, Dattani MT. Genetic regulation of pituitary gland development in human and mouse. Endocr Rev. 2009;30:790–829.
129. Cohen LE, Wondisford FE, Salvatoni A, Maghnie M, Brucker-Davis F, Weintraub BD, et al. A "hot spot" in the Pit-1 gene responsible for combined pituitary hormone deficiency: clinical and molecular correlates. J Clin Endocrinol Metab. 1995;80:679–84.
130. Okamoto N, Wada Y, Ida S, Koga R, Ozono K, Chiyo H, et al. Monoallelic expression of normal mRNA in the PIT1 mutation heterozygotes with normal phenotype and biallelic expression in the abnormal phenotype. Hum Mol Genet. 1994;3:1565–8.
131. de Zegher F, Pernasetti F, Vanhole C, Devlieger H, Van den Berghe G, Martial JA. The prenatal role of thyroid hormone evidenced by fetomaternal Pit-1 deficiency. J Clin Endocrinol Metab. 1995;80:3127–30.
132. Holl RW, Pfaffle R, Kim C, Sorgo W, Teller WM, Heimann G. Combined pituitary deficiencies of growth hormone, thyroid stimulating hormone and prolactin due to Pit-1 gene mutation: a case report. Eur J Pediatr. 1997;156:835–7.
133. Aarskog D, Eiken HG, Bjerknes R, Myking OL. Pituitary dwarfism in the R271W Pit-1 gene mutation. Eur J Pediatr. 1997;156:829–34.
134. Rodrigues Martineli AM, Braga M, De Lacerda L, Raskin S, Graf H. Description of a Brazilian patient bearing the R271W Pit-1 gene mutation. Thyroid. 1998;8:299–304.
135. Ward L, Chavez M, Huot C, Lecocq P, Collu R, Decarie JC, et al. Severe congenital hypopituitarism with low prolactin levels and age-dependent anterior pituitary hypoplasia: a clue to a PIT-1 mutation. J Pediatr. 1998;132:1036–8.
136. Oliveira NA, Cecchi CR, Higuti E, Oliveira JE, Jensen TG, Bartolini P, et al. Long-term human growth hormone expression and partial phenotypic correction by plasmid-based gene therapy in an animal model of isolated growth hormone deficiency. J Gene Med. 2010;12:580–5.

Chapter 6
Structural Abnormalities in Congenital Growth Hormone Deficiency

Andrea Secco, Natascia Di Iorgi, and Mohamad Maghnie

Abstract Until the advent of magnetic resonance imaging (MRI), only small advances were made in the field of pituitary imaging. MRI, however, led to an enormous increase in our detailed knowledge of pituitary morphology, thus improving the differential diagnosis of hypopituitarism. Indeed, MRI represents the examination method of choice for evaluating hypothalamic-pituitary-related endocrine diseases thanks to its ability to provide strongly contrasted high-resolution multiplanar and spatial images. Specifically, MRI allows for a detailed and precise anatomical study of the pituitary gland by differentiating between the anterior and posterior pituitary lobes. The MRI identification of pituitary hyperintensity in the posterior part of the *sella*, now considered a marker of neurohypophyseal functional integrity, has been the most striking finding for the diagnosis and understanding of some forms of "idiopathic" and permanent growth hormone deficiency (GHD).

Published data show a number of correlations between pituitary abnormalities as observed on MRI and a patient's endocrine profile. Indeed, several trends have emerged and have been confirmed: (1) normal MRI or anterior pituitary hypoplasia generally indicates isolated GHD which is transient and not confirmed after adult height achievement; (2) patients with MPHD seldom show a normal pituitary gland; (3) the classic triad of ectopic posterior pituitary gland, pituitary stalk hypoplasia/agenesis, and anterior pituitary gland hypoplasia is more frequently reported in MPHD patients and is generally associated with permanent GHD. Pituitary abnormalities have been reported in patients with GHD carrying mutations in several genes encoding transcription factors such as POU1F1, PROP1, HESX1, LHX3, LHX4, GLI2, PITX1, PITX2, SOX3, SOX2, and OTX2. Establishing endocrine and MRI phenotypes is extremely helpful in the selection and management of patients

---

M. Maghnie (✉)
Department of Pediatrics, IRCCS, Giannina Gaslini – University of Genova,
Largo Gerolamo Gaslini 5, 16147 Genova, Italy
e-mail: mohamadmaghnie@ospedale-gaslini.ge.it

K. Ho (ed.), *Growth Hormone Related Diseases and Therapy: A Molecular and Physiological Perspective for the Clinician*, Contemporary Endocrinology,
DOI 10.1007/978-1-60761-317-6_6, © Springer Science+Business Media, LLC 2011

with hypopituitarism, both in terms of possible genetic counseling, as well as that of early diagnosis of evolving anterior pituitary hormone deficiencies.

Keywords Pituitary organogenesis • Imaging of normal pituitary gland • Disorders of early pituitary development • Genes and structural hypothalamic-pituitary abnormalities • Pituitary cellular differentiation • Disorders of GH secretion (IGHD/CPHD)

Introduction

Establishing endocrine and magnetic resonance imaging (MRI) phenotypes is extremely helpful in the selection and management of patients with growth hormone deficiency (GHD), both in terms of possible genetic counselling and the early diagnosis of evolving anterior pituitary hormone deficiencies. Indeed, the advent of MRI has led to significant progress in the understanding of the pathogenesis of disorders that affect the hypothalamic-pituitary area and of several endocrine diseases. Today, there is convincing evidence to support the hypothesis that marked MRI differences in pituitary morphology indicate a range of disorders which affect the organogenesis and function of the anterior pituitary gland with different prognoses.

Specifically, MRI allows a detailed and precise anatomical study of the pituitary gland by differentiating between the anterior and posterior pituitary lobes. The MRI identification of pituitary hyperintensity in the posterior part of the sella, now commonly considered a marker of neurohypophyseal functional integrity, has been the most striking finding for the diagnosis and understanding of anterior and posterior pituitary diseases.

In the meantime, our understanding of the genetic regulation of pituitary gland development in humans and the mouse has increased considerably and mutations in a number of genes have been associated with pituitary dysfunction and abnormal pituitary gland development. The association of extrapituitary malformations accurately defined by MRI has led to a better definition of several conditions linked to pituitary hormone deficiencies and midline defects.

This chapter will discuss the current state of knowledge in our understanding of the etiology of hypopituitarism associated with structural hypothalamic-pituitary abnormalities. The role of transcription factors and genes implicated in abnormal pituitary development and pituitary cell differentiation will be underlined, and the lack of a genetic characterization in a high number of patients with congenital conditions will also be considered.

Pituitary Organogenesis

The development of the pituitary gland is strictly related to that of the forebrain in an early stage of embryogenesis. The three pituitary lobes have different embryological origins: anterior and intermediates lobes are derived from the oral ectoderm, while the posterior lobe develops from neural ectoderm (Fig. 6.1).

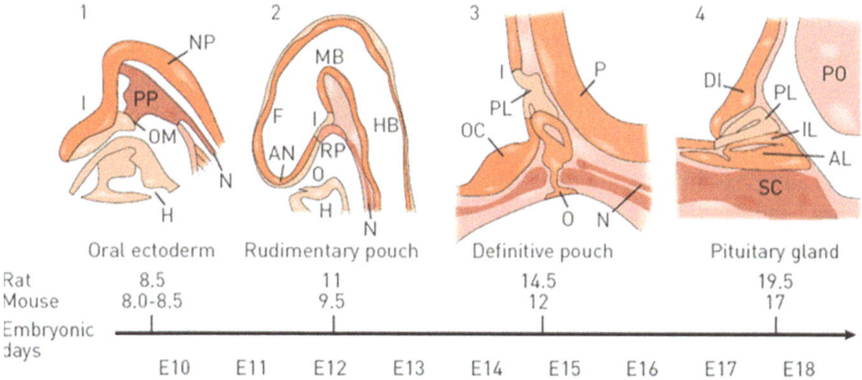

Fig. 6.1 Development of the pituitary gland in rodents. Midsagittal or para-sagittal section drawings of rat embryos showing pituitary development: (1) growth of the preinfundibular portion of the neural plate and establishment of the presumptive Rathke's pouch area; (2) formation of a rudimentary pouch with absence of the mesoderm between the pouch and the floor of the diencephalons; (3) formation of a definitive pouch and posterior lobe with invasion of the neural crest and mesenchymal tissue, and separation of the brain and oral cavities; (4) nascent pituitary gland. Corresponding stages in mouse and rat development indicated. *AL* anterior lobe; *AN* anterior neural pore; *DI* diencephalon; *E* embryonic day; *F* forebrain; *H* heart; *HB* hindbrain; *I* infundibulum; *IL* intermediate lobe; *MB* midbrain; *N* notochord; *NP* neural plate; *O* oral cavity; *OC* optic chiasm; *OM* oral membrane; *P* pontine flexure; *PL* posterior lobe; *PO* pons; *PP* prechordal plate; *RP* Rathke's pouch; *SC* sphenoid cartilage. Reprinted with permission from Sheng and Westphal [2]

The first marker of pituitary organogenesis, occurring in the mouse between 8 and 14 embryonic days, is the formation of an upward evagination of the ectoderm of the primitive oral cavity, known as Rathke's pouch. At the same time, another evagination originates from the neural ectoderm of the dorsal diencephalon, giving rise to an infundibulum (the future posterior pituitary) that maintains close contact with Rathke's pouch during the early stages of pituitary organogenesis. This relationship has a key role for inductive tissue interactions and for the influence exerted by neuroectodermal signaling on the primordial pituitary gland [1, 2].

In recent years, a significant understanding of the molecular mechanisms regulating these processes has been attained, especially in rodents, compared to the relatively smaller amount of existing knowledge about pituitary embryogenesis in humans. However, significant overlaps have been identified between the phenotypes of spontaneous and induced murine mutations and various human clinical syndromes, suggesting that pituitary development in rodents and humans shares some common mechanisms. The sequential expression of a complex cascade of transcription factors is mandatory for normal development of the pituitary gland. The lack of one or more of these mechanisms during different phases may lead to various degrees of pituitary dysmorphogenesis and/or function impairment. Early pituitary development is structured in three steps involving the Lim-homeobox genes Lhx3, Lhx4, and Isl, in response to inductive signals from the diencephalon [3].

The first step, occurring near embryonic day 8.5 in mice, is the formation of a placode in the oral ectoderm in response to the signaling molecule Bmp4 (bone morphogenic protein 4) from the diencephalon. At this stage, cells within the placode express Isl1, probably in response to Bmp4. The second step (at 9.5 dpc) consists of the transformation of the placode into the definitive Rathke's pouch, expressing Lhx3 and Lhx4 in response to a diencephalic signal molecule (believed to be Fgf8). It has been demonstrated that Rathke's pouch development requires the expression of either or both Lhx3 and Lhx4; when both factors are lacking, no pouch development can be recognized, whereas a single normal copy of a single gene is sufficient in mice for the formation of a definitive pouch [3].

During the third step, oral ectoderm cells in the pouch differentiate into the different pituitary cell types to form a nascent anterior pituitary gland and pars intermedia. This differentiation occurs in "zones" according to the relative activity levels of transcriptional factors.

Imaging of Normal Pituitary Gland

MRI Principles and Technical Requirements

MRI is the first choice modality for the evaluation of the hypothalamic-pituitary and parasellar regions due to its multiplanar capability, noninvasiveness, and tissue characterization potential. Special considerations when diagnosing children include incomplete cooperation and/or need for sedation/anesthesia, which may negatively impact scanning time and imaging quality. MRI contraindications include pacemakers, vascular clips or other metallic devices, foreign bodies, and cochlear implants.

The optimal amount of information is obtained with 2–3-mm thick, high-resolution T1- and T2-weighted images in the coronal and sagittal planes. Coronal scans allow for the visualization of the pituitary gland, stalk, chiasm, and parasellar regions. Sagittal images are best suited for the evaluation of the midline plane. Ideally, T1-weighted images should be obtained both before and after intravenous gadolinium-chelate administration, at dosages that depend on clinical suspicion. In fact, pituitary adenomas are better evaluated with half of the usual dose (i.e., 0.05 mmol/Kg body weight), showing as poorly enhanced areas, compared to other pituitary parenchyma. Other conditions, including diabetes insipidus and precocious puberty, require a full dose of gadolinium IV for better detection of lesions with pathological contrast enhancement at the level of the pituitary stalk, hypothalamus, and pineal gland. Postcontrast imaging can safely be dropped only in patients with isolated growth hormone defect, where an anatomic characterization of the pituitary gland and stalk is usually satisfactory. Axial T2-weighted images covering the entire brain should also be obtained in all cases in order to screen for additional abnormalities.

When MRI is unavailable or contraindicated, computerized tomography (CT) remains a valuable alternative for the study of the sellar and parasellar regions. The hazards of ionizing radiation (while minimal) should not be overlooked, especially in infants and young children. The coronal plane is the optimal imaging plane, but direct coronal imaging requires hyperextension of the neck which is difficult to obtain in sedated or uncooperative patients. Helical CT acquisitions may provide high-quality coronal (and sagittal) reformatted images from axial acquisitions. Helical imaging is also useful for minimizing imaging time and radiation dose administration. Examinations should be performed using 1.5–2 mm contiguous slices both before and after intravenous administration of an iodized contrast agent.

Normal Pituitary Anatomy

The pituitary gland consists of two major parts, which differ in origin, structure, and function: the posterior pituitary (or *neurohypophysis*) is a diencephalic downgrowth connected to the hypothalamus, whereas the anterior pituitary (or *adenohypophysis*) is an ectodermal derivative. Both portions include a part of the *infundibulum*, which has a central infundibular stem, called the *pituitary stalk* and is continuous with the *median eminence* of the hypothalamus. Surrounding the infundibular stem is the *pars tuberalis* or *infundibularis*, a component of the adenohypophysis. The latter can be divided into the *pars anterior* or *distalis* and *pars intermedia*, separated from each other in fetal and early postnatal life by the so-called hypophyseal cleft, a vestige of the Rathke's pouch from which the adenohypophysis develops. Usually obliterated in childhood, the hypophyseal cleft may sometimes persist as a variably sized cystic cavitation. In brief, the *neurohypophysis* includes the pars posterior, the infundibulum or pituitary stalk, and the median eminence, while the *adenoypophysis*, which makes up about 75% of the gland, includes the pars anterior, pars intermedia, and pars tuberalis [4] (Fig. 6.2a, b).

Prenatal Appearance

Prenatal imaging using MRI technology has given important insights into morphological abnormalities associated with congenital hypopituitarism. Structures of the hypothalamic-pituitary region, including the pituitary stalk, can be studied in utero by using single-shot fast spin-echo (SS-FSE) T2-weighted images, which are the main technique for fetal MRI diagnosis. The whole pituitary gland is recognizable during prenatal age as a round, ovoid, or irregular triangular hypointense structure on the sellar floor, without differentiation between anterior and posterior lobe. The pituitary stalk, represented by a linear isointense structure connecting the hypothalamic region to the floor of the sella turcica, was detected on coronal or sagittal section in 100% of fetuses with gestational age later than 26 weeks and in approximately 71% of those between 19 and 25 weeks.

Ultrasound is, however, the standard imaging tool for prenatal screening and it can make a significant contribution towards the early diagnosis of hypothalamic-pituitary disease. In particular, the absence of the septum pellucidum, detectable

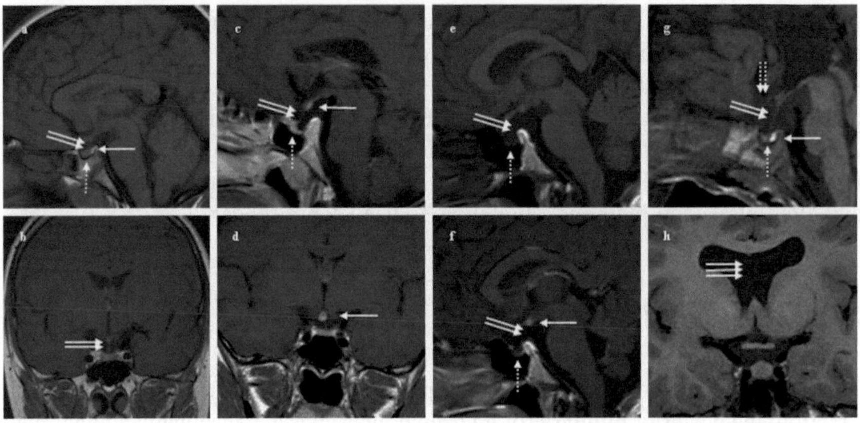

Fig. 6.2 MRI scans showing: (**a, b**) normal hypothalamic-pituitary anatomy in sagittal (**a**) and coronal (**b**) sections. (**c, d**) Ectopic posterior pituitary (EPP) and anterior pituitary hypoplasia with thin pituitary stalk after contrast-medium enhancement (Gd-DTPA) in sagittal (**c**) and coronal (**d**) sections. (**e, f**) Pituitary stalk agenesis with EPP and severe anterior pituitary hypoplasia before (**e**) and after (**f**) Gd-DTPA contrast injection. (**g, h**) Septo-optic dysplasia in sagittal (**g**) and coronal (**h**) sections. Please note that the *single arrow* indicates the posterior pituitary lobe, *double arrows* indicate the pituitary stalk, the *single dotted arrow* indicates the anterior pituitary gland, the *double dotted arrows* indicate absence of the optic chiasm, and the *triple arrows* indicate agenesis of the septum pellucidum

from a gestational age of 22 weeks as fusion of the frontal horns of the lateral ventricles, is a clue for the diagnosis of septo-optic dysplasia (SOD) [5]. In light of reported data, MRI may actually play a complementary role in diagnostic definition when associated abnormalities have been sonographically demonstrated previously and could be helpful both for the early management of neonatal hypopituitarism and for counselling prospective parents.

Postnatal Appearance

The pituitary gland undergoes dynamic changes in size and shape throughout life, reflecting its complex hormonal environment. In newborns, the gland is typically convex, sometimes pear-shaped, with very high signal intensity on T1-weighted images. This appearance changes during the second month of life as it progresses towards adult appearance, i.e., with a flat superior surface and isointensity of the anterior lobe on T1- and T2-weighted images. In addition, in preterm babies a bright anterior pituitary gland has been reported during the first age-corrected 2 months [6]. By the second month of life, the posterior neural lobe of the gland becomes progressively recognizable next to the dorsum sellae as the "bright spot," because of its marked hyperintensity on T1-weighted images. Regardless of its chemical origin, the bright spot serves as an important marker of neurohypophyseal function and, when present, documents integrity of the hypothalamic-neurohypophyseal tract. However, it is important to remember that the bright spot may be absent in 10% of normal individuals. The posterior pituitary does not undergo physiological variations in either size or signal intensity during childhood [4].

Following gadopentetate dimeglumine (Gd-DTPA) administration, marked enhancement of the adenohypophysis and of the infundibulo-tuberal region is clearly evident. The appearance of the posterior lobe blends with the anterior lobe due to its spontaneous hyperintensity. A normal pituitary stalk usually tapers smoothly along its course. It is approximately 3 mm in diameter near the optic chiasm and 2 mm where it inserts into the gland; it should not exceed basilar artery diameter. The pituitary stalk is divided into two parts: one is the neuronal component made up of the track of axons extending from the hypothalamic nuclei down to the axon terminals in the posterior pituitary pouch, while the other is the vascular component that provides blood supply to the anterior pituitary gland from the superior hypophyseal arteries through the pituitary portal system [4].

Normal Pituitary Size

Pituitary height is between 3 and 6 mm during prepubertal age. At puberty, the pituitary gland undergoes profound changes in size and shape, basically represented by marked enlargement. In girls, the gland may swell symmetrically to a height of 10 mm, appearing nearly spherical, whereas in pubertal boys it may reach 7–8 mm [4]. The maximum height of the anterior pituitary is measured (sagittal image) on a plane perpendicular to the sella turcica floor, whereas the volume is calculated using the Di Chiro formula [6]: $V = 1/2$ length x height x width (underestimated) or, alternatively, $V =$ area x width (overestimated). Most data on pituitary size and volume come from one- or two-dimensional indirect measurements [7], whereas a three-dimensional (3D) age-related reference range for pituitary volume is currently lacking. The estimated pituitary volumes calculated from 3D MRI sequences showed a wide variation in pituitary morphology and shape, as well as in the distribution of the signal intensity of both the anterior and posterior pituitary, independent of the subject's age [8]. The overall information provided by this study of 3D assessment of pituitary volume is a significant advance over the information collected using current methods. However, the weak correlation between pituitary height and measured pituitary volume, as well as between calculated and measured pituitary volume, makes the simple evaluation of pituitary height unreliable. This technique cannot be used with children over 10 years of age or in adolescents. The relative discrepancy between the results of this study and those found in two earlier reports [9, 10], as well as the absence of interobserver evaluation, render its results rather incomplete. Table 6.1 summarizes the main normal pituitary MRI characteristics.

Table 6.1 Hypothalamic-pituitary evaluation at MRI

Signal intensity	Anterior pituitary	Signal intensity equivalent to that of the gray matter on precontrast T1 images (hyperintense at age 0–6 weeks)	
	Posterior pituitary	Focal area of high signal at the posterior aspect of the sella on precontrast T1 images (best seen on thin midline sagittal scans)	
Shape	Concave	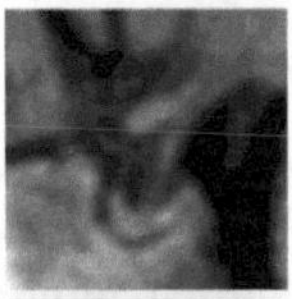	
	Flat	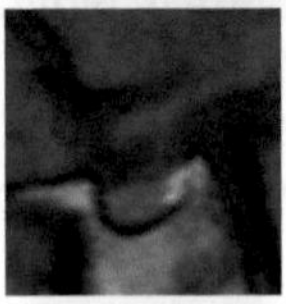	
	Convex	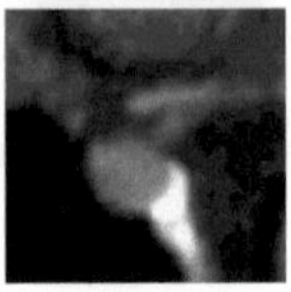	
Size/height	Pituitary height:	0–6 weeks: 2.6–5 mm >2 years: 3–6 mm Puberty: 6–8 mm	Measurement in the sagittal section on a plane perpendicular to the sella turcica 
	Pituitary volume:	V = (length × height × width)/2 (Di Chiro formula: underestimation) V = area x width (overestimation) 3D volumetric MRI	
Position and connections of pituitary stalk	Normal	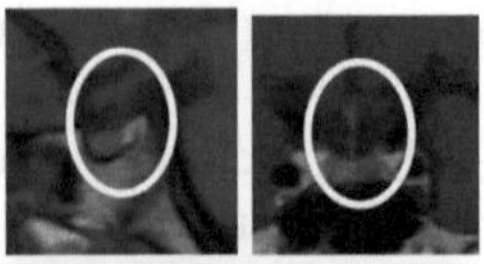	
	Deconnection	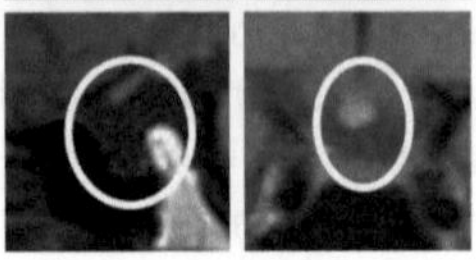	
CNS malformations	Arnold–Chiari type I/II, tentorial anomaly, septum pellucidum agenesis, septo-optic dysplasia, cortical dysplasia, absence of internal carotid artery, vermis dysplasia, periventricular heterotopia, corpus callosum dysgenesis, basilar impression, arachnoid cyst, syringomielia		

Genes and Structural Hypothalamic-Pituitary Abnormalities

Disorders of Early Pituitary Development (Syndromic GHD/CPHD/MPHD)

HESX1

HESX1 (homeobox gene expressed in ES Cells) is a member of the family of the class of paired-like homeobox genes. The HESX1 locus spans 1.7 kb (excluding the 3′ and 5′ untranslated regions), maps on chromosome 3p21.2-p21.1, and contains four coding exons [11]. It has significant repressor activity, playing a role in the regulation of PROP1 expression which takes place during the attenuation of HESX1 expression. Repression domains are found within the N-terminal region and the DNA-binding homeodomain [12, 13].

Hesx1 is an early marker of murine pituitary development. In fact, its expression begins during gastrulation in the primitive forebrain and is subsequently (by 9.0 dpc) restricted to the ventral diencephalon and the thickened layer of oral ectoderm that will give rise to Rathke's pouch. A progressive attenuation of Hesx1 expression takes place from 12 to 13.5 dpc, corresponding to initial pituitary cell differentiation.

Hesx1 null mice show a complex phenotype, with reduction in prospective forebrain tissue, absence of developing optic vesicles, microcephaly, craniofacial dysplasia with a short nose, absence of olfactory placodes, hypothalamic abnormalities, and aberrant morphogenesis of Rathke's pouch [11]. Other features of mutant subjects include hypoplastic nasal cavities and olfactory bulbs, microphthalmia and/or anophthalmia, and abnormalities of the septum pellucidum and corpus callosum. A small proportion of Hesx1-null mutants (5%) show complete anterior pituitary aplasia, with a detectable thickening of the oral ectoderm [12]. These characteristics confirm the strict interactions between the pituitary gland and forebrain development during early embryogenesis and resemble those of a human disease, known as SOD or de Morsier syndrome [14, 15].

SOD is characterized by various combinations of the triad of optic nerve hypoplasia (ONH), midline neuroradiological abnormalities (such as agenesis of the corpus callosum and absence of the septum pellucidum), and pituitary hypoplasia with combined pituitary hormone deficiency [16–19] (Fig. 6.2g, h). An incidence of 1/10,000 live births has been reported [20], without sex difference. This condition shows a significant prevalence in children born from young mothers (mean maternal age 22 years) [18, 21], especially if primigravida [22, 23]. SOD phenotype can be widely variable: 30% of cases have complete manifestations, 62% have hypopituitarism, and 60% have absence of the septum pellucidum [24].

Most cases of SOD are sporadic; however, a small percentage of familial cases have been identified, with autosomal recessive [25–27] or dominant [21, 28–30] inheritance.

The first HESX1 (OMIM 601802) analysis in patients affected by SOD revealed a homozygous missense mutation (R160C) in a consanguineous family with two

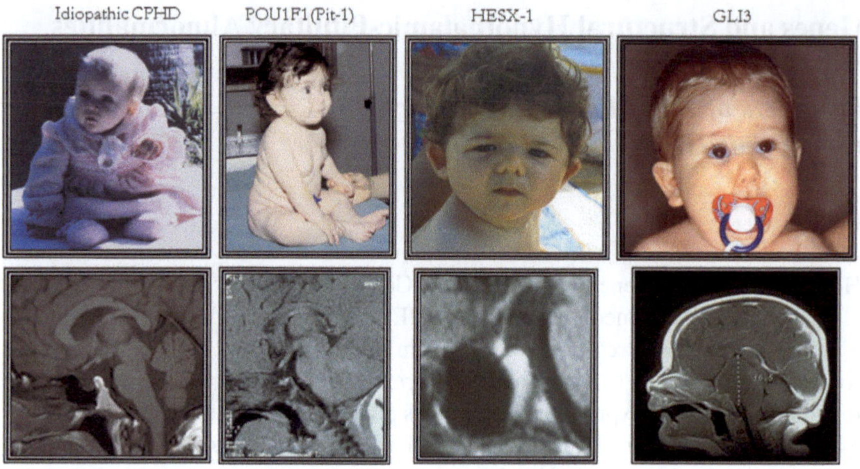

Fig. 6.3 Clinical and MRI phenotype of subjects with GHD and idiopathic combined pituitary hormone deficiency with EPP, anterior pituitary hypoplasia, and pituitary stalk agenesis; POU1F1 mutation and mild anterior pituitary hypoplasia; HESX1 mutation with normal optic chiasm and EPP along the pituitary stalk Pallister–Hall syndrome and GLI3 mutation associated with hypothalamic hamartoblastoma

siblings affected by ONH, hypoplastic corpus callosum, anterior pituitary hypoplasia, ectopic posterior pituitary (EPP), and consequent panhypopituitarism [11, 25]. This substitution leads to a loss of DNA-binding capability. The parents were heterozygous for the mutation and phenotypically normal, suggesting an autosomal recessive inheritance. Other homozygous mutations have been reported in subjects with heterogeneous phenotypes, ranging from evolving hypopituitarism in the absence of a midline and eye defects through SOD and pituitary aplasia [31, 32] (Fig. 6.3). Generally, milder clinical pictures have been reported in subjects with heterozygous mutations, with or without midline abnormalities [28, 29]. HESX1 penetrance may be variable and HESX1 mutations are generally rare even in SOD subjects (<1%), as recently reported [21]. Other genes were recently linked with an SOD phenotype, including SOX2 and OTX2 (See following paragraphs) (Table 6.2).

LHX3

Lhx3 is a member of the LIM family of homeobox genes, characterized by a unique cysteine/histidine-rich zinc-binding LIM domain. The protein contains two tandemly repeated LIM motifs between the N-terminus and the homeodomain which are likely involved in protein–protein interactions [33, 34].

Lhx3 expression in Rathke's pouch is involved, with other Lim-homeobox genes (Lhx4 and Isl1), in early pituitary development, according to a three-step model [2, 35]. During this stage, Lhx3 expression is restricted to the developing anterior and intermediate lobes, where it is maintained until adult age, and by 16.5 dpc, it is not

Table 6.2 Endocrine/pituitary phenotypes and associated features linked with mutations responsible for congenital GHD with structural hypothalamic-pituitary abnormalities

			Hormone deficiencies						
Genes	Inheritance	Locus	GH	PRL	TSH	LH/FSH	ACTH	Pituitary phenotypes	Associated features
GH-1 (IGHD IA)	AR	17q22-q24	■	□	□	□	□	Normal AP/EPP	No
GH1 (IGHD II)	AD	17q22-q24	■	□	■	□	■	Normal AP/small AP	Evolutive endocrinopathy
GHRHR (IGHD IB)	AR	7p15-p14	■	uncertain	□	□	□	Small AP	No
POU1F1	AR/AD	3p11	■	■	■/□	□	□	APH/small/normal AP	No
PROP1	AR	5q	■	■	■	■	■/□	Small/normal/large AP	ACTH deficiency in 1/3 of cases Risk of evolutive endocrinopathy
HESX1	AR/AD	3p21.2-p21.1	■	■/□	■/□	■/□	■/□	Normal AP/AP aplasia/ APH + EPP/isolated APH	SOD – SOD variants Coloboma – CC agensis or hypoplasia
LHX3	AR	9q34.3	■	■	■	■	□	APH/large AP	Stubby neck – anomalies of cervical vertebrae (including atlantic spina bifida occulta) Elevated and anteverted shoulders Limited head rotation in most cases Psychomotor delay – Short arms
LHX4	AD	1q25	■	□	■	■	■	APH/EPP	Chiari I – CC hypoplasia
GLI2	AD	2q14	■	uncertain	■/□	■/□	■/□	APH/AP aplasia	Holoprosencephaly-like features
SOX3	X-linked	Xq27	■	uncertain	■/□	■/□	■/□	APH/EPP	Mental retardation ± holoprosencephaly-like features

(continued)

Table 6.2 (continued)

			Hormone deficiencies						
Genes	Inheritance	Locus	GH	PRL	TSH	LH/FSH	ACTH	Pituitary phenotypes	Associated features
SOX2	AD	3q26.3-q27	■/□	□	■/□	■	■/□	APH/EPP/normal AP	Microphtalmia/anophtalmia, ONH, CNS abnormalities,[a] micropenis, cryptorchidism
OTX2	AD	14q22-23	■	□	■	■	■/□	Normal AP/APH/EPP/PSA	Bilateral anophtalmia, ONH, retinal distrophy, Chiari-I, cleft palate, short stature. Neonatal hypoglicemia, jaundice, and respiratory distress.
Kallmann syndrome genes	X-linked, AD, AR		■	□	■	■	■/□	EPP	

AD autosomal dominant; *AR* autosomal recessive; *filled square* hormone deficiency; *open square* no hormone deficiency; *AP* anterior pituitary; *EPP* ectopic posterior pituitary; *APH* anterior pituitary hypoplasia; *SOD* septo-optic dysplasia; *CC* corpus callosum; *ONH* optic nerve hypoplasia; *CNS* central nervous system; *PS* pituitary stalk; *PSA* pituitary stalk agenesis

[a]Hippocampal abnormalities, partial agenesis of the corpus callosum, absence of septum pellucidum, hypothalamic hamartoma, reduction of white matter

recognizable at the level of the posterior pituitary. This typical pattern suggests that Lhx3 could have significant regulatory activity in the differentiation and function of hormone-producing cell types [33, 36].

LHX3 (OMIM 600577) maps to human chromosome 9q and several mutations have been described to date in GHD patients with various associated clinical and neuroimaging findings, among them are: short and rigid cervical spine with limited head rotation and trunk movement, other anomalies of cervical vertebrae (especially atlantic spina bifida occulta, abnormalities of occipito-atlantoaxial joints in combination with a basilar impression of the dens axis), sensorineural deafness, small or enlarged anterior pituitary, and even a hypointense lesion within the anterior pituitary, compatible with microadenoma (Table 6.2) [37–41].

LHX4

Lhx4 is coexpressed with Lhx3 in Rathke's pouch during early pituitary organogenesis, but subsequently its expression will be limited to the anterior lobe, as demonstrated both in mice and in humans [42].

To date, seven heterozygous LHX4 (OMIM 602146) mutations have been reported in familial or sporadic cases of GH deficiency with variable pituitary and extrapituitary features. All pituitary lineages may be functionally affected, ranging from isolated GHD to combined pituitary hormone deficiency. The first mutation, described in 2001 by Machinis et al. [43], was associated with anterior pituitary hypoplasia, EPP, pituitary stalk agenesis, and extrapituitary malformations, including poorly formed sella turcica, persistent craniopharyngeal canal, and pointed cerebellar tonsils suggestive of Chiari malformation. New LHX4 mutations were identified in following years; they were characterized by various combinations of these clinical findings and their pathogenetic role has been confirmed in functional studies (Table 6.2) [43–47].

SOX3

SOX3 (OMIM 313430) is a member of the SOX (SRY-related high mobility group – HMG – box) family of transcription factors, which conserve the binding motif of the HMG class, typical of the mammalian sex-determining gene, SRY [48]. Among the various SOX genes, SOX3, SOX1, and SOX2 belong to the SOXB1 subfamily and have the highest degree of similarity to SRY.

SOX3 is encoded by a single exon (1.3 kb), mapping to chromosome Xq27. The SOX3 protein is constituted by a 66-amino acid N-terminal domain, a 79-amino acid DNA-binding HMG domain, and a longer C-terminal domain, containing four polyalanine stretches, involved in transcriptional activation [48].

In mice, Sox3 expression has been detected in the ventral diencephalon, including the infundibulum and presumptive hypothalamus, in an early developmental stage, suggesting that SOXB1 subfamily genes are some of the earliest neural markers involved in neuronal determination [49]. Furthermore, SOX3 is not expressed in Rathke's pouch, confirming the inductive interactions between neural structures and the developing pituitary [50].

Targeted SOX3 disruption in mice leads to different and complex phenotypes, with craniofacial abnormalities, midline central nervous system (CNS) defects, reduction in size and fertility, and various grades of pituitary defects [51, 52]. Heterozygous females generally appear normal. Pituitary histology reveals anterior pituitary hypoplasia with additional abnormal cleft between anterior and intermediate lobes, due to bifurcation of Rathke's pouch, while the presumptive hypothalamus is thinner and shorter.

In humans, SOX3 involvement has been described in association with X-linked hypopituitarism and mental retardation, generally due to duplications of various extent in the Xq26-Xq27 region [53–57] or due to polyalanine expansion within the gene [53, 58]. Male patients display various degrees of hypopituitarism, ranging from isolated GHD to CPHD, anterior pituitary and infundibular hypoplasia, EPP, corpus callosum abnormalities, and mental retardation. Females carriers are generally asymptomatic, due to preferential inactivation of the expanded X chromosome. However, five female subjects belonging to a single family were reported with GH deficiency, language disorders, hearing impairment, and craniofacial abnormalities probably related to a Xq26.2-Xq27.1 duplication [59].

Two siblings have also been described by Woods et al. [58] with X-linked hypopituitarism in the absence of significant mental retardation; they carried a submicroscopic duplication of about 690 kb in the Xq27.1 region, containing SOX3. Brain MRI revealed anterior pituitary hypoplasia, EPP, and infundibulum abnormalities in both subjects; the first also reported a cyst of the corpus callosum, while the second showed hypoplastic genitalia in the absence of corpus callosum anomalies.

In vitro functional studies of mutant SOX3 with polyalanine expansion [58, 60] show a partial loss of function of the mutant protein, which localizes almost exclusively within the cytoplasm, as compared to the nuclear localization of the wild-type protein. These data and evidence of Xq26-Xq27 expansions in subjects with infundibular hypoplasia associated with other complex hypothalamic-pituitary phenotypes seem to suggest that a "dose-dependent" effect of SOX3 expression is crucial for the correct development of the diencephalic and hypothalamic-pituitary areas (Table 6.2).

SOX2

The SRY-related HMG-box gene 2, known as SOX2 (OMIM 184429), is a member of the SOXB1 subfamily, like SOX3 and SOX1. In the mouse, Sox2 expression is already detected at 2.5 dpc (at the morula stage) and, by 9.5 dpc, its expression is prevalent in the CNS, sensory placode, branchial arches, and gut endoderm [61, 62]. While Sox2 homozygosis leads invariably to peri-implantation mortality, heterozygous mutants are viable, but show a reduction in size and male fertility and anophtalmia, especially when Sox2 expression decreases below 40% [63], as well as anterior pituitary hypoplasia and various degrees of CPHD [64].

The human counterpart SOX2 is a single exon gene, encoding 317 amino acids, with a structure similar to that of SOX3, mapping to chromosome 3q26.3-q27. SOX2 mutations have been described in subjects with bilateral anophtalmia or microphtalmia and developmental delay, esophageal atresia, and genital abnormalities, including micropenis and cryptorchidism [62, 65–68]. Indeed, a group of patients have been reported with additional features, such as anterior pituitary hypoplasia, hypogonadotropic hypogonadism, hippocampal abnormalities, defects of the corpus callosum, and sensorineural hearing loss [64, 69]. A missense mutation (L75Q) in the HMG domain was recently detected in a female subject with monolateral anopthtalmia and isolated central hypogonadism, without pituitary abnormalities or mental retardation [70]. These findings are consistent with a critical involvement of SOX2 in the regulation of eye development and central gonadotropic function [71] and suggest a strict endocrine follow-up in subjects with SOX-2 mutations in view of the high likelihood of central hypogonadism and evolving hormone deficiencies (Table 6.2).

GLI2

GLI2 (OMIM 165230, gene map locus 2q14) is a mediator of the Sonic hedgehog (SHH) signal transduction (OMIM 600725), involved in the activation or repression of specific target genes under the control of SHH activity. Mutation within SHH may be associated with holoprosencephaly [72], similar to GLI2 mutations, which have been reported in subjects with highly variable phenotypes (even in one subject with no obvious pathologic phenotype), depending on variable phenotypic penetrance [73]. However, GLI2 loss of function leads more commonly to anterior pituitary disorders (ranging from hypoplasia to complete absence) and various degrees of hypopituitarism and craniofacial abnormalities (Table 6.2).

OTX2

The Otx-family of homeobox genes are vertebrate orthologs to the Drosophila orthodenticle homeobox gene and include the Otx1 and Otx2 genes. After gastrulation, Otx2 is expressed in the prosencephalon, mesencephalon, and cerebellum of the developing CNS. In addition, OTX2 mRNA has been found in the retinal pigment epithelium of human fetal retina, and to a lesser extent, in the neural retina. Studies on heterozygous Otx2-deficient mice (Otx2+/–) report various phenotypes including eyes lacking a lens, cornea, or iris, as well as anophthalmia or microphthalmia. Indeed, mice homozygous for a null allele of Otx2 are embryonic lethal due to severe brain abnormalities [74].

In humans, OTX2 (OMIM 600037, gene map locus 14q21-q22) is a three-exon gene encoding a 297-aminoacids protein that includes a homeodomain and a proline, serine, threonine-rich C-terminal region involved in HESX1 transactivation. The role of OTX2 in forebrain and eye development is well known, but its relevance in hypothalamic-pituitary disease has only recently been characterized, based on a previous observation of the involvement of a 14q22-23 deletion (containing OTX2 and other candidate genes) in the development of anophtalmia and pituitary abnormalities [75, 76]. OTX2 mutations have been described in subjects displaying various degrees of congenital hypopituitarism (ranging from isolated GH deficiency (IGFD) to combined pituitary hormone deficiency), normal pituitary morphology at MRI [77], or structural hypothalamic-pituitary abnormalities (anterior pituitary hypoplasia, EPP, and pituitary stalk agenesis) [78], along with anophtalmia and other extrapituitary malformations. Furthermore, a N233S substitution was seen to be related to neonatal hypopituitarism with structural hypothalamic-pituitary abnormalities in the absence of ocular anomalies [79], and a heterozygous missense mutation (S138X) was identified in a boy with early-onset retinal dystrophy and GH deficiency [80].

In the presence of eye abnormalities and pituitary dysfunction, a differential diagnosis between OTX2 and SOX2 mutations and deletion of region q22-23of chromosome 14 should be ruled out. Whether OTX2 mutation might be associated with IGFD, as reported in one of the studies' patients [77], remains to be determined during follow-up for evolving pituitary hormone deficiencies (Table 6.2).

KAL1, FGFR1, FGF8, PROKR2 (Gene Overlap with Kallmann Syndrome)

Kallmann syndrome is characterized by hypogonadotropic hypogonadism with anosmia. In the past 20 years, many studies have shed light on the genetic basis of central hypogonadism [81, 82], but a recent report [83] has established a link between congenital hypopituitarism and mutations of genes involved in the pathogenesis of Kallman syndrome, including KAL1 (OMIM 308700, chromosome Xp22.3), FGFR1 (OMIM 136350, chromosome 8p11.2-p11.1), FGF8 (OMIM 600483, chromosome 10q24), and PROKR2 (OMIM 607123, chromosome 20p13). An EPP was found at brain MRI in one subject, while SOD has been identified in two subjects. No mutations of all other known genes responsible for congenital hypopituitarism have been found in any study subjects. Interestingly, KAL1 (p.V543I) and PROKR2 (p.R85H) mutations had already been described in subjects affected by Kallmann syndrome (Table 6.2). Although this interesting report expands the number of loci involved in congenital hypopituitarism, additional information from clinical, biochemical, neuroimaging, and molecular studies awaits further elucidation.

Disorders of Pituitary Cellular Differentiation

PROP1

PROP1 (acronym of "Prophet of Pit 1") belongs to the paired-like family of homeodomain transcription factors, like HESX1. It is expressed during early pituitary development (10–15.5 dpc in mice) within Rathke's pouch and possess both an activator and repressor function in the regulation of gene transcription.

The Ames dwarf mouse is the most significant animal model of Prop1 mutation [84], which displays DNA binding. Ames mice showed multiple pituitary hormone deficiency causing severe dwarfism, infertility, pituitary hyperplasia during embryonic period and hypoplasia in the adult age, anomalies of glandular vascularization, and absence of the intermediate lobe [85]. The switch from Hesx1 (repressor) to Prop1 (activator) expression leads to the emergence of Pou1f1 (Pit1) (GH, PRL and TSH) and gonadotropes lineages, corresponding to the first phases of pituitary cell differentiation [12].

The first human PROP1 (OMIM 601538, gene map locus 5q) mutation was described in 1998 in patients with GH, TSH, PRL, and gonadotropin deficiency [86], a phenotype partially resembling that of POU1F1 mutation. To date, at least 26 different PROP1 mutations (principally involving homeodomain and DNA-binding capability) have been described in more than 180 patients from 21 different countries. PROP1 mutation is the most common genetic cause of hereditary CPHD, accounting for about 50% of familial cases [87–89]; its incidence is lower in sporadic cases [87, 90, 91].

The clinical phenotype of CPHD in PROP1 defect varies considerably, in time of onset and the severity of the pituitary-derived hormone deficiencies. ACTH deficiency is a striking and unexpected associated feature because PROP1 is not expressed in corticotrophs; it is not linked to a given mutation and it has been reported in as many as one third of these patients. The mechanism of ACTH deficiency remains unclear and longitudinal analysis shows that the frequency of cortisol deficiency may be even higher than generally suspected [92].

In patients with PROP1 mutation, the clinical spectrum of gonadotropin deficiency may range from congenital hypogonadism with micropenis and cryptorchidism to delayed spontaneous puberty. The diverse spectrum suggests a role for PROP1 not only in gonadotroph differentiation (responsible for congenital forms of central hypogonadism) [90], but also in the maintenance of gonadotroph function in adolescence and adult age (responsible for acquired forms of central hypogonadism in carriers of PROP1 mutations) [92]. These considerations should lead to a strict endocrine follow-up in subjects affected by PROP1 mutations, because of the high likelihood of evolving hypopituitarism.

The most frequently observed MRI pituitary feature is a small anterior pituitary gland [93]. There have been reports of enlargement of the anterior pituitary [94]. In some cases, there has been progression from a large and full sella turcica to suprasellar extension of a pituitary mass, followed by cystic change, loss of contrast enhancement, and eventual regression to large empty sella [95–97]. The mass can even wax and wane before involuting [90].

The phenomenon of pituitary hyperplasia followed by shrinkage of the pituitary gland has not been observed in Ames mice and is still puzzling in humans. Several hypotheses have been proposed, including an abnormal proliferation of undifferentiated pituitary cells leading to a "pseudotumor" mass [96]. Recently, it has been shown that changes in the intermediate loby may underlie pituitary enlargement and subsequent spontaneous regression [98]. Indeed, a study by the Camper group provides insight into the understanding of the intriguing transition from pituitary hyperplasia to hypoplasia in humans with such a PROP1 mutation. With loss-of-function, progenitors are trapped and cause a hyperplastic overgrowth which subsequently undergoes apoptosis and disappear [99]. In PROP1 mutation, the posterior pituitary is in fact in its normal location (Table 6.2).

POU1F1

Pou1f1 (POU domain, class 1, transcription factor 1, previously named Pit1, pituitary-specific transcription factor 1) is a member of the POU homeodomain family of transcription factors, containing a POU-specific domain and a DNA-binding domain. Pou1f1 expression takes place in a relatively late phase of pituitary organogenesis (from 13.5 dpc) in future-developing somatotrophs, lactotrophs, and thyrotrophs [100], peaking at 16 dpc and persisting until adult age. Pou1f1 is required for GH, PRL, and TSH production and for Ghrhr expression [101, 102].

Two animal models harbor Pou1f1 mutations: Snell dwarf mice revealed a naturally occurring point substitution within the POU domain (p. W261C), while Jackson dwarf mice showed impaired Pou1f1 expression due to chromosomal rearrangement [103]. Both display embryologically normal GH-, TSH-, and PRL-secreting cell population, but a significant reduction in postnatal proliferation with accelerated apoptosis, suggesting that Pou1f1 is required both for embryological differentiation and for postnatal expansion of these cell types [85].

In addition, Pou1f1 regulate other pituitary cell lineages via inhibition of Gata2, a transcription factor required for specification of gonadotropes and thyrotropes [104].

Since the first report in humans, in 1992 [105], 28 different POU1F1 (OMIM 173110, gene map locus 3p11) mutations have been described in over 60 patients from diverse countries; of these, 23 mutations display autosomal recessive inheritance whereas 5 were autosomal dominant, mostly depending on the type and mutation of the POU1F1 gene [106].

The main features of patients with POU1F1 mutations are early GH and PRL deficiency, while TSH deficiency is highly variable. While it presents early in most cases, it can develop during dolescence [107–109].

Anterior pituitary size in subjects with POU1F1 mutations, as assessed by MRI, is commonly small, though sometimes normal. It appears that there is no relationship between the size of the pituitary and the age of the patient or the type of mutation. An analysis of the type of mutation and of age at the time of MRI study indicates that the most common C-to-T sporadic mutations, changing an arginine to a tryptophan in codon 271 (p.R271W) in one allele of the POU1F1 gene, is associated with normal pituitary size under the age of 2 years [110]. A lack of longitudinal studies currently hinders drawing a final conclusion, though sequential MRI may potentially be valuable for better understanding the role of POU1F1 gene in anterior pituitary cell survival. The pituitary stalk is normal and the posterior pituitary hyperintensity is located in its normal site. No extrapituitary abnormalities have been described to date (Table 6.2, Fig. 6.3).

Among recessive mutations, only four have been described in more than one pedigree: p.R172X [105, 109, 111], p.A158P [107], p.P239S [112], and p.E230K [113, 114].

Disorders of GH Secretion (IGHD/CPHD)

Four distinct familial types of isolated GHD have been defined on the basis of inheritance and other hormone deficiencies. This classification includes IGHD type IA, autosomal recessive with absent endogenous GH; type IB (GHRH receptor mutations), autosomal recessive with diminished GH [115, 116]; type II, autosomal dominant with diminished GH; and type III, X-linked with diminished GH [117].

Basically, genetic forms of isolated GHD, except for anterior pituitary hypoplasia, are not associated with major structural hypothalamic-pituitary abnormalities. Indeed, it is of interest to note that specific mutations related to type IB and to II IGHD are associated with a reduction of anterior pituitary size and evolving pituitary hormone deficiencies, including ACTH and TSH.

IGHD type II shows an autosomal dominant mode of inheritance and is mainly associated with genetic alterations involving loss of exon 3 of the GH-1 gene with subsequent impairment of tertiary structure, due to a "nonsense-mediated altered splicing." The altered splicing leads to the formation of a 17.5 kDa GH isoform, disrupting 22 kDa GH secretion through a dominant negative effect [118]. The first identified mutation at the basis of this mechanism was at the level of the first six bp of the intervening sequences 3 (5′IVS 3) [117]. Type II IGHD due to 5′IVS 3 mutations shows heterogeneous endocrine and MRI phenotypes, depending on which of the six bp is mutated. In a paper by Mullis et al. [119], 57 subjects with IGHD II, belonging to 19 pedigrees, were studied and followed up during growth and showed a phenotypic variability within and between families. Patients with splice site mutations 5′IVS-3 at +1/+2 bp present with a more severe phenotype of decreased cortisol levels and evolving endocrinopathy (mainly TSH and ACTH deficiency in postpubertal age). Pituitary size at diagnosis became smaller in adult age, compared to patients with 5′IVS-3 +5/+6 bp mutations, suggesting that patients developing additional pituitary hormone deficiencies showed a more hypoplastic pituitary gland over time. The authors hypothesized that a possible mechanism underlying the development of additional hormone deficiencies arises from damage secondary to macrophages activity in clearing dying somatotroph, as observed in a transgene mouse study [120].

Furthermore, evolving pituitary hormone deficits and pituitary hypoplasia have also been reported in patients with isolated GHD type II caused by a P89L GH missense mutation [119, 121, 122]. This confirms that misfolded proteins may have an adverse impact on anterior pituitary secretory pathways, leading to a local toxic effect with consequent "acquired" disease (Table 6.2).

Idiopathic GH Deficiency and Structural Hypothalamic-Pituitary Abnormalities

Epidemiology

Idiopathic isolated GHD (IGHD) is the most common sporadic form of hypopituitarism and its incidence is estimated to be between 1/4,000 and 1/10,000 among live births [123–125]. Although in 5–30% of cases there is also an affected first degree relative [126], most cases are sporadic and believed to be related to a combination of environmental and/or developmental risk factors, including viral infections, vascular or degenerative changes, drug exposure, and young maternal age, as is the case for SOD [20, 21, 127]. MRI variably detects structural hypothalamic-pituitary abnormalities in about 12–20% of IGHD cases [128].

Pathophysiology

Epidemiologic data seem to suggest that many of the so-called "idiopathic" cases of IGHD may actually have a genetic cause, as demonstrated by recent studies, but at present this remains unconfirmed. Indeed, in recent decades, various pathophysiological mechanisms have been suggested, including traumatic or vascular etiology.

Rare malformations involving the internal carotid artery system have been reported in association with hypopituitarism [129].

A noteworthy mechanism of anterior hypopituitarism on a possible vascular basis could be related to pituitary stalk diseases associated with impairment of the blood supply to the anterior pituitary. The term "pituitary stalk transection syndrome," applied in a high percentage of cases of birth trauma or breech delivery in patients with GHD and EPP and based on the results of experimental studies showing that surgical section of the pituitary stalk leads to posterior pituitary lobe regeneration at the hypothalamic level, does not appear appropriate [130]. The traumatic origin of such anatomical findings cannot be confirmed on the basis of birth trauma and consequent mechanical section of the pituitary stalk at the time of delivery, because two thirds of these patients are born with no reported adverse perinatal events, with cephalic delivery in approximately 50% and caesarean section in 15% of cases [131]. The precise cause of the high frequency of breech births in cases of agenesis of the stalk and severe hypoplasia of the pituitary remains largely unexplained, though a hypothesis of fetal hypotonia secondary to anatomical and pituitary dysfunction leading to breech presentation appears plausible [132, 133].

A congenital-genetic origin (still to be demonstrated) is supported by various factors such as: the presence of a phenotype appearance compatible with congenital defect; the absence of perinatal adverse events in a large number of cases; the presence of familial cases associated with the same anatomic-functional pituitary characteristics; the association of chromosome defects in some patients; the presence of CNS malformations; and findings of ectopia of the posterior pituitary, with absence of local signs secondary to trauma of the stalk in twins who were born with normal delivery but who died during the first month of life [134–143], as well as in patients with syndromes associated with hypothalamic-pituitary abnormalities. These include Fanconi's anemia [144], Poland's syndrome [145], Arthrogryposis multiplex congenita [146], Dubowitz syndrome [147], Worster-Drought syndrome, congenital bilateral perisylvian syndrome, neuronal migration defect, epilepsy, neuromotor retardation [148], situs inversus totalis [149], or mutations of transcription factors involved in pituitary development (HESX-1 and LHX-4), as previously reported.

Phenotype and Function

Clinical Phenotype

Idiopathic GH deficiency is a heterogeneous condition whose pathophysiology remains largely unknown. The typical GHD clinical phenotype in childhood is short stature and persistent growth failure associated with frontal bossing, midfacial hypoplasia, truncal adiposity, small genitalia in the male, and/or hypoglycemia and cholestatic jaundice. However, this presentation tends to be the exception rather than the rule and thus the clinical phenotype may not be particularly notable (Fig. 6.3).

A recent study analyzed the clinical phenotype of GHD photographs of 137 patients (73 IGHD and 64 MPHD). Standardized frontal and lateral digital pictures were taken of each patient and analyzed using specific software. The analysis revealed that Canthal Index (CI), the relative distance between the eyes, was related to pituitary morphology. Patients with an EPP had significantly higher CI values than patients without EPP, with a cut-off value being CI > 39 for identifying children with the highest probability of having EPP.

The combination of CI > 39 in the presence of hormonal deficiencies in addition to GHD strongly predicted EPP: 93% of the patients with a CI > 39 and additional hormonal deficiencies had EPP, compared to 77% of patients with additional hormonal deficiencies but a CI <39, and 29% of patients without any of these criteria. The association between Canthal Index, measured from digital pictures, and EPP could be caused by altered midline development, affecting both the pituitary gland and the facial structures of GHD patients [150].

MRI Evaluation of Hypothalamic-Pituitary Region and Associated CNS Abnormalities

These include (a) a normal or hypoplastic pituitary gland without anatomical abnormalities of the hypothalamus or pituitary stalk, and (b) moderate to severe (vertical height ≤3 mm) hypoplastic pituitary gland with EPP located anywhere from the median eminence to the distal stalk (Fig. 6.2c–f).

Table 6.3 Reports from the literature on MRI features in subjects with idiopathic GHD (number ≥ 35)

		MRI			Presenting pituitary functions	
References	Patients	EPP	HA	NA	IGHD (EPP)	CPHD (EPP)
Bressani et al. [115]	57	34 (59%)	23 (40%)	0	36 (37%)	21 (95%)
Maghnie et al. [134, 138]	45	20 (44%)	13 (29%)	12 (26%)	33 (24%)	8 (100%)
Abrahams et al. [153]	35	15 (43%)	20 (57%)	15 (43%)	2 (10%)	13 (87%)
Argyropoulou et al. [132]	46	29 (63%)	10 (21%)	7 (15%)	26 (38%)	20 (95%)
Triulzi et al. [131]	101	59 (48%)	ND	ND	67 (44%)	34 (85%)
Pinto et al. [137]	51	51 (100%)	41 (83%)	6 (13%)	16 (100%)	35 (100%)
Nagel et al. [154]	56	24 (42%)	11 (19%)	17 (30%)	43 (27%)	13 (92%)
Kornreich et al. [155]	51	41 (80%)	0	10 (20%)	23 (56%)	28 (100%)
Hamilton et al. [142]	35	25 (71%)	3 (<1%)	2 (<1%)	20 (70%)	15 (73%)
Arifa et al. [143]	100	20 (20%)	28 (28%)	41 (41%)	79 (ND)	19 (ND)
Bozzola et al. [156]	93	30 (32%)	69 (65%)	24 (35%)	5 (8%)	25 (76%)
Osorio et al. [95]	76	54 (71%)	36 (84%)	ND	33[a] (55%)	43 (57%)[b]
Arends et al. [157]	39	24 (61%)	22 (56%)	17 (44%)	11 (44%)	13 (87%)
Melo et al. [159]	62	48 (100%)[c]	0	0	10 (100%)[c]	38 (100%)[c]

EPP ectopic posterior pituitary; *HA* hypoplastic anterior pituitary with normal located posterior pituitary; *NA* normal anatomy; *ND* not determined

[a]1 patient with GH-1 deletion

[b]5 Prop1 and 1 Pit1 defects

[c]Posterior pituitary position not evaluated in 14 subjects

Isolated GHD is more commonly observed in the first category and multiple pituitary hormone deficiencies (MPHD) in the second. An empty sella occurs in 10% of IGHD [151]. MPHD may be associated with a spectrum of cerebral malformations such as Arnold–Chiari I, and II, agenesis of the septum pellucidum, SOD, vermis dysplasia, syringomyelia, absence of the internal carotid artery, dysgenesis of the corpus callosum, arachnoid cysts, and tentorium anomalies with basilar impression [7, 95, 131, 134, 135, 137, 143, 152–159]. The frequency of these radiological findings and their spectrum of pituitary hormone deficiencies is variable. In particular, MPHD is more frequently associated with EPP and with anterior pituitary hypoplasia that isolated GHD (Table 6.3). The variability between different studies could be attributed variously to the degree of restriction in the studies' diagnostic criteria, to the diagnostic limits of GHD (transitory deficits, recovery, false positives, etc.), and/or to the lack of a convincing standard for normal size of the pituitary gland among the prepubertal pediatric population. Idiopathic MPHD has been reported less frequently in association with anterior pituitary hypoplasia and more frequently with normal posterior pituitary and pituitary stalk; the high frequency of sporadic forms of idiopathic GHD associated with EPP in the absence of genetic identification remains intriguing, suggesting that other factors may play a role as well (Fig. 6.4).

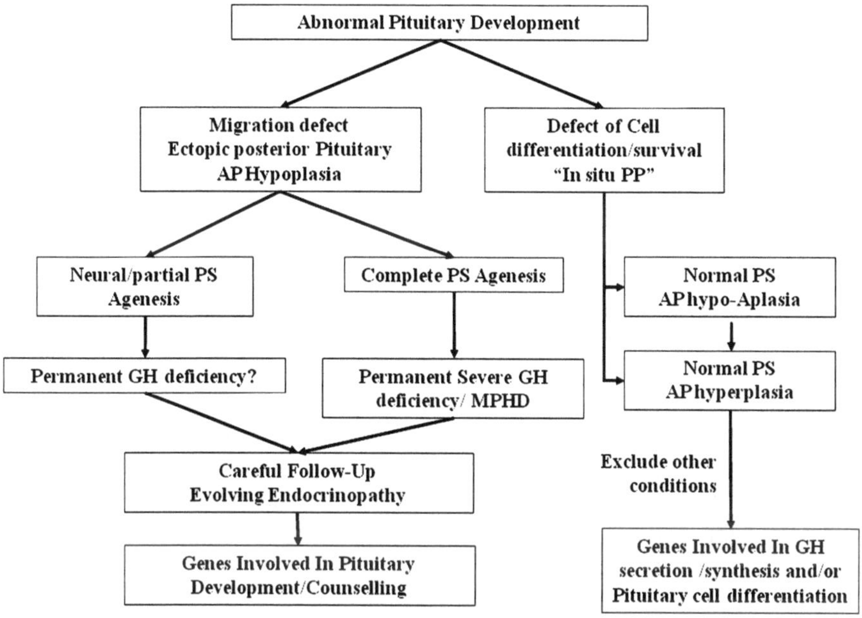

Fig. 6.4 Differential diagnosis and follow-up of congenital GHD with abnormal pituitary development. *AP* anterior pituitary; *PP* posterior pituitary; *PS* pituitary stalk; *MPHD* multiple pituitary hormone deficiency

Prognosis

MRI Evaluation of Risk of Evolving Hypopituitarism During Childhood

A detailed study of the pituitary stalk should be carried out with administration of Gd-DTPA. The presence of a vascular component of the stalk has prognostic significance since patients with agenesis of the pituitary stalk run a greater risk of developing multiple pituitary hormone deficits than those who show a vascular residue of the stalk (Fig. 6.2e, f). Patients in whom a pituitary stalk cannot be identified after Gd-DTPA have a risk of developing CPHD evolving to panhypopituitarism that is 27 times greater than those with a residual vascular pituitary stalk [132, 133].

Height Response to GH Treatment

Evidence of congenital developmental abnormalities at MR imaging is a stronger predictor than the maximual-stimulated GH response of height gain during the first 3 years of rhGH treatment in prepubertal children with GHD [160]. In addition, whereas patients with normal pituitary size treated with rhGH do not exhibit a difference in catch-up growth or in achievement of adult height compared to untreated subjects with idiopathic short stature, significantly higher final stature with a greater height gain was, however, observed in patients with pituitary gland abnormalities at MRI [7, 161].

Hormonal status does not significantly influence statural prognosis in terms of adult height and pubertal height gain, as recently demonstrated in two cohorts of subjects with childhood-onset IGHD and MPHD associated with structural hypothalamic-pituitary abnormalities [162].

Implications of MRI Pituitary Morphology After Adult Height Achievement in the Definition of Transient or Permanent GHD

The diagnosis of GHD in young adults is not straightforward and represents a major clinical challenge. The key predictors of persistent GHD are the severity of the original GH deficiency, the presence of additional pituitary hormone deficits, severely low IGF-I concentration, and structural HP abnormalities [163, 164]. We have shown that patients with GHD and congenital hypothalamic-pituitary abnormalities might not require reevaluation of GH secretion, whereas patients with isolated GHD and normal or small pituitary gland should be retested well before the attainment of adult height [163]. MRI findings of the hypothalamic-pituitary area in patients with GHD may be the most important criterion upon which the decision to reevaluate the patient should be based, rather than response to pharmacological stimulation.

These observations are in line with another recent study evaluating GH secretion after arginine stimulation test in two cohorts of subjects with childhood-onset GHD and ectopic or normally located posterior pituitary (EPP vs. NPP) at the attainment of adult height. The study demonstrates that the presence of an EPP compared to an NPP increases the likelihood of persistent GHD by 26% [165].

Two recent Consensus Statements on the management of young adults with childhood-onset GHD during transition phase [166, 167] state that patients with severe GHD in childhood with or without two or three additional hormone deficits, possibly due to a defined genetic cause, those with severe GHD due to structural hypothalamic-pituitary abnormalities, with CNS tumors, or patients having received high-dose cranial irradiation all have a high likelihood of permanent GHD after adult height attainment.

A subgroup of subjects with idiopathic GHD of childhood onset presenting with congenital structural hypothalamic-pituitary abnormalities confirms that GHD patients – defined "a priori" as those with GH response <5 μg/L and with anterior pituitary hypoplasia, pituitary stalk agenesis, and posterior pituitary ectopia at the level of the median eminence – are probable candidates for permanent GHD in adult life [168], while those with less severe MRI features have an uncertain diagnosis or a likelihood of normal GH response after stimulation tests. These findings have important clinical implications in the diagnosis and prognosis of GHD after adult height achievement. By applying the criterion of peak GH values of less than 3 or 5 μg/L, several misdiagnosed GHD subjects would be wrongly excluded from a potentially beneficial renewal of GH replacement treatment [169]. The comparison between a cohort of subjects with a high likelihood of permanent GHD (defined as those with hypothalamic-pituitary abnormalities at MRI) and a control group showed that the lowest values observed in normal subjects of peak GH after insulin tolerance test (ITT) of 6.1 μg/L and IGF-I of −1.7 SDS could identify GHD subjects with a sensitivity, respectively, of 96 and 77% and a sensibility of 100% for both groups [169]. An additional recent study that compared two groups of subjects with high and low likelihood of permanent GHD confirmed that a peak GH response after ITT of less than 5.62 μg/L can distinguish between the two groups, correctly classifying 87.34% of subjects (with sensitivity of 77.42% and sensibility of 93.75%). Indeed, IGF-I measurement is a useful marker of the degree of GHD, ensuring a high discrimination of patients with severe GHD for cut-off levels of ≤−2.8 SDS [170].

Pituitary function should be periodically assessed in subjects with pituitary stalk agenesis and IGHD or CPHD, as they may develop additional pituitary hormone deficiencies and deterioration of metabolic parameters [171]. In our recent study, ACTH deficiency characterized a subset of patients with idiopathic GHD and pituitary stalk abnormalities, revealing that several of them had undiagnosed subclinical ACTH deficiency [172].

References

1. Takuma N, Sheng HZ, Furuta Y, et al. Formation of Rathke's pouch requires dual induction from the diencephalon. Development. 1998;125(23):4835–40.
2. Sheng HZ, Westphal H. Early steps in pituitary organogenesis. Trends Genet. 1999;15(6):236–40.
3. Sheng HZ, Moriyama K, Yamashita T, et al. Multistep control of pituitary organogenesis. Science. 1997;278(5344):1809–12.
4. Tortori-Donati P, Rossi A, Biancheri R. Sellar and suprasellar disorders. In: Tortori-Donati P, editor. Pediatric neuroradiology. Berlin: Springer; 2005. p. 855–91.
5. Malinger G, Lev D, Kidron D, Heredia F, Hershkovitz R, Lerman-Sagie T. Differential diagnosis in fetuses with absent septum pellucidum. Ultrasound Obstet Gynecol. 2005;25(1):42–9.
6. Argyropoulou MI, Xydis V, Kiortsis DN, et al. Pituitary gland signal in pre-term infants during the first year of life: an MRI study. Neuroradiology. 2004;46(12):1031–5.
7. Maghnie M, Ghirardello S, Genovese E. Magnetic resonance imaging of the hypothalamus-pituitary unit in childrensuspected of hypopituitarism: who, how and when to investigate. J Endocrinol Invest. 2004;27(5):496–509.
8. Fink AM, Vidmar S, Kumbla S, et al. Age-related pituitary volumes in prepubertal children with normal endocrine function: volumetric magnetic resonance data. J Clin Endocrinol Metab. 2005;90(6):3274–8.
9. Takano K, Utsunomiya H, Ono H, Ohfu M, Okazaki M. Normal development of the pituitary gland: assessment with three-dimensional MR volumetry. AJNR Am J Neuroradiol. 1999;20(2):312–5.
10. Marziali S, Gaudiello F, Bozzao A, et al. Evaluation of anterior pituitary gland volume in childhood using three-dimensional MRI. Pediatr Radiol. 2004;34(7):547–51.
11. Dattani MT, Martinez-Barbera JP, Thomas PQ, et al. Mutations in the homeobox gene HESX1/Hesx1 associated with septo-optic dysplasia in human and mouse. Nat Genet. 1998;19(2):125–33.

12. Dasen JS, Barbera JP, Herman TS, et al. Temporal regulation of a paired-like homeodomain repressor/TLE corepressor complex and a related activator is required for pituitary organogenesis. Genes Dev. 2001;15(23):3193–207.
13. Brickman JM, Clements M, Tyrell R, et al. Molecular effects of novel mutations in Hesx1/HESX1 associated with human pituitary disorders. Development. 2001;128(24):5189–99.
14. De Morsier G. Studies on malformation of cranio-encephalic sutures. III. Agenesis of the septum lucidum with malformation of the optic tract. Schweiz Arch Neurol Psychiatr. 1956;77(1–2):267–92.
15. Kelberman D, Dattani MT. Hypothalamic and pituitary development: novel insights into the aetiology. Eur J Endocrinol. 2007;157 Suppl 1:S3–14.
16. St John JR, Reeves DL. Congenital absence of the septum pellucidum: a review of the literature with case report. Am J Surg. 1957;94(6):974–80.
17. Hoyt WF, Kaplan SL, Grumbach MM, Glaser JS. Septo-optic dysplasia and pituitary dwarfism. Lancet. 1970;1(7652):893–4.
18. Arslanian SA, Rothfus WE, Foley Jr TP, Becker DJ. Hormonal, metabolic, and neuroradiologic abnormalities associated with septo-optic dysplasia. Acta Endocrinol (Copenh). 1984;107(2):282–8.
19. Stanhope R, Preece MA, Brook CG. Hypoplastic optic nerves and pituitary dysfunction. A spectrum of anatomical and endocrine abnormalities. Arch Dis Child. 1984;59(2):111–4.
20. Patel L, McNally RJ, Harrison E, Lloyd IC, Clayton PE. Geographical distribution of optic nerve hypoplasia and septo-optic dysplasia in Northwest England. J Pediatr. 2006;148(1):85–8.
21. McNay DE, Turton JP, Kelberman D, et al. HESX1 mutations are an uncommon cause of septooptic dysplasia and hypopituitarism. J Clin Endocrinol Metab. 2007;92(2):691–7.
22. Acers TE. Optic nerve hypoplasia: septo-optic-pituitary dysplasia syndrome. Trans Am Ophthalmol Soc. 1981;79:425–57.
23. Izenberg N, Rosenblum M, Parks JS. The endocrine spectrum of septo-optic dysplasia. Clin Pediatr (Phila). 1984;23(11):632–6.
24. Morishima A, Aranoff GS. Syndrome of septo-optic-pituitary dysplasia: the clinical spectrum. Brain Dev. 1986;8(3):233–9.
25. Wales JK, Quarrell OW. Evidence for possible Mendelian inheritance of septo-optic dysplasia. Acta Paediatr. 1996;85(3):391–2.
26. Blethen SL, Weldon VV. Hypopituitarism and septooptic "dysplasia" in first cousins. Am J Med Genet. 1985;21(1):123–9.
27. Benner JD, Preslan MW, Gratz E, Joslyn J, Schwartz M, Kelman S. Septo-optic dysplasia in two siblings. Am J Ophthalmol. 1990;109(6):632–7.
28. Thomas PQ, Dattani MT, Brickman JM, et al. Heterozygous HESX1 mutations associated with isolated congenital pituitary hypoplasia and septo-optic dysplasia. Hum Mol Genet. 2001;10(1):39–45.
29. Cohen RN, Cohen LE, Botero D, et al. Enhanced repression by HESX1 as a cause of hypopituitarism and septooptic dysplasia. J Clin Endocrinol Metab. 2003;88(10):4832–9.
30. Tajima T, Hattorri T, Nakajima T, et al. Sporadic heterozygous frameshift mutation of HESX1 causing pituitary and optic nerve hypoplasia and combined pituitary hormone deficiency in a Japanese patient. J Clin Endocrinol Metab. 2003;88(1):45–50.
31. Sobrier ML, Netchine I, Heinrichs C, et al. Alu-element insertion in the homeodomain of HESX1 and aplasia of the anterior pituitary. Hum Mutat. 2005;25(5):503.
32. Sobrier ML, Maghnie M, Vie-Luton MP, et al. Novel HESX1 mutations associated with a life-threatening neonatal phenotype, pituitary aplasia, but normally located posterior pituitary and no optic nerve abnormalities. J Clin Endocrinol Metab. 2006;91(11):4528–36.
33. Bach I, Rhodes SJ, Pearse II RV, et al. P-Lim, a LIM homeodomain factor, is expressed during pituitary organ and cell commitment and synergizes with Pit-1. Proc Natl Acad Sci USA. 1995;92(7):2720–4.
34. Schmitt S, Biason-Lauber A, Betts D, Schoenle EJ. Genomic structure, chromosomal localization, and expression pattern of the human LIM-homeobox3 (LHX 3) gene. Biochem Biophys Res Commun. 2000;274(1):49–56.

35. Treier M, Gleiberman AS, O'Connell SM, et al. Multistep signaling requirements for pituitary organogenesis in vivo. Genes Dev. 1998;12(11):1691–704.
36. Zhadanov AB, Bertuzzi S, Taira M, Dawid IB, Westphal H. Expression pattern of the murine LIM class homeobox gene Lhx3 in subsets of neural and neuroendocrine tissues. Dev Dyn. 1995;202(4):354–64.
37. Rajab A, Kelberman D, de Castro SC, et al. Novel mutations in LHX3 are associated with hypopituitarism and sensorineural hearing loss. Hum Mol Genet. 2008;17(14):2150–9.
38. Netchine I, Sobrier ML, Krude H, et al. Mutations in LHX3 result in a new syndrome revealed by combined pituitary hormone deficiency. Nat Genet. 2000;25(2):182–6.
39. Bhangoo AP, Hunter CS, Savage JJ, et al. Clinical case seminar: a novel LHX3 mutation presenting as combined pituitary hormonal deficiency. J Clin Endocrinol Metab. 2006;91(3):747–53.
40. Pfaeffle RW, Savage JJ, Hunter CS, et al. Four novel mutations of the LHX3 gene cause combined pituitary hormone deficiencies with or without limited neck rotation. J Clin Endocrinol Metab. 2007;92(5):1909–19.
41. Kristrom B, Zdunek AM, Rydh A, Jonsson H, Sehlin P, Escher SA. A novel mutation in the LIM homeobox 3 gene is responsible for combined pituitary hormone deficiency, hearing impairment, and vertebral malformations. J Clin Endocrinol Metab. 2009;94(4):1154–61.
42. Sobrier ML, Attie-Bitach T, Netchine I, Encha-Razavi F, Vekemans M, Amselem S. Pathophysiology of syndromic combined pituitary hormone deficiency due to a LHX3 defect in light of LHX3 and LHX4 expression during early human development. Gene Expr Patterns. 2004;5(2):279–84.
43. Machinis K, Pantel J, Netchine I, et al. Syndromic short stature in patients with a germline mutation in the LIM homeobox LHX4. Am J Hum Genet. 2001;69(5):961–8.
44. Tajima T, Hattori T, Nakajima T, Okuhara K, Tsubaki J, Fujieda K. A novel missense mutation (P366T) of the LHX4 gene causes severe combined pituitary hormone deficiency with pituitary hypoplasia, ectopic posterior lobe and a poorly developed sella turcica. Endocr J. 2007;54(4):637–41.
45. Pfaeffle RW, Hunter CS, Savage JJ, et al. Three novel missense mutations within the LHX4 gene are associated with variable pituitary hormone deficiencies. J Clin Endocrinol Metab. 2008;93(3):1062–71.
46. Castinetti F, Saveanu A, Reynaud R, et al. A novel dysfunctional LHX4 mutation with high phenotypical variability in patients with hypopituitarism. J Clin Endocrinol Metab. 2008;93(7):2790–9.
47. Tajima T, Yorifuji T, Ishizu K, Fujieda K. A novel mutation (V101A) of the LHX4 gene in a Japanese patient with combined pituitary hormone deficiency. Exp Clin Endocrinol Diabetes. 2010;118:405–9.
48. Stevanovic M, Lovell-Badge R, Collignon J, Goodfellow PN. SOX3 is an X-linked gene related to SRY. Hum Mol Genet. 1993;2(12):2013–8.
49. Pevny L, Placzek M. SOX genes and neural progenitor identity. Curr Opin Neurobiol. 2005;15(1):7–13.
50. Rizzoti K, Lovell-Badge R. Early development of the pituitary gland: induction and shaping of Rathke's pouch. Rev Endocr Metab Disord. 2005;6(3):161–72.
51. Rizzoti K, Brunelli S, Carmignac D, Thomas PQ, Robinson IC, Lovell-Badge R. SOX3 is required during the formation of the hypothalamo-pituitary axis. Nat Genet. 2004;36(3):247–55.
52. Weiss J, Meeks JJ, Hurley L, Raverot G, Frassetto A, Jameson JL. Sox3 is required for gonadal function, but not sex determination, in males and females. Mol Cell Biol. 2003;23(22): 8084–91.
53. Laumonnier F, Ronce N, Hamel BC, et al. Transcription factor SOX3 is involved in X-linked mental retardation with growth hormone deficiency. Am J Hum Genet. 2002;71(6):1450–5.
54. Solomon NM, Nouri S, Warne GL, Lagerstrom-Fermer M, Forrest SM, Thomas PQ. Increased gene dosage at Xq26-q27 is associated with X-linked hypopituitarism. Genomics. 2002;79(4):553–9.

55. Hamel BC, Smits AP, Otten BJ, van den Helm B, Ropers HH, Mariman EC. Familial X-linked mental retardation and isolated growth hormone deficiency: clinical and molecular findings. Am J Med Genet. 1996;64(1):35–41.
56. Lagerstrom-Fermer M, Sundvall M, Johnsen E, et al. X-linked recessive panhypopituitarism associated with a regional duplication in Xq25-q26. Am J Hum Genet. 1997;60(4):910–6.
57. Hol FA, Schepens MT, van Beersum SE, et al. Identification and characterization of an Xq26-q27 duplication in a family with spina bifida and panhypopituitarism suggests the involvement of two distinct genes. Genomics. 2000;69(2):174–81.
58. Woods KS, Cundall M, Turton J, et al. Over- and underdosage of SOX3 is associated with infundibular hypoplasia and hypopituitarism. Am J Hum Genet. 2005;76(5):833–49.
59. Stankiewicz P, Thiele H, Schlicker M, et al. Duplication of Xq26.2-q27.1, including SOX3, in a mother and daughter with short stature and dyslalia. Am J Med Genet A. 2005;138(1):11–7.
60. Albrecht AN, Kornak U, Boddrich A, et al. A molecular pathogenesis for transcription factor associated poly-alanine tract expansions. Hum Mol Genet. 2004;13(20):2351–9.
61. Wood HB, Episkopou V. Comparative expression of the mouse Sox1, Sox2 and Sox3 genes from pre-gastrulation to early somite stages. Mech Dev. 1999;86(1–2):197–201.
62. Williamson KA, Hever AM, Rainger J, et al. Mutations in SOX2 cause anophthalmia-esophageal-genital (AEG) syndrome. Hum Mol Genet. 2006;15(9):1413–22.
63. Taranova OV, Magness ST, Fagan BM, et al. SOX2 is a dose-dependent regulator of retinal neural progenitor competence. Genes Dev. 2006;20(9):1187–202.
64. Kelberman D, Rizzoti K, Avilion A, et al. Mutations within Sox2/SOX2 are associated with abnormalities in the hypothalamo-pituitary-gonadal axis in mice and humans. J Clin Invest. 2006;116(9):2442–55.
65. Fantes J, Ragge NK, Lynch SA, et al. Mutations in SOX2 cause anophthalmia. Nat Genet. 2003;33(4):461–3.
66. Ragge NK, Lorenz B, Schneider A, et al. SOX2 anophthalmia syndrome. Am J Med Genet A. 2005;135(1):1–7; discussion 8.
67. Hagstrom SA, Pauer GJ, Reid J, et al. SOX2 mutation causes anophthalmia, hearing loss, and brain anomalies. Am J Med Genet A. 2005;138A(2):95–8.
68. Zenteno JC, Gascon-Guzman G, Tovilla-Canales JL. Bilateral anophthalmia and brain malformations caused by a 20-bp deletion in the SOX2 gene. Clin Genet. 2005;68(6):564–6.
69. Sisodiya SM, Ragge NK, Cavalleri GL, et al. Role of SOX2 mutations in human hippocampal malformations and epilepsy. Epilepsia. 2006;47(3):534–42.
70. Sato N, Kamachi Y, Kondoh H, et al. Hypogonadotropic hypogonadism in an adult female with a heterozygous hypomorphic mutation of SOX2. Eur J Endocrinol. 2007;156(2):167–71.
71. Kelberman D, de Castro SC, Huang S, et al. SOX2 plays a critical role in the pituitary, forebrain, and eye during human embryonic development. J Clin Endocrinol Metab. 2008;93(5): 1865–73.
72. Roessler E, Belloni E, Gaudenz K, et al. Mutations in the human Sonic Hedgehog gene cause holoprosencephaly. Nat Genet. 1996;14(3):357–60.
73. Roessler E, Du YZ, Mullor JL, et al. Loss-of-function mutations in the human GLI2 gene are associated with pituitary anomalies and holoprosencephaly-like features. Proc Natl Acad Sci USA. 2003;100(23):13424–9.
74. Larsen KB, Lutterodt M, Rath MF, Moller M. Expression of the homeobox genes PAX6, OTX2, and OTX1 in the early human fetal retina. Int J Dev Neurosci. 2009;27(5):485–92.
75. Lemyre E, Lemieux N, Decarie JC, Lambert M. Del(14)(q22.1q23.2) in a patient with anophthalmia and pituitary hypoplasia. Am J Med Genet. 1998;77(2):162–5.
76. Nolen LD, Amor D, Haywood A, et al. Deletion at 14q22-23 indicates a contiguous gene syndrome comprising anophthalmia, pituitary hypoplasia, and ear anomalies. Am J Med Genet A. 2006;140(16):1711–8.
77. Dateki S, Fukami M, Sato N, Muroya K, Adachi M, Ogata T. OTX2 mutation in a patient with anophthalmia, short stature, and partial growth hormone deficiency: functional studies using the IRBP, HESX1, and POU1F1 promoters. J Clin Endocrinol Metab. 2008;93(10):3697–702.

78. Tajima T, Ohtake A, Hoshino M, et al. OTX2 loss of function mutation causes anophthalmia and combined pituitary hormone deficiency with a small anterior and ectopic posterior pituitary. J Clin Endocrinol Metab. 2009;94(1):314–9.
79. Diaczok D, Romero C, Zunich J, Marshall I, Radovick S. A novel dominant negative mutation of OTX2 associated with combined pituitary hormone deficiency. J Clin Endocrinol Metab. 2008;93(11):4351–9.
80. Henderson RH, Williamson KA, Kennedy JS, et al. A rare de novo nonsense mutation in OTX2 causes early onset retinal dystrophy and pituitary dysfunction. Mol Vis. 2009;15: 2442–7.
81. Bhangoo A, Jacobson-Dickman E. The genetics of idiopathic hypogonadotropic hypogonadism:unraveling the biology of human sexual development. Pediatr Endocrinol Rev. 2009;6(3):395–404.
82. Semple RK, Topaloglu AK. The recent genetics of hypogonadotrophic hypogonadism – novel insights and new questions. Clin Endocrinol (Oxf). 2009;72:427–35.
83. Avbelij M, Romero C, Tziaferi V, et al. New loci for congenital hypopituitarism: overlap with Kallmann syndrome. Abstract presented at the LWPES/ESPE 8th Joint Meeting Paediatric Endocrinology. Horm Res. 2009. p. 30–1.
84. Sornson MW, Wu W, Dasen JS, et al. Pituitary lineage determination by the Prophet of Pit-1 homeodomain factor defective in Ames dwarfism. Nature. 1996;384(6607):327–33.
85. Ward RD, Stone BM, Raetzman LT, Camper SA. Cell proliferation and vascularization in mouse models of pituitary hormone deficiency. Mol Endocrinol. 2006;20(6):1378–90.
86. Wu W, Cogan JD, Pfaffle RW, et al. Mutations in PROP1 cause familial combined pituitary hormone deficiency. Nat Genet. 1998;18(2):147–9.
87. Cogan JD, Wu W, Phillips III JA, et al. The PROP1 2-base pair deletion is a common cause of combined pituitary hormone deficiency. J Clin Endocrinol Metab. 1998;83(9):3346–9.
88. Deladoey J, Fluck C, Buyukgebiz A, et al. "Hot spot" in the PROP1 gene responsible for combined pituitary hormone deficiency. J Clin Endocrinol Metab. 1999;84(5):1645–50.
89. Vieira TC, Boldarine VT, Abucham J. Molecular analysis of PROP1, PIT1, HESX1, LHX3, and LHX4 shows high frequency of PROP1 mutations in patients with familial forms of combined pituitary hormone deficiency. Arq Bras Endocrinol Metabol. 2007;51(7):1097–103.
90. Turton JP, Mehta A, Raza J, et al. Mutations within the transcription factor PROP1 are rare in a cohort of patients with sporadic combined pituitary hormone deficiency (CPHD). Clin Endocrinol (Oxf). 2005;63(1):10–8.
91. Kelberman D, Rizzoti K, Lovell-Badge R, Robinson IC, Dattani MT. Genetic regulation of pituitary gland development in human and mouse. Endocr Rev. 2009;30(7):790–829.
92. Bottner A, Keller E, Kratzsch J, et al. PROP1 mutations cause progressive deterioration of anterior pituitary function including adrenal insufficiency: a longitudinal analysis. J Clin Endocrinol Metab. 2004;89(10):5256–65.
93. Fofanova O, Takamura N, Kinoshita E, et al. MR imaging of the pituitary gland in children and young adults with congenital combined pituitary hormone deficiency associated with PROP1 mutations. AJR Am J Roentgenol. 2000;174(2):555–9.
94. Riepe FG, Partsch CJ, Blankenstein O, Monig H, Pfaffle RW, Sippell WG. Longitudinal imaging reveals pituitary enlargement preceding hypoplasia in two brothers with combined pituitary hormone deficiency attributable to PROP1 mutation. J Clin Endocrinol Metab. 2001;86(9):4353–7.
95. Osorio MG, Marui S, Jorge AA, et al. Pituitary magnetic resonance imaging and function in patients with growth hormone deficiency with and without mutations in GHRH-R, GH-1, or PROP-1 genes. J Clin Endocrinol Metab. 2002;87(11):5076–84.
96. Teinturier C, Vallette S, Adamsbaum C, Bendaoud M, Brue T, Bougneres PF. Pseudotumor of the pituitary due to PROP-1 deletion. J Pediatr Endocrinol Metab. 2002;15(1):95–101.
97. Vallette-Kasic S, Barlier A, Teinturier C, et al. PROP1 gene screening in patients with multiple pituitary hormone deficiency reveals two sites of hypermutability and a high incidence of corticotroph deficiency. J Clin Endocrinol Metab. 2001;86(9):4529–35.

98. Voutetakis A, Argyropoulou M, Sertedaki A, et al. Pituitary magnetic resonance imaging in 15 patients with Prop1 gene mutations: pituitary enlargement may originate from the intermediate lobe. J Clin Endocrinol Metab. 2004;89(5):2200–6.
99. Ward RD, Raetzman LT, Suh H, Stone BM, Nasonkin IO, Camper SA. Role of PROP1 in pituitary gland growth. Mol Endocrinol. 2005;19(3):698–710.
100. Bodner M, Castrillo JL, Theill LE, Deerinck T, Ellisman M, Karin M. The pituitary-specific transcription factor GHF-1 is a homeobox-containing protein. Cell. 1988;55(3):505–18.
101. Rhodes SJ, DiMattia GE, Rosenfeld MG. Transcriptional mechanisms in anterior pituitary cell differentiation. Curr Opin Genet Dev. 1994;4(5):709–17.
102. Andersen B, Rosenfeld MG. POU domain factors in the neuroendocrine system: lessons from developmental biology provide insights into human disease. Endocr Rev. 2001;22(1):2–35.
103. Li S, Crenshaw III EB, Rawson EJ, Simmons DM, Swanson LW, Rosenfeld MG. Dwarf locus mutants lacking three pituitary cell types result from mutations in the POU-domain gene pit-1. Nature. 1990;347(6293):528–33.
104. Dasen JS, O'Connell SM, Flynn SE, et al. Reciprocal interactions of Pit1 and GATA2 mediate signaling gradient-induced determination of pituitary cell types. Cell. 1999;97(5):587–98.
105. Tatsumi K, Miyai K, Notomi T, et al. Cretinism with combined hormone deficiency caused by a mutation in the PIT1 gene. Nat Genet. 1992;1(1):56–8.
106. Miyata I, Vallette-Kasic S, Saveanu A, et al. Identification and functional analysis of the novel S179R POU1F1 mutation associated with combined pituitary hormone deficiency. J Clin Endocrinol Metab. 2006;91(12):4981–7.
107. Pfaffle RW, DiMattia GE, Parks JS, et al. Mutation of the POU-specific domain of Pit-1 and hypopituitarism without pituitary hypoplasia. Science. 1992;257(5073):1118–21.
108. Pellegrini-Bouiller I, Belicar P, Barlier A, et al. A new mutation of the gene encoding the transcription factor Pit-1 is responsible for combined pituitary hormone deficiency. J Clin Endocrinol Metab. 1996;81(8):2790–6.
109. Pfaffle RW, Martinez R, Kim C, et al. GH and TSH deficiency. Exp Clin Endocrinol Diabetes. 1997;105 Suppl 4:1–5.
110. Cohen LE, Radovick S. Molecular basis of combined pituitary hormone deficiencies. Endocr Rev. 2002;23(4):431–42.
111. Brown MR, Parks JS, Adess ME, et al. Central hypothyroidism reveals compound heterozygous mutations in the Pit-1 gene. Horm Res. 1998;49(2):98–102.
112. Pernasetti F, Milner RD, al Ashwal AA, et al. Pro239Ser: a novel recessive mutation of the Pit-1 gene in seven Middle Eastern children with growth hormone, prolactin, and thyrotropin deficiency. J Clin Endocrinol Metab. 1998;83(6):2079–83.
113. Turton JP, Reynaud R, Mehta A, et al. Novel mutations within the POU1F1 gene associated with variable combined pituitary hormone deficiency. J Clin Endocrinol Metab. 2005;90(8):4762–70.
114. Gat-Yablonski G, Lazar L, Pertzelan A, Phillip M. A novel mutation in PIT-1: phenotypic variability in familial combined pituitary hormone deficiencies. J Pediatr Endocrinol Metab. 2002;15(3):325–30.
115. Maheshwari HG, Silverman BL, Dupuis J, Baumann G. Phenotype and genetic analysis of a syndrome caused by an inactivating mutation in the growth hormone-releasing hormone receptor: Dwarfism of Sindh. J Clin Endocrinol Metab. 1998;83(11):4065–74.
116. Netchine I, Talon P, Dastot F, Vitaux F, Goossens M, Amselem S. Extensive phenotypic analysis of a family with growth hormone (GH) deficiency caused by a mutation in the GH-releasing hormone receptor gene. J Clin Endocrinol Metab. 1998;83(2):432–6.
117. Mullis PE. Genetic control of growth. Eur J Endocrinol. 2005;152(1):11–31.
118. Ryther RC, McGuinness LM, Phillips III JA, et al. Disruption of exon definition produces a dominant-negative growth hormone isoform that causes somatotroph death and IGHD II. Hum Genet. 2003;113(2):140–8.
119. Mullis PE, Robinson IC, Salemi S, et al. Isolated autosomal dominant growth hormone deficiency: an evolving pituitary deficit? A multicenter follow-up study. J Clin Endocrinol Metab. 2005;90(4):2089–96.

120. McGuinness L, Magoulas C, Sesay AK, et al. Autosomal dominant growth hormone deficiency disrupts secretory vesicles in vitro and in vivo in transgenic mice. Endocrinology. 2003;144(2):720–31.
121. Besson A, Salemi S, Deladoey J, et al. Short stature caused by a biologically inactive mutant growth hormone (GH-C53S). J Clin Endocrinol Metab. 2005;90(5):2493–9.
122. Salemi S, Yousefi S, Baltensperger K, et al. Variability of isolated autosomal dominant GH deficiency (IGHD II): impact of the P89L GH mutation on clinical follow-up and GH secretion. Eur J Endocrinol. 2005;153(6):791–802.
123. Lacey KA, Parkin JM. Causes of short stature. A community study of children in Newcastle upon Tyne. Lancet. 1974;1(7846):42–5.
124. Rona RJ, Tanner JM. Aetiology of idiopathic growth hormone deficiency in England and Wales. Arch Dis Child. 1977;52(3):197–208.
125. Vimpani GV, Vimpani AF, Lidgard GP, Cameron EH, Farquhar JW. Prevalence of severe growth hormone deficiency. Br Med J. 1977;2(6084):427–30.
126. Phillips III JA, Cogan JD. Genetic basis of endocrine disease. 6. Molecular basis of familial human growth hormone deficiency. J Clin Endocrinol Metab. 1994;78(1):11–6.
127. Murray PG, Paterson WF, Donaldson MD. Maternal age in patients with septo-optic dysplasia. J Pediatr Endocrinol Metab. 2005;18(5):471–6.
128. Mullis PE. Genetics of growth hormone deficiency. Endocrinol Metab Clin North Am. 2007;36(1):17–36.
129. Kjellin IB, Kaiserman KB, Curran JG, Geffner ME. Aplasia of right internal carotid artery and hypopituitarism. Pediatr Radiol. 1999;29(8):586–8; discussion 585.
130. Kikuchi K, Fujisawa I, Momoi T, et al. Hypothalamic-pituitary function in growth hormone-deficient patients with pituitary stalk transection. J Clin Endocrinol Metab. 1988;67(4):817–23.
131. Triulzi F, Scotti G, di Natale B, et al. Evidence of a congenital midline brain anomaly in pituitary dwarfs: a magnetic resonance imaging study in 101 patients. Pediatrics. 1994;93(3):409–16.
132. Maghnie M, Genovese E, Villa A, Spagnolo L, Campan R, Severi F. Dynamic MRI in the congenital agenesis of the neural pituitary stalk syndrome: the role of the vascular pituitary stalk in predicting residual anterior pituitary function. Clin Endocrinol (Oxf). 1996;45(3):281–90.
133. Genovese E, Maghnie M, Beluffi G, et al. Hypothalamic-pituitary vascularization in pituitary stalk transection syndrome: is the pituitary stalk really transected? The role of gadolinium-DTPA with spin-echo T1 imaging and turbo-FLASH technique. Pediatr Radiol. 1997;27(1):48–53.
134. Maghnie M, Triulzi F, Larizza D, et al. Hypothalamic-pituitary dysfunction in growth hormone-deficient patients with pituitary abnormalities. J Clin Endocrinol Metab. 1991;73(1):79–83.
135. Argyropoulou M, Perignon F, Brauner R, Brunelle F. Magnetic resonance imaging in the diagnosis of growth hormone deficiency. J Pediatr. 1992;120(6):886–91.
136. Chen S, Leger J, Garel C, Hassan M, Czernichow P. Growth hormone deficiency with ectopic neurohypophysis: anatomical variations and relationship between the visibility of the pituitary stalk asserted by magnetic resonance imaging and anterior pituitary function. J Clin Endocrinol Metab. 1999;84(7):2408–13.
137. Pinto G, Netchine I, Sobrier ML, Brunelle F, Souberbielle JC, Brauner R. Pituitary stalk interruption syndrome: a clinical-biological-genetic assessment of its pathogenesis. J Clin Endocrinol Metab. 1997;82(10):3450–4.
138. Maghnie M, Larizza D, Triulzi F, Sampaolo P, Scotti G, Severi F. Hypopituitarism and stalk agenesis: a congenital syndrome worsened by breech delivery? Horm Res. 1991;35(3–4):104–8.
139. Maghnie M, Larizza D, Zuliani I, Severi F. Congenital central nervous system abnormalities, idiopathic hypopituitarism and breech delivery: what is the connection? Eur J Pediatr. 1993;152(2):175.
140. Siegel SF, Ahdab-Barmada M, Arslanian S, Foley Jr TP. Ectopic posterior pituitary tissue and paracentric inversion of the short arm of chromosome 1 in twins. Eur J Endocrinol. 1995;133(1):87–92.
141. Larizza D, Maraschio P, Maghnie M, Sampaolo P. Hypogonadism in a patient with balanced X/18 translocation and pituitary hormone deficiency. Eur J Pediatr. 1993;152(5):424–7.

142. Hamilton J, Chitayat D, Blaser S, Cohen LE, Phillips III JA, Daneman D. Familial growth hormone deficiency associated with MRI abnormalities. Am J Med Genet. 1998;80(2):128–32.
143. Arifa N, Leger J, Garel C, Czernichow P, Hassan M. Cerebral anomalies associated with growth hormone insufficiency in children: major markers for diagnosis? Arch Pediatr. 1999;6(1):14–21.
144. Dupuis-Girod S, Gluckman E, Souberbielle JC, Brauner R. Growth hormone deficiency caused by pituitary stalk interruption in Fanconi's anemia. J Pediatr. 2001;138(1):129–33.
145. Larizza D, Maghnie M. Poland's syndrome associated with growth hormone deficiency. J Med Genet. 1990;27(1):53–5.
146. Parano E, Trifiletti RR, Barone R, Pavone V, Pavone P. Arthrogryposis multiplex congenita and pituitary ectopia. A case report. Neuropediatrics. 2000;31(6):325–7.
147. Oguz KK, Ozgen B, Erdem Z. Cranial midline abnormalities in Dubowitz syndrome: MR imaging findings. Eur Radiol. 2003;13(5):1056–7.
148. Bas F, Darendeliler F, Yapici Z, et al. Worster-Drought syndrome (congenital bilateral perisylvian syndrome) with posterior pituitary ectopia, pituitary hypoplasia, empty sella and panhypopituitarism: a patient report. J Pediatr Endocrinol Metab. 2006;19(4):535–40.
149. Halasz Z, Bertalan R, Toke J, et al. Laterality disturbance and hypopituitarism. A case report of co-existing situs inversus totalis and combined pituitary hormone deficiency. J Endocrinol Invest. 2008;31(1):74–8.
150. de Graaff LC, Baan J, Govaerts LC, Hokken-Koelega AC. Facial and pituitary morphology are related in Dutch patients with GH deficiency. Clin Endocrinol (Oxf). 2008;69(1):112–6.
151. Cacciari E, Zucchini S, Ambrosetto P, et al. Empty sella in children and adolescents with possible hypothalamic-pituitary disorders. J Clin Endocrinol Metab. 1994;78(3):767–71.
152. Bressani N, di Natale B, Pellini C, Triulzi F, Scotti G, Chiumello G. Evidence of morphological and functional abnormalities in the hypothalamus of growth-hormone-deficient children: a combined magnetic resonance imaging and endocrine study. Horm Res. 1990;34(5–6):189–92.
153. Abrahams JJ, Trefelner E, Boulware SD. Idiopathic growth hormone deficiency: MR findings in 35 patients. AJNR Am J Neuroradiol. 1991;12(1):155–60.
154. Nagel BH, Palmbach M, Petersen D, Ranke MB. Magnetic resonance images of 91 children with different causes of short stature: pituitary size reflects growth hormone secretion. Eur J Pediatr. 1997;156(10):758–63.
155. Kornreich L, Horev G, Lazar L, Josefsberg Z, Pertzelan A. MR findings in hereditary isolated growth hormone deficiency. AJNR Am J Neuroradiol. 1997;18(9):1743–7.
156. Bozzola M, Mengarda F, Sartirana P, Tato L, Chaussain JL. Long-term follow-up evaluation of magnetic resonance imaging in the prognosis of permanent GH deficiency. Eur J Endocrinol. 2000;143(4):493–6.
157. Arends NJ, V d Lip W, Robben SG, Hokken-Koelega AC. MRI findings of the pituitary gland in short children born small for gestational age (SGA) in comparison with growth hormone-deficient (GHD) children and children with normal stature. Clin Endocrinol (Oxf). 2002;57(6):719–24.
158. Maghnie M, Loche S, Cappa M. Pituitary magnetic resonance imaging in idiopathic and genetic growth hormone deficiency. J Clin Endocrinol Metab. 2003;88(4):1911; author reply 1911–12.
159. Melo ME, Marui S, Carvalho LR, et al. Hormonal, pituitary magnetic resonance, LHX4 and HESX1 evaluation in patients with hypopituitarism and ectopic posterior pituitary lobe. Clin Endocrinol (Oxf). 2007;66(1):95–102.
160. Zenaty D, Garel C, Limoni C, Czernichow P, Leger J. Presence of magnetic resonance imaging abnormalities of the hypothalamic-pituitary axis is a significant determinant of the first 3 years growth response to human growth hormone treatment in prepubertal children with nonacquired growth hormone deficiency. Clin Endocrinol (Oxf). 2003;58(5):647–52.
161. Coutant R, Rouleau S, Despert F, Magontier N, Loisel D, Limal JM. Growth and adult height in GH-treated children with nonacquired GH deficiency and idiopathic short stature: the influence of pituitary magnetic resonance imaging findings. J Clin Endocrinol Metab. 2001;86(10):4649–54.

162. Maghnie M, Ambrosini L, Cappa M, et al. Adult height in patients with permanent growth hormone deficiency with and without multiple pituitary hormone deficiencies. J Clin Endocrinol Metab. 2006;91(8):2900–5.
163. Maghnie M, Strigazzi C, Tinelli C, et al. Growth hormone (GH) deficiency (GHD) of childhood onset: reassessment of GH status and evaluation of the predictive criteria for permanent GHD in young adults. J Clin Endocrinol Metab. 1999;84(4):1324–8.
164. Adan L, Souberbielle JC, Brauner R. Diagnostic markers of permanent idiopathic growth hormone deficiency. J Clin Endocrinol Metab. 1994;78(2):353–8.
165. Murray PG, Hague C, Fafoula O, et al. Likelihood of persistent GH deficiency into late adolescence: relationship to the presence of an ectopic or normally sited posterior pituitary gland. Clin Endocrinol (Oxf). 2009;71(2):215–9.
166. Clayton PE, Cuneo RC, Juul A, Monson JP, Shalet SM, Tauber M. Consensus statement on the management of the GH-treated adolescent in the transition to adult care. Eur J Endocrinol. 2005;152(2):165–70.
167. Ho KK. Consensus guidelines for the diagnosis and treatment of adults with GH deficiency II: a statement of the GH Research Society in association with the European Society for Pediatric Endocrinology, Lawson Wilkins Society, European Society of Endocrinology, Japan Endocrine Society, and Endocrine Society of Australia. Eur J Endocrinol. 2007;157(6):695–700.
168. Leger J, Danner S, Simon D, Garel C, Czernichow P. Do all patients with childhood-onset growth hormone deficiency (GHD) and ectopic neurohypophysis have persistent GHD in adulthood? J Clin Endocrinol Metab. 2005;90(2):650–6.
169. Maghnie M, Aimaretti G, Bellone S, et al. Diagnosis of GH deficiency in the transition period: accuracy of insulin tolerance test and insulin-like growth factor-I measurement. Eur J Endocrinol. 2005;152(4):589–96.
170. Secco A, di Iorgi N, Napoli F, et al. Reassessment of the growth hormone status in young adults with childhood-onset growth hormone deficiency: reappraisal of insulin tolerance testing. J Clin Endocrinol Metab. 2009;94(11):4195–204.
171. di Iorgi N, Secco A, Napoli F, et al. Deterioration of growth hormone (GH) response and anterior pituitary function in young adults with childhood-onset GH deficiency and ectopic posterior pituitary: a two-year prospective follow-up study. J Clin Endocrinol Metab. 2007;92(10):3875–84.
172. Maghnie M, Uga E, Temporini F, et al. Evaluation of adrenal function in patients with growth hormone deficiency and hypothalamic-pituitary disorders: comparison between insulin-induced hypoglycemia, low-dose ACTH, standard ACTH and CRH stimulation tests. Eur J Endocrinol. 2005;152(5):735–41.

Chapter 7
Genetic Causes of Familial Pituitary Adenomas

Silvia Vandeva, Sabina Zacharieva, Adrian F. Daly, and Albert Beckers

Abstract Pituitary adenomas are benign intracranial neoplasms and clinically apparent pituitary adenomas have a prevalence of approximately 1:1,000 individuals. They usually arise sporadically, but familial pituitary adenomas comprise about 5% of all cases, more than half of which include multiple endocrine neoplasia type 1 (MEN1) and Carney complex (CNC). The other half is represented by familial isolated pituitary adenomas (FIPA), a clinical entity described in the late 1990s. Recently, interest has been focused on the genetic pathophysiology of familial pituitary adenomas. MEN1 is due to mutations in *MEN1* gene. CNC is related to *PRKAR1A* mutations and still unknown disruptions on 2p16. Mutations of *CDKN1B* in MEN1-like patients without *MEN1* mutations allowed the differentiation of the condition as MEN4. About 15% of FIPA kindreds are associated with aryl hydrocarbon receptor-interacting protein (*AIP*) gene mutations, which suggests that this is a genetically heterogeneous condition. Overall, familial pituitary adenomas represent a small proportion of pituitary tumors, but are particularly significant as affected individuals may be younger, and adenomas may be relatively difficult to treat.

Keywords Multiple endocrine neoplasia type 1 • MEN4 • Carney complex • Familial isolated pituitary adenomas

Introduction

The pituitary gland has a vital role in maintaining metabolic homeostasis through rigorous function control of all glands with internal secretion. This balance could be disrupted by pituitary adenomas, manifesting clinically by hormonal hyperproduction

A. Beckers (✉)
Department of Endocrinology, C.H.U. de Liège, University of Liège,
Domaine Universitaire du Sart-Tilman, 4000 Liège, Belgium
e-mail: albert.beckers@chu.ulg.ac.be

K. Ho (ed.), *Growth Hormone Related Diseases and Therapy: A Molecular and Physiological Perspective for the Clinician*, Contemporary Endocrinology,
DOI 10.1007/978-1-60761-317-6_7,

Table 7.1 Genetic alterations in familial pituitary adenomas

Gene	Biological action	Pituitary adenoma	Syndrome
MEN1	Promoter regulation, histone modification, protein interaction with transcriptional factors	All tumor types	70–80% of Multiple endocrine neoplasia type 1 (MEN1)
CDKN1B	Cell cycle modulation	GH, ACTH	Less than 2% in MEN4 syndrome
PRKAR1	Modulation of cAMP-dependent PKA activity	Somatolactotrope hyperplasia and adenomas in Carney complex (CNC)	60% of CNC
2p16 locus	Unknown		
AIP	Cytoplasmic retention of AHR–hsp90–AIP complex, interaction with phosphodiesterases (PDEs)	All adenoma types	15–25% Familial isolated pituitary adenomas (FIPA), aggressive behavior

syndromes and/or tumor mass symptoms. Based on previous population studies, it was thought that pituitary adenomas were an uncommon pathology with a prevalence of 1:3,571-1:5,263 or less [1, 2]. These numbers, however, were highly discordant with autopsy and radiology series showing prevalence of 14.4 and 22.5%, respectively [3]. In 2006, a rigorous cross-sectional, population-based study was performed in the province of Liege, Belgium, revealing that clinically relevant pituitary adenomas are actually 3–5 times more frequent than previously reported – 1:1,064 [4]. Furthermore, pituitary tumors represent about one fifth of the intracranial neoplasms and are the second most common tumor type by histology in young adults between age of 20 and 34 years [5]. These data suggest that pituitary adenomas are a clinical entity with a significant clinical burden.

Evidence for genetic implication in the tumorigenesis of pituitary adenomas includes their monoclonality and their, albeit rare, familial predisposition. The greatest proportion of the pituitary adenomas arise in a sporadic setting and numerous possible culprit genes affecting cell cycle, growth factors, signal transduction pathways, and others are under intense investigation. Familial pituitary adenomas, representing only about 5% of all pituitary adenomas, are characterized by known genetic alterations in a significant proportion of the cases. Familial pituitary adenomas could be either part of multiple endocrine neoplasia (MEN) syndromes, such as multiple endocrine neoplasia type 1 (MEN1), MEN4, or Carney complex (CNC); alternatively, they can be isolated to the pituitary as occurs in the setting of familial isolated pituitary adenomas (FIPA; Table 7.1).

MEN1

MEN1 is an autosomal dominant disease, comprising two of the three following tumors: parathyroid, enteropancreatic, and pituitary. Primary hyperparathyroidism is usually the first and the most common manifestation of MEN1, reaching nearly

100% penetrance by the age of 50. Enteropancreatic tumors are present in about 30–80% of the cases, with gastrinomas and insulinomas being the most common types. Pituitary adenomas occur in 10–60% of the cases. Nonendocrine tumors, such as angiofibromas, collagenomas, and lipomas, have also been frequently described [6] The familial character of this condition was first suggested by Wermer in the 1950s who reported four sisters presenting with pituitary adenomas, hypercalcemia, and adenomatosis of the pancreas and gut [7]. The combination of a pituitary and parathyroid adenoma, however, was described 50 years earlier by Erdheim [8].

The primary genetic defect in more than two thirds of the cases with MEN1 is a mutation in the *MEN1* gene that encodes the 610 amino acid protein, menin. This tumor suppressor gene is located on chromosome 11q13 [9] and was first cloned in the 1990s [10]. More than 500 *MEN1* mutations have been described. About 85% of these mutations were germline, the rest being somatic. More than two thirds of these genetic alterations cause truncated forms of menin.

The biological role of menin seems to be quite complex and it is not completely understood. As mainly a nuclear protein, it is involved in gene transcription control by promoter regulation of a number of genes [11–14]. There is also evidence of menin interplay with histone modification [15]. Furthermore, menin is implicated in protein–protein interactions in various cell cycle and proliferation pathways [16–19]. Interestingly, menin expression is found in various tissues, although genetic disruptions are preferentially found in endocrine cells. Differential regulation of menin expression through upstream genetic elements could probably explain this phenomenon [20].

MEN1-related pituitary tumors comprise about 2.7% of all pituitary adenomas [21]. Pituitary adenomas have been observed in up to 60% of the MEN1 cases [6] and it is the first manifestation of the syndrome in about 17% of the affected subjects [22]. Somatic *MEN1* mutations do not play an important role in sporadic pituitary pathogenesis [23]. In a comprehensive French–Belgium study, the distribution by functional pituitary tumor type in MEN1 did not differ from that of sporadic tumors in that prolactinomas predominate in both groups. Other series report higher frequency of plurihormonality, especially prolactin and ACTH cosecretion, or PRL and ACTH double adenomas, combinations that are extremely rare in the sporadic setting [21]. MEN1 pituitary adenomas show some distinct clinical features as well. MEN1-related prolactinomas are almost invariably macroadenomas with poorer response to dopamine agonist therapy vs. sporadic adenomas [22]. Overall, MEN1 pituitary tumors are more aggressive and invasive compared to non-MEN1 pituitary adenomas [22, 24].

Within the MEN1 cohort, no genotype–phenotype relation has been found. However, pituitary disease was more frequent in familial than in sporadic MEN1 cases [22]. An interesting observation is the clustering of certain phenotypes, like the dominance of prolactinomas in $MEN1_{Burin}$, characterized by a nonsense founder mutation [20, 25] and $MEN1_{Tasman}$ characterized by a splice site founder mutation [26]. However, unlike the Burin kindred, prolactinomas were unevenly distributed in the different branches of the Tasman 1 kindred which could be possibly due to modifier genes, requiring further research.

More profound understanding of MEN1-related tumorigenesis requires further investigation and could be possibly facilitated by development of appropriate mouse models. Up to now, it has been shown that heterozygous *Men1+/–* mice develop pituitary adenomas in 36% of animals by the age of 26 months. Similarly to human MEN1, female preponderance and a greater proportion of aggressive adenomas were observed [27]. Interestingly, unlike homozygous *Men1–/–* mice that die early in their embryonic life, knockout animals that are conditionally restricted to pancreas and pituitary gland develop normally but have frequent findings of prolactinomas and pancreatic hyperplasia [28, 29], which mimics the human phenotype.

Despite clinical presentation of MEN1, about 20–30% of either familial or sporadic cases do not bear a *MEN1* mutation [30]. In such cases, possible genetic alterations could include mutations of the promoter region, intronic mutations, and disruptions in sequences related to menin pathway.

Clinical management of MEN1 starts with confirming the diagnosis. According to the consensus criteria, MEN1 phenotype is established by the presence of two out of the three following tumors: parathyroid, enteropancreatic, or pituitary. Afterwards in *MEN1* mutation positive adults, annual assessment of calcium, parathyroid hormone, gastrin, and prolactin levels is recommended, with regular screening for pituitary, pancreatic, and parathyroid disease in mutation carriers from early childhood [6].

MEN4

MEN4, a MEN1 like condition, has been described in a small number of human cases. This condition was originally described in rats with multiple neuroendocrine neoplasia and pituitary adenomas [31, 32]. The possible culprit gene was firstly mapped to chromosome 4 and was identified as the cyclin-dependent kinase inhibitor 1B gene [33]. The human analog, *CDKN1B,* is located on chromosome 12 and consists of 3 exons, encoding a 198 amino acid protein, p27. This protein belongs to the Cip/Kip family of cyclin-dependent kinase inhibitors and regulates cell cycle progression through modulation of cyclin-dependent kinases [34]. The implication of p27 in pituitary tumorigenesis has been demonstrated in mice models, where all p27–/– mice develop pituitary adenomas by the age of 12 months [35, 36]. In humans, p27 has been found underexpressed in sporadic pituitary adenomas compared to the normal gland [37, 38]. Additional insight on the biological role of p27 may be derived from potential involvement of this protein in the menin, Ret, and aryl hydrocarbon receptor (AHR) signaling pathways [39, 40].

In 2006, Pellegata et al. [33] reported the first human mutation of *CDKN1B* (TGG>TAG at codon 76) in a German family with MEN4 – acromegaly, primary hyperparathyroidism, renal angiomyolipoma, and testicular cancer within the kindred. Later, another mutation (19 bp heterozygous duplication in exon 1) was described in a Dutch female patient presenting with a pituitary adenoma (Cushing's disease), hyperparathyroidism, and a small-cell neuroendocrine cervical carcinoma [41]. These findings provoked an extensive research for other *CDKN1B* disruptions in MEN4 patients which led to the discovery of three new mutations of CDKN1B and other CDKI genes (*p15* N41D, *p15* L64R, *p18* V31L, *p21* R67L) by Agarwal et al. Out of these, only *p21* R67L was associated with pituitary adenomas (prolactinoma) [42]. Nevertheless, mutations in other CDKIs do not seem to contribute significantly to the pathology of cases of MEN syndromes, as they are found in less than 2% of the cases [42, 43].

Carney Complex

CNC is a rare autosomal dominant disease [44]. The combination of myxomas, spotty skin pigmentation (lentigines), and endocrine hyperactivity was reported by Carney [45]. Among CNC-related endocrine disorders, most frequent are primary pigmented nodular adrenocortical disease (PPNAD), testicular tumors (large cell calcifying Sertoli cell tumors (LCCSCT), Leydig cell tumors), thyroid tumors and nodules, and acromegaly. CNC is a familial trait in 70% of the cases and shows female prevalence and no racial predisposition. About 60% of the cases are positive for a germline mutation of protein kinase A regulatory subunit-1-alpha (*PRKAR1A*) gene located on chromosome 17q22-24 [46]. After the identification of *PRKAR1A,* more than 40 mutations have been found, most of them leading to an absent protein [16]. CNC tissue analyses showed presence of germline mutations and LOH of the 17q22-24 locus, hence confirming the role of *PRKAR1A* as a tumor suppressor gene [47]. Similarly to MEN1, sporadic pituitary tumors are not characterized by *PRKAR1A* gene mutations [48, 49]. It seems that the inactivation of the regulatory subunits in mutated cases leads to up-regulation of the other subunit types of the PKA complex, which has the effect of increased total cAMP activity [50]. Another locus associated with CNC is localized on chromosome 2p16 [51].

Pituitary pathology in CNC is represented by acromegaly, found in 10% of the cases. However, about 75% of patients have biochemical alterations: asymptomatic elevations of GH, IGF-1 or prolactin levels, or abnormal responses to dynamic pituitary testing [52]. In general, acromegaly is slowly progressing; GH-associated alterations are frequently found in adolescence, but somatotropinomas appear after the third decade. Development of aggressive pituitary adenomas has been observed after operation for Cushing's syndrome due to PPNAD [52]. On histology, pituitary tumors stain positively both for GH and prolactin, with a small proportion showing TSH, LH, or alpha subunit positivity [53]. A typical finding in CNC acromegaly is a multifocal somatomammotropic hyperplasia against a background of normal pituitary. These zones of hyperplasia are poorly demarcated and exhibit altered reticulin staining. Electron microscopy is characterized by a heterogenous intracellular structure [54].

Recently, mice models have been developed in which homozygous *Prkar1a−/−* is not compatible with life [55, 56]. Heterozygous *Prkar1a+/−* mice, on the other hand, did not present with alterations typical for CNC. Significant progress has been made with tissue-specific knockout *Prkar1a+/−* mice that are similar to human CNC phenotype [57].

Similarly to MEN1, treatment of CNC manifestation does not differ from their sporadic counterparts. However, there are some peculiarities concerning diagnosis. CNC is established as a diagnosis at the presence of two of the following: PPNAD, cardiac myxoma, cutaneous myxoma, blue naevi, LCCSCT, thyroid tumor, ovarian cysts, Schwann cell tumor, acromegaly, and osteochondromyxoma. Secondly, a search for a germline mutation in *PRKAR1A* is undertaken, and in negative cases, methods such as testing for large genomic deletions or duplications are taken into consideration. A relevant management option for *PRKAR1A* mutation positive subjects is an annual clinical, hormonal, and imaging screening for CNC manifestations.

Familial Isolated Pituitary Adenomas

Apart from MENs, published evidence for familial cases of pituitary pathology without any other syndromic disorders date back from the beginning of the twentieth century. Quite notable were the descriptions of isolated familial acromegaly, and especially acro-gigantism [58–65]. More rarely, selected historical instances of familial cases of isolated prolactinomas, Cushing's disease, and nonfunctioning pituitary adenomas had been published [66–69].

In the late 1990s, the concept of FIPA was first described by the University of Liège group as a distinct clinical entity to cover these disparate cases. By definition, FIPA kindreds present with two or more pituitary adenomas in related family members in the absence of known genetic causation (principally MEN1 and CNC). FIPA families were further divided into two groups depending on the functional type of the pituitary adenomas: homogenous (one clinical phenotype in the kindred) and heterogeneous (two or more distinct phenotypes in the kindred) [70]. A multicenter collaborative study among tertiary referral centers led to the gradual identification of 64 FIPA families that included about 140 individuals with full clinical, biochemical, radiological, and pathological analyses [71, 72]. To date, there are more than 130 FIPA families in our collaborative series [73] and FIPA families have been reported by other research groups [74, 75].

Compared to sporadic pituitary adenomas, FIPA kindreds show some differences. In distribution of pituitary adenomas by functional type, somatotropinomas are observed more frequently in FIPA compared to their frequency in sporadic cases or in MEN1: prolactinoma (41%), somatotropinoma (30%), nonfunctioning adenoma (13%), somatomammotropinoma (7%), gonadotropinoma (4%), Cushing's disease (4%), and thyrotropinoma (1%). First-degree relationship between affected members within the kindred is found in about 75% of FIPA kindreds. At the time of diagnosis, FIPA patients are 4 years younger than their sporadic counterparts, this difference being 8 years for nonfunctioning pituitary adenomas. In multigenerational families, younger generations were diagnosed 20 years earlier than their ancestors. Furthermore, macroadenomas are seen in approximately two thirds of FIPA patients. Prolactinomas belonging to heterogeneous FIPA kindreds seem to have higher rates of invasion and extension in comparison to sporadic adenomas.

Microprolactinomas are characteristic for women, while men almost invariably have macroadenomas. In regard to GH secreting adenomas, 50% of them belong to the homogenous acromegaly kindreds, which are diagnosed earlier (about 10 years) than heterogeneous FIPA or sporadic acromegaly patients. The rest of somatotropinomas occur in combination with other functional types among heterogeneous families. Nonfunctioning adenomas generally occur in a heterogeneous kindreds, while Cushing's disease and gonadotropinomas are rarely found in homogenous FIPA kindreds [62].

In 2006, Vierimaa et al. performed detailed genome-wide screen and DNA mapping predominantly in patients with familial cases of acromegaly and prolactinomas. They discovered germline mutations in the aryl hydrocarbon receptor-interacting protein (*AIP/Ara9*) gene [76, 77]. *AIP* was established as a tumor suppressor gene on the basis of LOH at the *AIP* locus in pituitary adenoma specimens. In the FIPA cohort, our collaborative group analyzed 73 families, of which 15% proved to harbor germline *AIP* mutations. Among homogenous somatotropinoma FIPA kindreds, 50% were found to have *AIP* mutations. A total of ten separate mutations were found, of which R304X was also reported by Vierimaa in an apparently unrelated family [78]. Up to now, more than 30 distinct *AIP* mutations have been reported [76, 78–84] (Fig. 7.1). Unselected sporadic pituitary adenoma series have also been analyzed, but *AIP* mutations do not seem to contribute significantly to tumorigenesis in this entity [79, 82, 84]. Concerning clinical presentation of sporadic pituitary adenomas in patients with *AIP* mutations, they are diagnosed at younger age, mainly with somatotropinomas, although other functional types do occur [82, 84]. Altogether, the contribution of *AIP* mutations in understanding FIPA pathogenesis is undisputable, but the majority of FIPA cohort still has an unknown genetic alteration.

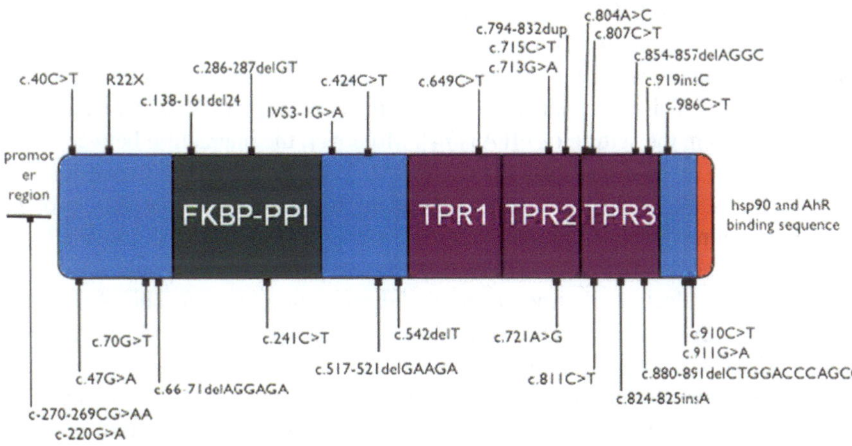

Fig. 7.1 Mutations in the aryl hydrocarbon receptor-interacting protein (*AIP*) *gene* found in familial isolated pituitary adenomas (FIPA) and sporadic adenomas. *AHR* Aryl hydrocarbon receptor; *FKBP-PPI* FK506-binding protein-type peptidyl-propyl *cis-trans* isomerase; *hsp90* heat-shock protein 90; *TPR* tetratricopeptide repeat domain

After the reports of *CDKN1B* mutations in MEN4, screening of a large cohort of FIPA families revealed several new *CDKN1B* variations. These, however, do not contribute significantly for FIPA tumorigenesis [85]

Structure of AIP protein could possibly give insight on its biological role and eventually AIP-dependent tumorigenesis. AIP is a 330-amino acid immunophilin protein, consisting of three tetratricopeptide repeat (TPR) domains and one domain characteristic for the immunophilin proteins – FK506-binding protein-type peptidyl-propyl *cis-trans* isomerase (FKBP-PPI). All four domains are highly conserved. The TPR domains have a key role in protein interaction, and without the last five amino acids of the C-terminus, binding with AHR is not possible [86, 87]. Most of the reported mutations lead to a premature stop codons and truncated protein with loss of the last TPR domain [88] AIP, together with a dimer of heat-shock protein 90 (hsp90), forms a complex with the AHR [89]. AHR is a transcriptional factor involved in the regulation of genes implicated in cellular growth and cell differentiation, liver and vascular system development, maturation of the immune system, and others. Through binding with environmental toxins, such as dioxin, AHR transduces a signal to response elements, which in turn leads to manifestation of toxic responses [90]. The role of AIP binding to the AHR-hsp90 complex is not completely understood. However, there is evidence that it is important for complex stabilization and cytoplasmic retention and consequently prevention from DNA interaction in the nucleus [91]. Furthermore, AIP functions extend to numerous protein–protein interactions. It interplays with hypoxia inducible factor-1alpha, nuclear factor-kB, and retinoblastoma protein, among others [62]. Modulation of phosphodiesterase (PDE) 4A5 and PDE2A activities are also very promising pathways for *AIP* mutation-related disease pathogenesis. PDEs are participants in the cAMP signal transduction pathway, which has a key role in pituitary adenoma pathogenesis. There is further evidence that AHR nuclear translocation is a cAMP-dependent process [90, 92, 93]. Leontiou et al. showed that wild-type AIP overexpression leads to reduction of cell proliferation in human embryonic kidney cell (HEK) 293, human embryonic lung fibroblast (TIG 3), and rat somatomammotroph (GH3) cell lines. This effect was negated after the expression of R304X and C238Y AIP (described in the setting of FIPA) [74]. Moreover, the interaction between AIP and PDE4A5 was disrupted also by C238Y, R271W, R81X, Q271X, and R304X *AIP* mutations. A recent immunohistochemical study suggests that AIP protein underexpression is related to a more aggressive course of somatotropinomas [94].

In murine knockout models, *Aip–/–* mice die early in their embryonic life due to cardiovascular defects [95]. Another model, displaying a hypomorphic *Aip* allele, has been recently developed. These mice had reduced expression of Aip protein and again cardiovascular disruption [96].

The risk of developing pituitary adenomas in *AIP* mutation bearers is not well established as the penetrance of the disease is divergent in *AIP* mutation positive FIPA kindreds. Nevertheless, our experience suggest an incomplete penetrance in the region of 33% in largest FIPA families [97].

Altogether, the presence of *AIP* mutations is related to a more aggressive course of pituitary disease. Furthermore, they prove to be more resistant to therapy, surgical results are often poor, and resistance to somatostatin analogs is observed [98].

It seems relevant to test for *AIP* mutations, at least one affected members of all families meeting criteria for FIPA. In *AIP* mutation bearers, it is reasonable to perform magnetic resonance imaging (MRI) and hormonal testing. In the absence of tumor on MRI, an annual follow-up based on clinical symptoms and basal hormonal tests (IGF-1 and prolactin) could be considered. In regard to sporadic pituitary adenomas, *AIP* mutations, in general, are uncommon. However, there is evidence that young patients with aggressive pituitary adenomas are more likely to harbor *AIP* mutations and *AIP* testing is likely justifiable in this subpopulation. To date, based on literature data, *AIP* is the only gene with a causative role for FIPA tumorigenesis. Nevertheless, *AIP* mutations are found in only 15% of FIPA kindreds, which renders FIPA pathogenesis still an issue for further investigation.

References

1. Ambrosi B, Faglia G. Epidemiology of pituitary tumors. In: Faglia G, Beck-Peccoz P, Ambrosi B, editors. Pituitary adenomas: new trends in basic and clinical research. Amsterdam: Elsevier; 1991. p. 159–68.
2. Clayton RN. Sporadic pituitary adenomas: from epidemiology to use of databases. Baillieres Best Pract Res Clin Endocrinol Metab. 1999;13:451–60.
3. Ezzat S, Asa SL, Couldwell WT, Barr CE, Dodge WE, Vance ML, et al. The prevalence of pituitary adenomas: a systematic review. Cancer. 2004;101:613–9.
4. Daly AF, Rixhon M, Adam C, Dempegioti A, Tichomirowa MA, Beckers A. High prevalence of pituitary adenomas: a crosssectional study in the province of Liege, Belgium. J Clin Endocrinol Metab. 2006;91:4769–75.
5. Central Brain Tumor Registry of the United States 2007–2008. Central Brain Tumor Registry of the United States Statistical Report; 2008.
6. Brandi ML, Gagel RF, Angeli A, et al. Guidelines for diagnosis and therapy of MEN type 1 and type 2. J Clin Endocrinol Metab. 2001;86:5658–71.
7. Wermer P. Genetic aspects of adenomatosis of endocrine glands. Am J Med. 1954;16:363–71.
8. Erdheim J. Zur normalen und pathologischen Histologie der Glandula thyreoidiea, parathyroidea und Hypophysis. Beitr Pathol Anat. 1903;33:158–65.
9. Larsson C, Skogseid B, Oberg K, Nakamura Y, Nordenskjold M. Multiple endocrine neoplasia type 1 gene maps to chromosome 11 and is lost in insulinoma. Nature. 1988;332:85–7.
10. Chandrasekharappa SC, Guru SC, Manickam P, Olufemi SE, Collins FS, Emmert-Buck MR, et al. Positional cloning of the gene for multiple endocrine neoplasia-type 1. Science. 1997;276:404–7.
11. Sayo Y, Murao K, Imachi H, Cao WM, Sato M, Dobashi H, et al. The multiple endocrine neoplasia type 1 gene product, menin, inhibits insulin production in rat insulinoma cells. Endocrinology. 2002;143:2437–40.
12. La P, Schnepp RW, Petersen D, Silva C, Hua X. Tumor suppressor menin regulates expression of insulin-like growth factor binding protein 2. Endocrinology. 2004;145:3443–50.
13. Namihira H, Sato M, Murao K, Cao WM, Matsubara S, Imachi H, et al. The multiple endocrine neoplasia type 1 gene product, menin, inhibits the human prolactin promoter activity. J Mol Endocrinol. 2002;29:297–304.
14. Scacheri PC, Davis S, Odom DT, Crawford GE, Perkins S, Halawi MJ, et al. Genome-wide analysis of menin binding provides insights into MEN1 tumorigenesis. PLoS Genet. 2006;2:e51.
15. Hughes CM, Rozenblatt-Rosen O, Milne TA, Copeland TD, Levine SS, Lee JC, et al. Menin associates with a trithorax family histone methyltransferase complex and with the hoxc8 locus. Mol Cell. 2004;13:587–97.

16. Horvath A, Stratakis CA. Clinical and molecular genetics of acromegaly: MEN1, Carney complex, McCune–Albright syndrome, familial acromegaly and genetic defects in sporadic tumors. Rev Endocr Metab Disord. 2008;9:1–11.
17. Pfarr CM, Mechta F, Spyrou G, Lallemand D, Carillo S, Yaniv M. Mouse JunD negatively regulates fibroblast growth and antagonizes transformation by ras. Cell. 1994;76:747–60.
18. Heppner C, Bilimoria KY, Agarwal SK, Kester M, Whitty LJ, Guru SC, et al. The tumor suppressor protein menin interacts with NF-kappaB proteins and inhibits NF-kappaB-mediated transactivation. Oncogene. 2001;20:4917–25.
19. Kaji H, Canaff L, Lebrun JJ, Goltzman D, Hendy GN. Inactivation of menin, a Smad3-interacting protein, blocks transforming growth factor type beta signaling. Proc Natl Acad Sci U S A. 2001;98:3837–42.
20. Farid N, Buehler S, Russel N, Maroun F, Allerdice P, Smyth H. Prolactinomas in familial multiple endocrine neoplasia syndrome type I. Relationship to HLA and carcinoid tumors. Am J Med. 1980;69:874–80.
21. Scheithauer BW, Laws Jr ER, Kovacs K, Horvath E, Randall RV, Carney JA. Pituitary adenomas of the multiple endocrine neoplasia type I syndrome. Semin Diagn Pathol. 1987;4:205–11.
22. Verges B, Boureille F, Goudet P, Murat A, Beckers A, Sassolas G, et al. Pituitary disease in MEN type 1 (MEN1): data from the France–Belgium MEN1 multicenter study. J Clin Endocrinol Metab. 2002;87:457–65.
23. Poncin J, Stevenaert A, Beckers A. Somatic MEN1 gene mutations does not contribute significantly to sporadic pituitary tumorigenesis. Eur J Endocrinol. 1999;140:573–6.
24. Trouillas J, Labat-Moleur F, Sturm N, Kujas M, Heymann M-F, Figarella-Branger D, et al. Pituitary tumors and hyperplasia in multiple endocrine neoplasia type 1 syndrome (MEN1): a case-control study in a series of 77 patients versus 2509 non-MEN1 patients. Am J Surg Pathol. 2008;32:534–43.
25. Olufemi SE, Green J, Manickam P, Guru SC, Agarwal SK, Kester M, et al. Common ancestral mutation in the MEN1 gene is likely responsible for the prolactinoma prolactinoma variant of MEN1 (MEN1Burin) in four kindreds from Newfoundland. Hum Mutat. 1998;11:264–9.
26. Burgess JR, Shepherd JJ, Parameswaran V, Hoffman L, Greenaway TM. Spectrum of pituitary disease in multiple endocrine neoplasia type 1 (MEN 1): clinical, biochemical, and radiological features of pituitary disease in a large MEN 1 kindred. J Clin Endocrinol Metab. 1996;81:2642–6.
27. Bertolinio P, Tong WM, Galendo D, Wang ZQ, Zhang CX. Heterozygous MEN1 mutant mice develop a range of endocrine tumors mimicking multiple endocrine neoplasia type 1. Mol Endocrinol. 2003;17:1880–92.
28. Bertolinio P, Radanovic I, Casse H, Aguzzi A, Wang ZQ, Zhang CX. Genetic ablation of the tumor suppressor menin causes lethality at mid-gestation with defects in multiple organs. Mech Dev. 2003;120:549–60.
29. Biondi CA, Garside MG, Waring P, Loffler KA, Stark MS, Magnuson MA, et al. Conditional inactivation of the MEN1 gene leads to pancreatic and pituitary tumorigenesis but does not affect normal development of these tissues. Mol Cell Biol. 2004;24:3125–31.
30. Hai N, Aoki N, Shimatsu A, Mod T, Kosugi S. Clinical features of multiple endocrine neoplasia type 1 (MEN1) phenocopy without germline MEN1 gene mutations: analysis of 20 Japanese sporadic casis with MEN1. Clin Endocrinol. 2000;52:509–18.
31. Fritz A, Walch A, Piotrowska K, et al. Recessive transmission of a multiple endocrine neoplasia syndrome in the rat. Cancer Res. 2002;62:3048–51.
32. Piotrowska K, Pellegata NS, Rosemann M, Fritz A, Graw J, Atkinson MJ. Mapping of a novel MEN-like syndrome locus to rat chromosome 4. Mamm Genome. 2004;15:135–41.
33. Pellegata NS, Quintanilla-Martinez L, Siggelkow H, Samson E, Bink K, Hofler H, et al. Germ-line mutations in p27Kip1 cause a multiple endocrine neoplasia syndrome in rats and humans. Proc Natl Acad Sci U S A. 2006;103:15558–63.
34. Quereda V, Malumbres M. Cell cycle control of the pituitary. J Mol Endocrinol. 2009;42:75–86.

35. Nakayama K, Ishida N, Shirane M, Inomata A, Inoue T, et al. Mice lacking p27Kip1 display increased body size, multiple organ hyperplasia, retinal dysplasia and pituitary tumors. Cell. 1996;85:707–20.
36. Kiyokawa H, Kineman RD, Manova-Todorova KO, Soares VC, Hoffman ES, et al. Enhanced growth of mice lacking the cyclin-dependent kinase inhibitor function of p27Kip1. Cell. 1996;85:721–32.
37. Jin L, Qian X, Kulig E, et al. Transforming growth factor-beta, transforming growth factor-beta receptor II and p27Kip1 expression in nontumorous and neoplastic human pituitaries. Am J Pathol. 1997;151:509–19.
38. Bamberger CM, Fehn M, Bamberger AM, et al. Reduced expression levels of the cell-cycle inhibitor p27Kip1 in human pituitary adenomas. Eur J Endocrinol. 1999;140:250–5.
39. Drosten M, Hilken G, Bockmann M, Rodicker F, Mise N, Cranston AN, et al. Role of MEN2A-derived RET in maintenance and proliferation of medullary thyroid carcinoma. J Natl Cancer Inst. 2004;96:1231–9.
40. Kolluri SK, Weiss C, Koff A, Gottlicher M. p27Kip1 induction and inhibition of proliferation by the intracellular Ah receptor in developing thymus and hepatoma cells. Genes Dev. 1999;13:1742–53.
41. Georgitsi M, Raitila A, Karhu A, van der Lujit RB, Aalfs CM, Sane T, et al. Germline CDKN1B/p27kip1 mutation in multiple endocrine neoplasia. J Clin Endocrinol Metab. 2007;92:3321–5.
42. Agarwal SK, Mateo CM, Marx SJ. Rare germline mutations in cyclin-dependent kinase inhibitor genes in MEN1 and related states. J Clin Endocrinol Metab. 2009;94:1826–34.
43. Igreja SC, Chahal HS, Akker SA, Gueorguiev M, Popovic V, Damjanovic S, et al. Assessment of p27 (cyclin-dependent kinase inhibitor 1B) and AIP (aryl hydrocarbon receptor-interacting protein) genes in MEN1 syndrome patients without any detectable MEN1gene mutations. Clin Endocrinol. 2009;70:259–64.
44. Stratakis CA, Bertherat J, Carney JA. Mutation of perinatal myosin heavy chain. N Engl J Med. 2004;351:2556–8.
45. Bain F. Carney complex [letter to editor]. Mayo Clin Proc. 1986;61:508.
46. Boikos SA, Stratakis CA. Carney complex: the first 20 years. Curr Opin Oncol. 2007;19:24–9.
47. Kirschner LS, Carney JA, Pack SD, Taymans SE, Giatzakis C, Cho YS, et al. Mutations of the gene encoding the protein kinase A type I-alpha regulatory subunit in patients with the Carney complex. Nat Genet. 2000;26:89–92.
48. Kaltsas GA, Kola B, Borboli N, Morris DG, Gueorguiev M, Swords FM, et al. Sequence analysis of the PRKAR1A gene in sporadic somatotroph and other pituitary tumours. Clin Endocrinol (Oxf). 2002;57:443–8.
49. Sandrini F, Kirschner LS, Bei T, Farmakidis C, Yasufuku-Takano J, Takano K, et al. PRKAR1A, one of the Carney complex genes, and its locus (17q22-24) are rarely altered in pituitary tumours outside the Carney complex. J Med Genet. 2002;39:e78.
50. Bossis I, Stratakis CA. PRKAR1A: normal and abnormal functions. Endocrinology. 2004;145:5452–8.
51. Stratakis CA, Carney JA, Lin JP, Papanicolaou DA, Karl M, Kastner DL, et al. Carney complex, a familial multiple neoplasia and lentiginosis syndrome. Analysis of 11 kindreds and linkage to the short arm of chromosome 2. J Clin Investig. 1996;97:699–705.
52. Boikos SA, Stratakis CA. Pituitary pathology in patients with Carney complex: growth-hormone producing hyperplasia or tumors and their association with other abnormalities. Pituitary. 2006;9:203–9.
53. Pack SD, Kirschner LS, Pak E, Zhuang Z, Carney JA, Stratakis CA. Genetic and histological studies of somatomammotropic tumors in patients with the 'Complex of spotty skin pigmentation, myxomas, endocrine overactivity and schwannomas' (Carney complex). J Clin Endocrinol Metab. 2000;85:3860–5.
54. Kurtkaya-Yapicier O, Scheithauer B, Carney JA, et al. Pituitary adenoma in Carney complex: an immunohistochemical, ultrastructural, and immunoelectron microscopic study. Ultrastruct Pathol. 2002;26:345–53.

55. Griffin KJ, Kirschner LS, Matyakhina L, Stergiopoulos SG, Robinson-White A, Lenherr SM, et al. A transgenic mouse bearing an antisense construct of regulatory subunit type 1A of protein kinase A develops endocrine and other tumours: comparison with Carney complex and other PRKAR1A induced lesions. J Med Genet. 2004;41:923–31.
56. Griffin KJ, Kirschner LS, Matyakhina L, Stergiopoulos S, Robinson-White A, Lenherr S, et al. Down-regulation of regulatory subunit type 1A of protein kinase A leads to endocrine and other tumors. Cancer Res. 2004;64:8811–5.
57. Yin Z, Williams-Simons L, Parlow AF, Asa S, Kirschner LS. Pituitary-specific knockout of the Carney complex gene PRKAR1A leads to pituitary tumorigenesis. Mol Endocrinol. 2008;22:380–7.
58. Bailey P, Cushing H. The microscopic structure of the adenomas in acromegalic dyspituitarism (fugitive acromegaly). Am J Pathol. 1928;4:545–63.
59. Daly AF, Jaffrain-Rea ML, Ciccarelli A, Valdes-Socin H, Rohmer V, Tamburrano G, et al. Clinical characterization of familial isolated pituitary adenomas. J Clin Endocrinol Metab. 2006;91:3316–23.
60. Verloes A, Stevenaert A, Teh BT, Petrossians P, Beckers A. Familial acromegaly: case report and review of the literature. Pituitary. 1999;1:273–7.
61. Frohman LA, Eguchi K. Familial acromegaly. Growth Horm IGF Res. 2004;14:S90–6.
62. Beckers A, Daly A. The clinical, pathological, and genetic features of familial isolated pituitary adenomas. Eur J Endocrinol. 2007;157:371–82.
63. Teh BT, Kytola S, Farnebo F, Bergman L, Wong FK, Weber G, et al. Mutation analysis of the MEN1 gene in multiple endocrine neoplasia type 1, familial acromegaly and familial isolated hyperparathyroidism. J Clin Endocrinol Metab. 1998;83:2621–6.
64. Gadelha MR, Une KN, Rohde K, Vaisman M, Kineman RD, Frohman LA. Isolated familial somatotropinomas: establishment of linkage to chromosome 11q13.1-11q13.3 and evidence for a potential second locus at chromosome 2p16-12. J Clin Endocrinol Metab. 2000;85:707–14.
65. Luccio-Camelo DC, Une KN, Ferreira RE, Khoo SK, Nickolov R, Bronstein MD, et al. A meiotic recombination in a new isolated familial somatotropinoma kindred. Eur J Endocrinol. 2004;150:643–8.
66. Linquette M, Herlant M, Laine E, Fossati P, Dupont-Lecompte J. Adenome a prolactine chez une jeune fine dont la mere etait porteuse d'un adenome hypophysaire avec amenorrhee galactorrhee. Ann Endocrinol. 1997;28:773–80.
67. Berezin M, Karasik A. Familial prolactinoma. Clin Endocrinol. 1995;42:483–6.
68. Salti IS, Mufarrij IS. Familial Cushing disease. Am J Med Genet. 1981;8:91–4.
69. Yuasa H, Tokito S, Nakagaki H, Kitamura K. Familial pituitary adenoma – report of four cases from two unrelated families. Neurol Med Chir (Tokyo). 1990;30:1016–9.
70. Valdes-Socin H, Poncin J, Stevens V, Stevenaert A, Beckers A. Adenomes hypophysaires familiaux isoles non lies avec la mutation somatique NEM-1. Suivi de 27 patients. Ann Endocrinol. 2000;61:301.
71. Valdes-Socin H, Jaffrain Rea ML, Tamburrano G, Cavagnini F, Ciccarelli A, Colao A, et al. Familial isolated pituitary tumors: clinical and molecular studies in 80 patients. In: Endocrine Society's 84th Annual Meeting; 2002. P3-663 647.
72. Beckers A. Familial isolated pituitary adenomas. The Ninth International Workshop on multiple endocrine neoplasia (MEN2004). J Intern Med. 2004;255:696–730.
73. Daly AF, Tichomirowa MA, Beckers A. Genetic, molecular and clinical features of familial isolated pituitary adenomas. Horm Res. 2009;71 Suppl 2:116–22.
74. Leontiou CA, Gueorguiev M, van der Spuy J, Quinton R, Lolli F, Hassan S, et al. The role of the aryl hydrocarbon receptor-interacting protein gene in familial and sporadic pituitary adenomas. J Clin Endocrinol Metab. 2008;93:2390–401.
75. Villa C, Magri F, Morbini P, et al. Silent familial isolated pituitary adenomas: histopathological and clinical case report. Endocr Pathol. 2008;19:40–6.
76. Vierimaa O, Georgitsi M, Lehtonen R, Vahteristo P, Kokko A, Raitila A, et al. Pituitary adenoma predisposition caused by germline mutations in the AIP gene. Science. 2006;312:1228–30.

77. Yu R, Bonert V, Saporta I, Raffel LJ, Melmed S. Aryl hydrocarbon receptor interacting protein variants in sporadic pituitary adenomas. J Clin Endocrinol Metab. 2006;91:5126–9.
78. Daly AF, Vanbellinghen JF, Khoo SK, Jaffrain-Rea ML, Naves LA, Guitelman MA, et al. Aryl hydrocarbon receptorinteracting protein gene mutations in familial isolated pituitary adenomas: analysis in 73 families. J Clin Endocrinol Metab. 2007;92:1891–6.
79. Barlier A, Vanbellinghen JF, Daly AF, Silvy M, Jaffrain-Rea ML, Trouillas J, et al. Mutations in the aryl hydrocarbon receptor interacting protein gene are not highly prevalent among subjects with sporadic pituitary adenomas. J Clin Endocrinol Metab. 2007;92:1952–5.
80. Toledo RA, Lourenco Jr DM, Liberman B, Cunha-Neto MB, Cavalcanti MG, Moyses CB, et al. Germline mutation in the aryl hydrocarbon receptor interacting protein gene in familial somatotropinoma. J Clin Endocrinol Metab. 2007;92:1934–7.
81. Iwata T, Yamada S, Mizusawa N, Golam HM, Sano T, Yoshimoto K. The aryl hydrocarbon receptor-interacting protein gene is rarely mutated in sporadic GH-secreting adenomas. Clin Endocrinol. 2007;66:499–502.
82. Georgitsi M, Raitila A, Karhu A, Tuppurainen K, Makinen MJ, Vierimaa O, et al. Molecular diagnosis of pituitary adenoma predisposition caused by aryl hydrocarbon receptor-interacting protein gene mutations. Proc Natl Acad Sci U S A. 2007;104:4101–5.
83. Georgitsi M, Karhu A, Winqvist R, Visakorpi T, Waltering K, Vahteristo P, et al. Mutation analysis of aryl hydrocarbon receptor interacting protein (AIP) gene in colorectal, breast, and prostate cancers. Br J Cancer. 2007;96:352–6.
84. Cazabat L, Libe R, Perlemoine K, Rene-Corail F, Burnichon N, Gimenez-Roquiplo AP, et al. Germline inactivating mutations of the aryl hydrocarbon receptor-interacting protein gene in a large cohort of sporadic acromegaly: mutations are found in a subset of young patients with macroadenomas. Eur J Endocrinol. 2007;157:1–8.
85. Tichomirowa MA, Daly AF, Pujol J, Naves LA, Rodien P, Vanbellinghen JF, et al. An analysis of the role of cyclin dependent kinase inhibitor 1B (CDKN1B) gene mutations in 86 families with familial isolated pituitary adenomas (FIPA). In: The Endocrine Society's 91st Annual Meeting, 10–13 June, Washington; 2009.
86. Carver LA, LaPress JJ, Dunham EE, Bradfield CA. Characterization of the Ahreceptor-associated protein, ARA9. J Biol Chem. 1998;273:33580–7.
87. Bell DR, Poland A. Binding of aryl hydrocarbon receptor (AhR) to AhR-ineracting protein. The role of hsp90. J Biol Chem. 2000;275:36407–14.
88. Tichomirowa MA, Daly AF, Beckers A. Familial pituitary adenomas. J Intern Med. 2009;266:5–18.
89. Carver LA, Bradfield CA. Ligand-dependent interaction of the aryl hydrocarbon receptor with a novel immunophilin homolog in vivo. J Biol Chem. 1997;272:11452–6.
90. Oesch-Bartlomowicz B, Oesch F. Role of cAMP in mediating AHR signaling. Biochem Pharmacol. 2009;77:627–41.
91. Lees MJ, Peet DJ, Whitelaw ML. Defining the role of XAP2 in stabilization of dioxin receptor. J Biol Chem. 2003;278:35878–88.
92. Bolger GB, Peden AH, Steele MR, MacKenzie C, McEwan DG, Wallace DA, et al. Attenuation of the activity of the cAMP-specific phosphodiesterasePDE4A5 by interaction with the immunophilin XAP2. J Biol Chem. 2003;278:33351–63.
93. de Oliveira SK, Hoffmeister M, Gambaryan S, Muller-Esterl W, Guimaraes JA, Smolenski AP. Phosphodiesterase 2a forms a complex with the co-chaperone XAP2 and regulates nuclear translocation of the aryl hydrocarbon receptor. J Biol Chem. 2007;282:13656–63.
94. Jaffrain-Rea ML, Angelini M, Gargano D, Tichomirowa MA, Daly AF, Vanbellinghen JF, et al. Expression of aryl hydrocarbon receptor (AHR) and AHR-interacting protein in pituitary adenomas: pathological and clinical implications. Endocr Relat Cancer. 2009;16:1029–43.
95. Lin BC, Sullivan R, Lee Y, Moran S, Glover E, Bradfield CA. Deletion of the aryl hydrocarbon receptor-associated protein 9 leads to cardiac malformation and embryonic lethality. J Biol Chem. 2007;282:35924–32.
96. Lin BC, Nguyen LP, Walisser JA, Bradfield CA. A hypomorphic allele of aryl hydrocarbon receptor-associated protein-9 produces a phenocopy of the AHR-null mouse. Mol Pharmacol. 2008;74:1367–71.

97. Naves LA, Daly AF, Vanbellinghen JF, Casulari LA, Spilioti C, Magalhaes AV, et al. Variable pathological and clinical features of a large Brazilian family harboring a mutation in the aryl hydrocarbon receptor-interacting protein gene. Eur J Endocrinol. 2007;157:383–91.
98. Daly AF., Tichomirowa MA, Petrossians P, Heliövaara E, Jaffrain-Rea ML, Barlier A, et al. Clinical characteristics and therapeutic responses in patients with germ-line AIP mutations and pituitary adenomas: an international collaborative study. The Journal of Clinical Endocrinology and Metabolism. 2010;95(11):E373–83.

Part III
Growth Hormone Deficiency

Chapter 8
The Epidemiology of Growth Hormone Deficiency

Kirstine Stochholm and Jens Sandahl Christiansen

Abstract The definition of growth hormone deficiency (GHD) has been disputed for decades. Pituitary disorders or the consequences of the treatment hereof can lead to GHD; however, numerous extrapituitary diseases or syndromes can result in GHD. Guidelines and clarification have been published in order to shed more light on the issue. However, no standardized approach, which can be applied for currently diagnosed persons as well as recently diagnosed persons, exists.

When focusing on the diagnosis of GHD, it seems important to differentiate between childhood-onset (CO) and adult-onset (AO) GHD. No arbitrary cut-off age exists, and in the guidelines, this is a clinical definition. Epidemiologically, this is not manageable and often 18 years is used as cut-off age.

Data on incidence and prevalence are scarce; however, generally males seem to be diagnosed more often than.

Morbidity is increased in patients with GHD; this applies both when focusing on clinical markers and in terms of admissions to hospitals. Similarly, studies from different countries have identified an increased mortality in various subgroups of GHD patients and in total. The increased mortality was especially due to cardiovascular diseases or cancer.

The long-term consequences of growth hormone treatment on morbidity and mortality are not fully known; however, the treatment is considered safe except with regard to patients with active malignancy, in whom the treatment is contraindicated.

Keywords Growth hormone deficiency • Incidence • Prevalence • Childhood onset • Adult onset • Morbidity mortality • Children • Adults • Cancer • Cardiovascular • Hypopituitarism

---

J.S. Christiansen (✉)
Department of Internal Medicine and Endocrinology, Aarhus University Hospital, Noerrebrogade 42, 8000 Aarhus, Denmark
e-mail: jsc@ki.au.dk

K. Ho (ed.), *Growth Hormone Related Diseases and Therapy: A Molecular and Physiological Perspective for the Clinician*, Contemporary Endocrinology, DOI 10.1007/978-1-60761-317-6_8, © Springer Science+Business Media, LLC 2011

Introduction

Growth hormone (GH) deficiency (GHD) has been recognized as a clinical entity for decades, but confusion regarding diagnostic criteria still exists [1]. Consensus reports for diagnosing GHD in children [2] and in adults [3] have facilitated a more uniform diagnostic approach worldwide, and a clinical guideline was published in 2006 [4]. Recently, revisions of the guidelines and clarifications have been published [5, 6].

However, there is still no standardized approach to the diagnosis of hypopituitarism or GHD which impedes epidemiologic assessment, and a definition of GHD which can be applied currently as well as previously does not exist. This problem is reflected by the variability of criteria applied in previous epidemiologic studies. In order to understand GHD epidemiologically, it is important to recognize all aspects of the heterogeneity of the disease.

Causes of GHD

Childhood-onset (CO) GHD: There is little doubt that numerous diseases can lead to GHD. In a Belgium prevalence study of children who were candidates for human GH (hGH) treatment, 41% were classified as having idiopathic GHD, 35% acquired GHD, and 20% congenital GHD [7]. All were younger than 18 years at diagnosis. In a Danish incidence study, 16% of the persons identified with CO GHD had benign tumors in or close to the pituitary, 11% had idiopathic GHD, and 10% craniopharyngeomas [8]. In this study, CO GHD was defined as age at onset younger than 18 years. Interestingly, 33% had a defined, but rare cause of GHD, such as aplasia/hypoplasia of the pituitary, neurofibromatosis Recklinghausen, or birth trauma and 20% had GHD due to unknown causes.

Adult-onset (AO) GHD: In adults, pituitary adenomas or the treatment hereof are the most common causes of GHD [4]. The adenomas are typically nonfunctioning; however, they may secrete one or more of the pituitary hormones. In a study in a part of United Kingdom (West Midlands, representing about 10% of the UK population) including prevalent *hypopituitary* persons, 57% had a nonfunctioning pituitary adenoma, 9% a prolactinoma, 1% a gonadotropinoma, and <1% a thyrotropin-secreting adenoma [9]; acromegalic patients as well as patients with Cushing's syndrome were excluded. Danish nationwide data in *GHD* patients identified 28% with a nonfunctioning adenoma, 14% with hormonally active adenoma (including former GH-producing and ACTH-producing), and 17% with either hemorrhage in adenoma or unspecified [8]. The symptoms of the nonfunctioning adenoma are often due to growth of the adenoma, thus visual disturbance is a common first symptom due to pressure on the optic chiasm. Surgery rarely takes place before eventual visual disturbance occurs.

Craniopharyngiomas are relatively common causes of GHD, also in adults. One study identified 108 persons with GHD due to a craniopahryngeoma out of

1,329 (8%) [8]. Anatomically, craniopharyngiomas are sellar or suprasellar tumors, and the consequences of the craniopharyngioma or of therapy of any kind are closely related to the function of hypothalamus and the pituitary [10]. Thus, hyperphagia, obesity, diabetes insipidus, and pubertas tarda are often observed. Furthermore, loss of releasing hormones of the pituitary hormones is observed.

The long-term consequences of irradiation of the brain, resulting in hypopituitarism, are now widely recognized [11]. Furthermore, it is known that the impact of radiation on the pituitary function is dose-dependent [12]. Due to the increased survival of children with cancer [13], it is anticipated that a majority of these patients will develop clinical hypopituitarism.

Acquired GHD due to traumatic brain injury (TBI) has recently been recognized as a clinical entity [14], with a steeply increasing number of publications during the last decade. One of the first more detailed reports in a group of head-injured patients was published in 2000 [15]. Thus, it is obvious that the clinical surveillance for GHD/hypopituitarism in these TBI patients was limited in the twentieth century. Incidence and prevalence data including patients from 1999 and before are therefore to some degree biased towards lower numbers. Thus, the TBI patients elegantly demonstrate one of the pitfalls in historical epidemiological data.

As one of the few clinical entities, the number of women with classical Sheehan's syndrome originally described by Sheehan [16] is hopefully decreasing at least in the part of the world with modern obstetric birth clinics.

Genetic disorders in pituitary transcription factors (Pit) cause multiple pituitary hormone deficits (including GHD), such as caused by mutation in Pit-1 [17] (now called POU1F1) and prophet of Pit-1 (PROP-1) [18]. Another type of genetic disorders is the growth hormone-1 (GH-1) gene defect, which causes isolated GHD [19]. Various genetic disorders leading to congenital hypopituitarism has been described in detail [20]. All these conditions are rare.

This heterogeneity in causes of GHD is important to keep in mind whenever scrutinizing reports regarding mortality and morbidity. Data in selected cohorts should only be regarded as valid in relation to the specific cohorts studied.

Testing GH Secretion

As the release of GH is pulsatile, testing is necessary to identify possible disturbances in GH secretion [2]. In routine clinical practice, a considerable number of different stimulatory tests have been performed during the last decades: Insulin tolerance test (ITT) [21], L-dopa [22], clonidine [23], glucagon [24, 25], arginine [26], exercise [27], heat tolerance [28], growth hormone-releasing hormone (GHRH) [29], pyridostigmine and GHRH [30], urinary growth hormone [31], and combinations hereof [32, 33], including also combinations with for instance propranolol, ornithine, and betaxolol, as described in a European survey by Juul et al. [1]. Another review stated that at least 34 provocative tests with 189 combinations have been found [34]. The struggle of defining the "gold standard" is obvious considering the problem of achieving a normal response from a normal subject under a test [35] and as the reproducibility of the tests is highly variable [36]. The problem is further accentuated as cranial irradiation can cause hypothalamic damage and confuse the interpretation of the tests [37, 38]. Since the hypothalamus is considered more radio-sensitive than the pituitary [39], a consequence of irradiation may be deficiency in the normal pattern of GHRH secretion, but no deficiency in the ability of the pituitary to produce GH when stimulated appropriately. This ability diminishes over time. Some patients may be able to secrete GH on GHRH stimulation within the first 5–10 years after irradiation, but lack the ability to secrete GH during an ITT [40].

GH is usually the most sensitive hormone to irradiation and surgery of the pituitary [11, 41]. A Swedish group found after stimulation with GHRH a specificity of 94% and a sensitivity of 66% of the GHD diagnosis among young adults with acute lymphoblastic leukemia (ALL) treated with cranial irradiation using 7.5 μg/L as cut-off and ITT as gold standard [42]. These patients were irradiated 21 years (range: 8.9–28) before the test. Other factors which influence tests are fasting and physical activity [43] and hot baths [44]. By contrast, some suggest that no testing should be undertaken in children [45], and instead, diagnosis should be based on a combination of auxological, biochemical, neuroradiological, and genetic considerations.

The variability of assays to determine GH concentrations and in cut-off limits is considerable [46]. There have been various international standards for human GH with different conversion factors, leading to erroneous use of standards [47]. The consensus guidelines recommend the use of the World Health Organization's (WHO) definition: International reference preparation (IRP) 88/624 [3], where 1 mg of protein (somatropin and somatropin-related impurities) corresponds to 3 International Unit (IU) of biological activity. The IRP 88/624 is still considered the standard of choice; however, the problem of standardization is clearly expressed in the latest amendment [5]. The first international standard was established recently for IGF-I [48] with the preparation coded 02/254.

Once again, the heterogeneity of tests hinders a clear-cut definition of GHD, and as a consequence, epidemiological studies investigating the growth hormone-deficient patients are extremely difficult to compare. In conclusion, it is important to stress that the changes in the definition of GHD necessitate knowledge of former diagnostic criteria. Furthermore, the pitfalls when using these criteria must be accepted, if used as diagnostic evaluation of patients historically.

Childhood-Onset vs. Adult-Onset GHD

The first guideline describing the diagnosis and treatment in children with short stature was printed in 1995 [49], but the diagnosis and treatment had already taken place for decades. An interesting aspect is the missing distinct cut-off between CO GHD and AO GHD *combined* with the change in definition of CO and AO GHD. due to the highly variable time of attainment of puberty, there is no consequent arbitrary age at which this occurs. At the end of linear growth or, better, after attainment of peak bone mass [50]? Or as stated in the consensus guidelines "After the attainment of final height, retesting of the GH-IGF axis, using the adult GHD diagnostic criteria,...should be undertaken by the pediatric endocrinologist...." However, in the epidemiological studies, an unambiguous number is much more applicable than this relevant but unmanageable definition, and it seems like the most consistently used cut-off age has been 18 years.

Some children with CO GHD may respond normally to a GH stimulation test when reevaluated after completion of linear growth and puberty [51, 52]. The mechanism behind this phenomenon is poorly understood. A gender-specific increase in spontaneous (unstimulated) GH-release was found in healthy children during puberty [53], and an increased percentage of normal responses to stimulation tests has been identified in pubertal compared to nonpubertal patients [54]. However, these apparent differences might simply be due to lack of reproducibility of the GH stimulation tests used [55]. The percentage of persistent severe GHD in adults who had CO GHD and who formerly were treated with GH is stated to be between 12.5 and 90% [56].

Marin et al. [57] suggested the use of sex hormone priming in prepubertal and pubertal stage 2 and 3 children in order to reduce the number of false positive results during GH testing. This should be undertaken in order to distinguish between genuinely GHD children and children with constitutional delay in growth. However, data are still scarce, and sex hormone priming is considered controversial [58], although generally recommended [2].

Intuitively, there is little doubt that the impact on morbidity and mortality varies in GHD children and adults. Interestingly, a study in patients with craniopharyngioma by Kendall-Taylor et al. [59] demonstrates reduced body mass index and reduced cholesterol levels in CO patients compared with AO patients. A cut-off age of 18 years was applied. Differences between CO and AO GHD have also been found in cohorts of mixed etiology [60]. Cut-off age was not given.

The percentage of CO GHD persons who remain growth hormone-deficient when tested after attainment of final height is highly dependable on the specific cause of the deficiency in the first place. Thus, a high number of other pituitary deficiencies yield a high risk of staying growth hormone-deficient.

The Incidence and Prevalence of GHD

Data on incidence and prevalence of GHD in children and adults are scarce. Various approaches have been undertaken [7, 8, 61–65], including large screening studies of elementary school children. See Table 8.1 for further details. Generally, the incidence and prevalence of GHD are increased in boys compared to girls and in men compared to women.

In adults (≥18 years at diagnosis), there is a population-based estimate on the incidence rate (IR) and the prevalence of *hypopituitarism* from the northwestern Spain [66]. Here the incidence was estimated to be 4.2 per 100,000 per year, and the prevalence was estimated to be 46 per 100,000 in 1999. The Growth Hormone Research Society (GRS) estimated an annual incidence of 1.0 patient with AO *GHD*

Table 8.1 Descriptions of epidemiological findings regarding incidence and prevalence of growth hormone deficiency (GHD)

Author	CO/AO	Number of patients[a]	Number of background, per year	Patients' age (years, range)	Criteria for GH test	GH assay	Incidence per 100,000
Incidence							
Parkin [61]	CO[b]	12 (na)	Approximately 48,000	7–22 at Evaluation	No information	No information	3.3
Sassolas et al. [65]	AO[b]	16 (na)	Approximately 1,319,550	na	GH < 3 μg/L in one or two tests	No information	1.2
Stochholm et al. [8]	CO	494 (303)	Not given	0–18 at Diagnosis	Detailed criteria given	No information	2.58 (boys), 1.70 (girls)
Stochholm et al. [8]	AO	1,329 (744)	Not given	18 Years or more at diagnosis	Detailed criteria given	No information	1.90 (men), 1.42 (women)

Author	CO/AO[b]	Number of patients[a]	Number of background[a]	Patients' age (years, range)	Criteria for GH test	GH assay	Prevalence per 100,000 males	Prevalence per 100,000 females
Prevalence								
Vimpani et al. [62]	CO[b]	13 (8)	48,221 (24,670)	6–9	GH ≤ 9 mU/L (ITT)	No information	32.4	21.2
Bao et al. [63]	CO[b]	12 (9)	103,753 (51,994)	6–15	GH < 10 μg/L in two tests	No information	17.3	5.8
Lindsay et al. [64]	CO[b]	33 (24)	114,881 (59,087)	Kindergarten to fifth grade	GH < 10 ng/mL in two tests	No information	40.6	16.1
Thomas et al. [7]	CO[b]	386 (na)	2,200,000 (na)	0–18	GH < 10 ng/mL in two tests (GH<5 ng/mL)(na)	No information	18 (Both genders)	
Sassolas et al. [65]	AO[b]	Estimated 82	1,759,400 (na) Includes children	na	GH < 3 μg/L in one or two tests	No information	4.6 (Both genders)	

[a]In *parentheses* number of males
[b]The definitions of childhood onset (CO) and adult onset (AO) are not described in the studies and are here applied as the most probable definition
na: not available

per 100,000 citizens. This estimate is based on the incidence of pituitary tumors and the age of onset in the specific patients is not defined [67].

Morbidity in GHD

In general, an aspect different from the clinical perspective must be undertaken when performing epidemiological studies in morbidity and mortality in GHD patients. As the previously mentioned changes in diagnostic tools and criteria occur with time, many patients have not been investigated corresponding to today's standard and many have never been tested for eventual GHD. It is important to realize that the results may be even more biased if only patients diagnosed according to today's criteria are included. Thus, it is relevant to include patients as GHD when they lack hormones from the other pituitary axes [68, 69], have been irradiated [11], or have a history which suggest a high risk of suffering from GHD.

In none of the following morbidity and mortality studies, a description of the GH assays can be found, probably due to the historical changes in the use of assays (see above). Moreover, it is noteworthy that many clinical and epidemiological studies analyze data on CO and AO GHD pooled together or do not state the age at diagnosis.

Morbidity of patients with GHD has been described in a number of clinical papers, typically focusing on clinical or laboratory markers for a disease, for instance, cardiovascular risk markers. Data can be supplemented with data on hypopituitary patients. In female patients with hypopituitarism, an increased prevalence of cardiovascular risk factors such as lower high density lipoprotein cholesterol compared to age- and gender-matched controls was reported [70]. Colao et al. [71] studied the cardiovascular impairment in GHD adolescents and found that GH replacement improved cardiovascular risk factors. In adult GHD patients with and without treatment with GH, the metabolic changes have been described in detail [72, 73] and an induction of, for example, anabolic effects have been identified [74]. An increased severity of cardiovascular risk factors in children compared to adults was found in a subgroup of patients suffering from craniopharyngioma [59]. In adults with CO GHD, there is an increased risk of fractures [75] and an increased risk of reduced bone mineral density (BMD) [76]. Importantly, BMD has been reported to be significantly higher in those who were treated with growth hormone compared to the untreated persons during their adolescence. In GHD adults, an increased risk of fractures was reported [77]. Furthermore, the treatment (e.g., irradiation) of children with pituitary disease who might become GHD enhances these disturbances and may to some extent overshadow the relative impact of GHD [78].

An increased morbidity in GHD patients seems to be a natural consequence of these clinical findings and has been identified in a cohort of CO and AO GHD patients compared to the background population, where time from diagnosis to first admittance at a hospital is used as a indication of morbidity [79]. Here a significantly increased morbidity was identified in both genders in CO and AO GHD. Using the international classification of diseases, all admittances were divided into 20 chapters and an increased morbidity was identified in the vast majority of chapters. Interestingly, the only chapter with a significantly decreased morbidity was due to pregnancy and puerperal complications in both CO and AO, possibly caused by a decreased number of pregnancies. However, as stated, "It remains to be clarified whether this was due to lack of interest, lack of partner, or reduced fertility."

Cancer has been of major concern in children with GHD, especially those treated with human GH (hGH) and GH. The recurrence of cancer has been studied in children (the vast majority was younger than 19 years at first treatment) treated with hGH [80] in the UK. A significantly increased mortality due to cancer was present, with a significantly increased mortality in colon cancer and Hodgkin's disease. These findings were evident even though only a limited number of deaths were identified. In the US, a significantly increased relative risk of secondary leukemia among children treated with hGH was found [81]. Sklar et al. [82] found a significantly increased risk of development of a secondary neoplasia among patients treated with hGH compared to nontreated patients, primarily among patients with primary acute leukemia (all were diagnosed with cancer before the age of 21 years). Here it is extremely important to emphasize that the authors stress that there is also "may have been inherent, but unrecognized, biases in the selection of patients who received therapy with GH." They found no difference in relative risk of *death* in GH-treated patients compared to nontreated patients and concluded that the risk of secondary cancer during GH treatment is not increased, but precaution should be taken in regard to patients with primary acute leukemia. In contrast, Swerdlow et al. [83] found a significantly decreased relative risk of mortality and risk of recurrence of tumor in hGH-treated compared to nontreated children (all younger than 16 years at tumor diagnosis); both groups primarily had had radiotherapy due to a brain tumor. This finding was supported by Mills who found no increased risk of cancer among hGH-treated children compared to the background population when high risk children were included [84]. Similarly, no increased risk of de novo neoplasms in recipients of recombinant human GH has been found [85]. Here the cohort of interest was 42% with idiopathic GHD, 26% with organic GHD, 14% with idiopathic short stature, 10% with Turner syndrome, and 8% with other etiologies.

Thus, pros and cons regarding the consequences of GH treatment exist. Overall, the treatment is considered safe, with the precaution of not treating patients with active malignancy [4].

Mortality

The specific consequences on mortality of the increased morbidity of GHD are not fully described.

One Swedish study analyzed 289 GHD patients treated with GH and found a normal mortality [86], whereas mortality and morbidity were increased in an older group of hypopituitary patients not treated with GH, both groups were compared with the background population. Danish data on GHD patients identified nationwide found a significantly increased hazard ratio (HR) of 8.3 (4.5–15.1) in CO boys,

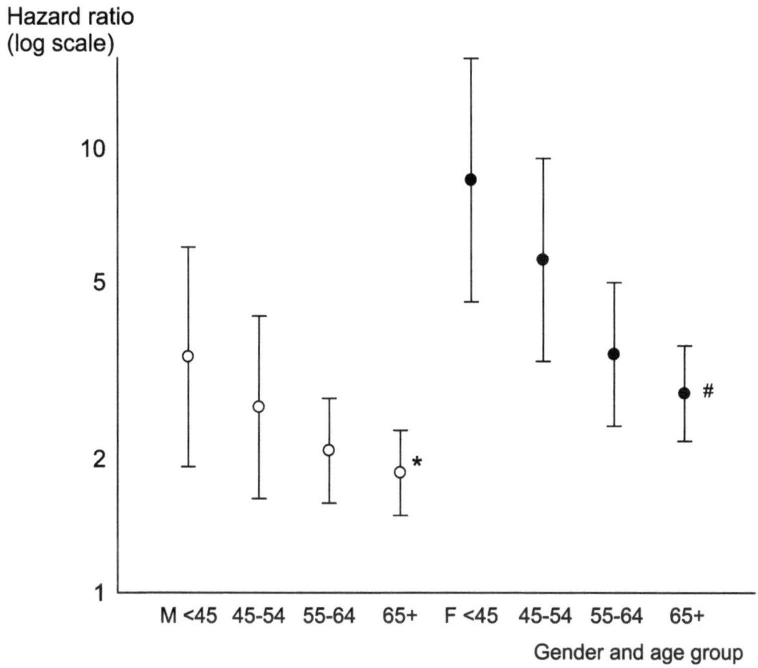

Fig. 8.1 Hazard ratios of total mortality adult-onset GHD, 1980–2004, subdivided into four age groups according to age at entry and gender. *Open circles* males, and *black circles* females. Mortality risk falls with increasing age at entry *($p<0.05$) and gender #($p<0.001$)

9.4 (4.6–19.4) in CO girls, 1.9 (1.7–2.2) in AO men, and 3.4 (2.9–4.0) in AO women, all compared with age- and gender-matched background population [87]. For details on mortality with increasing age at entry in study, see Fig. 8.1. Furthermore, mortality in AO women was significantly increased compared to AO men. In Canadian GHD children (age: 0–29 years at onset) treated with hGH, mortality was comparable to the background population, except in a subgroup of boys younger than 2 years at diagnosis [88].

Most epidemiologic studies in adults have focused on patients with hypopituitarism defined by at least one deficient pituitary axis [9, 89, 90] or on patients after a pituitary operation [91, 92], and an increased mortality compared to the background population has been reported.

An interesting small study of mortality in patients with isolated and untreated GHD due to a mutation in the GH-1 gene found a significantly increased mortality compared to siblings and to age-matched persons from the same regional area [93]. Furthermore, a doubled mortality compared to the background population was found in reoperated and irradiated patients with a macroadenoma [94]. In patients with pituitary adenoma, a Swedish group identified a doubled standardized mortality ratio (SMR) in the total cohort and a significantly increased SMR in females compared to males [95]. In a recent much smaller Danish study in patients with transsphenoidally operated nonfunctioning adenomas, total SMR was not increased;

however, the gender difference was still present with a significantly increased SMR in females, but not in males [96].

It is important to emphasize that the consequences of GH treatment on mortality are not known. The appropriate randomized study to identify an eventual effect is ethically not acceptable. This is for instance due to the known improvement in quality of life and cardiovascular risk markers when treated, as well as the contraindication of treatment in patients with active malignancy.

Comments on Epidemiological Studies in Persons with Growth Hormone Deficiency

In summary, while scrutinizing *epidemiological* reports it is important to consider whether focus of the report is hypopituitarism or GHD or other. Check how the identification of the patients was performed – which diagnostic criteria were applied and are they acceptable? Check whether age at onset of GHD or hypopituitarism is stated – is CO and AO (GHD) patients mixed? Does the report contain both prevalent and incident patients? This typically takes place when recruiting patients from the outpatients' clinic. Check whether the patients for instance were identified in university hospitals with special interest in rare pituitary diseases? Or are peripheral hospitals included, where other forms of GHD/hypopituitarism may be present?

In screening surveys, beware of the percentage of participants – could it possibly be the sick and hypopituitary persons who are never tested because they do not attend normal school? Or are only the small persons tested, with subsequent loss of the persons with hypopituitarism for only a short while?

Conclusion

In conclusion, the incidence and prevalence of GHD are highly variable with findings from 1.2–33 per 100,000 citizens per year to 4.6–40.6 per 100,000 citizens, respectively. Morbidity was significantly increased in the GHD cohorts. This applies to as different measures of morbidity as cardiovascular profiles, BMD, or when using number of admissions at hospitals. Mortality was significantly increased in all cohorts of GHD persons and was especially due to cancer or cardiovascular diseases.

References

1. Juul A, Bernasconi S, Clayton PE, Kiess W, DeMuinck-Keizer SS. European audit of current practice in diagnosis and treatment of childhood growth hormone deficiency. Horm Res. 2002;58(5):233–41.
2. GH Research Society. Consensus guidelines for the diagnosis and treatment of growth hormone (GH) deficiency in childhood and adolescence: summary statement of the GH Research Society. J Clin Endocrinol Metab. 2000;85(11):3990–3.

3. Sonksen PH, Christiansen JS; Growth Hormone Research Society. Consensus guidelines for the diagnosis and treatment of adults with growth hormone deficiency. Growth Horm IGF Res. 1998;8(Suppl B):89–92.
4. Molitch ME, Clemmons DR, Malozowski S, Merriam GR, Shalet SM, Vance ML. Evaluation and treatment of adult growth hormone deficiency: an Endocrine Society Clinical Practice Guideline. J Clin Endocrinol Metab. 2006;91(5):1621–34.
5. Ghigo E, Aimaretti G, Corneli G. Diagnosis of adult GH deficiency. Growth Horm IGF Res. 2008;18(1):1–16.
6. Ho KKY. Consensus guidelines for the diagnosis and treatment of adults with GH deficiency II: a statement of the GH Research Society in association with the European Society for Pediatric Endocrinology, Lawson Wilkins Society, European Society of Endocrinology, Japan Endocrine Society, and Endocrine Society of Australia. Eur J Endocrinol. 2007;157(6):695–700.
7. Thomas M, Massa G, Craen M, et al. Prevalence and demographic features of childhood growth hormone deficiency in Belgium during the period 1986-2001. Eur J Endocrinol. 2004;151(1):67–72.
8. Stochholm K, Gravholt CH, Laursen T, et al. Incidence of GH deficiency – a nationwide study. Eur J Endocrinol. 2006;155:61–71.
9. Tomlinson JW, Holden N, Hills RK, et al. Association between premature mortality and hypopituitarism. West Midlands Prospective Hypopituitary Study Group. Lancet. 2001; 357(9254):425–31.
10. Halac I, Zimmerman D. Endocrine manifestations of craniopharyngioma. Childs Nerv Syst. 2005;21(8–9):640–8.
11. Littley MD, Shalet SM, Beardwell CG, Ahmed SR, Applegate G, Sutton ML. Hypopituitarism following external radiotherapy for pituitary tumours in adults. Q J Med. 1989;70(262):145–60.
12. Littley MD, Shalet SM, Beardwell CG, Robinson EL, Sutton ML. Radiation-induced hypopituitarism is dose-dependent. Clin Endocrinol (Oxf). 1989;31(3):363–73.
13. Bleyer A. Older adolescents with cancer in North America deficits in outcome and research. Pediatr Clin North Am. 2002;49(5):1027–42.
14. Ghigo E, Masel B, Aimaretti G, et al. Consensus guidelines on screening for hypopituitarism following traumatic brain injury. Brain Inj. 2005;19(9):711–24.
15. Kelly DF, Gonzalo IT, Cohan P, Berman N, Swerdloff R, Wang C. Hypopituitarism following traumatic brain injury and aneurysmal subarachnoid hemorrhage: a preliminary report. J Neurosurg. 2000;93(5):743–52.
16. Sheehan H. Postpartum necrosis of the anterior pituitary. J Pathol Bacteriol. 1939;45:189.
17. Camper SA, Saunders TL, Katz RW, Reeves RH. The Pit-1 transcription factor gene is a candidate for the murine Snell dwarf mutation. Genomics. 1990;8(3):586–90.
18. Sornson MW, Wu W, Dasen JS, et al. Pituitary lineage determination by the Prophet of Pit-1 homeodomain factor defective in Ames dwarfism. Nature. 1996;384(6607):327–33.
19. Akinci A, Kanaka C, Eble A, Akar N, Vidinlisan S, Mullis PE. Isolated growth hormone (GH) deficiency type IA associated with a 45-kilobase gene deletion within the human GH gene cluster. J Clin Endocrinol Metab. 1992;75(2):437–41.
20. Kelberman D, Dattani MT. The role of transcription factors implicated in anterior pituitary development in the aetiology of congenital hypopituitarism. Ann Med. 2006;38(8):560–77.
21. Hoffman DM, O'Sullivan AJ, Baxter RC, Ho KK. Diagnosis of growth-hormone deficiency in adults. Lancet. 1994;343(8905):1064–8.
22. Lim YJ, Kwan E, Low LC. Screening test for growth hormone deficiency: usefulness of L-dopa-propranolol provocative test. J Paediatr Child Health. 1994;30(4):328–30.
23. Gil-Ad I, Topper E, Laron Z. Oral clonidine as a growth hormone stimulation test. Lancet. 1979;2(8154):1242.
24. Rao RH, Spathis GS. Intramuscular glucagon as a provocative stimulus for the assessment of pituitary function: growth hormone and cortisol responses. Metabolism. 1987;36(7):658–63.
25. Orme SM, Price A, Weetman AP, Ross RJ. Comparison of the diagnostic utility of the simplified and standard i.m. glucagon stimulation test (IMGST). Clin Endocrinol (Oxf). 1998;49(6):773–8.
26. Merimee TJ, Lillicrap DA, Rabinowitz D. Effect of arginine on serum-levels of human growth-hormone. Lancet. 1965;2(7414):668–70.

27. de San LC, Parkin JM, Turner SJ. Treadmill exercise test in short children. Arch Dis Child. 1984;59(12):1179–82.
28. Christensen SE, Jorgensen OL, Moller N, Orskov H. Characterization of growth hormone release in response to external heating. Comparison to exercise induced release. Acta Endocrinol (Copenh). 1984;107(3):295–301.
29. Rogol AD, Blizzard RM, Johanson AJ, et al. Growth hormone release in response to human pancreatic tumor growth hormone-releasing hormone-40 in children with short stature. J Clin Endocrinol Metab. 1984;59(4):580–6.
30. Andersen M, Hansen TB, Stoving RK, et al. The pyridostigmine-growth-hormone-releasing-hormone test in adults. The reference interval and a comparison with the insulin tolerance test. Endocrinol Metab. 1996;3(3):197–206.
31. Skinner AM, Clayton PE, Price DA, Addison GM, Soo A. Urinary growth hormone excretion in the assessment of children with disorders of growth. Clin Endocrinol (Oxf). 1993;39(2):201–6.
32. Liberman B, Cesar FP, Wajchenberg BL. Human growth hormone (hGH) stimulation tests: the sequential exercise and L-dopa procedure. Clin Endocrinol (Oxf). 1979;10(6):649–54.
33. Weldon VV, Gupta SK, Klingensmith G, et al. Evaluation of growth hormone release in children using arginine and L-dopa in combination. J Pediatr. 1975;87(4):540–4.
34. Sizonenko PC, Clayton PE, Cohen P, Hintz RL, Tanaka T, Laron Z. Diagnosis and management of growth hormone deficiency in childhood and adolescence. Part 1: diagnosis of growth hormone deficiency. Growth Horm IGF Res. 2001;11(3):137–65.
35. Ghigo E, Bellone J, Aimaretti G, et al. Reliability of provocative tests to assess growth hormone secretory status. Study in 472 normally growing children. J Clin Endocrinol Metab. 199; 81(9):3323–7.
36. Rochiccioli P, Enjaume C, Tauber MT, Pienkowski C. Statistical study of 5473 results of nine pharmacological stimulation tests: a proposed weighting index. Acta Paediatr. 1993;82(3):245–8.
37. Romshe CA, Zipf WB, Miser A, Miser J, Sotos JF, Newton WA. Evaluation of growth hormone release and human growth hormone treatment in children with cranial irradiation-associated short stature. J Pediatr. 1984;104(2):177–81.
38. Ahmed SR, Shalet SM. Hypothalamic growth hormone releasing factor deficiency following cranial irradiation. Clin Endocrinol (Oxf). 1984;21(5):483–8.
39. Chrousos GP, Poplack D, Brown T, O'Neill D, Schwade J, Bercu BB. Effects of cranial radiation on hypothalamic-adenohypophyseal function: abnormal growth hormone secretory dynamics. J Clin Endocrinol Metab. 1982;54(6):1135–9.
40. Darzy KH, Aimaretti G, Wieringa G, Gattamaneni HR, Ghigo E, Shalet SM. The usefulness of the combined growth hormone (GH)-releasing hormone and arginine stimulation test in the diagnosis of radiation-induced GH deficiency is dependent on the post-irradiation time interval. J Clin Endocrinol Metab. 2003;88(1):95–102.
41. Clayton PE, Shalet SM. Dose dependency of time of onset of radiation-induced growth hormone deficiency. J Pediatr. 1991;118(2):226–8.
42. Bjork J, Link K, Erfurth EM. The utility of the growth hormone (GH) releasing hormone-arginine test for diagnosing GH deficiency in adults with childhood acute lymphoblastic leukemia treated with cranial irradiation. J Clin Endocrinol Metab. 2005;90(11):6048–54.
43. Fisker S, Jorgensen JO, Christiansen JS. Variability in growth hormone stimulation tests. Growth Horm IGF Res. 1998;8(Suppl A):31–5.
44. Fisker S, Jorgensen JO, Orskov H, Christiansen JS. GH stimulation tests: evaluation of GH responses to heat test versus insulin-tolerance test. Eur J Endocrinol. 1998;139(6):605–10.
45. Badaru A, Wilson DM. Alternatives to growth hormone stimulation testing in children. Trends Endocrinol Metab. 2004;15(6):252–8.
46. Juul A, Bernasconi S, Chatelain P, et al. Diagnosis of growth hormone (GH) deficiency and the use of GH in children with growth disorders. Horm Res. 1999;51(6):284–99.
47. Bristow AF. International standards for growth hormone. Horm Res. 1999;51 Suppl 1:7–12.
48. Burns C, Rigsby P, Moore M, Rafferty B. The First International Standard for Insulin-like Growth Factor-1 (IGF-1) for immunoassay: preparation and calibration in an international collaborative study. Growth Horm IGF Res. 2009;19(5):457–62.

49. The Drug and Therapeutics Committee of the Lawson Wilkins Pediatric Endocrine Society. Guidelines for the use of growth hormone in children with short stature. A report by the Drug and Therapeutics Committee of the Lawson Wilkins Pediatric Endocrine Society. J Pediatr. 1995;127(6):857–67.
50. Shalet S. Stepping into adulthood: the transition period. Horm Res. 2004;62 Suppl 4:15–22.
51. Wacharasindhu S, Cotterill AM, Camacho-Hubner C, Besser GM, Savage MO. Normal growth hormone secretion in growth hormone insufficient children retested after completion of linear growth. Clin Endocrinol (Oxf). 1996;45(5):553–6.
52. Juul A, Kastrup KW, Pedersen SA, Skakkebak NE. Growth hormone (GH) provocative retesting of 108 young adults with childhood-onset GH deficiency and the diagnostic value of insulin-like growth factor I (IGF-I) and IGF-binding protein-3. J Clin Endocrinol Metab. 1997;82(4): 1195–201.
53. Albertsson-Wikland K, Rosberg S, Karlberg J, Groth T. Analysis of 24-hour growth hormone profiles in healthy boys and girls of normal stature: relation to puberty. J Clin Endocrinol Metab. 1994;78(5):1195–201.
54. Cacciari E, Tassoni P, Cicognani A, et al. Value and limits of pharmacological and physiological tests to diagnose growth hormone (GH) deficiency and predict therapy response: first and second retesting during replacement therapy of patients defined as GH deficient. J Clin Endocrinol Metab. 1994;79(6):1663–9.
55. Tauber M, Moulin P, Pienkowski C, Jouret B, Rochiccioli P. Growth hormone (GH) retesting and auxological data in 131 GH-deficient patients after completion of treatment. J Clin Endocrinol Metab. 1997;82(2):352–6.
56. Gasco V, Corneli G, Beccuti G, et al. Retesting the childhood-onset GH-deficient patient. Eur J Endocrinol. 2008;159 Suppl 1:S45–52.
57. Marin G, Domene HM, Barnes KM, Blackwell BJ, Cassorla FG, Cutler GB. The effects of estrogen priming and puberty on the growth-hormone response to standardized treadmill exercise and arginine-insulin in normal girls and boys. J Clin Endocrinol Metab. 1994;79(2):537–41.
58. Wyatt DT, Mark D, Slyper A. Survey of growth hormone treatment practices by 251 pediatric endocrinologists. J Clin Endocrinol Metab. 1995;80(11):3292–7.
59. Kendall-Taylor P, Jonsson PJ, Abs R, et al. The clinical, metabolic and endocrine features and the quality of life in adults with childhood-onset craniopharyngioma compared with adult-onset craniopharyngioma. Eur J Endocrinol. 2005;152(4):557–67.
60. Attanasio AF, Lamberts SW, Matranga AM, et al. Adult growth hormone (GH)-deficient patients demonstrate heterogeneity between childhood onset and adult onset before and during human GH treatment. Adult Growth Hormone Deficiency Study Group. J Clin Endocrinol Metab. 1997;82(1):82–8.
61. Parkin JM. Incidence of growth hormone deficiency. Arch Dis Child. 1974;49(11):904–5.
62. Vimpani GV, Vimpani AF, Lidgard GP, Cameron EH, Farquhar JW. Prevalence of severe growth hormone deficiency. Br Med J. 1977;2(6084):427–30.
63. Bao XL, Shi YF, Du YC, Liu R, Deng JY, Gao SM. Prevalence of growth hormone deficiency of children in Beijing. Cochrane Database Syst Rev (Online). 1992;105(5):401–5.
64. Lindsay R, Feldkamp M, Harris D, Robertson J, Rallison M. Utah Growth Study: growth standards and the prevalence of growth hormone deficiency. J Pediatr. 1994;125(1):29–35.
65. Sassolas G, Chazot FB, Jaquet P, et al. GH deficiency in adults: an epidemiological approach. Eur J Endocrinol. 1999;141(6):595–600.
66. Regal M, Paramo C, Sierra SM, Garcia-Mayor RV. Prevalence and incidence of hypopituitarism in an adult Caucasian population in northwestern Spain. Clin Endocrinol (Oxf). 2001; 55(6):735–40.
67. Carroll PV, Christ E, Bengtsson BA, et al.; the members of Growth Hormone Research Society Scientific Committee. Growth hormone deficiency in adulthood and the effects of growth hormone replacement: a review. J Clin Endocrinol Metab. 1998;83(2):382–95.
68. Hartman ML, Crowe BJ, Biller BM, Ho KK, Clemmons DR, Chipman JJ. Which patients do not require a GH stimulation test for the diagnosis of adult GH deficiency? J Clin Endocrinol Metab. 2002;87(2):477–85.

69. Toogood AA, Beardwell CG, Shalet SM. The severity of growth hormone deficiency in adults with pituitary disease is related to the degree of hypopituitarism. Clin Endocrinol (Oxf). 1994;41(4):511–6.
70. Bulow B, Hagmar L, Eskilsson J, Erfurth EM. Hypopituitary females have a high incidence of cardiovascular morbidity and an increased prevalence of cardiovascular risk factors. J Clin Endocrinol Metab. 2000;85(2):574–84.
71. Colao A, Di Somma C, Salerno M, Spinelli L, Orio F, Lombardi G. The cardiovascular risk of GH-deficient adolescents. J Clin Endocrinol Metab. 2002;87(8):3650–5.
72. Jorgensen JO, Pedersen SA, Thuesen L, et al. Beneficial effects of growth hormone treatment in GH-deficient adults. Lancet. 1989;1(8649):1221–5.
73. Salomon F, Cuneo RC, Hesp R, Sonksen PH. The effects of treatment with recombinant human growth-hormone on body-composition and metabolism in adults with growth-hormone deficiency. N Engl J Med. 1989;321(26):1797–803.
74. Lucidi P, Lauteri M, Laureti S, et al. A dose-response study of growth hormone (GH) replacement on whole body protein and lipid kinetics in GH-deficient adults. J Clin Endocrinol Metab. 1998;83(2):353–7.
75. de Boer H, Blok GJ, van Lingen A, Teule GJ, Lips P, van der Veen EA. Consequences of childhood-onset growth hormone deficiency for adult bone mass. J Bone Miner Res. 1994;9(8):1319–26.
76. Hyer SL, Rodin DA, Tobias JH, Leiper A, Nussey SS. Growth hormone deficiency during puberty reduces adult bone mineral density. Arch Dis Child. 1992;67(12):1472–4.
77. Rosen T, Wilhelmsen L, Landin-Wilhelmsen K, Lappas G, Bengtsson BA. Increased fracture frequency in adult patients with hypopituitarism and GH deficiency. Eur J Endocrinol. 1997;137(3):240–5.
78. Murray RD, Brennan BM, Rahim A, Shalet SM. Survivors of childhood cancer: long-term endocrine and metabolic problems dwarf the growth disturbance. Acta Paediatr Suppl. 1999;88(433):5–12.
79. Stochholm K, Laursen T, Green A, et al. Morbidity and GH deficiency: a nationwide study. Eur J Endocrinol. 2008;158(4):447–57.
80. Swerdlow AJ, Higgins CD, Adlard P, Preece MA. Risk of cancer in patients treated with human pituitary growth hormone in the UK, 1959-85: a cohort study. Lancet. 2002;360(9329):273–7.
81. Fradkin JE, Mills JL, Schonberger LB, et al. Risk of leukemia after treatment with pituitary growth hormone. JAMA. 1993;270(23):2829–32.
82. Sklar CA, Mertens AC, Mitby P, et al. Risk of disease recurrence and second neoplasms in survivors of childhood cancer treated with growth hormone: a report from the Childhood Cancer Survivor Study. J Clin Endocrinol Metab. 2002;87(7):3136–41.
83. Swerdlow AJ, Reddingius RE, Higgins CD, et al. Growth hormone treatment of children with brain tumors and risk of tumor recurrence. J Clin Endocrinol Metab. 2000;85(12):4444–9.
84. Mills JL, Schonberger LB, Wysowski DK, et al. Long-term mortality in the United States cohort of pituitary-derived growth hormone recipients. J Pediatr. 2004;144(4):430–6.
85. Tuffli GA, Johanson A, Rundle AC, Allen DB. Lack of increased risk for extracranial, nonleukemic neoplasms in recipients of recombinant deoxyribonucleic acid growth hormone. J Clin Endocrinol Metab. 1995;80(4):1416–22.
86. Svensson J, Bengtsson BA, Rosen T, Oden A, Johannsson G. Malignant disease and cardiovascular morbidity in hypopituitary adults with or without growth hormone replacement therapy. J Clin Endocrinol Metab. 2004;89(7):3306–12.
87. Stochholm K, Gravholt CH, Laursen T, et al. Mortality in growth hormone deficiency – a nationwide study. Eur J Endocrinol. 2007;157:9–18.
88. Taback SP, Dean HJ. Mortality in Canadian children with growth hormone (GH) deficiency receiving GH therapy 1967-1992. The Canadian Growth Hormone Advisory Committee. J Clin Endocrinol Metab. 1996;81(5):1693–6.
89. Rosen T, Bengtsson BA. Premature mortality due to cardiovascular disease in hypopituitarism. Lancet. 1990;336(8710):285–8.

90. Bates AS, Van't Hoff W, Jones PJ, Clayton RN, et al. The effect of hypopituitarism on life expectancy. J Clin Endocrinol Metab. 1996;81(3):1169–72.
91. Bulow B, Hagmar L, Mikoczy Z, Nordstrom CH, Erfurth EM. Increased cerebrovascular mortality in patients with hypopituitarism. Clin Endocrinol (Oxf). 1997;46(1):75–81.
92. Bates AS, Bullivant B, Sheppard MC, Stewart PM. Life expectancy following surgery for pituitay tumours. Clin Endocrinol (Oxf). 1999;50(1999):315–9.
93. Besson A, Salemi S, Gallati S, et al. Reduced longevity in untreated patients with isolated growth hormone deficiency. J Clin Endocrinol Metab. 2003;88(8):3664–7.
94. Erfurth EM, Bulow B, Svahn-Tapper G, et al. Risk factors for cerebrovascular deaths in patients operated and irradiated for pituitary tumors. J Clin Endocrinol Metab. 2002;87(11): 4892–9.
95. Nilsson B, Gustavsson-Kadaka E, Bengtsson BA, Jonsson B. Pituitary adenomas in Sweden between 1958 and 1991: incidence, survival, and mortality. J Clin Endocrinol Metab. 2000;85(4):1420–5.
96. Lindholm J, Nielsen EH, Bjerre P, et al. Hypopituitarism and mortality in pituitary adenoma. Clin Endocrinol (Oxf). 2006;65(1):51–8.

Chapter 9
Diagnosis of Growth Hormone Deficiency in Adults

Sandra Pekic and Vera Popovic

Abstract Adult growth hormone deficiency (GHD) has a wide spectrum of clinical presentations including abnormal body composition, increased blood pressure, increased body weight, adverse lipid profiles, increased coagulability, and increased markers of inflammation, all associated with an increased cardiovascular morbidity and mortality. Detection of GHD is important, especially as successful GH replacement therapy improves the adult GHD syndrome and is accompanied by a significant increase in quality of life. Health authorities in many countries have approved the therapeutic use of GH in adult hypopituitary patients in whom GHD has been identified. The identification of adults with GHD remains challenging due to the episodic nature of GH secretion (pulsatile secretion) and multiple GH samples over 24 h are not practical as diagnostic procedure. For this reason in adults with known pituitary disease, diagnostic testing involves provocative tests of GH secretion. The insulin tolerance test (ITT) is regarded as the standard test of choice. Reproducibility, demanding surveillance during the test, and safety issues under some circumstances may be of some concern. Other more powerful provocative tests like the GH-releasing hormone (GHRH) + arginine and GHRH + GH-releasing hexapeptide 6 (GHRP-6) have gained acceptance after being evaluated in a large series of patients and controls. Glucagon test has yet to be evaluated in large series of patients and controls, but has already gained popularity particularly in the US after shortage of GHRH supplies. In adult patients with known pituitary disease, the presence and severity of GHD are related to the number of additional pituitary hormone deficits so that in panhypopituitary patients GH stimulation tests may be unnecessary. In that situation, insulin-like growth factor 1 (IGF-1) may be a useful diagnostic tool. Otherwise, considerable overlap exists for IGF-1 levels between normal subjects and those with GHD.

V. Popovic (✉)
Neuroendocrine Unit, Institute of Endocrinology, University Clinical Center,
Dr Subotic 13, 11000 Belgrade, Serbia
e-mail: popver@EUnet.yu

K. Ho (ed.), *Growth Hormone Related Diseases and Therapy: A Molecular and Physiological Perspective for the Clinician*, Contemporary Endocrinology,
DOI 10.1007/978-1-60761-317-6_9,

Therefore, IGF-1 is not a reliable marker for diagnosis of GHD. When focused on childhood-onset GHD, these young adults who had previously received GH replacement in childhood need to be reevaluated for GH status.

Keywords Growth hormone • Growth hormone deficiency • Hypopituitarism

The frequency of adult growth hormone deficiency (GHD) is reported to be 1/3,000 to 1/4,000 [1]. The incidence of isolated GHD (IGHD) relative to multiple pituitary hormone deficiencies varies from 10 to 20%, due to different criteria and tests used [2, 3]. Idiopathic IGHD occurring de novo in the adult is not a recognized diagnostic entity [4].

Who Should Be Evaluated for Adult GHD?

GH is secreted episodically and this is modified by age and sex. Because of this pulsatile secretion, only multiple sampling over 24 h may accurately reflect GH status, but this is not practical in clinical practice. Since GH and insulin-like growth factor 1 (IGF-1) concentrations are insufficient to diagnose adult GHD, many pharmacological provocative tests were investigated, and currently, the diagnosis of GHD is based on an inadequate response of the GH to the provocative testing [4–16]. Current consensus recommendations define patient groups which require GH testing: (1) patients with signs and symptoms of hypothalamic-pituitary disease, (2) patients who have received cranial irradiation or tumor treatment, and (3) patients who survived traumatic brain injury (TBI) or subarachnoid hemorrhage (SAH) [4, 16].

All provocative tests for diagnosing GHD are imperfect, and their accuracy is strongly affected by the underlying appropriate clinical setting. A large number of structural lesions in the hypothalamic-pituitary region underlie adult-onset GHD (AO-GHD) [17, 18]. Acquired GHD in adults is most commonly secondary to hypothalamic-pituitary tumors who may present with hypopituitarism. Generally, growth hormone (GH) is the first hormone to be lost, followed by the gonatotrophins, adrenocorticotrophic hormone, and thyroid-stimulating hormone [19]. The greater the number of pituitary hormone deficiencies, the likelihood of GHD increases [20, 21]. Data emerging after 2,000 demonstrate the relevance of TBI and SAH as the cause of GHD, whereas postpartum pituitary necrosis, lymphocytic hypophysitis, infections, and neoplastic lesions in the sella remain rarer diagnoses of AO-GHD [4, 8, 18, 22–27]. Testing for GHD should not be sooner than 1 year after TBI due to the possibility of its reversal [16]. In adults with childhood-onset GHD (CO-GHD), the most common entity is idiopathic GHD [28].

The consensus also recommended that one stimulation test is sufficient for the diagnosis of adult GHD [4]. Patients with three or more pituitary hormone deficiencies and an IGF-1 level below the reference range have >97% chance of being GHD and therefore do not need a GH stimulation test [4, 16, 21].

Interpretation and Validation of a GH Stimulation Test

In real life, there is a spectrum of GHD from severe, partial to not at all. The diagnosis of GHD is based on an inadequate response of GH to stimulation. It is common in clinical practice to classify subjects as normal or abnormal with regard to the response to a stimulation test as an aid to decision-making and thus GH treatment. The situation is complicated by the availability of many stimulation tests and many assays for its measurement [29–31] (Table 9.1). In general, there are two kinds of GH stimulation tests: those less and those more potent (see Fig. 9.1). The combined use of two stimulation factors in one test has promoted the use of powerful tests for GH release such as GHRH+arginine, GHRH+hexarelin, GHRH+GHRP-6,

Table 9.1 Procedures for different GH stimulation tests to diagnose adult GH deficiency

Test	Procedure	Sampling
ITT	0.15–0.3 U/kg of soluble insulin IV bolus at 0 min	0, 30, 60, 90, 120 min
Glucagon	1 mg (1.5 mg if >90 kg) IM at 0 min	0, 90, 120, 150, 180, 210, 240 min
GHRH+ARG	GHRH (1 μg/kg IV bolus at 0 min GEREF 1-29 NH_2)+ARG (ARG hydrochloride, 0.5 g/kg IV infusion over 30 min)	0, 15, 30, 45, 60, 90, 120 min
GHRH+GHRP-6	GHRH (1 μg/kg IV bolus at 0 min GEREF 1-29 NH_2)+GHRP-6 (1 μg/kg IV bolus at 0 min His-D-Trp-Ala-Trp-D-Phe-Lys-NH_2)	0, 15, 30, 45, 60, 90, 120 min

ITT Insulin tolerance test; *GHRH* growth hormone-releasing hormone; *ARG* arginine; *GHRP-6* growth hormone-releasing peptide 6; *IV* intravenously; *IM* intramuscularly

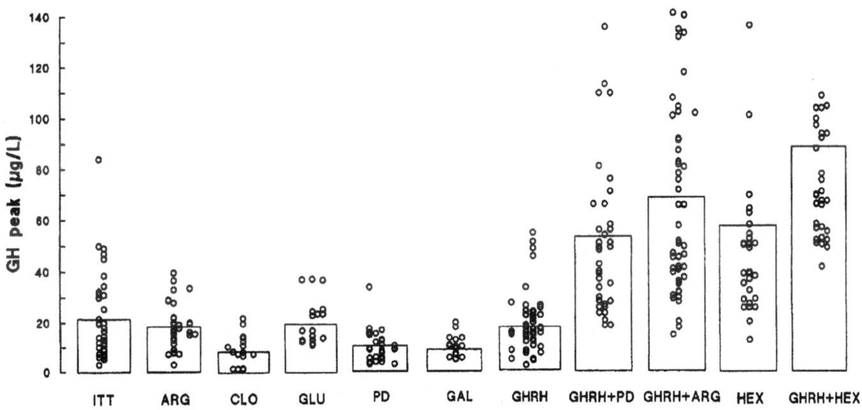

Fig. 9.1 Mean and individual peak GH responses to classical stimuli, GH-releasing hormone (GHRH), GHRH+pyridostigmine (PD), GHRH+arginine (ARG), hexarelin (HEX), and GHRH+HEX in normal adult subjects (from Aimaretti et al. [29]; © Society of the European Journal of Endocrinology (2000). Reproduced by permission)

Table 9.2 Results from studies on the effects of body weight on GH responses in different tests to diagnose adult GH deficiency

References	Test		GH value
Hoffman et al. [36]	ITT	Adult	<3 μg/L
Ho [4]	Consensus Guidelines	Transition period	<6 μg/L
Gomez et al. [49]	Glucagon		<3 μg/L
Corneli et al. [12]	GHRH + arginine	BMI <25 kg/m^2	<11.5 μg/L
		BMI 25–30 kg/m^2	<8.0 μg/L
		BMI >30 kg/m^2	<4.2 μg/L
Popovic et al. [56]	GHRH + GHRP-6	BMI <35 kg/m^2	<10 μg/L
Kelestimur et al. [61]		BMI >35 kg/m^2	<5 μg/L

BMI Body mass index; *GHRH* growth hormone-releasing hormone; *GHRP* growth hormone-releasing peptide; *ITT* insulin tolerance test

and GHRH + GHRP-2. In patients with defined pathology in the hypothalamic-pituitary region, one test will suffice.

The stimulated GH threshold levels (cut-off values) are somewhat arbitrary, since in reality GH secretion as mentioned is a continuum between deficient, insufficient, and sufficient GH state. The GH thresholds defining severe GHD in adults 3 or 5 μg/L have often been used in clinical practice when less potent provocative tests for GH secretion were adopted [30, 32, 33]. These GH thresholds are thought to best discriminate between patients and healthy subjects irrespective of age, gender, and BMI. Depending on the stimulus, peak GH responses to stimulation tests show interindividual variability and great inconsistency. Thus, following criteria for the validation of a GH stimulation test were proposed: (1) should be potent and reproducible, with reproducibility assessed in normal subjects, (2) the influence of gender, age, and adiposity should be validated in control subjects, (3) effectiveness needs to be assessed in controls and patients, evaluated by ROC curve analysis, and (4) the cut-off point for the test should be established by ROC curve analysis being the value that provides the best pairing of sensitivity and specificity.

Furthermore, threshold GH levels are variable as a function of the assay used and therefore not generalizable to other assays [34]. In 1997, the polyclonal radioimmunoassays were replaced by highly sensitive two-site monoclonal assays. A universal calibrator is essential and GRS advocates the use of recombinant 22 kDa GH calibrator (International Reference Preparation (IRP) 98/574) in all GH assays [4].

Another problem is interpretation of the results of most GH stimulation tests in adult GHD patients in the presence of obesity. Nearly half of the patients with acquired hypothalamic-pituitary disease are overweight or obese, indicating the need to define BMI-related cut-off limits of the GH responses to the stimulation test, because obesity per se is a state of functional hyposomatotropism. Obese patients have enhanced GH clearance, shorter half-life of GH, and significant decrease in the production and secretion of GH. Each one unit increase in BMI has been shown to be associated with a 6% decrease in 24-h GH secretion [35]. Thus, a BMI-related GH threshold levels in a GH stimulation test have to be taken into account when interpreting the results (Table 9.2).

Stimulation Tests for GH Secretion

Insulin Tolerance Test

Among the stimulation tests, the ITT is the diagnostic test of choice with peak GH of less than 3 μg/L during adequate hypoglycemia (blood glucose <2.2 mmol/L) indicating severe GHD [36] (Tables 9.1 and 9.2). When compared with a series of multiple GH samples over 24 h, ITT accurately reflected the GH status and this test was shown to be able to reliably distinguish patients with GHD from healthy volunteers [36]. On the other hand, recently it has been shown that under certain circumstances (in some cranially irradiated patients), failure to pass ITT in isolation may not reflect severe GHD [37, 38]. It appears that within partially damaged somatotroph axis, there is a compensatory increase in hypothalamic stimulatory input which may result in restoration of spontaneous GH secretion.

The mechanism by which hypoglycemia stimulates GH secretion is thought to be suppression of somatostatin tone and stimulation of α-adrenergic receptors. The advantage of ITT is that it stimulates both GH and ACTH release through the hypothalamus. Both ACTH and GH rise in response to stressful situation such as hypoglycaemia.

None of the tests including ITT fulfill the requirements of an ideal test being convenient and economical, with high discriminatory power, reproducible, safe, not dependent on confounding factors (age, gender, nutritional status, obesity, etc.). ITT demands surveillance and some patients fail to achieve adequate hypoglycemia following a standard dose of intravenous insulin during the ITT [39]. For such patients, a second, usually larger dose is administered, thus prolonging the procedure with some risks. Factors associated with insulin resistance, such as obesity and fasting hyperglycemia, both influence the ability to achieve adequate hypoglycemia following insulin administration. Lee et al. performed a study with the aim to identify the factors that predict failure to achieve adequate hypoglycemia during an ITT after pituitary surgery. In that study, fasting blood glucose (FBG) was an important determinant of the dose of insulin required to achieve adequate hypoglycemia during an ITT. In patients with FBG >5.5 mmol/L, a standard insulin dose of OD 0.1 U/kg is insufficient for adequate hypoglycemia. Their data indicate that doubling the insulin dose to 0.2 U/kg is appropriate for patients with FBG of 5.5–5.9 mmol/L and that 0.3 U/kg may be necessary for patients with FBG ≥6 mmol/L. Another factor causing inadequate hypoglycemia during the ITT is baseline and peak cortisol concentration. Cortisol enhances hepatic glucose production and is associated with decreased insulin sensitivity.

Although ITT demonstrates good sensitivity, reproducibility is a major problem [40]. Up to a sixfold difference in peak GH response has been demonstrated on different days in healthy adults undergoing ITTs, regardless of the degree of hypoglycemia [33]. Inadequate GH response may sometimes occur in normal subjects who have had a recent spontaneous pulse of GH. Inadequate GH response may also occur in obese patients frequently overlapping with GH response in nonobese adult

GHD patients [41]. The ITT is unpleasant and certain precautions are necessary. ECG must be normal, cortisol >100 nmol/L, and thyroxine levels normal. ITT is contraindicated in epilepsy, ischemic heart disease, and age <2 years. The demanding surveillance during ITT stimulated the search for other more practical procedures.

GHRH–Arginine Stimulation Test

Among alternative stimulation tests, GHRH + arginine has gained wide popularity defining severe GHD with a cut-off point for GH <9 μg/L [42, 43] (Table 9.1). GH-releasing hormone (GHRH) + arginine is very potent and reproducible test [44]. Arginine, an amino acid, potentiates the response to the GHRH probably via inhibition of hypothalamic somatostatin release [45, 46]. The GHRH + arginine test has excellent sensitivity and specificity and is unaffected by gender [19]. The GH response to GHRH + arginine is independent of age, and there is less inter- and intraindividual variability than with other stimulation tests. The reliability of GHRH + arginine was tested in obese hypopituitary patients [47]. The cut-off limits for peak GH response to the GHRH + arginine test obtained by ROC curve analysis were 11.5, 8.0 and 4.2 μg/L for the lean (BMI <25 kg/m^2), overweight (BMI 25–30 kg/m^2), and obese population (BMI >30 kg/m^2), respectively (Table 9.2). Inverse correlation between peak GH responses to the GHRH + arginine test and BMI was found in control subjects, but not in hypopituitary patients [30].

GHRH + arginine explores the maximum GH secretory reserve of the pituitary and is a powerful test that may yield false-negative responses in young patients with GHD, particularly in those with radiation-induced GHD [48]. There are situations where discordant GH responses to ITT and GHRH + arginine may be encountered. Hypothalamic dysfunction occurs early following cranial irradiation, while somatotroph dysfunction later. This may cause an inadequate GH response to ITT and adequate GH response to GHRH + arginine test. The discordant results largely depend on the total radiation dose and length of follow-up time [37].

In a large study by Biller et al. [30] in which utility, sensitivity, and specificity of six different tests for adult GHD were investigated, the greatest diagnostic accuracy occurred with the ITT and the GHRH + arginine test.

The Glucagon Test

The rationale behind the choice of various provocative tests may be related not only to specific personnel experience, but also to the availability of drugs like GHRH. Glucagon test has gained popularity particularly in the US, due to lack of recombinant GHRH [31]. It assesses ACTH/cortisol and GH reserve in patients in whom ITT is contraindicated. This test has very few side effects, allows simultaneous evaluation of pituitary-adrenal axis function, and has few contraindications (pheochromocytoma, insulinoma, glycogen storage disease, severe cortisol deficiency <100 nmol/L).

The interpretation is as for ITT defining severe GHD with a cut-off for GH <3 μg/L [49] (Tables 9.1 and 9.2). GH response to glucagon in 73 adult patients with GHD and in 46 controls compared with the ITT demonstrated that glucagon test was reliable for assessing GH status [49]. Unfortunately, a negative correlation between GH response to glucagon and age and body mass index was observed in healthy controls. Thus, glucagon test may be less reliable than ITT as it does not discriminate between healthy subjects and adult GHD patients above 50 years of age [50]. Recently, the glucagon test was thought to be valuable in young patients [51].

The mechanism of stimulation of GH secretion is still poorly understood. It is unlikely that the glucagon-induced GH release is mediated by changes in glucose concentrations. One possible mechanism by which glucagon stimulates GH secretion is via activation of central noradrenergic pathways. Glucagon has been shown to induce noradrenaline release in healthy subjects and this may trigger GH secretion via α-adrenergic receptors. Glucagon test needs to be evaluated in large series of patients and controls and this has not yet been reported.

GHRH + Growth Hormone Secretagogues

Growth hormone secretagogues (GHS) are peptidyl (GH-releasing peptides, GHRP) and nonpeptidyl, nonnatural molecules that possess potent dose-dependent and specific stimulatory effect on somatotrope secretion both in animals and in humans [52, 53]. GHS also acutely stimulate prolactin, ACTH, and cortisol secretion, without any effects on LH, FSH, or TSH secretion [54]. They act as functional antagonists of somatostatin via binding to specific hypothalamic and pituitary receptors (GHSR-1a) for which ghrelin has been shown to be the natural ligand [55]. Thus, GHS are considered to be analogs of ghrelin. GHS and GHRH act through different receptors and have synergistic effect on GH secretion. Analogs of ghrelin studied so far include GHRP-2, GHRP-6, hexarelin, and MK-677.

The combined GHRH + GHRP-6 test has been shown to be well tolerated, sensitive, and reproducible in diagnosing GHD [56–58] (Tables 9.1 and 9.2). It has been shown that GHRH + GHRP-6 test is a very potent and reproducible stimulation test which has been evaluated in a large series of patients and controls [56]. Severe GHD was defined with GH peak <10 μg/L to GHRH + GHRP-6 tests, while a normal GH response to the combined test was considered with GH peak >20 μg/L [56] (Table 9.2). The GH stimulating effect of GHRH + GHRP-6 test is blunted in hypothalamic-pituitary disconnection [59] (see Fig. 9.2). The GH-releasing effect in adults is independent of gender, age, sex, or obesity except for BMI >35 kg/m^2 [60, 61]. In a recent study by Kelestimur et al. [61], the cut-off level after GHRH + GHRP-6 test in patients with BMI >35 kg/m^2 defining severe GHD is <5 μg/L (Table 9.2). The availability of BMI-related cut-off limits is mandatory in view of the fact that many hypopituitary patients with suspected GHD are overweight or truly obese.

Furthermore, physical activity or food intake prior to testing did not affect the reproducibility of GH secretion elicited by GHRH + GHRP-6 test in normal adult subjects, making it convenient for everyday clinical practice [62]. In order to further

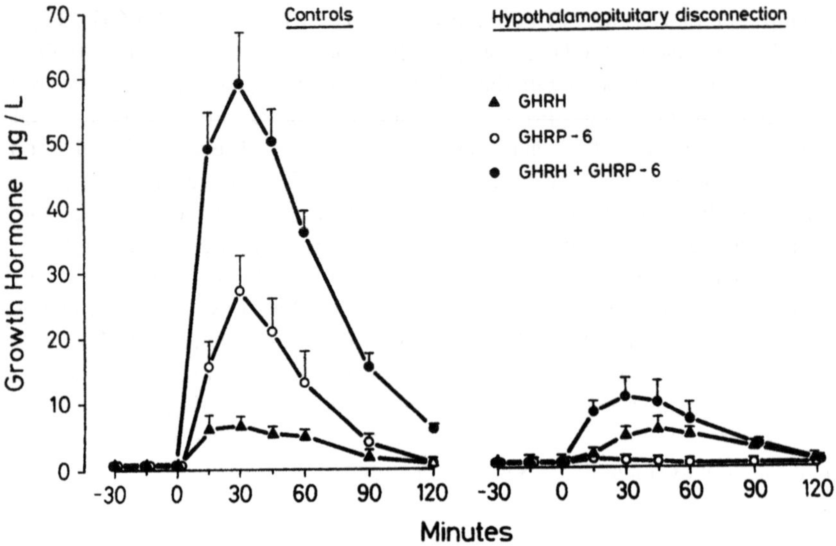

Fig. 9.2 Mean ± SE of GH release in 11 normal subjects and in a group of 12 patients with hypothalamopituitary disconnection after the administration on separate occasions of GH-releasing hormone (GHRH) (100 µg IV, *filled triangle*), growth hormone-releasing peptide (GHRP-6) (90 µg IV, *open circle*), and GHRH + GHRP-6 at the above-mentioned doses (*filled circle*) (reproduced with permission from Popovic et al. [59])

simplify the test in a study with hypopituitary patients and healthy subjects, a single 30 min GH sample after GHRH + GHRP-6 was sufficient for diagnosing GHD [63].

It has been shown that short-term glucocorticoid deprivation does not have a major impact on GH responsiveness to GHRH + GHRP-6, so hydrocortisone replacement is not necessary when testing GH secretion in hypopituitary patients. However, in patients with long-standing hypocortisolism, GH response is decreased but still in normal limits, suggesting the presence of subtle changes in somatotroph function [64]. The consensus of the Growth Hormone Research Society included this test as an alternative for diagnosis of GHD in adults [4, 65].

Both the GHRH + arginine and GHRH + GHRP-6 tests are equally reliable for establishing the diagnosis of GHD in adults [66] (see Fig. 9.3).

The possible problems with powerful combined tests such as GHRH + arginine or GHRH + GHRP-6 tests are false-negative results and this problem is probably not completely solved. Since GHRH and GHRP-6 act directly on the pituitary, it is possible that their administration can restore GH secretion in patients who lack hypothalamic stimulation. This possibility is supported by the finding that some patients with idiopathic GHD, identified by a poor response to ITT, show an exuberant response to GHRH + GHRP-6 [57]. Furthermore, in cranially irradiated patients who have acquired hypothalamic dysfunction, discordant GH results may be obtained after comparing ITT with both combined tests (GHRH + arginine and GHRH + GHRP-6) [48, 67]. Hypothalamic dysfunction occurs earlier than pituitary dysfunction, so

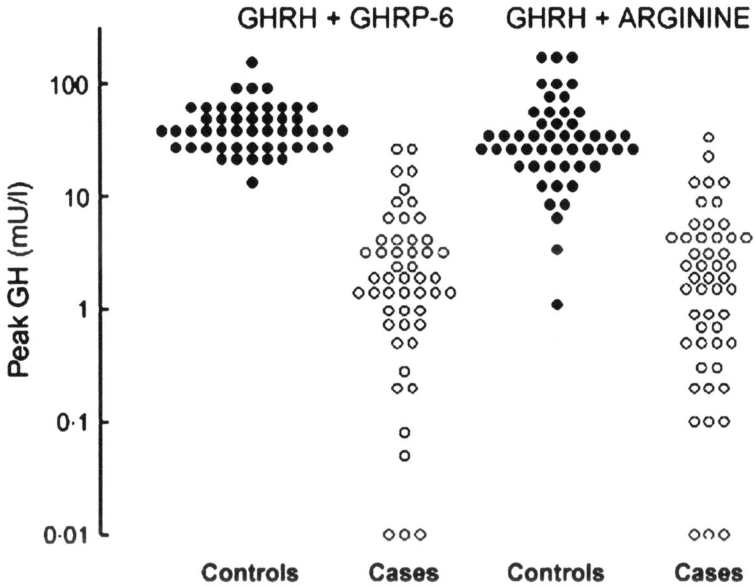

Fig. 9.3 Individual GH peak responses to GHRH + GHRP-6 and arginine + GHRH in controls and hypopituitary patients (logarithmic scale) (reproduced with permission from Popovic et al. [66])

patients can pass the combined test and fail ITT in the first 5 years after radiotherapy. Thus, the combined tests may be preferably used in patients with pituitary rather than hypothalamic causes of GH deficiency. Among GHD patients who may have substantial hypothalamic dysfunction, the ITT remains the preferred test [65].

Ghrelin and GH Secretagogues (GHRP-2, GHRP-6, hexarelin, MK-677)

Ghrelin is the natural GH secretagogue. Ghrelin's capability to release GH was studied in humans and was compared with hexarelin, a nonnatural peptidyl GH secretagogue, and GHRH [68]. Ghrelin administration in healthy subjects induced a prompt and marked increase in GH levels, clearly higher than after GHRH administration and even significantly higher than after hexarelin. Ghrelin administration also induced an increase in PRL, ACTH, and cortisol levels, higher than after hexarelin. Coadministration of ghrelin and GHRH had a synergistical effect on GH secretion. GH response to ghrelin in comparison with ITT and GHRH + arginine test in adult patients was studied in patients with isolated severe CO-GHD [69]. Ghrelin administration induced GH responses higher than those induced by an ITT or GHRH + arginine test, but significantly lower compared with healthy subjects. The main action of ghrelin is exerted at the hypothalamic level and requires the integrity of hypothalamic-pituitary connection [70]. Ghrelin test needs to be evaluated in large series of patients and controls and this has not been done as yet.

In several studies, synthetic GH-releasing peptide GHRP-2 in combination with GHRH was tested in hypopituitary patients and healthy subjects [57, 71]. GHRP-2 stimulation test had a cut-off GH value of 9 μg/L for diagnosis of GHD [72]. The test showed good reproducibility, but the test was affected by age and obesity.

GHRP-6 administered alone in comparison with ITT in the diagnosis of GHD in adults showed high specificity with limited sensitivity [73]. ROC analysis suggested GH cut-point of 3.5 μg/L with 80% sensitivity and 95% specificity. In almost one third of their patients, further testing by ITT was necessary.

GH response to hexarelin, a synthetic hexapeptide analog to GHRP-6 in patients with different hypothalamic-pituitary abnormalities, confirmed that the integrity of the hypothalamic-pituitary connection was essential for hexarelin to express its GH-releasing activity [74].

Chapman et al. [75] investigated oral administration of GHRP-mimetic MK-677 on somatotropic axis in adult patients with CO-GHD. These authors showed that once daily oral administration of such GHRP-mimetic compounds can stimulate 24-h mean GH concentration by enhancing the pulsatile pattern of GH release, i.e., increasing GH pulse amplitude with no effect on pulse frequency. Concomitantly, serum IGF-1 and IGF-BP3 levels increased. The MK-677 was generally well tolerated.

Other Stimulation Tests

Clonidine is a specific α-adrenergic receptor stimulant which increases serum GH concentration, ascribed to stimulation of GHRH release. It was frequently used in children and represents a very useful screening measure for the detection of GHD [76]. Unfortunately, it is not a reliable test in adults, is unpleasant, lowers blood pressure, and can induce somnolence.

The L-dopa test is less frequently used due to its side effects (nausea, vomiting, vertigo) and a high incidence of false-negative results. L-Dopa increases GH secretion through dopaminergic and α-adrenergic pathways.

Arginine alone is also less potent and less used. It stimulates GH secretion by reducing somatostatin tone and possibly by stimulation of α-adrenergic receptors with GHRH release.

Pyridostigmine plus GHRH (PD+GHRH) and clonidine plus GHRH (CLO+GHRH) tests were compared with ITT in patients with hypothalamic-pituitary disease, showing that the diagnostic value of the PD+GHRH test was equal to that of the ITT, but pyridostigmine+GHRH cannot be used across life span, because the potentiating effect of pyridostigmine on GH release declines with age [33].

There are some attempts to introduce other simple and cost-effective approaches to assess pituitary GH and ACTH reserve with combined use of the overnight metyrapone test and IGF-1 SDS [77]. The overnight metyrapone test is an inexpensive and convenient test without contraindications that can be carried out in the outpatient setting. It, like ITT, examines the hypothalamic-pituitary-adrenal axis. By this approach, ACTH and GH reserve can be diagnosed accurately in approximately 50% of patients with organic pituitary disease without provocative testing.

Studies of Pulsatile GH Secretion (24 and 12 h Spontaneous Nocturnal GH Secretion)

The GH pulse frequency and amplitude can be assessed by serial blood sampling for GH measurement usually performed at 20–30 min intervals, either during 24 h or overnight (12 h). Such studies require hospitalization and necessitate multiple sampling. Therefore, these are expensive investigations, which require specialized teams and set-up, of limited value in the diagnosis of GHD.

Some consider measurement of spontaneous nocturnal GH over 12 h to be comparable with the results of GH secretion during ITT, particularly in a group of young adult patients with CO-GHD [78]. These authors demonstrated similar sensitivity and specificity for ITT and 12-h spontaneous nocturnal GH. Despite the fact that 12-h spontaneous nocturnal GH measurements need specific conditions, it may provide some additional useful information. Some patients who passed the ITT have abnormal 12 h-nocturnal GH profiles and IGF-1 levels. Thus, passing ITT does not exclude the possibility of GHD [78, 79].

Pulsatility of GH secretion differs with age, with marked increase in amplitude of pulsatile GH secretion in puberty and decrease approximately 10–15% per decade during adulthood. Reproducibility of the GH secretory profiles over time is poor and pulsatile secretion of GH remains clinical research tool.

Urinary GH

GH is present in urine in very low concentrations. Despite the development of very sensitive assays, the measurement of 24 h or overnight GH excretion, total or corrected for urinary creatinine, remains controversial [80]. There is a marked overlap between normal and abnormal GH secretion and the reproducibility of the urinary GH measurement is poor. Up to five overnight collections are advised to overcome this issue. Urinary GH determination may be useful for monitoring GH replacement therapy.

IGF-1 Measurement

The IGF-1 is synthesized in the liver and in the periphery and is an important mediator of GH actions. It circulates bound to a number of different proteins, of which IGF-binding protein-3 (IGF-BP3) is the most important. Growth hormone induces production of IGF-1 and this GH/IGF system is dynamic. Its activity changes with age and sexual maturation and is influenced by body composition and many factors like nutrition [81]. IGF-1 levels are not reliable for diagnostic screening of adult GHD, even though GHD patients have lower serum IGF-1 concentrations than GH-sufficient patients [81]. IGF-1 has poor discriminative power, especially in patients with IGHD, GHD + one other pituitary hormone deficiency, and especially in older patients [6].

Serum IGF-1 levels are interpreted based on age- and sex-specific reference values from the reference population using specific assays [82]. Unfortunately, there is a considerable overlap for IGF-1 levels between normal subjects and those with GHD, even among patients with the most severe degree of GHD according to the stimulation tests [36]. Several studies have shown that about one third of patients with GHD diagnosed by stimulated GH levels have IGF-1 levels in the normal range [21, 30, 36]. Low level of GH secretion may be sufficient to maintain normal IGF-1 levels.

However, IGF-1 can be of some diagnostic assistance if levels are well below the age-adjusted normal range and if factors known to lower IGF-1 levels (liver disease, starvation, diabetes mellitus, hypothyroidism) are taken into consideration [81]. The timing of the onset of GHD may also influence the interpretation of IGF-1 levels. There are profound differences in IGF-1 levels between patients with CO-GHD and patients with AO-GHD [83, 84]. Patients with CO-GHD (a study from KIMS database) had IGF-1 values on average 1.43 SDS lower than those with AO-GHD [84].

Patients with three or four other pituitary hormone deficiencies and low IGF-1 levels (less than 84 ng/ml; approximating to an SDS of −3) are highly likely to be GH-deficient and thus do not need GH stimulation test [21].

Reevaluation of the GH Status in Young Adults with Childhood-Onset GHD

Patients with CO-GHD may need to continue GH replacement after the attainment of adult height [85]. The diagnosis of GHD in the transition period represents a major clinical challenge to either pediatric or adult endocrinologist [86]. Several studies have shown that many patients with CO-GHD are no longer GHD when retested in early or late adolescence [87, 88]. The 2007 Consensus guidelines for the diagnosis and treatment of adults with GHD recommended that patients with CO-GHD should be reevaluated, with the exception of those with identified transcription factor or gene mutations or those with severe long-standing multiple pituitary hormone deficiencies (MPHD) [4]. Thus, GH retesting is not required for those with a transcription factor mutation (e.g., POU1F1, PROP1, HESX1, LHX3, and LHX4), those with three pituitary hormone deficits, and those with IGHD associated with an identified mutation (e.g., GH, GHRHR, SOX3). MRI findings of structural hypothalamic-pituitary abnormalities may predict persistence of GHD and be important when deciding who to reevaluate in CO-GHD. It has been shown that patients with IGHD and a normal or small pituitary gland should be retested [87].

The reason for reevaluation for GHD in patients with CO-GHD is to identify those who may profit from further replacement therapy [78, 89–91]. This raises the question about the most appropriate cut-off value for diagnosing GHD after completion of growth and puberty. The diagnosis of GHD in adults, based on peak GH response of less than 3 μg/L after ITT, was performed in adult patients and healthy subjects at a mean age >45 years [36]. The negative correlation between age at reevaluation and peak GH response after a stimulation test in patients with CO-GHD suggested that the cut-off value for peak GH response should be reestablished for the age range between 16 and 25 years. In several studies this was evaluated [78, 91] supporting the 2007 Consensus Statement recommendation of a peak GH cut-off to ITT of less than 6 μg/L in the diagnosis of permanent GHD in young adults with CO-GHD (Table 9.2).

Conclusions

The clinical features of adult GHD are recognizable but not distinctive, so clinical suspicion must be confirmed by biochemical tests. Within an appropriate underlying clinical context, GHD in adults may be diagnosed by a single stimulation test, provided it is reproducible and with a high discriminatory power between health and disease. Consensus guidelines for the diagnosis and treatment of adult GHD recommend stimulation testing of GH secretion for patients with hypothalamic-pituitary disease or with CO-GHD, and moreover, patients who have undergone cranial irradiation or have a history of head trauma or subarachnoid hemorrhage. The ITT is a test of choice and the GH threshold level in adults is 3 μg/L, while in young adults (adolescence) it is 6 μg/L. The GHRH + arginine and GHRH + GHRP-6 tests represent the best alternatives to ITT provided appropriate cut-off limits are considered. These tests have been validated and considered convenient, safe, and reliable for the diagnosis of GHD. Glucagon test is reliable, but further validation is necessary. Ghrelin mimetics are currently under evaluation. Retesting adolescents who received treatment with GH in childhood (CO-GHD) remain a hot topic. The 2007 Consensus guidelines for the diagnosis and treatment of adults with GHD recommend that patients with CO-GHD should be reevaluated, with the exception of those with identified transcription factor or gene mutations or those with severe long-standing MPHD.

References

1. Dattani M, Preece M. Growth hormone deficiency and related disorders: insights into causation and treatment. Lancet. 2004;363:1977–87.
2. Sassolas G, Borson Chazot F, Jaquet P, et al. GH deficiency in adults: an epidemiological approach. Eur J Endocrinol. 1999;141:595–600.

3. Abs R, Mattsson A, Bengtsson BA, et al. On behalf of the KIMS study group isolated growth hormone (GH) deficiency in adult patients: baseline clinical characteristics and responses to GH replacement in comparison with hypopituitary patients. Growth Horm IGF Res. 2005;15:349–59.
4. Ho KKY. Consensus guidelines for the diagnosis and treatment of adults with GH deficiency II: a statement of the GH Research Society in association with the European Society for Pediatric Endocrinology, Lawson Wilkins Society, European Society of Endocrinology, Japan Endocrine Society, and Endocrine Society of Australia. Eur J Endocrinol. 2007;157:695–700.
5. Lisset CA, Thompson EGE, Rahim A, Brennan BMD, Shalet SM. How many tests are required to diagnose growth hormone (GH) deficiency in adults? Clin Endocrinol. 1999;51:551–7.
6. Molitch M. Diagnosis of GH deficiency in adults-how good do the criteria need to be? J Clin Endocrinol Metab. 2002;87:473–6.
7. Abs R. Update on the diagnosis of GH deficiency in adults. Eur J Endocrinol. 2003;148:S3–8.
8. Ghigo E, Masel B, Aimaretti G, et al. Consensus guidelines on screening for hypopituitarism following traumatic brain injury. Brain Inj. 2005;19:711–24.
9. Doga M, Bonadonna S, Gola M, Mazziotti G, Giustina A. Growth hormone deficiency in the adult. Pituitary. 2006;9:305–11.
10. Molitch ME. Evaluation and treatment of adult growth hormone deficiency: an Endocrine Society Clinical Practice Guideline. J Clin Endocrinol Metab. 2006;91:1621–34.
11. Prabhakar VKB, Shalet SM. Aetiology, diagnosis, and management of hypopituitarism in adult life. Postgrad Med J. 2006;82:259–66.
12. Corneli G, Gasco V, Prodam F, Grottoli S, Aimaretti G, Ghigo E. Growth hormone levels in the diagnosis of growth hormone deficiency in adulthood. Pituitary. 2007;10:141–9.
13. Kaushal K, Shalet SM. Defining growth hormone status in adults with hypopituitarism. Horm Res. 2007;68:185–94.
14. Gasco V, Corneli G, Rovere S, et al. Diagnosis of adult GH deficiency. Pituitary. 2008;11:121–8.
15. Ghigo E, Aimaretti G, Corneli G. Diagnosis of adult GH deficiency. Growth Horm IGF Res. 2008;18:1–16.
16. Casanueva FF, Catro AI, Micic D, Kelestimur F, Dieguez C. New guidelines for the diagnosis of growth hormone deficiency in adults. Horm Res. 2009;71 Suppl 1:112–5.
17. Ho KY. Growth hormone deficiency in adults. In: DeGroot LJ, Jameson JL, editors. Endocrinology. Philadelphia: Elsevier; 2006. p. 755–65.
18. Schneider HJ, Aimaretti G, Kreitschmann-Andermahr I, Stalla GK, Ghigo E. Hypopituitarism. Lancet. 2007;369:1461–70.
19. Growth Hormone Research Society. Consensus guidelines for the diagnosis and treatment of adults with growth hormone deficiency. J Clin Endocrinol Metab. 1998;83:379–81.
20. Toogood A, Beardwell C, Shalet S. The severity of growth hormone deficiency in adults with pitutiary disease is related to the degree of hypopituitarism. Clin Endocrinol. 1994;41:511–6.
21. Hartman M, Crowe B, Biller B, Ho K, Clemmons D, Chipman J; On behalf of the HypoCCS Advisory Board and the US HypoCCS Study Group. Which patients do not require a GH stimulation test for the diagnosis of GH deficiency? J Clin Endocrinol Metab. 2002;87:477–85.
22. Kelly DF, Gonzalo IT, Cohan P, et al. Hypopituitarism following traumatic brain injury and aneurysmal subarachnoid hemorrhage: a preliminary report. J Neurosurg. 2000;93:743–52.
23. Benvenga S, Campenni A, Ruggeri RM, Trimarchi F. Hypopituitarism secondary to head trauma. J Clin Endocrinol Metab. 2000;85:1353–61.
24. Lieberman SA, Oberoi AL, Gilkison CR, et al. Prevalence of neuroendocrine dysfunction in patients recovering from traumatic brain injury. J Clin Endocrinol Metab. 2001;86:2752–6.
25. Aimaretti G, Ambrosio MR, Di Somma C, et al. Traumatic brain injury and subarachnoid haemorrhage are conditions at high risk for hypopituitarism: screening study at 3 months after the brain injury. Clin Endocrinol. 2004;61:320–6.
26. Popovic V, Pekic S, Pavlovic D, et al. Hypopituitarism as a consequence of traumatic brain injury (TBI) and its possible relation with cognitive and mental distress. J Endocrinol Invest. 2004;27:1048–54.

27. Popovic V. GH deficiency as the most common pituitary defect after TBI: clinical implications. Pituitary. 2005;8:239–43.
28. Dattani MT, Hindmarsh PC. Growth hormone deficiency in children. In: DeGroot LJ, Jameson JL, editors. Endocrinology. Philadelphia: Elsevier; 2006. p. 733–54.
29. Aimaretti G, Baffoni C, DiVito L, et al. Comparison among old and new provocative tests of GH secretion in 178 normal adults. Eur J Endocrinol. 2000;142:347–52.
30. Biller BMK, Samuels MH, Zagar A, et al. Sensitivity and specificity of six tests for the diagnosis of adult GH deficiency. J Clin Endocrinol Metab. 2002;87:2067–79.
31. Yuen KC, Biller BM, Molitch ME, Cook DM. Is lack of recombinant growth hormone (GH)-releasing hormone in the United States a setback or time to consider glucagon testing for adult GH deficiency? J Clin Endocrinol Metab. 2009;94:2702–7.
32. Shalet SM, Toogood A, Rahim A, Brennan BM. The diagnosis of growth hormone deficiency in children and adults. Endocr Rev. 1998;19:203–23.
33. Hoeck H, Vestergaard P, Jakobsen PE, Falhof J, Laurberg P. Diagnosis of growth hormone (GH) deficiency in adults with hypothalamic-pituitary disorders: comparison of test results using pyridostigmine plus GH-releasing hormone (GHRH), clonidine plus GHRH, and insulin-induced hypoglycemia as GH secretagogues. J Clin Endocrinol Metab. 2000;85:1467–72.
34. Bidlingmaier M, Freda PU. Measurement of human growth hormone by immunoassays: current status, unsulved problems and clinical consequences. Growth Horm IGF Res. 2009. doi:10.1016/j.ghir.2009.09.005.
35. Iranmanesh A, Lizzarralde G, Veldhuis JD. Age and relative adiposity are specific negative determinants of the frequency and amplitude of growth hormone (GH) secretory bursts and the half-life of endogenous GH in healthy men. J Clin Endocrinol Metab. 1991;73:1081–8.
36. Hoffman DM, O'Sullivan AJ, Baxter RC, Ho KK. Diagnosis of growth hormone deficiency in adults. Lancet. 1994;343:1064–8.
37. Darzy KH, Thorner MO, Shalet SM. Cranially irradiated adult cancer survivors may have normal spontaneous GH secretion in the presence of discordant peak GH responses to stimulation tests (compensated GH deficiency). Clin Endocrinol. 2009;70:287–93.
38. Darzy KH, Shalet SM. Hypopituitarism following radiotherapy. Pituitary. 2009;12:40–50.
39. Lee P, Greenfield JR, Ho KKY. Factors determining inadequate hypoglycaemia during insulin tolerance testing (ITT) after pituitary surgery. Clin Endocrinol. 2009;71:82–5.
40. Pfeifer M, Kanc K, Verhovec R, Kocijancic A. Reproducibility of the insulin tolerance test (ITT) for assessment of growth hormone and cortisol secretion in normal and hypopituitary adult men. Clin Endocrinol. 2001;54:17–22.
41. Cordido F, Alvarez-Castro P, Isidro ML, Casanueva FF, Dieguez C. Comparison between insulin tolerance test, growth hormone (GH)-releasing hormone (GHRH), GHRH plus acipimox and GHRH plus GH-releasing peptide-6 for the diagnosis of adults GH deficiency in normal subjects, obese and hypopituitary patients. Eur J Endocrinol. 2003;149:117–22.
42. Aimaretii G, Corneli G, Razzore P, et al. Comparison between insulin-induced hypoglycaemia and growth hormone (GH)-releasing hormone+arginine as provocative tests for the diagnosis of GH deficiency in adults. J Clin Endocrinol Metab. 1998;83:1615–8.
43. Ghigo E, Aimaretti G, Arvat E, Camanni F. Growth hormone-releasing hormone combined with arginine or growth hormone secretagogues for the diagnosis of growth hormone deficiency in adults. Endocrine. 2001;15:29–38.
44. Ghigo E, Aimaretti G, Gianotti L, Bellone J, Arvat E, Cammani F. New approach to the diagnosis of growth hormone deficiency in adults. Eur J Endocrinol. 1996;134:352–6.
45. Casanueva FF. Physiology of growth hormone secretion and action. In: Melmed S, editor. Endocrinology and metabolism clinics of North America, vol. 21. Philadelphia: W.B. Saunders; 1992. p. 483–517.
46. Ghigo E, Miola C, Aimaretti G, et al. Arginine abolishes the inhibitory effect of glucose on the GH response to GHRH in man. Metabolism. 1992;41:1000–3.
47. Corneli G, Di Somma C, Baldelli R, et al. The cut-off limits of the GH response to GH-releasing hormone-arginine test related to body mass index. Eur J Endocrinol. 2005;153:257–64.

48. Darzy KH, Aimaretti G, Wieringa G, Gattamaneni HR, Ghigo E, Shalet SM. The usefulness of the combined growth hormone (GH)-releasing hormone and arginine stimulation test in the diagnosis of radiation-induced GH deficiency is dependent on the post-irradiation time interval. J Clin Endocrinol Metab. 2003;88:95–102.
49. Gomez JM, Espadero RM, Escobar-Jimenez F, et al. Growth hormone release after glucagon as a reliable test of growth hormone assessment in adults. Clin Endocrinol. 2002;56:329–34.
50. Micmacher E, Assumpcao RP, Redorat R, et al. Growth hormone secretion in response to glucagon stimulation test in healthy middle-aged men. Arq Bras Endocrinol Metabol. 2009;53:853–8.
51. Secco A, di Iorgi N, Napoli F, et al. The glucagon test in the diagnosis of growth hormone deficiency in children with short stature younger than 6 years. J Clin Endocrinol Metab. 2009;94:4251–7.
52. Bowers CY. GH releasing peptides-structure and kinetics. J Pediatr Endocrinol. 1993;6:21–31.
53. Bowers CY. Synergistic release of growth hormone by GHRP and GHRH: scope and implication. In: Bercu BB, Walker RF, editors. Growth hormone secretagoguges in clinical practice. New York: Marcel Dekker; 1998. p. 1–25.
54. Frieboes RM, Murck H, maier P, Schier T, Holsboer F, Steiger A. Growth hormone-releasing peptide-6 stimulates sleep, growth hormone, ACTH and cortisol release in normal man. Neuroendocrinology. 1995;61:584–9.
55. Kojima M, Hosoda H, Date Y, Nakazoto M, Matsuo H, Kangawa K. Ghrelin is a growth hormone-releasing acylated peptide from stomach. Nature. 1999;402:656–60.
56. Popovic V, Leal A, Micic D, et al. GH-releasing hormone and GH-releasing peptide-6 for diagnostic testing in GH-deficient adults. Lancet. 2000;356:1137–42.
57. Mahajan T, Lightman SL. A simple test for growth hormone deficiency in adults. J Clin Endocrinol Metab. 2000;85:1473–6.
58. Popovic V, Pekic S, Micic D, et al. Evaluation of the reproducibility of the GHRH plus GHRP-6 test of growth hormone reserve in adults. Clin Endocrinol. 2004;60:185–91.
59. Popovic V, Damjanovic S, Micic D, Djurovic M, Dieguez C, Casanueva FF. Blocked growth hormone-releasing peptide (GHRP-6)-induced GH secretion and absence of the synergic action of GHRP-6 plus GH-releasing hormone in patients with hypothalamic-pituitary disconnection: evidence that GHRP-6 main action is exerted at the hypothalamic level. J Clin Endocrinol Metab. 1995;80:942–7.
60. Micic D, Popovic V, Doknic M, Macut D, Dieguez C, Casanueva FF. Preserved growth hormone (GH) secretion in aged and very old subjects after testing with the combined stimulus GH-releasing hormone plus GH-releasing hexapeptide-6. J Clin Endocrinol Metab. 1998;83:2569–72.
61. Kelestimur F, Popovic V, Leal A, et al. Effect of obesity and morbid obesity on the growth hormone (GH) secretion elicited by the combined GHRH+GHRP-6 test. Clin Endocrinol. 2006;64:667–71.
62. Popovic V, Pekic S, Simic M, et al. Physical activity or food intake prior to testing did not affect the reproducibility of GH secretion elicited by GH releasing hormone plus GH-releasing hexapeptide in normal adult subjects. Clin Endocrinol. 2002;56:89–94.
63. Leal A, Lage M, Popovic V, et al. A single growth hormone (GH) determination is sufficient for the diagnosis of GH-deficiency in adult patients using the growth hormone releasing hormone plus growth hormone releasing peptide-6 test. Clin Endocrinol. 2002;57:377–84.
64. Pekic S, Doknic M, Djurovic M, et al. The influence of serum cortisol levels on growth hormone (GH) responsiveness to GH-releasing hormone plus GH-releasing peptide-6 in patients with hypocortisolism. Hormones. 2003;4:243–9.
65. Ho K. Commentary: diagnosis of adult GH deficiency. Lancet. 2000;356:1125–6.
66. Popovic V, Pekic S, Doknic M, et al. The effectiveness of arginine+GHRH test compared with GHRH+GHRP-6 test in diagnosing growth hormone deficiency in adults. Clin Endocrinol. 2003;59:251–7.
67. Popovic V, Pekic S, Golubicic I, Doknic M, Dieguez C, Casanueva F. The impact of cranial irradiation on growth hormone responsiveness to GH-releasing hormone plus GH-releasing peptide-6. J Clin Endocrinol Metab. 2002;87:2095–9.

68. Arvat E, Maccario M, DiVito L, et al. Endocrine activities of ghrelin, a natural growth hormone secretagogue (GHS), in humans: comparison and interactions with hexarelin, a nonnatural peptidyl GHS, and GH-releasing hormone. J Clin Endocrinol Metab. 2001;86:1169–74.
69. Aimaretti G, Baffoni C, Broglio F, et al. Endocrine responses to ghrelin in adult patients with isolated childhood-onset growth hormone deficiency. Clin Endocrinol. 2002;56: 765–71.
70. Maghnie M, Pennati MC, Civardi E, et al. GH response to ghrelin in subjects with congenital GH deficiency: evidence that ghrelin action requires hypothalamic-pituitary connections. Eur J Endocrinol. 2007;156:449–54.
71. Tiulpakov AN, Brook CG, Pringle PJ, Peterkova VA, Volevodz NN, Bowers CY. GH responses to intravenous bolus infusions of GH releasing hormone and GH releasing peptide 2 separately and in combination in adult volunteers. Clin Endocrinol. 1995;43:347–50.
72. Chihara K, Shimatsu A, Hizuka N, Tanaka T, Seino Y, Kato Y; for the KP-102 Study Group. A simple diagnostic test using GH-releasing peptide-2 in adult GH deficiency. Eur J Endocrinol. 2007;157:19–27.
73. Alaioubi B, Mann K, Petersenn S. Diagnosis of growth hormone deficiency in adults: provocatine testing with GHRP-6 in comparison to the insulin tolerance test. Horm Metab Res. 2009;41:238–43.
74. Maghnie M, Spica-Russotto V, Cappa M, et al. The growth hormone response to hexarelin in patients with different hypothalamic-pituitary abnormalities. J Clin Endocrinol Metab. 1998;83:3886–9.
75. Chapman IM, Pescovitz OH, Murphy G, et al. Oral administration of growth hormone (GH) releasing peptide-mimetic MK-677 stimulates the GH/insulin-like growth factor-I axis in selected GH-deficient adults. J Clin Endocrinol Metab. 1997;82:3455–63.
76. Gil-Ad I, Topper E, Laaron Z. Oral clonidine as a growth hormone stimulation test. Lancet. 1979;11:278–9.
77. Gibney J, Healy M, Smith TP, McKenna TJ. A simple and cost-effective approach to assessment of pituitary adrenocorticotropin and growth hormone reserve: combined use of the overnight metyrapone test and insulin-like growth factor-I standard deviation scores. J Clin Endocrinol Metab. 2008;93:3763–8.
78. Secco A, di Iorgi N, Napoli F, et al. Reassessment of the growth hormone status in young adults with childhood-onset growth hormone deficiency: reappraisal of insulin tolerance test. J Clin Endocrinol Metab. 2009;94:4195–204.
79. Radetti G, di Iorgi N, Paganini C, et al. The advantage of measuring spontaneous growth hormone (GH) secretion compared with the insulin tolerance test in the diagnosis of GH deficiency in young adults. Clin Endocrinol. 2007;67:78–84.
80. Bates AS, Evans AJ, Jones P, Clayton RN. Assessment of GH status in adults with GH deficiency using serum growth hormone, serum insulin-like growth factor-I and urinary growth hormone excretion. Clin Endocrinol. 1995;42:425–30.
81. Mukherjee A, Shalet SM. The value of IGF1 estimation in adults with GH deficiency. Eur J Endocrinol. 2009;161:S33–9.
82. Brabant G, von zur Muhlen A, Wuster C, et al. Serum insulin-like growth factor I reference values for an automated chemiluminescence immunoassay system: results from a multicenter study. Horm Res. 2003;60:53–60.
83. Attanasio AF, Lamberts SW, Matranga AM, et al. Adult growth hormone (GH)-deficient patients demonstrate heterogeneity between childhood onset and adult onset before and during human GH treatment. Adult Growth Hormone Deficiency Study Group. J Clin Endocrinol Metab. 1997;82:82–8.
84. Lissett CA, Jonsson P, Monson JP, Shalet SM. Determinants of IGF-I status in a large cohort of growth hormone-deficient (GHD) subjects: the role of timing of onset of GHD. Clin Endocrinol. 2003;59:773–8.
85. Savage MO, Drake WM, Carroll PV, Monson JP. Transitional care of GH deficiency: when to stop GH therapy. Eur J Endocrinol. 2004;151:S61–5.

86. Clayton PE, Cuneo RC, Juul A, Monson JP, Shalet SM, Tauber M. Consensus statement on the management of the GH-treated adolescent in the transition to adult care. Eur J Endocrinol. 2005;152:165–70.
87. Maghnie M, Strigazzi C, Tinelli C, et al. Growth hormone (GH) deficiency (GHD) of childhood onset: reassessment of GH status and evaluation of the predictive criteria for permanent GHD in young adults. J Clin Endocrinol Metab. 1999;84:1324–8.
88. Attanasio AF, Howell S, Bates PC, et al. Confirmation of severe GH deficiency after final heigh in patients diagnosed as GH deficient during childhood. Clin Endocrinol. 2002;56:503–7.
89. Bonfig W, Bechtold S, Bachmann S, et al. Reassessment of the optimal growth hormone cut-off level in insulin tolerance testing for growth hormone secretion in patients with childhood-onset growth hormone deficiency during transition to adulthood. J Pediatr Endocrinol Metab. 2008;21:1049–56.
90. Gasco V, Corneli G, Beccuti G, et al. Retesting the childhood-onset GH-deficient patient. Eur J Endocrinol. 2008;159:S45–52.
91. Maghnie M, Aimaretti G, Bellone S, et al. Diagnosis of GH deficiency in the transition period: accuracy of insulin tolerance test and insulin-like growth factor-I measurement. Eur J Endocrinol. 2005;152:589–96.

Chapter 10
Transition from Puberty to Adulthood

Helena Gleeson

Abstract Growth hormone deficiency (GHD) has implications throughout the lifespan. In childhood, reduced linear growth is the primary consequence of untreated GHD, while in adulthood, a number of abnormalities are associated with the phenotype of untreated GHD. However, the phenotype of GHD in adulthood differs depending on timing of onset. Childhood-onset (CO) patients have reduced bone mineral content and lean body mass despite GHRT during childhood. These differences suggest the CO GHD phenotype is developmental. It is these developmental deficits that have led to a focus on GHRT in patients with CO GHD in the transition period, spanning from the completion of puberty into young adulthood. The aim of GHRT during this time is to normalise body composition and cardiovascular health with the long-term aim of reducing the morbidity and mortality observed in patients with hypopituitarism.

There is evidence that GHRT at this time aids to reduce this deficit in muscle and bone. This may have the effect of improving physical performance and reducing fractures later in life; however, the epidemiological evidence is currently lacking. There is inconsistent evidence that cardiovascular risk is increased and quality of life (QoL) is reduced and that these improve with GHRT. It is possible, and there is some evidence to support this, that certain patient groups may demonstrate a worsening cardiovascular risk profile and QoL after a more sustained period of time off GHRT.

More studies are required to provide the endocrinologist and the patient with adequate information to inform treatment decisions. It is essential that both paediatric and adult endocrinologists recognise that young people with GHD are young people first and foremost. Addressing the psychological, social, educational, and vocational needs of young people is essential in engaging them in their ongoing management with GHD.

H. Gleeson (✉)
Department of Endocrinology, Leicester Royal Infirmary, Leicester, LE1 5WW, UK
e-mail: helena.gleeson@uhl-tr.nhs.uk

K. Ho (ed.), *Growth Hormone Related Diseases and Therapy: A Molecular and Physiological Perspective for the Clinician*, Contemporary Endocrinology,
DOI 10.1007/978-1-60761-317-6_10,

Keywords Growth hormone deficiency • Transition • Adolescence • Aetiology • Diagnosis • Management • Bone health • Cardiovascular risk • Quality of life

Introduction

Growth hormone deficiency (GHD) has implications throughout the lifespan. In childhood, reduced linear growth is the primary consequence of untreated GHD, while in adulthood, a number of abnormalities (abnormal body composition, altered lipid metabolism, and poor quality of life (QoL)) are associated with the phenotype of untreated GHD [1]. There is clear evidence that these abnormalities improve with growth hormone replacement therapy (GHRT) in adulthood [1]. However, the phenotype of GHD in adulthood differs depending on timing of onset. Childhood-onset (CO) GHD patients have a lower serum IGF-1 and IGFBP3, body mass index (BMI), and better QoL compared with adult-onset (AO) GHD patients [2]. In addition, CO patients also have reduced bone mineral content (BMC) and lean body mass (LBM) despite GHRT during childhood [3]. These differences suggest the CO GHD phenotype is developmental.

It is these developmental deficits that have led to a focus on GHRT in patients with CO GHD in the transition period, spanning from the completion of puberty into young adulthood. The aim of GHRT during this time is to normalise body composition and cardiovascular health with the long-term aim of reducing the morbidity and mortality observed in patients with hypopituitarism [4, 5]. The recognition that the transition period may represent an important time for GHRT in a patient with GHD has raised many issues for paediatric and adult endocrinologists: diagnosis and management of GHD in adolescence, evidence of short- and long-term benefit of GHRT, and delivery of care.

This chapter will review the following areas:

Consequences of GHD in adolescence and evidence of benefit for GHRT including bone health and fracture risk, cardiovascular risk, QoL

Diagnosis and management of GHD in adolescence including aetiology of GHD in childhood and likelihood of persistent GHD in adolescence, making the diagnosis of GHD in adolescence, management strategy, transition

Consequences of GHD in Adolescence and Evidence of Benefit for GHRT

Bone Health and Fracture Risk

Background

Bone health is central to the discussion about the role of GHRT in adolescents with persistent CO GHD. GH and IGF-1 in collaboration with sex steroids form the hormonal control of linear growth, bone mineralisation, and augmentation of muscle mass [6]. Attanasio et al. [3] compared body composition in 92 CO with 35 age-matched untreated AO GHD patients; the mean age of those with severe GHD was 21 years and they had been off GHRT for a mean of 1.6 years. Corrected for height, CO compared with AO GHD patients had a lower BMC (2.1 vs. 2.4 kg, $p<0.001$) and LBM (38.5 vs. 50 kg, $p<0.001$) [3]. The closer the CO GHD patients were to achieving their genetic target height, the higher the BMC and LBM [3]. There is therefore a significant developmental deficit (muscle and bone) in CO GHD (16–20% less compared with AO GHD) treated with GHRT during childhood [3]. The hypothesis is that this deficit represents failure to complete somatic development up to young adult levels due to inadequate GHRT in childhood and/or absence of GHRT after the completion of linear growth. The concern is that skeletal gains during this time may be an important determinant of an individual's risk of fractures later in life.

"Peak Bone Mass Concept" vs. "Mechanostat Theory"

An understanding of the two differing approaches taken in the literature is required to discuss this deficit and its impact on fracture risk: one is based around the "peak bone mass concept", the other the "mechanostat theory" (reviewed in [7]).

"Peak bone mass concept" proposes that, although the majority (40%) of bone mass is accrued during puberty, only 90% of bone mass has been accrued by the end of the second decade with gains continuing into the third decade to achieve "peak bone mass". The "peak bone mass" achieved has been linked to risk of osteoporosis and fractures in later life. Therefore in the setting of CO GHD, the hypothesis is that discontinuation of GHRT at the end of linear growth results in a reduced "peak bone mass" and an increased risk of osteoporosis and fractures in later life.

"Mechanostat theory" proposes that bones adapt their strength to keep the strain caused by physiological loads close to a set point, and as the largest physiological loads are caused by muscle contractions, physical size and muscle mass and forces dictate bone strength. In the setting of CO GHD, the hypothesis is that bones are adapted to the reduced physiological load caused by relative short stature and reduced muscle mass and force observed in adolescents with GHD and therefore the risk of fractures is low.

Assessment of Bone

An understanding of methods of achieving an accurate assessment of bone mineral density (BMD) in adolescents is essential to interpret the literature. This is particularly the case in GHD when the majority of adolescents may have relative short stature. Techniques used to assess BMD should be size-adjusted, as failure to do so could result in reporting inappropriate low levels of BMD. DEXA is the most frequently available assessment of BMD. If DEXA is used, volumetric (g/cm^3) rather than areal BMD (g/cm^2) should be measured. pQCT is another technique of assessing volumetric BMD and also bone structure, another important component of bone strength. As bone length and muscle size are important for assessing bone accurately, this, in some centres, is also measured. Size-adjusted data should be reported as Z-scores as this takes into consideration age and sex rather than T-scores which compare the patient with young adults.

Bone and Muscle in Adolescents with CO GHD and Response to GHRT

Using the appropriate assessment technique of BMD, there is little evidence that volumetric BMD is reduced in children, adolescents, or adults with CO GHD (reviewed in [7]). Studies in children [8] and adults [9] with CO GHD using pQCT have found normal cortical and trabecular BMD but altered structure and normal trabecular area but reduced cortical thickness (by 20%) and area (periosteal expansion) (by 23%) in mid-shaft long bone diaphyses. However, no account, in these studies, is taken of the length of the bones. It is possible that the cortical area and thickness are proportional to the length of the bone. In contrast to BMD, muscle size is reduced at diagnosis in GHD.

GHRT has an anabolic effect on both muscle and bone in children, adolescents, and adults (reviewed in [7]). The primary effect of muscle mass and force on bone is to increase cortical thickness through periosteal expansion, which has been observed to be reduced in pQCT studies in children and adults with CO GHD [8, 9]. There is evidence that cortical thickness which would be affected by muscle increases in the metacarpals, but no similar evidence has been observed in the radius [8].

The studies of the effect of GHD and GHRT from the transition period on LBM and bone are reviewed (Table 10.1). One study comparing the effect of discontinuation of GHRT at final height in adolescents retesting severely GHD with those retesting GH sufficient (GHS) demonstrated no difference in measurements of bone between the two groups either at baseline or after 2 years of observation [10]. The same group, however, demonstrated a marked reduction in LBM (−8%) in comparison to no change in GHS patients [11]. Other studies examining discontinuation at the end of linear growth found that total body BMC increases at a rate of 2.5% per year and lumbar spine BMD at a rate of 0.5–2% per year [10, 12–14]. This is consistent with ongoing increases in BMD and BMC after discontinuation of GHRT; it is not known how long this effect continues.

In studies comparing the effect of "seamless" continuation with discontinuation, there were contrasting results. Drake et al. [12] in a 1-year study reported a greater increase in total body BMC (6%) and lumbar spine BMD (5%) in GHD subjects on GHRT compared to those off GHRT (Fig. 10.1). The same group reported an improvement in body composition with a 4–6% increase in LBM over 1–2 years [15, 16]. The increase in BMC and LBM is similar to the increase in healthy adolescents over a similar time period. Mauras et al. [17] in a similar design of study over 2 years found no effect of GHRT on total or lumbar spine BMD and a decrease in percentage LBM in three groups (GHD treated with GHRT; GHD off GHRT; and children retesting GHS), but no significant differences between the groups. When considering the results of these studies, one needs to take into consideration that the

Table 10.1 Effect of discontinuation, continuation, and recommencement of GHRT on bone health and body composition in studies of growth hormone deficiency (GHD) patients in the transition period

References	Number of GHD subjects	Age (years) (mean (SD))	Duration (years)	Treatment (dose (mcg/kg/day))	TB BMC	LS BMD	LBM	FM	FM%
					% Change from baseline				
[10, 11, 31]	21	19 (2)	2	Off GHRT	+5	+4	−8		+7
[12, 15]	24	17 (1)	1	Off GHRT	+2	+3	−2	+10	+3
			1	Continued GHRT (17)	+6	+5	+4	−7	−1
[17]	45	16 (2)	2	Off GHRT		ND			+5
			2	Continued GHRT (20)		ND			+5
[16, 35]	19	20 (1)	1	Off GHRT			+2	+17	
			2	Continued GHRT (18)			+6	−6	
			1	Recommenced GHRT (20)			+14	−25	
[13, 18]	149	20 (3)	2	Off GHRT	+6	+3	+2	+13	
			2	Recommenced GHRT					
			2	Paediatric dose (25)	+8	+5	+14	−6	
				Adult dose (12.5)	+10	+6	+13	−7	
[14]	64	24 (4)	2	Off GHRT		+1	+3	+11	
			2	Recommenced GHRT					
			2	Paediatric dose (25)		+5	+13	−18	
				Adult dose (12.5)		+3	+13	−1	
Average % change/year				Off GHRT	+2.5	+1.7	−0.3	+9.7	+3.0
				Continued GHRT	+6.0	+5.0	+3.5	−5.0	+0.7
				Recommenced GHRT	+4.5	+2.4	+8.1	−8.2	

TB Total body; *BMC* bone mineral content; *LS* lumbar spine; *BMD* bone mineral density; *FM* fat mass; *FM%* percentage fat mass; *ND* data not available

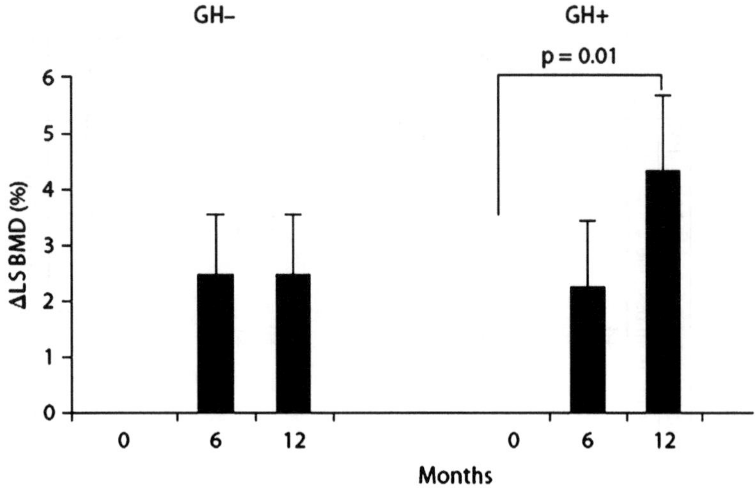

Fig. 10.1 Continuation (GH+) compared with discontinuation (GH–) of GHRT at final height in 24 adolescents with severe GHD: percentage change in lumbar spine BMD (ΔLSBMD(%)) over 6 and 12 months [12]

patients in the study by Mauras et al. [17] had less severe GHD (GH < 5 mcg/L) and were treated with higher doses of GHRT (42 mcg/kg/day) up to final height resulting in supraphysiological IGF-1 levels, while the patients in the study by Drake et al. [12] were more severely GHD (GH <3 mcg/L) and received a lower dose of GHRT (25 mcg/kg/day) up to final height as evidenced by an IGF-1 SDS of −1 at final height. The higher dose of GHRT may have resulted in increased accrual of bone mass during growth.

Studies examining the effect of recommencement of GHRT demonstrated improvement in total body and lumbar spine BMC and BMD [13, 14]. However, the increase in total body BMC and lumbar spine BMD after 2 years of GHRT [13, 14] was of similar magnitude to the increase seen after 1 year of "seamless" continuation [12]. This is due in part to an initial reduction in BMC and BMD when GHRT is restarted after a period of discontinuation with bone resorption being at first greater than bone mineralisation [13, 14]. Recommencement of GHRT did, however, result in a marked improvement in body composition with an increase in LBM (13–14%) regardless of the duration off GHRT (Fig. 10.2) [14, 16, 18]. The effect of GHRT on muscle is known to exceed and precede the effect on bone; this difference reduces with time as evidenced by the studies examining continuation. This increase in LBM equates to 65–85% of the deficit observed in young adults with CO compared with age-matched AO subjects [3].

Two studies have examined the effect of a paediatric (25 mcg/kg/day) vs. an adult dose (12.5 mcg/kg/day) of GHRT on bone mass accrual [13, 14]. Attanasio et al. [18] and Shalet et al. [13] found no dose effect on LBM (Fig. 10.2), BMC or BMD. Underwood et al. [14], in a similarly designed study, identified a dose effect in improvement in lumbar spine BMD and a trend in improvement in total body BMD.

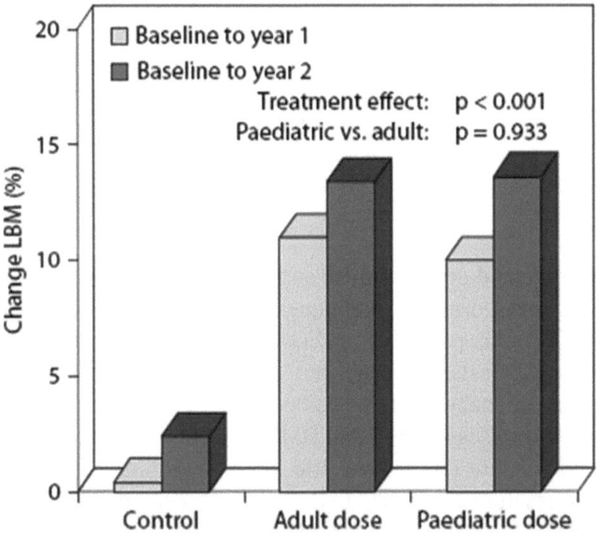

Fig. 10.2 Recommencement of GHRT in young people with severe GHD with paediatric ($n=58$) and adult ($n=59$) doses of GHRT compared with untreated controls ($n=32$): percentage (%) change in LBM over 1 and 2 years [18]

Several studies have examined whether the alteration in LBM translates to improved muscle strength with conflicting results. Hulthen et al. [11] showed an increase in isometric knee flexion and handgrip strength (7–15%) in GHS patients and normal controls, while GHD patients off GHRT demonstrated no such improvement over the 2 years. However, other studies have showed no difference in grip strength or isometric quadriceps strength on or off GHRT [16, 17].

Fractures

Several studies report increased risk of fractures in GHD [19, 20] and increased fracture rate in adults with hypopituitarism [21–26]. However, the majority of subjects in these studies were AO GHD and had MPHD secondary to pituitary adenomas with small numbers of patients with either CO GHD or isolated GHD. In large studies from the KIMS database, the prevalence of fractures was 28–34% in MPHD patients over the age of 60, and interestingly, wrist fracture was lower in CO GHD patients (6%) compared with AO GHD patients (11%) [21]. This study identified no difference in fracture prevalence between isolated GHD and number of additional hormone deficiencies, but a greater prevalence in men with DI [21]. In a follow-up study, the incidence in CO GHD patients was 20% and AO GHD patients was 25% [24], similar to the normal population [27]. Another large-scale study reported a lower incidence in men with AO MPHD and a greater incidence in women with CO MPHD on complete hormone replacement [26]. Studies of cohorts of patients with CO GHD demonstrated no increased risk of fracture in isolated GHD [28–30], but an increased risk in CO MPHD in untreated patients [28]. There is no evidence that CO or AO isolated GHD increases the risk of fractures. There is some evidence to support an increase risk of fractures in AO MPHD.

Summary

There is a developmental deficit in bone and muscle in CO GHD patients at the end of growth. Using size-corrected techniques to assess BMD bone structure rather than density is affected in patients with CO GHD. Regardless of whether you adopt the "peak bone mass" or the "mechanostat" approach, or a combination of the two, GHRT clearly has an anabolic effect on both muscle and bone. However, it is not clear whether the increases in BMC and BMD are sustained and whether they translate to a reduced fracture rate in later life. Studies to date do not demonstrate an increased rate of fractures in isolated GHD and the evidence for increased risk of fractures in MPHD is predominantly in AO patients.

Cardiovascular Risk

Background

Adults with hypopituitarism have a twofold higher risk of death from cardiovascular disease compared with healthy controls. This has been attributed to untreated GHD. The risk is higher in certain groups, for example, females are more at risk than males and AO GHD patients are more at risk than CO GHD patients. GHD mediates negative effects both directly on the heart and endothelium and indirectly via abdominal obesity, insulin resistance, and high total and LDL-cholesterol and low HDL-cholesterol. Although GHRT reduces cardiovascular risk and improves cardiovascular abnormalities in adults, it is not clear whether it reduces the excess in morbidity and mortality.

Body Fat

As with bone and muscle, a hormonal interplay between GH, IGF-1, and sex steroids dictates body fat and body fat distribution [6]. GHD is associated with increased fat mass, truncal adiposity, and intra abdominal fat which is associated with insulin resistance and increased cardiovascular risk. The studies of the effect of GHD and GHRT from the transition period on body fat are reviewed (Table 10.1).

Discontinuation of GHRT in GHD subjects over 1–2 years is associated with a marked increase in FM (+10–17%) [14–16, 18] and truncal body fat of about 22% [31]. In the "seamless" placebo controlled continuation studies, there were contrasting results. Two studies demonstrated a 6–8% decrease in FM over 1–2 years consistent with normal changes in body composition [15, 16]. Mauras et al. [17] demonstrated an increase in %FM in all three groups (GHD treated with GHRT; GHD off GHRT; and children retesting GHS), but no significant differences between the groups. The adolescents studied by Carroll et al. [15] and Vahl et al. [16] had no significant change in IGF-1 in those continuing on GHRT compared with a reduction in IGF-1 in the study by Mauras et al. [17], consistent with the reduction in GHRT dosage (42–20 mcg/kg/day). The fall in IGF-1 levels may have influenced the apparent lack of effect of continued GHRT on body composition [17]. With recommencement of GHRT, a less consistent effect was seen for FM reduction (1–25%) [14, 16, 18]. Differences in FM response may be related in part to GHRT dose and in part to a gender effect. Underwood et al. [14] demonstrated a dose effect on FM reduction, while in the study of Attanasio et al. [18] females in the first year of GHRT had a greater decrease in FM on the higher paediatric GHRT dose.

Lipid Profile

Some studies have reported an adverse lipid profile (elevated total and LDL-cholesterol, triglycerides and apoliporotein B, and reduced HDL-cholesterol) at final height despite conventional GHRT [31–33]. On discontinuation of GHRT, the lipid profile deteriorates [18, 31–33]. In contrast, some studies comparing discontinuation with continuation of GHRT have shown no difference in lipid profile changes [15, 17], while another study demonstrated an increase in HDL-cholesterol on GHRT [16]. Recommencement of GHRT after a period of discontinuation resulted in improvements in total cholesterol/HDL-cholesterol ratio and triglyceride levels [18, 32, 33]; other studies failed to demonstrate an improvement [14, 16]. The recent paper from the KIMS database identified that patients with acquired GHD had a more adverse lipid profile (elevated total cholesterol and triglyceride levels) than patients with idiopathic GHD. There was also a correlation between lipid profile and duration of time off GHRT in patients with acquired GHD. This implies that lipid profile worsens significantly with longer the time off GHRT (Fig. 10.3) [34]. After a mean GHRT interruption of 4.4 years, almost half the cohort had total cholesterol, LDL-cholesterol, and triglyceride levels exceeding the target values. Only those who had been off GHRT for longer than 2 years demonstrated an improvement in HDL-cholesterol levels after 1 year of GHRT [34].

Insulin Resistance

There are few data regarding effects of GHRT on insulin resistance during the transition period. Mauras et al. [17] found no significant differences in global measures of insulin resistance (HOMA-IR) or insulin sensitivity (QUICKI) in

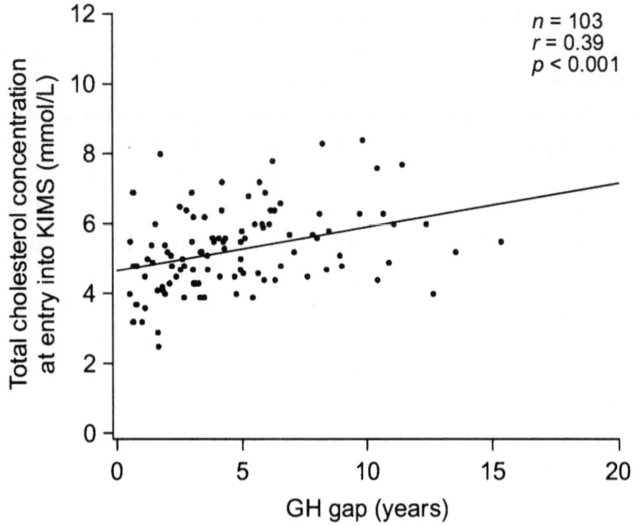

Fig. 10.3 Correlation of length of time between the end of paediatric GHRT and the start of GHRT in adulthood (GH gap) with total cholesterol concentration (millimoles per litre) in 103 patients with acquired CO GHD [34]

their study examining effects of GHRT vs. placebo during the transition period. One study found a decrease in glucose levels [16] in GHD patients receiving placebo but no change in insulin sensitivity despite fat accumulation [35] and an increase in glucose [16] and a non-significant decrease in insulin sensitivity, despite reduction in fat mass and increase in fat free mass [35] when these patients were switched to GHRT in adult doses. It is possible that insulin resistance-inducing effects of GH are counteracted by reductions in trunk fat, which would reduce insulin resistance.

Fibrinolytic Activity, Inflammatory Markers

Elevated fibrinogen levels and reduced fibrinolytic activity are associated with increased cardiovascular risk. Plasminogen activator inhibitor (PAI-1) is the main physiologic inhibitor of tissue plasminogen activator (tPA). Low levels of PAI-1 indicate reduced fibrinolytic activity. Adolescents with untreated GHD have impaired fibrinolytic activity [36]; however, the loss of circadian variation observed in adult GHD [37] has not been reported in this age group. In addition, Colao et al. [32] observed an elevated fibrinogen level in adolescents treated with GHRT during childhood at the end of growth, which increases further after GHRT has been discontinued for 6 months and remained elevated compared with controls after reintroduction of GHRT for 6 months.

Adiponectin is a cytokine expressed by adipose tissue known to have anti-inflammatory and antiathrogenic properties. It is low in untreated adolescents compared to treated adolescents [38]. There is no data about the effect of discontinuation and recommencement of GHRT on adiponectin levels.

Cardiac and Common Carotid Artery Morphology and Function

The relationship between GH and IGF-1 and the cardiovascular system has been demonstrated by numerous experimental studies (reviewed in [39]). The studies of Colao et al. [32, 33] have assessed cardiac and common carotid artery function during discontinuation of GHRT for 6 months before recommencement of GHRT for 6 months in GHD subjects compared with normal controls.

Cardiac structure and function have been assessed by echocardiography. At baseline, GHD subjects had a lower early-to-late mitral flow velocity ratio (*E/A*) (marker of diastolic function), but a normal left ventricular (LV) mass index and ejection fraction compared with normal controls. Six months after GHRT withdrawal, both LV mass index and *E/A* decreased, although remaining within the normal range. Six months after restarting GHRT, cardiac parameters were brought back to the levels measured at study entry, with LV ejection fraction and *E/A* remaining lower than normal controls [32]. These results have been confirmed by other studies in children and adults with CO GHD (reviewed in [39]). Two other studies have identified no echocardiographic abnormalities [14, 17], although concern may be raised at higher doses of GHRT which are associated with significant increases in LV mass index [14].

The clinical relevance of these cardiac changes is not clear. Reassuringly, continuation, discontinuation, and recommencement of GHRT have not been shown to have any effect on exercise capacity [16, 17].

Intima media thickness (IMT) at the common carotid arteries was similar in GHD and controls at baseline. In GHD adolescents, despite changes in IGF-1 levels, lipid profile, and insulin resistance, 6 months of GHRT withdrawal and 6 months of GHRT reinstitution did not change IMT or systolic and diastolic peak velocities at the common carotid arteries. Thus, the discontinuation of GHRT was not followed by significant alterations within the common carotid arteries [33]. This is in contrast to what was observed in adult GHD patients [40].

Summary

Markers of increased cardiovascular risk have been detected in children, adolescents, and young adults with CO GHD. In contrast to adults with GHD, there is no consistent evidence that GHRT in the transition period improves these. Reassuringly, there were only subtle effects on cardiac morphology and function which improved with recommencement, but no effect was observed on IMT in the common carotid arteries. With more prolonged discontinuation of GHRT, negative effects on the cardiovascular system may occur. More studies are required to explore this and whether cardiovascular risk and morbidity and mortality from cardiovascular disease are affected by aetiology of CO GHD.

QoL

QoL is reduced in adults with GHD regardless of the timing of onset; it has been reported, however, that patients with GHD of AO have a worse QoL than those with CO [2]. As CO GHD is a heterogeneous condition, it is important to consider that certain groups of patients may have a worse QoL. Two papers from the KIMS database using QoL-AGHDA (QoL Assessment of GHD in Adults) to assess QoL in CO GHD identified that QoL was worse in those patients with an acquired compared with idiopathic cause for GHD [34] and in those patients with GHD secondary to brain tumours compared with other causes despite higher serum IGF-1 levels off GHRT [41].

Both these papers report improvements in QoL after GHRT recommencement [34, 41]. The most recent paper from the KIMS database also reported that there was a correlation between duration off GHRT and a poorer QoL in the whole group. This was also evident in the group of patients with acquired GHD; the longer the duration off GHRT, the worse the QoL (more vs. less than 2 years off GHRT: score 10.2 ± 6.7 vs. 6.9 ± 5.6, $p<0.01$ (higher score indicates a poorer QoL)) [34]. Other studies have reported the negative effect of GHRT cessation and positive effect of recommencement of CO GHD on QoL and other psychological aspects [42, 43].

However, randomised controlled studies have been less convincing in demonstrating the positive effects of GHRT on QoL in the transition period. One study examined the effect of seamless continuation of GHRT compared with discontinuation and found no effect on QoL at baseline or after 2 years on or off GHRT [17]. In a study looking at recommencement of GHRT, although QoL, measured with an adult-specific GHD questionnaire, QLS(M)-H (Questions of life satisfaction – Hypopituitarism), was significantly lower than normal controls (females −0.35 (1.17) SDS; males −0.7 (1.05) SDS), no change in overall score was identified with GHRT despite significant improvement in individual parameters that were low at baseline, including sexual arousal and body shape [44]. The affected dimensions which did improve with GHRT were related to age-specific behavioural and psychological issues which may worsen with age [44]. No such effect on QoL was confirmed by a similar study design [14].

Summary

Although there is evidence of reduced QoL in patients with CO GHD, GHRT for GHD in the transition period does not consistently demonstrate an improvement. Data suggest that there are groups of patients who have a more impaired QoL and therefore may in turn benefit more from GHRT. There is evidence that the longer duration of time off GHRT may have a negative impact on QoL particularly in patients with GHD due to acquired causes. One study in childhood cancer survivors with GHD in adulthood selected for poor QoL demonstrated a clear improvement in QoL measured by AGHDA-QoL at 3 and 12 months after the commencement of GHRT [45]. Further studies are required to clarify the effect of aetiology, severity of GHD, and presence of additional pituitary hormone deficits on QoL and the effect of duration off GHRT after the completion of linear growth and subsequent benefit of recommencing GHRT. In addition, there is a need to develop or identify an appropriate instrument to measure QoL in adolescents and young adults with CO GHD and to translate any change in QoL, together with physical benefits, into economic cost-benefit or cost-effectiveness analyses [46, 47].

Diagnosis and Management of GHD in Adolescence

Aetiology of GHD in Childhood and Likelihood of Persistent GHD in Adolescence

CO GHD is a heterogeneous condition. At one end of the spectrum, patients with poor growth with no identifiable congenital or acquired cause and a normal magnetic resonance imaging (MRI) of the hypothalamic pituitary axis may be diagnosed biochemically with isolated idiopathic GHD and receive treatment with GHRT, while at the other, patients have GHD in association with destructive hypothalamic pituitary lesions such as a craniopharyngioma with MPHD. CO GHD can be divided into congenital, idiopathic, and acquired (Table 10.2). Patients with congenital GHD may have structural defects such as septo optic dysplasia or a defined genetic mutation. Patients with idiopathic CO GHD make up the largest group and may have isolated GHD or GHD in association with other pituitary deficits with or without a structural hypothalamic pituitary abnormality. Patients with acquired CO GHD may have a destructive hypothalamic pituitary lesion or CO GHD caused by cranial irradiation.

The aetiology of CO GHD in adolescence differs from that during childhood as a significant number of patients treated during childhood recover "normal" GH secretion in late adolescence and therefore the number of patients with idiopathic and/or isolated GHD reduces in the cohort, whereas those with congenital or acquired remain relatively unchanged. Identifying which patients are likely to have persistent GHD is important for making the final biochemical diagnosis of GHD in adolescence (Table 10.2). The studies examining this are discussed.

In the idiopathic group, a significant proportion of patients with and without structural abnormalities of the hypothalamic pituitary axis on MRI will have "normal" GH secretion on the completion of puberty and at the end of linear growth. One study identified that 61% of subjects with an ectopic posterior pituitary were diagnosed with severe GHD in late adolescence and young adulthood [48]. In a similar study, 66% of

Table 10.2 Aetiology of CO GHD and likelihood of persistent GHD in adolescence

High likelihood
Congenital
Septo optic dysplasia and other midline defects
Defined genetic mutation
Idiopathic
With structural abnormalities of the hypothalamic pituitary axis affecting positioning of the posterior pituitary gland or the pituitary stalk or associated with three or more pituitary hormone deficits
Acquired
Destructive lesion affecting the hypothalamic pituitary axis
Cranial irradiation
Low likelihood
Idiopathic
Isolated growth hormone deficiency (GHD) or GHD with two or less pituitary hormone deficits without structural abnormalities of the hypothalamic pituitary axis affecting positioning of the posterior pituitary gland or the pituitary stalk

patients with an ectopic posterior pituitary had persistent GHD compared with 40% of those with a normally placed pituitary [49]. In addition, location of the ectopic posterior pituitary at the median eminence, absence of a visible stalk, and MPHD were predictors of severe GHD on retesting in this cohort [48]. There is also evidence that patients with the presence of an ectopic posterior pituitary are at risk of progressing to reduced GH secretion and therefore require further retesting in the future [50].

In the acquired group, many of the patients with destructive hypothalamic pituitary lesions will have MPHD and therefore are likely to have persistent severe GHD. Patients treated with cranial irradiation frequently have isolated GHD unless very high doses of radiation are administered when other pituitary hormone deficiencies may occur [51]. Retesting at the end of growth has identified that patients treated with higher doses of cranial irradiation at a younger age are more likely to have severe GHD [52]. In the same study, they identified that only 64% of patients with severe GHD in childhood had persistent severe GHD in adolescence [52]. Hypothalamic pituitary damage secondary to radiation evolves with time and therefore patients will require further testing in the future [51].

Making the Diagnosis of GHD in Adolescence

The evidence of recovery of "normal" GH secretion after the completion of puberty and linear growth in a significant number of patients and the differing peak GH cut-offs during childhood and adulthood necessitate biochemical reassessment of GH status in the majority of adolescents with CO GHD. However, the biochemical diagnosis of GHD in adolescence is challenging and there is an expanding literature in this area to inform practice. The challenge can be divided into two areas: the most appropriate testing algorithms and the validation of GH provocative tests; and biochemical GH level cut-off below which GHD is diagnosed.

Consensus statements and practice guidelines [53–56] on diagnosing GHD in late adolescence suggest testing algorithms in which aetiology of CO GHD, other pituitary hormone deficits, severity of GHD in childhood, and serum IGF-1 SDS level are used to predict likelihood of persistent GHD and make recommendations about GH provocative testing and biochemical GH level cut-offs. The literature supporting the various approaches to the biochemical diagnosis of GHD in adolescence is reviewed.

Biochemical Diagnosis of GHD in Adolescence

The GH–IGF-1 axis varies throughout life. GH secretion increases markedly during puberty [57] and then decreases with ageing at a rate of 14% per decade [58, 59]. Therefore, it follows that biochemical diagnosis of GHD based on the measurement of serum IGF-1 levels and peak GH responses during provocative testing with various pharmacological stimuli should reflect this variation with age.

Utility of Serum IGF-1 Levels

IGF-1 can be measured on a single, randomly obtained blood sample and compared with age- and sex-adjusted reference ranges. From the adult literature, if a patient has a high likelihood of GHD, an IGF-1 level <–2 SDS (or has 3 or more additional pituitary hormone deficits), GHD is confirmed in 90% and therefore no GH provocative testing is required [60]. However, because hepatic production of IGF-1 can be decreased for other reasons than GHD and these include hepatic and renal failure, untreated hypothyroidism, and protein and/or calorie malnutrition, it is important to exclude these before making the diagnosis of severe GHD on a low IGF-1 alone. Similarly, a normal IGF-1 does not exclude severe GHD in approximately 20% of young adult patients [61].

GH Provocative Testing

In contrast, a similar age- and sex-matched guide for peak GH cut-off during different provocative tests is not available. The peak GH cut-off during provocative testing below which GHD is diagnosed is higher in children than in adults when only severe GHD is treated. Therefore, the position was taken that peak GH cut-off during provocative testing should be lower than in childhood, but higher than in adulthood. The first consensus statement arbitrarily identified a peak GH of 5 mcg/L as a suitable cut-off [56]. Since that time, the most appropriate peak GH cut-off has been an area of ongoing discussion. In addition, an exhaustive validation of the ITT and other GH provocative tests is still lacking in adolescent subjects with CO GHD.

The ITT is considered the gold standard test in the diagnosis of adult GHD. Two studies have examined peak GH cut-off levels during an ITT in late adolescence and

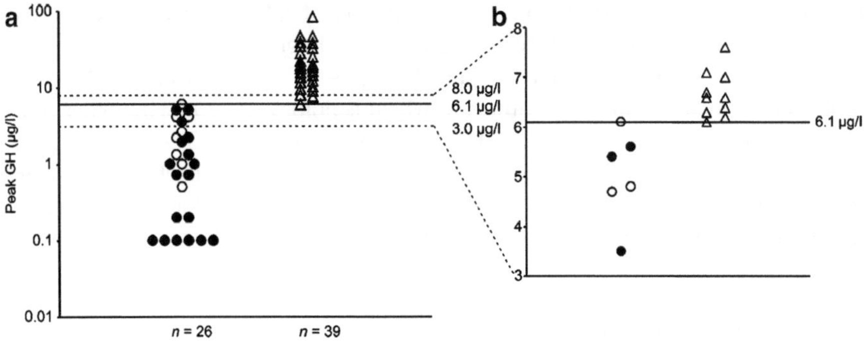

Fig. 10.4 Results of ITT in 26 young people with idiopathic GHD with either MRI abnormalities of the hypothalamic pituitary axis or other pituitary hormone deficiency and in 39 age- and sex-matched controls [62] (*n* = number of subjects)

young adulthood. Maghnie et al. [62] identified a peak GH cut-off of 6.1 mcg/L and a serum IGF-1 SDS level of <−1.7 based on 26 subjects with idiopathic CO GHD with a high likelihood of persistent GHD with either structural abnormalities or other pituitary hormone deficits compared with 39 controls (Fig. 10.4). Secco et al. [63] have recently validated this finding in a larger group of patients with idiopathic CO GHD either with MRI abnormalities or other pituitary hormone deficits. This second study using receiver-operating characteristic analysis found best diagnostic accuracy for peak GH cut-off during an ITT of 5.62 mcg/L and for serum IGF-1 SDS level of −2.83 SDS. This peak GH cut-off during an ITT of 6 mcg/L has been adopted by the recent consensus recommendations [54].

In recent years, the GHRH-AST has been identified as a robust alternative to the ITT in the diagnosis of adult GHD. The utility of the GHRH-AST in 62 patients with CO GHD in late adolescence and young adulthood has been studied and a peak GH cut-off of <10 mcg/L has been identified during two provocative tests [64]. Using a peak GH cut-off of <9 mcg/L (first centile), 94% of those with idiopathic GHD with other pituitary hormone deficits and/or structural abnormalities and 52% of those with idiopathic isolated GHD and no structural abnormalities have been identified, all patients with a peak GH <9 mcg/L during a GHRH-AST also had a peak GH of <3 mcg/L during an ITT [64]. There is a growing recognition that GH response during provocative testing is reduced with increasing BMI, evidence from GHRH-AST studies in adults [65–67], and has led to reducing peak GH cut-offs being recommended for increasing levels of BMI by the GH Research Society [54, 55].

In certain groups of patients with GHD in childhood, a normal response to a GHRH-AST may be misleading. A study comparing ITT with GHRH + AST in adult survivors of brain tumours and leukaemia showed that, within 5 years of cranial irradiation, there was a reduction in peak GH response observed during an ITT and not during a GHRH + AST, whereas after a longer duration the discordancy between the two tests reduced [68]. Therefore, a normal response to a GHRH-AST in a patient with CO GHD which is potentially secondary to hypothalamic dysfunction, for instance, secondary to cranial irradiation, does not exclude GHD.

Recently, availability of GHRH has been limited and therefore alternatives to the ITT and GHRH-AST have been sought. In adults, the GST is as specific and sensitive as the ITT at identifying severe GHD with a lower peak GH cut-off of 3 mcg/L [69, 70]. The AST with a peak GH cut-off of <0.4 mcg/L is also an alternative with a sensitivity of 87% and specificity of 91% [71]. Validation is required of both the GST and AST in CO GHD in late adolescence. Other pharmacological stimuli are not recommended.

Summary

All patients with CO GHD should be considered for reassessment at the end of linear growth. The likelihood of persistent severe GHD in a patient is predicted by aetiology and further increased if MRI abnormalities or other pituitary hormone deficits are present or severe GHD was identified during childhood.

Although there are differences between consensus statements and practice guidelines [53–56], they reflect similar evidence-based themes for reassessment.

Those patients with MPHD (three or more) can continue GHRT seamlessly (Fig. 10.5). Those remaining patients with high likelihood of persistent GHD (con genital or acquired GHD or idiopathic GHD with significant structural abnormalities (EPP or absent/thin stalk) particularly if severe GHD in childhood) should be screened with a serum IGF-1 level having been off GHRT for at least a month. If the IGF-1 is <–2 SDS off GHRT (as long as other causes of low IGF-1 levels have been excluded), the patient can forgo any GH provocative testing as persistent severe GHD is a very high likelihood. Patients in the high likelihood group with an IGF-1 level off GHRT for at least a month within the normal range require GH provocative testing and initiation of GHRT if biochemical GHD is confirmed. It is important to emphasise that patients in the high likelihood group with either partial GHD or the potential of an evolving hypothalamic pituitary dysfunction should have ongoing follow-up.

Those patients with low likelihood of persistent GHD (idiopathic GHD with no significant structural abnormalities either isolated GHD or with one or two additional pituitary hormone deficits particularly if GHD was not severe in childhood) all require to stop GHRT for a month and then check a serum IGF-1 level and one GH provocative test. If discordant results occur, then consideration should be given as to whether a second test is required or whether the patient should be kept under observation. If both tests are normal and there are no concerns about evolving hypothalamic pituitary pathology, the patient can be discharged.

GH provocative testing should be with a validated test and peak GH cut-off in late adolescence and young adulthood, therefore a peak GH during an ITT of <6 mcg/L is recommended and a peak GH during a GHRH-AST varying with BMI as recommended for AO GHD is recommended [54]. If these tests are not recommended or unavailable, the alternatives with some evidence of validation in older adults are the GST and AST; however, there are no current recommendations for

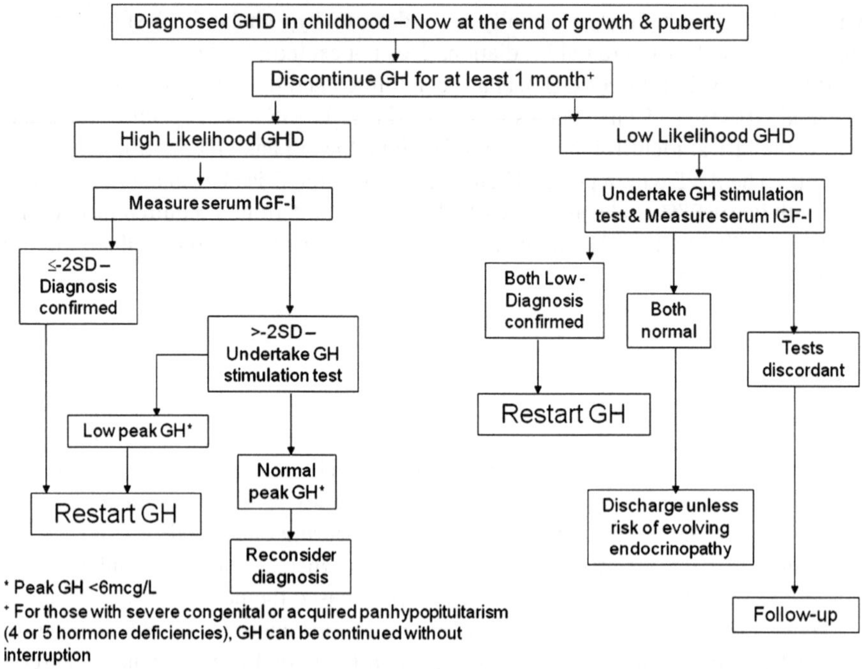

Fig. 10.5 Algorithm for the reassessment of growth hormone deficiency (GHD) at the end of growth in adolescents with CO GHD (adapted from [56])

peak GH cut-off for these tests in this age group. A normal GHRH-AST test in the setting of hypothalamic GHD, for instance, cranial irradiation, would require additional evaluation with an alternative provocative test.

Management Strategy

At final height, the adolescent treated with GHRT for GHD should undergo re-evaluation of the underlying diagnosis, adequacy of other hormone replacements, and reassessment of GH status (Fig. 10.6).

In adolescents with confirmed severe GHD on reassessment in the transition period, GHRT should be restarted. GHRT should be continued until the mid-20s, the expected time of adult somatic development, at which point further review of GH status and the need for GHRT in adulthood is required.

There is no clear evidence of the optimum dose regimen to produce maximal benefit in adolescence. It is recommended that the dose of GHRT should be altered using IGF-1-based titration using robust gender- and age-matched normative data rather than a weight-based approach as in paediatric care. There is evidence that this approach is better tolerated in adults. The current consensus [56] is to start at a dose

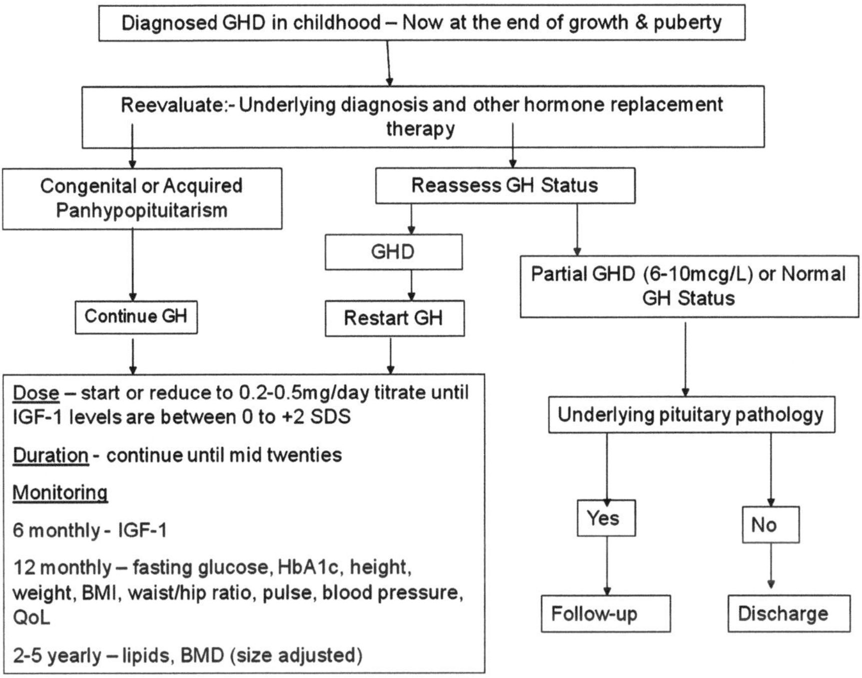

Fig. 10.6 Algorithm for the management at the end of growth of adolescents with CO GHD (adapted from [56])

of 0.2–0.5 mg/day and aim for an IGF-1 between 0 and +2 SDS, with measurements taken every 6 months.

The assessment of volumetric BMD by DEXA, if available, should be performed every 2–5 years, with measurement of Z-scores. The aim is to achieve a Z-score of greater than −1 SDS. In addition, the current consensus [56] has recommended the following minimum observations be performed for adolescents on GHRT: yearly – height, weight, BMI, waist/hip ratio, blood pressure, heart rate, fasting glucose, HbA1c, and QoL; and 2–5 yearly –lipids.

Transition

The transition from puberty to adulthood with GHD has more implications than the physical effects of GHD detailed above. Young people during this time should be becoming increasingly independent from their parents and considering their educational, training, and vocational options to achieve independent living. It is possible that growing up with a chronic condition, such as GHD, has affected their ability to achieve their potential socially, educationally, and vocationally. Endocrine care through the transition period should be holistic and consider the needs of the young person as well as the physical benefits of GHRT for GHD. The process of transition should start young and should include the encouragement of self advocacy, disease knowledge, and self-management. Engagement of a young person during this time is key to improving concordance with GHRT, and more importantly, engagement with adult services. This is the joint responsibility of both paediatric and adult endocrinologists. Transition only stops when a young person is fully engaged with adult services not at the achievement of peak bone mass.

Conclusion

The recognition that GHD has potential implications, if untreated in the transition period, has resulted in an emerging area of evidence-based practice. These areas include the consequence of GHD and the effect of GHRT and the diagnosis and management of GHD in this age group.

There is a clear developmental deficit in CO GHD patients at the end of growth compared to age- and sex-matched AO controls. There is evidence that GHRT at this time aids to reduce this deficit in muscle and bone. This may have the effect of improving physical performance and reducing fractures later in life; however, the epidemiological evidence is currently lacking. There is inconsistent evidence that cardiovascular risk is increased and QoL is reduced and that these improve with GHRT. It is possible, and there is some evidence to support this, that certain patient groups may demonstrate a worsening cardiovascular risk profile and QoL after a more sustained period of time off GHRT.

More studies are required to provide the endocrinologist and the young person with adequate information about long-term consequences and benefits to inform treatment decisions.

It is essential that both paediatric and adult endocrinologists also recognise that young people with GHD are young people first and foremost. Addressing the psychological, social, educational, and vocational needs of this group is essential in engaging them in their ongoing management with GHD.

References

1. Simpson H, Savine R, Sonksen P, Bengtsson BA, Carlsson L, Christiansen JS, et al. Growth hormone replacement therapy for adults: into the new millennium. Growth Horm IGF Res. 2002;12:1–33.
2. Attanasio AF, Lamberts SW, Matranga AM, Birkett MA, Bates PC, Valk NK, et al. Adult growth hormone (GH)-deficient patients demonstrate heterogeneity between childhood onset and adult onset before and during human GH treatment. Adult Growth Hormone Deficiency Study Group. J Clin Endocrinol Metab. 1997;82:82–8.
3. Attanasio AF, Howell S, Bates PC, Frewer P, Chipman J, Blum WF, et al. Body composition, IGF-I and IGFBP-3 concentrations as outcome measures in severely GH-deficient (GHD)

patients after childhood GH treatment: a comparison with adult onset GHD patients. J Clin Endocrinol Metab. 2002;87:3368–72.
4. Stochholm K, Gravholt CH, Laursen T, Laurberg P, Andersen M, Kristensen LO, et al. Mortality and GH deficiency: a nationwide study. Eur J Endocrinol. 2007;157:9–18.
5. Stochholm K, Laursen T, Green A, Laurberg P, Andersen M, Kristensen LO, et al. Morbidity and GH deficiency: a nationwide study. Eur J Endocrinol. 2008;158:447–57.
6. Veldhuis JD, Roemmich JN, Richmond EJ, Rogol AD, Lovejoy JC, Sheffield-Moore M, et al. Endocrine control of body composition in infancy, childhood, and puberty. Endocr Rev. 2005;26:114–46.
7. Hogler W, Shaw N. Childhood growth hormone deficiency, bone density, structures and fractures: scrutinizing the evidence. Clin Endocrinol (Oxf). 2010;72(3):281–9.
8. Schweizer R, Martin DD, Schwarze CP, Binder G, Georgiadou A, Ihle J, et al. Cortical bone density is normal in prepubertal children with growth hormone (GH) deficiency, but initially decreases during GH replacement due to early bone remodeling. J Clin Endocrinol Metab. 2003;88:5266–72.
9. Murray RD, Adams JE, Shalet SM. A densitometric and morphometric analysis of the skeleton in adults with varying degrees of growth hormone deficiency. J Clin Endocrinol Metab. 2006;91:432–8.
10. Fors H, Bjarnason R, Wirent L, Albertsson-Wikland K, Bosaeust L, Bengtsson BA, et al. Currently used growth-promoting treatment of children results in normal bone mass and density. A prospective trial of discontinuing growth hormone treatment in adolescents. Clin Endocrinol (Oxf). 2001;55:617–24.
11. Hulthen L, Bengtsson BA, Sunnerhagen KS, Hallberg L, Grimby G, Johannsson G. GH is needed for the maturation of muscle mass and strength in adolescents. J Clin Endocrinol Metab. 2001;86:4765–70.
12. Drake WM, Carroll PV, Maher KT, Metcalfe KA, Camacho-Hubner C, Shaw NJ, et al. The effect of cessation of growth hormone (GH) therapy on bone mineral accretion in GH-deficient adolescents at the completion of linear growth. J Clin Endocrinol Metab. 2003;88:1658–63.
13. Shalet SM, Shavrikova E, Cromer M, Child CJ, Keller E, Zapletalova J, et al. Effect of growth hormone (GH) treatment on bone in postpubertal GH-deficient patients: a 2-year randomized, controlled, dose-ranging study. J Clin Endocrinol Metab. 2003;88:4124–9.
14. Underwood LE, Attie KM, Baptista J. Growth hormone (GH) dose-response in young adults with childhood-onset GH deficiency: a two-year, multicenter, multiple-dose, placebo-controlled study. J Clin Endocrinol Metab. 2003;88:5273–80.
15. Carroll PV, Drake WM, Maher KT, Metcalfe K, Shaw NJ, Dunger DB, et al. Comparison of continuation or cessation of growth hormone (GH) therapy on body composition and metabolic status in adolescents with severe GH deficiency at completion of linear growth. J Clin Endocrinol Metab. 2004;89:3890–5.
16. Vahl N, Juul A, Jorgensen JO, Orskov H, Skakkebaek NE, Christiansen JS. Continuation of growth hormone (GH) replacement in GH-deficient patients during transition from childhood to adulthood: a two-year placebo-controlled study. J Clin Endocrinol Metab. 2000;85:1874–81.
17. Mauras N, Pescovitz OH, Allada V, Messig M, Wajnrajch MP, Lippe B. Limited efficacy of growth hormone (GH) during transition of GH-deficient patients from adolescence to adulthood: a phase III multicenter, double-blind, randomized two-year trial. J Clin Endocrinol Metab. 2005;90:3946–55.
18. Attanasio AF, Shavrikova E, Blum WF, Cromer M, Child CJ, Paskova M, et al. Continued growth hormone (GH) treatment after final height is necessary to complete somatic development in childhood-onset GH-deficient patients. J Clin Endocrinol Metab. 2004;89:4857–62.
19. Giustina A, Mazziotti G, Canalis E. Growth hormone, insulin-like growth factors, and the skeleton. Endocr Rev. 2008;29:535–59.
20. Canalis E. Insulin-like growth factors and osteoporosis. Bone. 1997;21:215–6.
21. Wuster C, Abs R, Bengtsson BA, Bennmarker H, Feldt-Rasmussen U, Hernberg-Stahl E, et al. The influence of growth hormone deficiency, growth hormone replacement therapy, and other

aspects of hypopituitarism on fracture rate and bone mineral density. J Bone Miner Res. 2001;16:398–405.
22. Rosen T, Wilhelmsen L, Landin-Wilhelmsen K, Lappas G, Bengtsson BA. Increased fracture frequency in adult patients with hypopituitarism and GH deficiency. Eur J Endocrinol. 1997;137:240–5.
23. Vestergaard P, Jorgensen JO, Hagen C, Hoeck HC, Laurberg P, Rejnmark L, et al. Fracture risk is increased in patients with GH deficiency or untreated prolactinomas – a case-control study. Clin Endocrinol (Oxf). 2002;56:159–67.
24. Abs R, Mattsson AF, Bengtsson BA, Feldt-Rasmussen U, Goth MI, Koltowska-Haggstrom M, et al. Isolated growth hormone (GH) deficiency in adult patients: baseline clinical characteristics and responses to GH replacement in comparison with hypopituitary patients. A sub-analysis of the KIMS database. Growth Horm IGF Res. 2005;15:349–59.
25. Mazziotti G, Bianchi A, Cimino V, Bonadonna S, Martini P, Fusco A, et al. Effect of gonadal status on bone mineral density and radiological spinal deformities in adult patients with growth hormone deficiency. Pituitary. 2008;11:55–61.
26. Holmer H, Svensson J, Rylander L, Johannsson G, Rosen T, Bengtsson BA, et al. Fracture incidence in GH-deficient patients on complete hormone replacement including GH. J Bone Miner Res. 2007;22:1842–50.
27. Holroyd C, Cooper C, Dennison E. Epidemiology of osteoporosis. Best Pract Res Clin Endocrinol Metab. 2008;22:671–85.
28. Bouillon R, Koledova E, Bezlepkina O, Nijs J, Shavrikhova E, Nagaeva E, et al. Bone status and fracture prevalence in Russian adults with childhood-onset growth hormone deficiency. J Clin Endocrinol Metab. 2004;89:4993–8.
29. Maheshwari HG, Bouillon R, Nijs J, Oganov VS, Bakulin AV, Baumann G. The Impact of congenital, severe, untreated growth hormone (GH) deficiency on bone size and density in young adults: insights from genetic GH-releasing hormone receptor deficiency. J Clin Endocrinol Metab. 2003;88:2614–8.
30. Baroncelli GI, Bertelloni S, Sodini F, Saggese G. Lumbar bone mineral density at final height and prevalence of fractures in treated children with GH deficiency. J Clin Endocrinol Metab. 2002;87:3624–31.
31. Johannsson G, Albertsson-Wikland K, Bengtsson BA. Discontinuation of growth hormone (GH) treatment: metabolic effects in GH-deficient and GH-sufficient adolescent patients compared with control subjects. Swedish Study Group for Growth Hormone Treatment in Children. J Clin Endocrinol Metab. 1999;84:4516–24.
32. Colao A, Di Somma C, Salerno M, Spinelli L, Orio F, Lombardi G. The cardiovascular risk of GH-deficient adolescents. J Clin Endocrinol Metab. 2002;87:3650–5.
33. Colao A, Di Somma C, Rota F, Di Maio S, Salerno M, Klain A, et al. Common carotid intima-media thickness in growth hormone (GH)-deficient adolescents: a prospective study after GH withdrawal and restarting GH replacement. J Clin Endocrinol Metab. 2005;90: 2659–65.
34. Koltowska-Haggstrom M, Geffner ME, Jonsson P, Monson JP, Abs R, Hana V, et al. Discontinuation of growth hormone (GH) treatment during the transition phase is an important factor determining the phenotype of young adults with nonidiopathic childhood-onset GH deficiency. J Clin Endocrinol Metab. 2010;95(6):2646–54.
35. Norrelund H, Vahl N, Juul A, Moller N, Alberti KG, Skakkebaek NE, et al. Continuation of growth hormone (GH) therapy in GH-deficient patients during transition from childhood to adulthood: impact on insulin sensitivity and substrate metabolism. J Clin Endocrinol Metab. 2000;85:1912–7.
36. Lanes R, Paoli M, Carrillo E, Villaroel O, Palacios A. Peripheral inflammatory and fibrinolytic markers in adolescents with growth hormone deficiency: relation to postprandial dyslipidemia. J Pediatr. 2004;145:657–61.
37. Devin JK, Blevins Jr LS, Verity DK, Chen Q, Bloodworth Jr JR, Covington J, et al. Markedly impaired fibrinolytic balance contributes to cardiovascular risk in adults with growth hormone deficiency. J Clin Endocrinol Metab. 2007;92:3633–9.

38. Lanes R, Soros A, Gunczler P, Paoli M, Carrillo E, Villaroel O, et al. Growth hormone deficiency, low levels of adiponectin, and unfavorable plasma lipid and lipoproteins. J Pediatr. 2006;149:324–9.
39. Colao A. The GH-IGF-I axis and the cardiovascular system: clinical implications. Clin Endocrinol (Oxf). 2008;69:347–58.
40. Colao A, Di Somma C, Cuocolo A, Spinelli L, Acampa W, Spiezia S, et al. Does a gender-related effect of growth hormone (GH) replacement exist on cardiovascular risk factors, cardiac morphology, and performance and atherosclerosis? Results of a two-year open, prospective study in young adult men and women with severe GH deficiency. J Clin Endocrinol Metab. 2005;90:5146–55.
41. Hoybye C, Jonsson P, Monson JP, Koltowska-Haggstrom M, Hana V, Geffner M, et al. Impact of the primary aetiology upon the clinical outcome of adults with childhood-onset GH deficiency. Eur J Endocrinol. 2007;157:589–96.
42. Stouthart PJ, Deijen JB, Roffel M, Delemarre-van de Waal HA. Quality of life of growth hormone (GH) deficient young adults during discontinuation and restart of GH therapy. Psychoneuroendocrinology. 2003;28:612–26.
43. Sartorio A, Molinari E, Riva G, Conti A, Morabito F, Faglia G. Growth hormone treatment in adults with childhood onset growth hormone deficiency: effects on psychological capabilities. Horm Res. 1995;44:6–11.
44. Attanasio AF, Shavrikova EP, Blum WF, Shalet SM. Quality of life in childhood onset growth hormone-deficient patients in the transition phase from childhood to adulthood. J Clin Endocrinol Metab. 2005;90:4525–9.
45. Murray RD, Darzy KH, Gleeson HK, Shalet SM. GH-deficient survivors of childhood cancer: GH replacement during adult life. J Clin Endocrinol Metab. 2002;87:129–35.
46. Bullinger M, Koltowska-Haggstrom M, Sandberg D, Chaplin J, Wollmann H, Noeker M, et al. Health-related quality of life of children and adolescents with growth hormone deficiency or idiopathic short stature – part 2: available results and future directions. Horm Res. 2009;72:74–81.
47. Brutt AL, Sandberg DE, Chaplin J, Wollmann H, Noeker M, Koltowska-Haggstrom M, et al. Assessment of health-related quality of life and patient satisfaction in children and adolescents with growth hormone deficiency or idiopathic short stature – part 1: a critical evaluation of available tools. Horm Res. 2009;72:65–73.
48. Leger J, Danner S, Simon D, Garel C, Czernichow P. Do all patients with childhood-onset growth hormone deficiency (GHD) and ectopic neurohypophysis have persistent GHD in adulthood? J Clin Endocrinol Metab. 2005;90:650–6.
49. Murray PG, Hague C, Fafoula O, Gleeson H, Patel L, Banerjee I, et al. Likelihood of persistent growth hormone deficiency into late adolescence: relationship to the presence of an ectopic or normally sited posterior pituitary gland. Clin Endocrinol (Oxf). 2009;71(2):215–9.
50. di Iorgi N, Secco A, Napoli F, Tinelli C, Calcagno A, Fratangeli N, et al. Deterioration of growth hormone (GH) response and anterior pituitary function in young adults with childhood-onset GH deficiency and ectopic posterior pituitary: a two-year prospective follow-up study. J Clin Endocrinol Metab. 2007;92:3875–84.
51. Darzy KH, Shalet SM. Hypopituitarism following radiotherapy. Pituitary. 2009;12:40–50.
52. Gleeson HK, Gattamaneni HR, Smethurst L, Brennan BM, Shalet SM. Reassessment of growth hormone status is required at final height in children treated with growth hormone replacement after radiation therapy. J Clin Endocrinol Metab. 2004;89:662–6.
53. Molitch ME, Clemmons DR, Malozowski S, Merriam GR, Shalet SM, Vance ML, et al. Evaluation and treatment of adult growth hormone deficiency: an Endocrine Society Clinical Practice Guideline. J Clin Endocrinol Metab. 2006;91:1621–34.
54. Ho KK. Consensus guidelines for the diagnosis and treatment of adults with GH deficiency II: a statement of the GH Research Society in association with the European Society for Pediatric Endocrinology, Lawson Wilkins Society, European Society of Endocrinology, Japan Endocrine Society, and Endocrine Society of Australia. Eur J Endocrinol. 2007;157:695–700.
55. Cook DM, Yuen KC, Biller BM, Kemp SF, Vance ML. American Association of Clinical Endocrinologists medical guidelines for clinical practice for growth hormone use in growth

hormone-deficient adults and transition patients – 2009 update. Endocr Pract. 2009;15 Suppl 2:1–29.
56. Clayton PE, Cuneo RC, Juul A, Monson JP, Shalet SM, Tauber M. Consensus statement on the management of the GH-treated adolescent in the transition to adult care. Eur J Endocrinol. 2005;152:165–70.
57. Martha Jr PM, Gorman KM, Blizzard RM, Rogol AD, Veldhuis JD. Endogenous growth hormone secretion and clearance rates in normal boys, as determined by deconvolution analysis: relationship to age, pubertal status, and body mass. J Clin Endocrinol Metab. 1992;74:336–44.
58. Iranmanesh A, Lizarralde G, Veldhuis JD. Age and relative adiposity are specific negative determinants of the frequency and amplitude of growth hormone (GH) secretory bursts and the half-life of endogenous GH in healthy men. J Clin Endocrinol Metab. 1991;73:1081–8.
59. Zadik Z, Chalew SA, McCarter Jr RJ, Meistas M, Kowarski AA. The influence of age on the 24-hour integrated concentration of growth hormone in normal individuals. J Clin Endocrinol Metab. 1985;60:513–6.
60. Hartman ML, Crowe BJ, Biller BM, Ho KK, Clemmons DR, Chipman JJ. Which patients do not require a GH stimulation test for the diagnosis of adult GH deficiency? J Clin Endocrinol Metab. 2002;87:477–85.
61. Attanasio AF, Howell S, Bates PC, Blum WF, Frewer P, Quigley C, et al. Confirmation of severe GH deficiency after final height in patients diagnosed as GH deficient during childhood. Clin Endocrinol (Oxf). 2002;56:503–7.
62. Maghnie M, Aimaretti G, Bellone S, Bona G, Bellone J, Baldelli R, et al. Diagnosis of GH deficiency in the transition period: accuracy of insulin tolerance test and insulin-like growth factor-I measurement. Eur J Endocrinol. 2005;152:589–96.
63. Secco A, di Iorgi N, Napoli F, Calandra E, Calcagno A, Ghezzi M, et al. Reassessment of the growth hormone status in young adults with childhood-onset growth hormone deficiency: reappraisal of insulin tolerance testing. J Clin Endocrinol Metab. 2009;94:4195–204.
64. Aimaretti G, Baffoni C, Bellone S, Di Vito L, Corneli G, Arvat E, et al. Retesting young adults with childhood-onset growth hormone (GH) deficiency with GH-releasing-hormone-plus-arginine test. J Clin Endocrinol Metab. 2000;85:3693–9.
65. Bonert VS, Elashoff JD, Barnett P, Melmed S. Body mass index determines evoked growth hormone (GH) responsiveness in normal healthy male subjects: diagnostic caveat for adult GH deficiency. J Clin Endocrinol Metab. 2004;89:3397–401.
66. Corneli G, Di Somma C, Baldelli R, Rovere S, Gasco V, Croce CG, et al. The cut-off limits of the GH response to GH-releasing hormone-arginine test related to body mass index. Eur J Endocrinol. 2005;153:257–64.
67. Qu XD, Gaw Gonzalo IT, Al Sayed MY, Cohan P, Christenson PD, Swerdloff RS, et al. Influence of body mass index and gender on growth hormone (GH) responses to GH-releasing hormone plus arginine and insulin tolerance tests. J Clin Endocrinol Metab. 2005;90:1563–9.
68. Darzy KH, Aimaretti G, Wieringa G, Gattamaneni HR, Ghigo E, Shalet SM. The usefulness of the combined growth hormone (GH)-releasing hormone and arginine stimulation test in the diagnosis of radiation-induced GH deficiency is dependent on the post-irradiation time interval. J Clin Endocrinol Metab. 2003;88:95–102.
69. Conceicao FL, da Costa e Silva A, Leal Costa AJ, Vaisman M. Glucagon stimulation test for the diagnosis of GH deficiency in adults. J Endocrinol Invest. 2003;26:1065–70.
70. Gomez JM, Espadero RM, Escobar-Jimenez F, Hawkins F, Pico A, Herrera-Pombo JL, et al. Growth hormone release after glucagon as a reliable test of growth hormone assessment in adults. Clin Endocrinol (Oxf). 2002;56:329–34.
71. Biller BM, Samuels MH, Zagar A, Cook DM, Arafah BM, Bonert V, et al. Sensitivity and specificity of six tests for the diagnosis of adult GH deficiency. J Clin Endocrinol Metab. 2002;87:2067–79.

Chapter 11
Issues in Long-Term Management of Adults with Growth Hormone Deficiency

Anne McGowan and James Gibney

Abstract The first clinical trials of recombinant growth hormone (GH) replacement in GH-deficient (GHD) adults were carried out in the 1980s. Since some of the original participants in these trials have continued to the present day on GH replacement, there are now patients who have been treated with GH in adult life for more than 20 years and many more who have been treated for more than a decade. It is timely to consider the long-term effects of GH replacement both from the point of view of efficacy and safety. Assessment of long-term effects of GH replacement is difficult due to lack of placebo-controlled data, but insights can be gained from a number of sources including follow-up studies of early trials, cohort studies from individual centres, and surveillance studies undertaken by pharmaceutical companies.

Keywords Growth hormone • Long-term • Growth hormone deficiency

Background

The widespread availability of recombinant human Growth Hormone (r-hGH) in the 1980s gave rise to an extensive series of studies investigating the role of GH in adult life, and in particular, the effects of GH replacement in GHD adults. These studies demonstrated that GH deficiency (GHD) in adults was associated with abnormal body composition, dyslipidaemia, reduced exercise capacity and strength, increased surrogate markers of cardiovascular risk, and impaired quality of life (for reviews, see [1–6]). In view of the association of GHD with atherosclerosis,

J. Gibney (✉)
Department of Endocrinology and Diabetes, Adelaide and Meath Hospital, Tallaght, Dublin 24, Ireland
e-mail: James.Gibney@amnch.ie

K. Ho (ed.), *Growth Hormone Related Diseases and Therapy: A Molecular and Physiological Perspective for the Clinician*, Contemporary Endocrinology, DOI 10.1007/978-1-60761-317-6_11,

it has also been suggested that GHD might contribute to the increase in cardiovascular and cerebrovascular mortality that is observed in hypopituitarism [7].

GH replacement has been clearly shown in placebo-controlled studies lasting up to 3 years to increase lean body mass (LBM) and bone mineral density (BMD), reduce total body fat (TBF), and reduce total and LDL-cholesterol [1, 2, 5, 6]. Some short-term placebo-controlled studies have also demonstrated improved quality of life [8, 9]. Data from meta-analysis support an effect of GH to increase exercise capacity [10], improve some indices of cardiac function [11], and improve surrogate markers of cardiovascular risk [12]. Large placebo-controlled studies and meta-analysis results generally have not supported an effect of short-term GH replacement to improve muscle strength or quality of life, although these findings may be limited by methodological issues [13, 14].

Evaluation of more long-term aspects of GH replacement is complex. Placebo-controlled studies have not been carried out for duration of more than 3 years. Uncontrolled data are difficult to interpret and are confounded by issues including the effects of ageing, effects of recovery from previous illness, and the effects of other pituitary hormone deficiencies and their replacement. Results from short-term controlled trials cannot necessarily be extrapolated to imply long-term efficacy (or lack of efficacy) since some short-term effects may not persist, whereas other effects (e.g. muscle strength) may only emerge after long-term treatment. Data regarding clinically relevant endpoints such as cardiovascular events, fractures, and overall mortality are lacking but such outcomes could only be definitively established in large long-term placebo-controlled studies, which are clearly unrealistic.

Potential benefits of long-term GH replacement must also be balanced against possible risks. Although identical to the native human GH molecule, for a number of reasons safety issues surrounding replacement with r-hGH have been subjected to intense scrutiny. First, the pathophysiologic model of acromegaly, in which supra-physiologic levels of GH result from a somatotroph adenoma of the pituitary gland, provides persuasive evidence that sustained exposure to supraphysiologic levels GH is harmful [15]. Second, treatment of patients in intensive care units with high doses of GH was associated with doubling of overall mortality [16]. Third, retrospective analysis of a cohort of GHD children in whom pituitary-derived GH (pit-GH) was used to increase linear growth revealed unexpectedly high rates of both cancer and type 2 diabetes mellitus [17, 18]. Finally, components of the insulin-like growth factor (IGF)/IGF-binding protein (IGFBP) axis have been implicated in increased cancer risk in normal subjects [19]. These concerns are particularly relevant in GHD children and adults, who may be more vulnerable to mitogenic stimuli, both because of the underlying cause of GHD (genetic, residual tumour) and also because of previous treatment such as radiotherapy and chemotherapy [7].

GH replacement can start at any age from infancy onward, and since there is no age at which it is automatically discontinued, treatment is potentially lifelong. It is an expensive treatment, which some patients find inconvenient and therefore it is important to both the patient and the healthcare provider to provide evidence for benefits of long-term treatment and also to address whether there are genuine safety concerns. Evidence regarding the long-term efficacy and safety of GH replacement can be derived principally from three sources. First, follow-up studies of subjects who had taken part in early trials of GH replacement remain important as they include control groups of GHD subjects who did not receive long-term GH replacement. Second, a cohort of extensively characterised GHD subjects in a single centre has now been followed up for more than a decade. Change over time in this cohort has been compared with age-related normative data from the same population, although not with GHD patients who have not received GH replacement. Finally, post-marketing surveillance studies sponsored by pharmaceutical companies include much larger subject numbers and also provide useful although uncontrolled data.

In this review, we will summarise data from studies of GH replacement in adulthood lasting more than 3 years. More short-term data have already been extensively reviewed. We will address whether there is evidence to support beneficial long-term effects of GH replacement and also whether genuine safety concerns exist and how they should be addressed. We will consider mortality in GHD patients and how it might be influenced by GH replacement. Finally, we will consider practical aspects of long-term GH replacement including deciding which patients are suitable for treatment, appropriate dose adjustment, and treatment monitoring, and when consideration should be given to discontinuation of treatment.

Long-Term Efficacy of GH

Body Composition

Lean Body Mass and Body Fat

LBM is reduced by approximately 7–8%, or 4 kg, in GHD adults compared with age- and gender-matched normal subjects [1–4]. Using a four-compartment model, it has been demonstrated that this difference is proportionately accounted for by reduction in both extracellular water (ECW) and body cell mass (BCM) [20]. TBF is increased in GHD adults compared to age- and gender-matched normal subjects. The excess fat is predominantly centrally distributed in both abdominal subcutaneous and abdominal visceral compartments [21]. Short-term GH replacement has been consistently shown to increase LBM and reduce total body and centrally distributed fat [1, 2, 4]. A meta-analysis including 473 GHD adults revealed a mean increase of 2.8 kg in LBM and a mean reduction in TBF of 3.05 kg following GH replacement for 6–12 months [12].

There is more long-term data concerning effects of GH on body composition than most other effects of GH replacement. Two studies have reported long-term outcomes in subjects who had taken part in early double blind placebo controlled trials (DBPCTs) of GH replacement and continued taking GH for up to 10 years. These studies, although small, remain important as they include control groups of GHD subjects who did not receive long-term GH replacement. Gibney et al. reviewed 21 adults, 10 of whom had been treated with GH for 10 years following a DBPCT and 11 of whom had not [22] (see Fig. 11.1). Twelve matched normal controls were also studied at baseline and at the 10-year time point. There was a

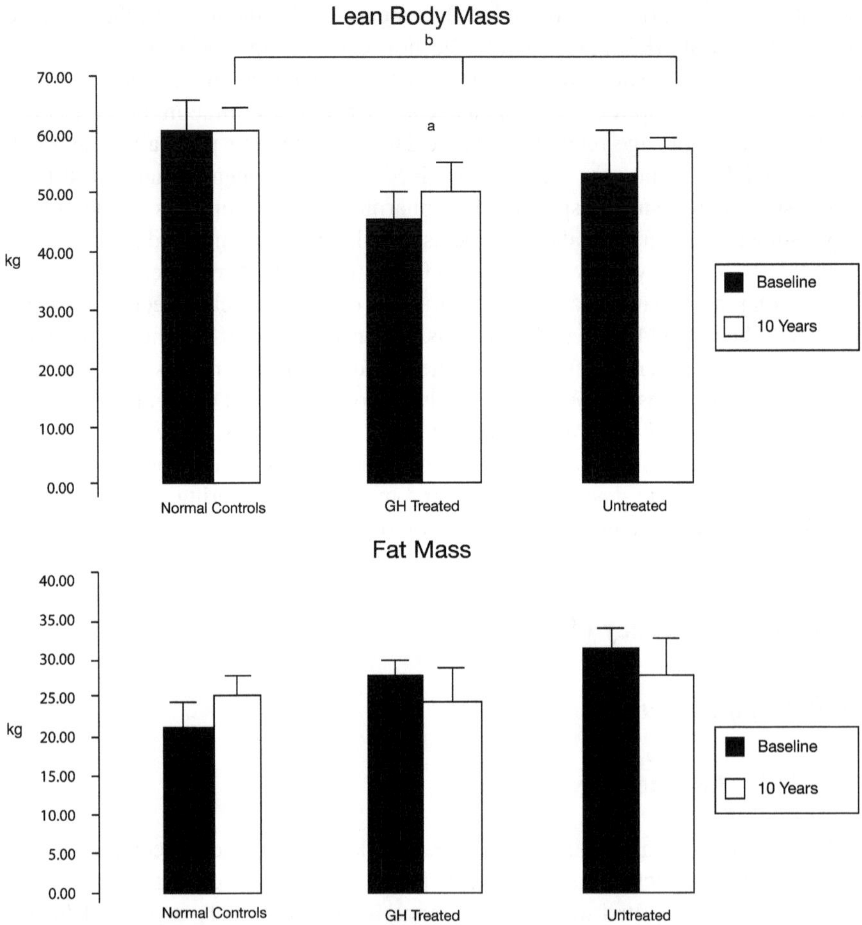

Fig. 11.1 Lean body mass and fat mass assessed by total body potassium measurement at baseline and 10 years in growth hormone-treated (GH-treated) and -untreated (untreated) patients and normal control subjects (controls). $a=p<0.02$ versus baseline and for GH treatment effect. (Reproduced with permission from Gibney et al. [22])

significant increase in LBM in the GH-treated group compared to the untreated group and to the normal control group, but no change in TBF in any of the groups. Muscle area, as measured by CT, increased significantly in the GH-treated group, whereas there was no change in the untreated GHD patients. Chrisoulidou et al. compared changes over 7 years in 12 GHD patients who had received GH replacement over this time, with two matched GHD groups, one of which had received GH replacement for a maximum of 18 months (finishing at least 5 years prior to follow-up) and one who had never received GH replacement [23]. Compared to control subjects, treatment with GH led to a sustained increase in total body water and fat free mass and reduction in TBF. Waist–hip ratio and subscapular skinfold thickness increased significantly over the 7 years in untreated GHD subjects, but not in those who received GH replacement.

More detailed, although uncontrolled, data from a larger cohort of GHD patients have been reported in stages from a centre in Sweden. Most recently, Gotherstrom et al. reported data in 87 of these patients followed up during 10 years of GH replacement [24]. Sustained increases in LBM, BCM, and ECW were observed, while the effect of GH replacement on TBF varied according to methods used to estimate it. TBF estimated by DXA or as part of a five-compartment model was not reduced compared to baseline, but was reduced when estimated as part of a four-compartment model and adjusted for age and sex. In a smaller study from the same centre, body composition was estimated before and following 7 years of GH replacement in GHD adults with results compared to normal subjects [25]. A reduction in body fat and an increase in free fat mass were observed following GH replacement, but no differences were observed between GHD and normal subjects at baseline or at the end of the study.

Two other studies have reported long-term changes in body composition. In an uncontrolled study, ter Maaten et al. demonstrated sustained increase in LBM following 3–5 years of GH replacement in 38 men with childhood-onset GHD [26], while a 4-year study by Fideleff et al. demonstrated non-significant between-group changes in LBM and TBF in 48 GHD adults treated with GH compared to 23 GHD adults who did not receive GH replacement [27].

In summary, the available data generally indicates sustained improvement in LBM, BCM, and ECW for up to 10 years of GH replacement. Data concerning long-term effects on body fat are unclear with some, but not all, studies suggesting sustained improvement.

Bone Mineral Density

Hypopituitarism is associated with reduced BMD [1, 2, 5] and increased fracture rate [28]. The contribution of GHD to these observations is unclear since over-replacement with glucocorticoid and thyroid hormone and untreated or inadequately treated hypogonadism could exert similar effects. However, in childhood-onset (CO-) GHD, the degree of reduction in BMD is similar in patients with isolated GHD compared to patients with multiple pituitary hormone defects suggesting that GHD plays a major role [5]. Markers of bone turnover are normal and decreased in adults with GHD and although variable, several studies have demonstrated no difference compared to controls [5]. CO-GHD is clearly associated with reduced BMD because in the absence of GH, normal peak bone mass is not achieved, but it is less clear whether onset of GHD after peak bone mass is achieved is also associated with reduction in BMD. In fact, BMD has been reported to be normal in middle-aged GHD adults and may actually be increased in older GHD adults [29]. GH replacement in subjects with GHD results in an initial increase in bone resorption leading to a reduction, but a later increase, in BMD [5]. Histomorphorphometric studies have demonstrated that 6–12 months of GH replacement leads to increased cortical thickness and increased bone formation [6].

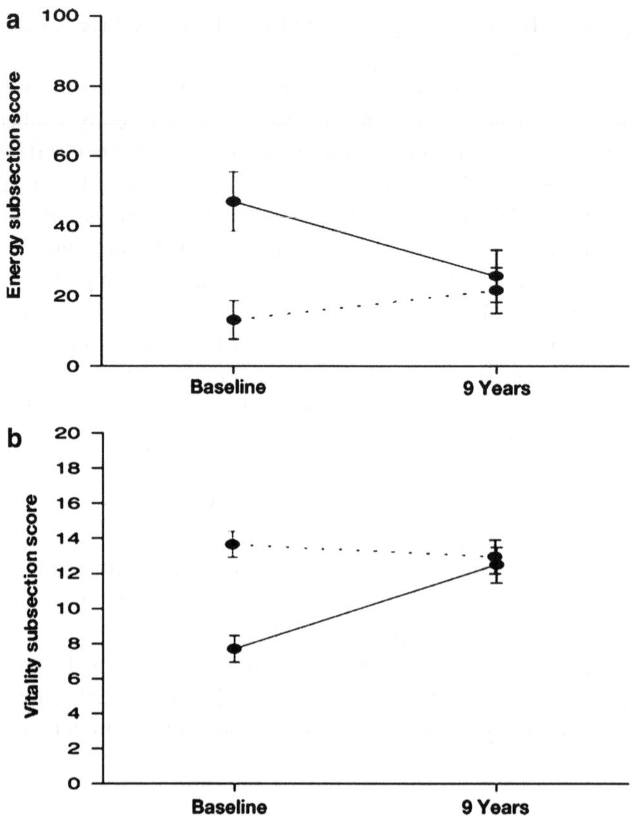

Fig. 11.2 The effect of GH replacement on bone mineral density in 23 adults with childhood-onset GHD followed up over 10 years. Bone mineral density was measured at the lumbar spine, femoral neck, and trochanter. (Reproduced with permission from Arwert et al. [30])

A number of studies have addressed the long-term effect of GH replacement on BMD. These studies have not included unreplaced GHD subjects as controls, but most have compared changes in BMD to normal subjects or to age-related normative data. The longest duration of follow-up was reported by Arwert et al. who studied BMD during 10 years of GH replacement in 23 adults with CO-GHD, and over the same period, in 19 healthy matched controls [30] (see Fig. 11.2). BMD increased continuously in the lumbar spine over the 10 years of the study in GHD subjects, but declined in controls. BMD at the femoral neck and trochanter increased over the first 5 years of GH replacement, but declined thereafter, resulting in values that did not differ at the 10-year time point compared to baseline. Femoral neck BMD in normal subjects declined over the 10 years. Changes in both lumbar spine and femoral neck BMD differed significantly in GHD subjects compared with changes observed in normal subjects. Gotherstrom et al. reported the effects of 5 years of GH replacement in 118 GH-deficient adults [31]. T- and Z-scores for BMD increased progressively in the lumbar spine and femoral neck over the course of the study, more markedly in men. Biermasz et al. studied the effect of GH replacement for 4 years on 30 adults with GHD, 15 of whom had osteoporosis and 15 had low bone mass [32]. Lumbar spine BMD increased significantly in both groups following 4 years of GH replacement, although the osteoporotic group remained in the osteoporotic range. Further increases were observed in a later phase of the study when alendronate was administered in addition to GH replacement. A number of other studies have demonstrated similar results and are summarised in Table 11.1.

In summary, therefore, there is consistent evidence that GH replacement increases BMD. The most marked effect seems to occur in the first 5 years and is greater in men. There is relatively little data available beyond 5 years and the natural history of change in BMD in untreated GHD patients is not known. Further improvements in BMD can be attained with the addition of a bisphosphonate. There is no long-term data available regarding fracture risk.

Exercise Capacity and Strength

Exercise Capacity

Exercise capacity is reduced in GHD adults. Maximal oxygen consumption (VO_{2max}, aerobic capacity, or the maximum ability to take in and use oxygen) in GHD adults has been consistently shown to be reduced by estimates ranging from 17 to 27% compared to values predicted for age, gender, and height [3]. The majority of studies reported to date have demonstrated increased maximum work-rate and VO_{2max} [3] following GH replacement in subjects with both childhood- and adult-onset GHD and a recent meta-analysis strongly supports these effects [10].

There is little data available concerning whether these effects are sustained. In the study described above by ter Maaten et al., significant increases in VO_{2max} and maximal workload were observed over 3–5 years of GH replacement. Although the study was uncontrolled, results were compared to age-related normative data and mean values improved from lesser than to greater than predicted values by the study end [26]. In contrast, in the study by Chrisoulidou et al. no change in exercise performance was observed [23].

Muscle Strength

Reduced muscle mass in GHD is associated with reduced isometric muscle strength, while some but not all studies have also demonstrated reduced isokinetic strength [3, 33]. The effects of short-term GH replacement on strength are less clear than effects on exercise performance with some but not all studies suggestive of a beneficial effect and no effect being demonstrated in a recent meta-analysis [13].

Table 11.1 Summary of studies of effects of GH replacement in GHD adults in studies of at least 3 years duration

Study	Design	Length of study	Number	End-point	Effect in GHD adults at study end	Effect in controls at study end compared to GHD subjects
Thuesen [42]	Open controlled trial	38 months	21/21	Blood pressure Doppler electrocardiography	↑ Heart rate ↑ Blood pressure ↑ Fractional Shortening of left ventricle and cardiac index	
Johannsson [43]	Open controlled trial	42 months	7/21	Blood pressure Exercise electrocardiography Doppler electrocardiography Body weight	↔ Blood pressure ↑ Exercise capacity – age adjusted ↑ Left ventricular wall mass ↔ Body weight	↔ Blood pressure ↔ Exercise capacity – age adjusted ↔ Left ventricular wall mass ↔ Body weight
Kann [82]	Prospective, controlled trial	42 months	20/20	Bone mineral density	↑ BMD ↔ Bone elasticity	
Al-Shoumer [83]	Prospective uncontrolled trial	4 years	13	Metabolic indices	↔ Fasting glucose ↓ Total and LDL-cholesterol ↔ Triglycerides and HDL cholesterol	
Valimaki [84]	Prospective uncontrolled trial	42 months	32–36 months 20–42 months	Bone mineral density	↑ BMD at 36 months ↔ BMD at 42 months ↔ Markers of bone turnover	

Gibney [22]	Double-blind randomised controlled trial	10 years	10/11	Metabolic indices Body composition Strength CV investigations QOL	↓ LDL-cholesterol ↑ Lean body mass ↔ Strength ↔ CIMT ↑ QOL	↔ LDL-cholesterol ↔ Lean body mass ↓ Strength ↑ CIMT ↔ QOL
ter Maaten [26]	Prospective, uncontrolled	3–5 years	38	Body composition Cardiac investigations	↑ Body weight ↑ Bone mineral density ↑ Exercise capacity ↑ Stroke volume and cardiac output	
Chrisoulidou [23]	Randomised controlled trial	7 years	12/11/10	Metabolic indices Body composition Anthropometry CV investigations	↓ LDL-cholesterol ↓ Body fat mass ↓ WHR ↔ Echocardiographic data	↓ LDL-cholesterol ↔ Body fat mass ↔ WHR ↔ Echocardiographic data
Götherström [31]	Prospective, uncontrolled	5 years	118	Metabolic indices Body composition	↑ Lean mass and ↓ Body fat ↔ Total bone mineral density ↑ Total bone mineral content ↓ Total and LDL-cholesterol	
Drake [85]	Prospective uncontrolled trial	44–72 months	13	Bone mineral density	↑ BMD ↔ Bone alkaline phosphatase	
Biermasz [32]	Prospective controlled trial	7 years	30/15	Bone mineral density 4 years – all GH 3 years – ± Alendronate	↑ BMD at 4 years ↔ BMD at 7 years	

(continued)

Table 11.1 (continued)

Study	Design	Length of study	Number	End-point	Effect in GHD adults at study end	Effect in controls at study end compared to GHD subjects
Arwert [30]	Prospective controlled trial	10 years	23/19	Metabolic indices Body composition	↓ LDL-cholesterol ↑ Fasting glucose ↑ BMI and WHR ↑ Lumbar spine BMD ↔ Femoral neck BMD	↓ Lumbar spine BMD
Götherström [24]	Prospective, uncontrolled	10 years	87	Metabolic indices Body composition	↑ Lean mass ↑ Fasting glucose ↓Cholesterol concentration ↓ Body fat – if measured by four-compartment model, ↔ if DEXA used	
Svensson [25]	Prospective, controlled	7 years	11/11	Metabolic profile Body composition Clamp data	↓ LDL-cholesterol ↔ Body fat ↔ Insulin Sensitivity	↔ LDL-cholesterol ↔ Body fat Trend towards ↓
Colao [39]	Prospective controlled trial	5 years	35/35	CIMT Insulin resistance	↓ CIMT ↓ Insulin Resistance, slight ↓ in GHD, not treated with GH	↑ CIMT (but lower overall than those with GHD) Slight ↑ in controls
Fideleff [27]	Prospective controlled trial	4 years	48/23	Metabolic indices Body composition Anthropometry CV investigations	↓ Total cholesterol ↓ Bone mineral content ↔ BMI ↓ Diastolic BP	↔ Total cholesterol ↔ Bone mineral content ↔ BMI ↔ Diastolic BP

↑ Increase; ↓ decrease; ↔ no change; *BMD* bone mineral density; *CV* cardiovascular; *CIMT* carotid intima media thickness; *BP* blood pressure; *QOL* quality of life

However, there are theoretical reasons why a detectable increase in muscle strength would require GH replacement of longer duration, and evidence to support this has been provided by an open-label study, which has now been reported up to 10 years. At baseline, compared with a reference population of normal subjects, GHD adults exhibited reductions in isometric and isokinetic muscle strength and local muscle endurance. Following 2 years of GH replacement, isometric and isokinetic strength increased into the normal range, although a reduction was seen in muscle endurance [34]. Most measures of strength continued to increase in absolute terms for the first 5 years of the study [35] and declined over the second 5 years, although isometric knee flexor strength continued to increase throughout the 10 years [36]. However, following correction for age and gender using predicted ratios, sustained increases in all variables reflecting muscle performance were seen. There are few other studies of long-term GH effects on muscle strength. In the 10-year study by Gibney et al., muscle strength declined significantly in unreplaced GHD patients, but remained unchanged in the GH-replaced subjects [22].

In summary, there is strong short-term evidence to suggest that GH replacement improves exercise performance and very limited information concerning more long-term treatment. While one study has shown promising data, it did not include a control group. In contrast, short-term GH replacement does not seem to improve muscle strength, but available evidence does suggest long-term improvement and ultimately normalisation compared to a reference population.

Lipids, Atherosclerosis, and Cardiovascular Effects

Lipids and Atherosclerosis

GHD adults are characterised by an atherogenic lipid profile, with increased total (TC) and low-density lipoprotein-cholesterol (LDL-C), increased apolipoprotein B (apo B), and in some reports, increased triglyceride (TG) and reduced high-density lipoprotein-cholesterol (HDL-C) [1, 2, 6]. GHD adults are also characterised by other features, which either predispose to or represent early stages in the progression of atherogenesis. Abnormal circulating markers of cardiovascular risk include high-sensitivity CRP, IL-6, fibrinogen, sialic acid [37], and plasminogen activator inhibitor (PAI)-I activity [38], while vascular studies have revealed endothelial dysfunction and increased carotid intima media thickness (CIMT) [6].

Short-term GH replacement reduces TC, LDL-C, and Apo B levels, and in some studies, increases HDL-C [1, 4]. In a meta-analysis [12], Maison et al. demonstrated significant changes in total (mean reduction 0.3 mmol/L) and LDL- (mean reduction 0.5 mmol/L) cholesterol. Improvements have also been demonstrated in most of the above surrogate markers of cardiovascular risk. A number of uncontrolled studies have reported effects on lipid profile for durations up to 10 years, most showing an improvement. The largest long-term study is that reported by Gotherstrom et al. in which significant improvements were observed in total, LDL-, and HDL- cholesterol, but not triglyceride concentration [24]. In the two long-term follow-up studies of early trials, no between-group changes in plasma lipids were observed, although in one of the studies a significant reduction in LDL-C was observed in GH-replaced patients only [22, 23] . Notably, in both of these studies, long-term improvements in lipid profile were observed in both replaced and unreplaced GHD patients. Similarly, in a 5-year study by Colao et al. which included GHD subjects who received GH replacement and a smaller group who refused GH replacement, significant improvements in total, LDL-, and HDL- cholesterol were observed in both groups with no between-group differences apparent [39]. These observations likely reflect lifestyle changes over this time and possible long-term recovery from original illness, but also cautions against over-interpretation of uncontrolled data.

Carotid intima media thickness (CIMT) has become established as a robust surrogate marker of atherosclerosis [40]. CIMT is increased in adults with GHD [6]. Short-term GH replacement decreases, but does not normalise CIMT [41]. The most persuasive long-term data regarding the effect of GH replacement on CIMT comes from a study by Colao et al., in which 35 GHD adults were matched with 35 normal subjects and studied over 5 years [39]. Eighteen of the 35 GHD patients received GH replacement, while 13 refused replacement. CIMT was greater in subjects with GHD compared to normal subjects. CIMT decreased (mean change −7.3%) in GH-replaced patients and the change was significantly different to unreplaced GHD patients in whom there was no significant change (mean change +2.8%) and normal subjects in whom CIMT increased (mean change +5%) [39]. Consistent with these findings, CIMT was greater at the end of the 10-year study reported by Gibney et al., although baseline measurements were not available [22].

Cardiac Function

Most, although not all, studies using echocardiography or equilibrium radionucleide angiography have demonstrated reduced left ventricular mass (LVM) and ejection fraction (EF) in GHD adults compared to normal subjects [11]. Reports of the effects of GH replacement on cardiac structure and function are inconsistent, but a meta-analysis of placebo-controlled trials demonstrated a significant effect of GH replacement to increase left ventricular posterior wall thickness and LVM [11].

There is relatively little long-term data concerning the effects of GH replacement on cardiac function. In the 10-year study reported by Gibney et al., no changes were observed in echocardiographic indices of cardiac function [22]. In the study by ter Maaten et al., an initial increase in LV mass and septal wall thickness was observed which subsequently returned to baseline values [26]. In contrast, increases in stroke volume and cardiac output were sustained over the median 55 months of study follow-up. Two other studies of 38 and 42 months duration have demonstrated increased cardiac index and left ventricular wall thickness, respectively [42, 43]. Notably, increases in left ventricular wall thickness in the latter study and also in the first year of the study by ter Maaten et al. occurred in response to doses of GH that would now be considered supraphysiologic.

In summary, although there is substantial data demonstrating improved lipid profile in GHD subjects following long-term GH replacement, studies that have included unreplaced GHD patients have shown statistically similar improvements. Data concerning reduced CIMT following GH replacement are more convincing as improvement was seen even when compared to unreplaced GHD patients, although this has only been demonstrated in one study with small subject numbers. There is also a paucity of evidence concerning long-term effects of GH replacement on cardiac structure and function. Importantly, supraphysiologic increases in LV thickness do not appear to occur following GH replacement doses that are close to the physiologic range.

Quality of Life

GHD is associated with low energy levels, social isolation, greater emotional lability, impaired socio-economic performance, and difficulties forming relationships [4, 33]. In a 21-month cross-over double-blind study, Burman et al. demonstrated a significant improvement in quality of life in adults with GHD treated with GH with improvements also observed by their partners [8]. However, not all placebo-controlled trials of GH replacement indicate an improvement in these variables [6, 33]. Meta-analysis data have shown improvement in quality of life following GH replacement in open-label, but not placebo-controlled studies [14]. Evaluation of whether GH replacement improves quality of life has been limited by use of a number of different questionnaires, some of which are considered inappropriate for this patient group [44]. It has also been suggested that effects would be more clearly observed if placebo-controlled studies were limited to patients in whom quality of life was markedly abnormal at baseline.

It is likely that studies evaluating psychosocial variables are more susceptible to placebo effect compared to those evaluating physical variables and long-term data, none of which is placebo-controlled and must be evaluated with this in mind. Several studies have included GHD patients who did not receive GH replacement (see Fig. 11.3). In the study of Gibney et al., significant improvements were observed in energy and emotional reaction in GH replaced subjects following 10 years of GH replacement compared to unreplaced GHD subjects [22]. Those who had not received GH replacement over the subsequent 9 years had a small but significant decline in physical mobility and general health and well-being, while those who were treated had improved energy and vitality.

Uncontrolled data have also been reported in the post-marketing surveillance observational studies, HypoCCS and KIMS. Both have shown sustained improvement in various aspects of quality of life in studies lasting up to 4 years and from 4–6 years, respectively [45, 46]. Both studies demonstrated normalisation of values relative to population-based data.

In summary, therefore, GH replacement in both short-term and long-term open-label studies improves aspects of quality of life with resulting measurements that no

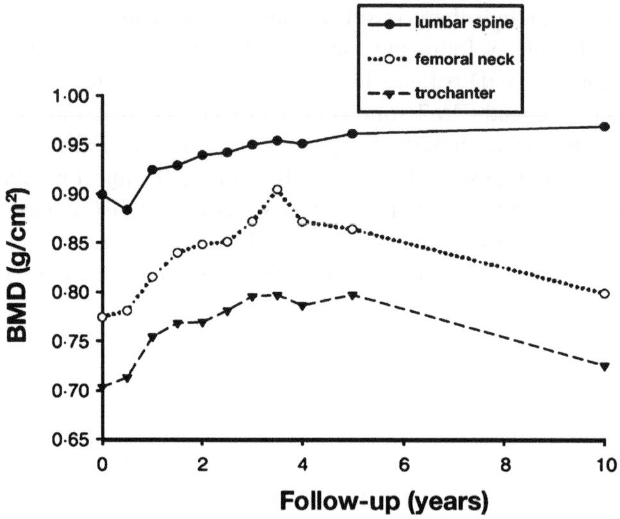

Fig. 11.3 The effect of GH replacement on Energy and Vitality, measured by Nottingham Health Profile in adults with GHD; receiving GH (*solid line*) and not receiving GH (*dotted line*) over 9 years. (Reproduced with permission from Gilchrist et al. [86])

longer differ from those of the normal population. Two long-term studies including small numbers of GHD patients who have and have not received long-term GH replacement have also shown significant between-group improvements. The relative paucity of short-term placebo-controlled data underpinning these long-term effects, however, remains a concern.

Safety

The effects of supraphysiologic circulating levels of GH in patients with acromegaly provide evidence of the harmful effects of excessive GH. Features of acromegaly include facial disfigurement, cardiomyopathy, arthropathy, obstructive sleep apnoea, impaired glucose tolerance, and diabetes mellitus [15]. Mortality rates are increased in acromegaly and are predominately due to cardio-respiratory causes, but can be normalised on correction of GH excess [47]. Clinical features of GH excess do not result from GH replacement when appropriate methods of dose titration are used [48], but the serious adverse consequences of acromegaly emphasise the importance of avoidance of over-treatment with GH through careful monitoring of IGF-I.

The main safety concerns associated with administration of doses of GH that more closely approximate physiologic secretion rates are possible increases in rates of diabetes and cancer. These are discussed in detail below.

Insulin Resistance and Diabetes

In view of the short-term insulin-antagonistic effect of GH and the well-documented association of insulin resistance with acromegaly, the observation that insulin sensitivity was reduced in GHD adults was initially surprising [49–51]. This finding has been confirmed in numerous studies and is probably explained by increased central adiposity, reduced muscle mass, and low levels of serum IGF-I. While short-term GH replacement leads to a further decrease in insulin sensitivity, most, although not all, available data suggest a return to baseline values following more prolonged treatment [6]. A study by Svensson et al. was of particular importance as it used the gold standard technique of the hyperinsulinaemic euglycaemic clamp and included a matched group of normal subjects [25]. Eleven GHD adults were studied at baseline and frequent intervals up to 7 years after starting GH replacement. No absolute change in insulin sensitivity occurred in the GH-replaced individuals during the 7 years of the study, but insulin sensitivity expressed as a percentage of that in the control subjects increased from 45 to 71%. The balance of evidence therefore suggests that long-term physiologic replacement exerts a neutral effect on glucose metabolism, probably reflecting a balance between the counter-regulatory effect of GH and the insulin sensitising effect of reduced visceral fat. This interpretation is supported by a recent report from the HypoCCS study, which demonstrates that the prevalence of the metabolic syndrome is higher in adults with GHD compared to the general population in the United States, and 3 years of GH replacement did not alter the prevalence [52].

There is some evidence, however, that rates of development of diabetes are increased following GH replacement in children and adults. Cutfield et al. reported data from an international surveillance study including more than 20,000 subjects, in which pituitary human GH treatment in childhood was associated with a sixfold increase in new cases of type 2 diabetes [18]. Data from the National Cooperative Growth Study (NCGS) in which the safety of recombinant human GH was studied in almost 55,000 children (42.5% GHD) between 1985 and 2006, however, did not clearly demonstrate any increase in diabetes, although it was not possible to determine the relative risk of type 2 diabetes due to lack of normative data in the age group studied [53]. In a randomised, double-blind, placebo-controlled, multicentre study of 1-year GH replacement in adult-onset GHD, there was a greater incidence of impaired glucose tolerance and type 2 diabetes in the GH-treated compared to the placebo-treated group [54]. In the 7-year study reported by Svensson et al., 2 of 11 patients on GH replacement developed diabetes although the authors noted that the overall figure for development of diabetes during GH replacement in their centre was 3.5% [25]. A retrospective study carried out in Sweden reviewed 750 adults with GHD and 2,314 matched controls [55]. The majority of subjects with GHD had been receiving GH replacement for a median duration of 6 years. The prevalence of type 2 diabetes was not significantly different between GHD and normal men, but the prevalence of diabetes in GHD women was double that in normal women. This increase remained significant following adjustment for BMI, smoking, and physical activity. Data from 5,120 GHD subjects from the KIMS database were reviewed, all of whom had at least 1 return visit. New cases of type 2 diabetes were identified in 26 men and 17 women and 16 of these were in the first year of GH treatment. The standard incidence ratio was calculated using a Swedish population-based cohort of 138,000 people, who were followed for prospectively for 3 years. The incidence of type 2 diabetes was not increased in people with GHD who had a normal BMI compared to the background population, but was increased in those with an increased BMI.

In summary, therefore, there is some evidence that patients on GH replacement are more likely to develop diabetes. The risk appears greater in subjects who are overweight and possibly also in women.

GH and Neoplasia

The mitogenic effect of GH and the relationship between the IGF/IGFBP system and neoplasia provide a theoretical basis by which GH treatment could increase cancer risk [19, 56]. Prolonged exposure to greatly supraphysiologic levels of GH in acromegalic patients has been thought for many years to be associated with cancer risk, although this remains controversial. Possible associations between GH replacement and neoplasia include recurrence of pituitary tumours, recurrence of other neoplasia, and emergence of de novo neoplasia. When assessing cancer rates, it must be taken into consideration that subjects with hypopituitarism secondary to genetic or tumour-related causes might have an inherently higher risk of neoplasia compared with normal subjects, and furthermore, that treatment modalities including radiotherapy and chemotherapy might also increase the risk of subsequent malignancy [7]. A large study including all individuals with pituitary tumours included in the Swedish Cancer Registry between 1958 and 1991 revealed an excess mortality from all tumours (pituitary excluded) with a SMR of 1.4 (CI, 1.21–1.7) and malignant tumours of the brain with a SMR of 7.1 (CI, 4.2–11.3) [57].

Recurrence of Pituitary Tumours

Determining whether GH replacement increases risk of recurrence of pituitary tumours is made difficult by the paucity of data regarding the natural history of pituitary tumours, whether previously untreated or treated with surgery or radiotherapy. The progression-free survival in patients with pituitary tumour has been reported as between 12 and 69% after surgery and between 72 and 92% following surgery and radiotherapy [58]. The implication of these widely differing results is that the influence of GH on tumour recurrence can only be considered in the context of an appropriate control group.

Two small studies have reported incidence of tumour re-growth in patients with pituitary adenomas, some of whom received GH replacement. Hatrick et al. carried out interval radiological scanning in 47 GHD patients who had received GH for a mean duration of 3.6 years and 28 who had not [59]. There was no difference in pituitary tumour recurrence rate between the two groups. Arnold et al. retrospectively reviewed subjects with non-functioning pituitary adenomas treated only by surgery over 6.8 years, the mean duration of GH replacement being 4.6 years [60]. Of the 130 subjects included in the study, 23 received GH replacement. Tumour re-growth was detected in 35% of GH-replaced and 36% of unreplaced GHD patients. Following adjustment for sex, age at tumour diagnosis, cavernous sinus invasion at diagnosis, and type of tumour removal, GH treatment was not a significant independent predictor of recurrence. In a prospective study, Frajese et al. carried out serial scanning in 100 consecutive GHD subjects who had been treated for pituitary or peri-pituitary tumour prior to commencing GH replacement [61]. There was one case of slight intrasellar tissue enlargement at 6 months – GH replacement was continued and there was no further enlargement by 12 months. The appearances remained unchanged in the other subjects. There was also no evidence of increased risk of tumour recurrence in data from the KIMS study [62]. It should be emphasised, however, that patients with residual hypothalamic-pituitary tumours were less likely to be included in this study and also that tumour recurrence or progression was not routinely monitored by imaging techniques.

Other Neoplasia

IGF-I/IGFBPs and Cancer

Prospective studies have demonstrated that greater IGF-I levels within the normal range are predictive of cancers including breast, prostate, and colon, while there is also evidence of associations between levels of IGFBPs and cancer risk [19, 56]. Renehan et al. carried out a systematic review and meta-regression analysis investigating the association between serum IGF-1 and IGFBP-3 levels and common cancers [63]. Included in the analysis were 21 studies, comprising 3,609 cases and 7,137 controls. High-normal concentrations of IGF-1 levels were associated with increased rates of prostate and premenopausal but not postmenopausal breast cancer, while high concentrations of IGFBP3 were also associated with increased risk of premenopausal breast cancer. A meta-analysis of prospective studies including 3,700 cases and 5,200 controls confirmed the association between IGF-1 levels and prostate cancer [64]. Although there was also an association between IGFBP3 levels and prostate cancer, this was not independent of IGF-1. The Endogenous Hormones and Breast Cancer Collaborative Group reviewed pooled individual data from 17 prospective studies in 12 countries, including an additional four trials and 1,500 subjects compared to the study carried out by Renehan et al [65]. Serum IGF-1 levels were predictive of breast cancer risk in oestrogen-receptor positive tumours only and were not modified by menopausal status or serum IGFBP-3 levels. Meta-analysis data support an association between IGF-I levels and colon cancer [63], while there are also smaller reports of associations between IGF-I and lung, bladder, and endometrial carcinoma [19].

GH Treatment and Malignancy

GH Treatment in Childhood

In addition to use in children with hypothalamic-pituitary tumours, GH replacement has also been extensively used in survivors of childhood malignancy such as brain tumours and leukaemia, in which cranial irradiation almost invariably leads to GHD. Much of the data surrounding tumour re-growth are complicated by the fact that childhood survivors of cancer may have an inherently higher risk of developing second tumours and cranial irradiation itself increases the rate of malignancy. Detailed analysis of the safety data from these populations is beyond the scope of this review. Briefly, rates of recurrence of primary malignancy including brain tumours and leukaemia do not appear to be increased by GH [53, 66, 67]. The risk of mortality from cancer (specifically Hodgkin's disease and colorectal cancer) was increased approximately threefold in a large cohort of children who were treated with pituitary-derived GH in the UK between 1959 and 1985 and followed up between 1995 and 2000 [17]. While this data has caused concern, some important points need to be addressed. Firstly, the absolute number of deaths was very small. Secondly, the incidence of cancer deaths in hypopituitary patients who did not receive GH is unknown. Finally, dosing schedules used at that time probably resulted in higher IGF-I levels than those currently recommended, although this cannot be retrospectively confirmed. However, both the Childhood Cancer Survivor Study (CCSS) [67] and the NCGS [53] have also demonstrated greater than predicted rates of second neoplasm following GH treatment. The relative risk in the CCSS for second solid tumour (most commonly meningioma) was 3.21 compared to cancer survivors who did not receive GH in the initial evaluation, but decreased to 2.15 during extended follow-up. A similar incidence of second neoplasms was reported in the NCGS.

GH Replacement in Adults

There is very limited data available concerning cancer rates in GHD adults receiving GH replacement. Cancer rates, particularly colorectal cancer, have been reported to be increased in hypopituitary patients not receiving GH [68]. A recent report by Popovic et al. using data from KIMS has evaluated the relationship between serum IGF-1, IGFBP-2, and IGFBP-3 levels in GH-replaced GHD patients and the risk of developing de novo malignancy [69]. Results from 100 patients who developed malignancy while receiving GH replacement were compared to 325 GH-replaced patients who did not develop malignancy. No association between IGF-1 standard deviation scores and relative risk was observed, but elevated levels of IGFBP-2 and IGFBP-3 SDSs were significantly associated with an increased relative risk of developing malignancy. The significance of this is unclear.

Summary

The overall body of evidence is reassuring that GH treatment does not increase the risk of recurrence of pituitary tumours or other cancers. There is evidence, however, that GH treatment in childhood increases the risk of second neoplasms. There is no evidence that GH increases cancer risk in adults. However, in view of the association between elevated IGF-I levels and risk of common cancers, surveillance studies carried out over many years in large numbers of patients are essential.

GH and Mortality

The relationship between pituitary disorders and mortality has recently been extensively reviewed by Sherlock et al [7]. Briefly, mortality rates are increased in hypopituitarism. A meta-analysis concluded that the all-cause standardised mortality ratio was 2.06 in men and 2.80 in women. The cause of this is unknown and probably multifactorial. It has been suggested that GHD contributes to increased mortality in view of the well-recognised effects described above on cardiovascular risk factors and surrogate markers of atherosclerosis and that GH replacement might reduce mortality rates. Further complexity is added by recent observations linking the IGF/IGFBP system with mortality rates in the general population.

IGF/IGFBP and Mortality

There is conflicting evidence surrounding the relationship between serum IGF-1, IGFBP-3, IGFBP-1 levels, and mortality. Low IGF-1 levels are associated with increased all-cause mortality [70, 71] in some studies and only cardiovascular mortality in others [72]. Mortality from IHD was studied in relation to IGF-1 and IGFBP-1 levels in the Rancho Bernardo Study. Low serum IGF-1 and IGFBP-1 levels were both jointly and independently related to IHD mortality. This effect seems to be cardiac specific, as neither IGF-1 nor IGFBP-1 was associated with non-IHD CVD mortality or all-cause mortality [73]. Similar conflicting evidence exists between serum IGFBP-3, IGFBP-1 levels, and mortality [71, 72, 74]. The discrepancy between these studies is likely due to the cohort sampled and their age at time of sampling. The assays used in measuring IGFBPs are variable and have also been suggested as a reason for the discrepancy between studies.

GH Replacement and Mortality

It is not practical to carry out DBPCTs of sufficient size and duration to evaluate effects of GH replacement on mortality. Useful data can be obtained from large observational studies and post-marketing surveillance studies carried out by pharmaceutical companies, although outcomes are potentially influenced by treatment bias and the well-documented phenomenon that patients' lifestyles are altered by taking part in clinical trials.

Svensson et al. compared mortality rates in 289 hypopituitary patients receiving growth hormone (studied prospectively) and 1,411 hypopituitary patients not receiving GH replacement (studied retrospectively) to mortality rates in the normal population [68]. Overall mortality, rates of cerebrovascular events, myocardial infarctions, and malignancies were significantly increased in hypopituitary patients not receiving GH replacement compared to the normal population. In contrast, mortality rates and rates of malignancy were similar in the GH-treated group compared to the normal population. In addition to problems created by comparing retrospective with prospective data, patient characteristics differed between the two hypopituitary cohorts and therefore these findings must be interpreted with caution. Early reports from the KIMS database have also reported that mortality rates in GH-replaced patients do not differ from the normal population. Overall, it is reassuring that mortality rates in GHD subjects who are on GH replacement and under long-term follow-up in major centres or as part of large surveillance studies appear not to differ from the normal population. It is not known, however, whether GH replacement contributes to this apparent improvement compared to historical data. Other possibilities include improved management of pituitary tumours and pituitary hormone deficiencies other than GH and more stringent attention to cardiovascular risk factors.

Practical Aspects of Long-Term GH Replacement

Before prescribing GH replacement, the likely benefits should be considered. It can be explained to the patient that potential benefits include changes in body composition and improved energy and strength. It can also be explained that it is likely that fracture risk will be reduced, although the importance of this will depend on the patient's age, baseline BMD, and overall fracture risk. It is important to avoid creating unrealistic expectations about currently unsubstantiated benefits of GH such as improved quality of life. It is premature to assume that excess cardiovascular risk in hypopituitary adults is due to GHD and also that GH replacement might reduce this risk. GH replacement is contraindicated in patients with previous and active malignancy and it may be prudent to avoid GH in patients considered at increased cancer risk, particularly regarding cancers in which a link with IGF-I has been observed.

Guidelines regarding dose titration regimes and treatment monitoring have been published by expert groups and should be followed [6, 75, 76]. Patients should be reviewed on a regular basis. Variables to be monitored should include at least BMI, BP, fasting plasma glucose and lipid levels, HbA1c, and IGF-I. In general, IGF-I levels should be maintained in the middle of the age and gender-related normal range. The frequency of pituitary imaging and assessment of other pituitary hormone replacement can be individualised depending on the underlying cause of GHD and known pituitary hormone deficiencies. It must be remembered that GH replacement can alter the requirement for glucocorticoid and thyroid hormone replacement [77, 78]. In women requiring oestrogen replacement, transdermal compared to oral administration reduces the dose of GH necessary to achieve the same IGF-I response [79].

If a patient is seen to be benefiting from GH replacement, there is no indication to discontinue treatment. In the absence of any apparent benefit after reasonable trial duration, typically 6–12 months, it is reasonable to discontinue. There is little data available concerning the effects of withdrawal of long-term GH replacement [80]. In one study, patients receiving GH replacement for a mean duration of 5 years were randomised to continuing GH replacement or receiving identical placebo for 6 months. Compared to those who remained on GH replacement, there was a significant decrease in LBM and increases in total body and trunk fat mass and LDL-cholesterol [81]. Significant changes in a small number of quality of life variables were also observed. Patients should therefore be advised when discontinuing GH replacement that these effects might occur.

Summary and Conclusions

Despite more than two decades of GH replacement, there is a relatively small body of evidence regarding its long-term efficacy. There is reasonable evidence that improvements in LBM, BMD, and strength persist in the long-term, but little evidence for sustained effect of other short-term benefits. Long-term GH replacement does not reduce insulin sensitivity, but diabetes risk may be increased. There is no evidence that GH replacement is associated with an excess incidence of cancer, but the association of IGF-I levels in the general population with certain cancers remains of concern. The association between hypopituitarism, GH replacement, other pituitary hormone replacements, and mortality is complex. Observational studies have demonstrated that the mortality rate in GH-replaced GHD subjects is similar to that of the normal population, which might suggest reduced mortality compared to GHD patients not receiving GH replacement in whom mortality rates are increased.

It is the opinion of the authors that the principal evidence-based indications for long-term GH replacement are to increase LBM, BMD, and strength. Continuation of GH replacement also seems reasonable where there is a demonstrable short-term improvement in variables influenced by GH replacement such as exercise capacity. Currently, the evidence-base to support prescribing GH for the indications of improving quality of life or reducing mortality rates is limited.

To enhance the evidence-base from which decisions about long-term GH replacement are made, large-scale placebo-controlled trials of longer duration (e.g. up to 5 years) are needed. Such trials would not be unethical given the paucity of long-term data currently available and should be feasible. Particular importance should be attached to providing evidence regarding quality of life using psychological instruments known to be sensitive in hypopituitary subjects and also to addressing safety concerns. It is likely that in the future healthcare providers will expect such long-term data to be available before funding expensive and lifelong treatment.

References

1. Carroll PV, Christ ER, Bengtsson BA, et al. Growth hormone deficiency in adulthood and the effects of growth hormone replacement: a review. Growth Hormone Research Society Scientific Committee. J Clin Endocrinol Metab. 1998;83(2):382–95.
2. de Boer H, Blok G-J, van der Veen EA. Clinical aspects of growth hormone deficiency in adults. Endocr Rev. 1995;16:63–86.
3. Gibney J, Healy ML, Sonksen PH. The growth hormone/insulin-like growth factor-I axis in exercise and sport. Endocr Rev. 2007;28(6):603–24.
4. Drake WM, Howell SJ, Monson JP, Shalet SM. Optimizing GH therapy in adults and children. Endocr Rev. 2001;22(4):425–50.
5. Ohlsson C, Bengtsson BA, Isaksson OG, Andreassen TT, Slootweg MC. Growth hormone and bone. Endocr Rev. 1998;19(1):55–79.
6. Molitch ME, Clemmons DR, Malozowski S, Merriam GR, Shalet SM, Vance ML. Evaluation and treatment of adult growth hormone deficiency: an endocrine society clinical practice guideline. J Clin Endo Metab. 2006;91(5):1621–34.
7. Sherlock M, Ayuk J, Tomlinson JW, et al. Mortality in patients with pituitary disease. Endocr Rev. 2010;31(3):301–42.
8. Burman P, Broman JE, Hetta J, et al. Quality of life in adults with growth hormone (GH) deficiency: response to treatment with recombinant human GH in a placebo-controlled 21-month trial. J Clin Endocrinol Metab. 1995;80(12):3585–90.
9. Wallymahmed ME, Foy P, Shaw D, Hutcheon R, Edwards RH, MacFarlane IA. Quality of life, body composition and muscle strength in adult growth hormone deficiency: the influence of growth hormone replacement therapy for up to 3 years. Clin Endocrinol (Oxf). 1997;47(4):439–46.
10. Widdowson WM, Gibney J. The Effect of Growth Hormone Replacement on Exercise Capacity in Patients with GH-deficiency: A Meta-Analysis. J Clin Endocrinol Metab. 2008;93(11):4413–7.
11. Maison P, Chanson P. Cardiac effects of growth hormone in adults with growth hormone deficiency: a meta-analysis. Circulation. 2003;108(21):2648–52.
12. Maison P, Griffin S, Nicoue-Beglah M, Haddad N, Balkau B, Chanson P. Impact of growth hormone (GH) treatment on cardiovascular risk factors in GH-deficient adults: a metaanalysis of blinded, randomized, placebo-controlled trials. J Clin Endocrinol Metab. 2004;89(5):2192–9.
13. Widdowson WM, Gibney J. The effect of growth hormone (GH) replacement on muscle strength in patients with GH-deficiency: a meta-analysis. Clin Endocrinol (Oxf). 2010;72(6):787–92.
14. Arwert LI, Deijen JB, Witlox J, Drent ML. The influence of growth hormone (GH) substitution on patient-reported outcomes and cognitive functions in GH-deficient patients: a meta-analysis. Growth Horm IGF Res. 2005;15(1):47–54.
15. Nabarro JD. Acromegaly. Clin Endocrinol (Oxf). 1987;26(4):481–512.
16. Takala J, Ruokonen E, Webster NR, et al. Increased mortality associated with growth hormone treatment in critically ill adults. N Engl J Med. 1999;341:785–92.
17. Swerdlow AJ, Higgins CD, Adlard P, Preece MA. Risk of cancer in patients treated with human pituitary growth hormone in the UK, 1959–85: a cohort study. Lancet. 2002;360:273–7.
18. Cutfield WS, Wilton P, Bennmarker H, et al. Incidence of diabetes mellitus and impaired glucose tolerance in children and adolescents receiving growth-hormone treatment. Lancet. 2000;355(9204):610–3.

19. Khandwala HM, McCutcheon IE, Flyvbjerg A, Friend KE. The effects of insulin-like growth factors on tumorigenesis and neoplastic growth. Endocr Rev. 2000;21(3):215–44.
20. Hoffman DM, O'Sullivan AJ, Freund J, Ho KK. Adults with growth hormone deficiency have abnormal body composition but normal energy metabolism. J Clin Endocrinol Metab. 1995;80(1):72–7.
21. Bengtsson BA, Eden S, Lonn L, et al. Treatment of adults with growth hormone (GH) deficiency with recombinant human GH. J Clin Endocrinol Metab. 1993;76(2):309–17.
22. Gibney J, Wallace JD, Spinks T, et al. The effects of 10 years of recombinant human growth hormone (GH) in adult GH-deficient patients. J Clin Endocrinol Metab. 1999;84(8):2596–602.
23. Chrisoulidou A, Beshyah SA, Rutherford O, et al. Effects of 7 years of growth hormone replacement therapy in hypopituitary adults. J Clin Endocrinol Metab. 2000;85(10):3762–9.
24. Gotherstrom G, Bengtsson BA, Bosaeus I, Johannsson G, Svensson J. A 10-year, prospective study of the metabolic effects of growth hormone replacement in adults. J Clin Endocrinol Metab. 2007;92(4):1442–5.
25. Svensson J, Fowelin J, Landin K, Bengtsson BA, Johansson JO. Effects of seven years of GH-replacement therapy on insulin sensitivity in GH-deficient adults. J Clin Endocrinol Metab. 2002;87(5):2121–7.
26. ter Maaten JC, de Boer H, Kamp O, Stuurman L, van der Veen EA. Long-term effects of growth hormone (GH) replacement in men with childhood-onset GH deficiency. J Clin Endocrinol Metab. 1999;84(7):2373–80.
27. Fideleff HL, Boquete HR, Stalldecker G, Giaccio AV, Sobrado PG. Comparative results of a 4-year study on cardiovascular parameters, lipid metabolism, body composition and bone mass between untreated and treated adult growth hormone deficient patients. Growth Horm IGF Res. 2008;18(4):318–24.
28. Wuster C, Abs R, Bengtsson BA, et al. The influence of growth hormone deficiency, growth hormone replacement therapy, and other aspects of hypopituitarism on fracture rate and bone mineral density. J Bone Miner Res. 2001;16(2):398–405.
29. Murray RD, Columb B, Adams JE, Shalet SM. Low bone mass is an infrequent feature of the adult growth hormone deficiency syndrome in middle-age adults and the elderly. J Clin Endocrinol Metab. 2004;89:1124–30.
30. Arwert LI, Roos JC, Lips P, Twisk JW, Manoliu RA, Drent ML. Effects of 10 years of growth hormone (GH) replacement therapy in adult GH-deficient men. Clin Endocrinol (Oxf). 2005;63(3):310–6.
31. Gotherstrom G, Svensson J, Koranyi J, et al. A prospective study of 5 years of GH replacement therapy in GH-deficient adults: sustained effects on body composition, bone mass, and metabolic indices. J Clin Endocrinol Metab. 2001;86(10):4657–65.
32. Biermasz NR, Hamdy NA, Pereira AM, Romijn JA, Roelfsema F. Long-term skeletal effects of recombinant human growth hormone (rhGH) alone and rhGH combined with alendronate in GH-deficient adults: a seven-year follow-up study. Clin Endocrinol (Oxf). 2004;60(5):568–75.
33. Woodhouse LJ, Mukherjee A, Shalet SM, Ezzat S. The influence of growth hormone status on physical impairments, functional limitations, and health-related quality of life in adults. Endocr Rev. 2006;27(3):287–317.
34. Johannsson G, Grimby G, Sunnerhagen KS, Bengtsson BA. Two years of growth hormone (GH) treatment increase isometric and isokinetic muscle strength in GH-deficient adults. J Clin Endocrinol Metab. 1997;82(9):2877–84.
35. Svensson J, Stibrant Sunnerhagen K, Johannsson G. Five years of growth hormone replacement therapy in adults: age- and gender-related changes in isometric and isokinetic muscle strength. J Clin Endocrinol Metab. 2003;88(5):2061–9.
36. Gotherstrom G, Elbornsson M, Stibrant-Sunnerhagen K, Bengtsson BA, Johannsson G, Svensson J. Ten years of growth hormone (GH) replacement normalizes muscle strength in GH-deficient adults. J Clin Endocrinol Metab. 2009;94(3):809–16.
37. Christ ER, Cummings MH, Lumb PJ, Crook MA, Sonksen PH, Russell-Jones DL. Growth hormone (GH) replacement therapy reduces serum sialic acid concentrations in adults with

GH-deficiency: a double-blind placebo- controlled study. Clin Endocrinol (Oxf). 1999;51(2):173–9.
38. Johannsson JO, Landin K, Tengborn L, Rosen T, Bengtsson BA. High fibrinogen and plasminogen activator inhibitor activity n growth hormone-deficient adults. Arterioscler Thromb. 1994;14:434–7.
39. Colao A, Di Somma C, Spiezia S, et al. Growth hormone treatment on atherosclerosis: results of a 5-year open, prospective, controlled study in male patients with severe growth hormone deficiency. J Clin Endocrinol Metab. 2008;93(9):3416–24.
40. Persson J, Formgren J, Israelsson B, Berglund G. Ultrasound-determined intima-media thickness and atherosclerosis. Direct and indirect validation. Arterioscler Thromb. 1994;14(2):261–4.
41. Colao A, Di Somma C, Cuocolo A, et al. Does a gender-related effect of growth hormone (GH) replacement exist on cardiovascular risk factors, cardiac morphology, and performance and atherosclerosis? Results of a two-year open, prospective study in young adult men and women with severe GH deficiency. J Clin Endocrinol Metab. 2005;90(9):5146–55.
42. Thuesen L, Jørgensen JOL, Müller JR, et al. Short and long-term cardiovascular effects of growth hormone therapy in growth hormone deficient adults. Clin Endocrinol. 1994;41:615–20.
43. Johannsson G, Bengtsson B-Å, Andersson B, Isgaard J, Caidahl K. Long-term cardiovascular effects of growth hormone treatment in GH-deficient adults: preliminary data from a small group of patients. Clin Endocrinol (Oxf). 1996;45:305–14.
44. Deijen JB, Arwert LI, Witlox J, Drent ML. Differential effect sizes of growth hormone replacement on Quality of Life, well-being and health status in growth hormone deficient patients: a meta-analysis. Health Qual Life Outcomes. 2005;3:63.
45. Rosilio M, Blum WF, Edwards DJ, et al. Long-term improvement of quality of life during growth hormone (GH) replacement therapy in adults with GH deficiency, as measured by questions on life satisfaction-hypopituitarism (QLS-H). J Clin Endocrinol Metab. 2004;89(4):1684–93.
46. Koltowska-Haggstrom M, Mattsson AF, Monson JP, et al. Does long-term GH replacement therapy in hypopituitary adults with GH deficiency normalise quality of life? Eur J Endocrinol. 2006;155(1):109–19.
47. Sherlock M, Aragon Alonso A, Reulen RC, et al. Monitoring disease activity using GH and IGF-I in the follow-up of 501 patients with acromegaly. Clin Endocrinol (Oxf). 2009;71(1):74–81.
48. Gibney J, Johannsson G. Long-term monitoring of insulin-like growth factor I in adult growth hormone deficiency: a critical appraisal. Horm Res. 2004;62 Suppl 1:66–72.
49. Beshyah SA, Henderson A, Niththyanathan R, Sharp P, Richmond W, Johnston DG. Metabolic abnormalities in growth hormone deficient adults: II. Carbohydrate tolerance and lipid metabolism. Endocrinol Metab. 1994;1:173–80.
50. Hew FL, Koschmann M, Christopher M, et al. Insulin resistance in growth hormone-deficient adults: defects in glucose utilization and glycogen synthase activity. J Clin Endocrinol Metab. 1996;81(2):555–64.
51. Johansson JO, Fowelin J, Landin K, Lager I, Bengtsson BA. Growth hormone-deficient adults are insulin-resistant. Metabolism. 1995;44(9):1126–9.
52. Attanasio AF, Mo D, Erfurth EM, et al. Prevalence of metabolic syndrome in adult hypopituitary growth hormone (GH)-deficient patients before and after GH replacement. J Clin Endocrinol Metab. 2010;95(1):74–81.
53. Bell J, Parker KL, Swinford RD, Hoffman AR, Maneatis T, Lippe B. Long-term safety of recombinant human growth hormone in children. J Clin Endocrinol Metab. 2010;95(1):167–77.
54. Hoffman AR, Kuntze JE, Baptista J, et al. Growth hormone (GH) replacement therapy in adult-onset GH deficiency: effects on body composition in men and women in a double-blind, randomized, placebo-controlled trial. J Clin Endocrinol Metab. 2004;89(5):2048–56.
55. Holmer H, Svensson J, Rylander L, et al. Nonfatal stroke, cardiac disease, and diabetes mellitus in hypopituitary patients on hormone replacement including growth hormone. J Clin Endocrinol Metab. 2007;92(9):3560–7.
56. Jogie-Brahim S, Feldman D, Oh Y. Unraveling insulin-like growth factor binding protein-3 actions in human disease. Endocr Rev. 2009;30(5):417–37.

57. Nilsson B, Gustavasson-Kadaka E, Bengtsson BA, Jonsson B. Pituitary adenomas in Sweden between 1958 and 1991: incidence, survival, and mortality. J Clin Endocrinol Metab. 2000;85(4):1420–5.
58. Brada M, Rajan B, Traish D, et al. The long-term efficacy of conservative surgery and radiotherapy in the control of pituitary adenomas. Clin Endocrinol (Oxf). 1993;38(6):571–8.
59. Hatrick AG, Boghalo P, Bingham JB, Ayres AB, Sonksen PH, Russell-Jones DL. Does GH replacement therapy in adult GH-deficient patients result in recurrence or increase in size of pituitary tumours? Eur J Endocrinol. 2002;146(6):807–11.
60. Arnold JR, Arnold DF, Marland A, Karavitaki N, Wass JA. GH replacement in patients with non-functioning pituitary adenoma (NFA) treated solely by surgery is not associated with increased risk of tumour recurrence. Clin Endocrinol (Oxf). 2009;70(3):435–8.
61. Frajese G, Drake WM, Loureiro RA, et al. Hypothalamo-pituitary surveillance imaging in hypopituitary patients receiving long-term GH replacement therapy. J Clin Endocrinol Metab. 2001;86(11):5172–5.
62. Abs R, Bengtsson BA, Hernberg-Stahl E, et al. GH replacement in 1034 growth hormone deficient hypopituitary adults: demographic and clinical characteristics, dosing and safety. Clin Endocrinol (Oxf). 1999;50(6):703–13.
63. Renehan AG, Zwahlen M, Minder C, O'Dwyer ST, Shalet SM, Egger M. Insulin-like growth factor (IGF)-I, IGF binding protein-3, and cancer risk: systematic review and meta-regression analysis. Lancet. 2004;363(9418):1346–53.
64. Roddam AW, Allen NE, Appleby P, et al. Insulin-like growth factors, their binding proteins, and prostate cancer risk: analysis of individual patient data from 12 prospective studies. Ann Intern Med. 2008;149(7):461–71, W483–68.
65. Key TJ, Appleby PN, Reeves GK, Roddam AW. Insulin-like growth factor 1 (IGF1), IGF binding protein 3 (IGFBP3), and breast cancer risk: pooled individual data analysis of 17 prospective studies. Lancet Oncol. 2010;11(6):530–42.
66. Jostel A, Mukherjee A, Hulse PA, Shalet SM. Adult growth hormone replacement therapy and neuroimaging surveillance in brain tumour survivors. Clin Endocrinol (Oxf). 2005; 62(6):698–705.
67. Sklar CA, Mertens AC, Mitby P, et al. Risk of disease recurrence and second neoplasms in survivors of childhood cancer treated with growth hormone: a report from the Childhood Cancer Survivor Study. J Clin Endocrinol Metab. 2002;87(7):3136–41.
68. Svensson J, Bengtsson BA, Rosen T, Oden A, Johannsson G. Malignant disease and cardiovascular morbidity in hypopituitary adults with or without growth hormone replacement therapy. J Clin Endocrinol Metab. 2004;89(7):3306–12.
69. Popovic V, Mattsson AF, Gaillard RC, Wilton P, Koltowska-Haggstrom M, Ranke MB. Serum insulin-like growth factor I (IGF-I), IGF-binding proteins 2 and 3, and the risk for development of malignancies in adults with growth hormone (GH) deficiency treated with GH: Data from KIMS (Pfizer International Metabolic Database). J Clin Endocrinol Metab. 2010;95(9):4449–54.
70. Roubenoff R, Parise H, Payette HA, et al. Cytokines, insulin-like growth factor 1, sarcopenia, and mortality in very old community-dwelling men and women: the Framingham Heart Study. Am J Med. 2003;115(6):429–35.
71. Friedrich N, Haring R, Nauck M, et al. Mortality and serum insulin-like growth factor (IGF)-I and IGF binding protein 3 concentrations. J Clin Endocrinol Metab. 2009;94(5):1732–9.
72. Saydah S, Graubard B, Ballard-Barbash R, Berrigan D. Insulin-like growth factors and subsequent risk of mortality in the United States. Am J Epidemiol. 2007;166(5):518–26.
73. Laughlin GA, Barrett-Connor E, Criqui MH, Kritz-Silverstein D. The prospective association of serum insulin-like growth factor I (IGF-I) and IGF-binding protein-1 levels with all cause and cardiovascular disease mortality in older adults: the Rancho Bernardo Study. J Clin Endocrinol Metab. 2004;89(1):114–20.
74. Harrela M, Qiao Q, Koistinen R, et al. High serum insulin-like growth factor binding protein-1 is associated with increased cardiovascular mortality in elderly men. Horm Metab Res. 2002;34(3):144–9.

75. Ho KK. Consensus guidelines for the diagnosis and treatment of adults with GH deficiency II: a statement of the GH Research Society in association with the European Society for Pediatric Endocrinology, Lawson Wilkins Society, European Society of Endocrinology, Japan Endocrine Society, and Endocrine Society of Australia. Eur J Endocrinol. 2007;157(6):695–700.
76. Cook DM, Yuen KC, Biller BM, Kemp SF, Vance ML. American Association of Clinical Endocrinologists medical guidelines for clinical practice for growth hormone use in growth hormone-deficient adults and transition patients – 2009 update. Endocr Pract. 2009;15 Suppl 2:1–29.
77. Giavoli C, Libe R, Corbetta S, et al. Effect of recombinant human growth hormone (GH) replacement on the hypothalamic-pituitary-adrenal axis in adult GH-deficient patients. J Clin Endocrinol Metab. 2004;89(11):5397–401.
78. Porretti S, Giavoli C, Ronchi C, et al. Recombinant human GH replacement therapy and thyroid function in a large group of adult GH-deficient patients: when does L-T(4) therapy become mandatory? J Clin Endocrinol Metab. 2002;87(5):2042–5.
79. Cook DM, Ludlam WH, Cook MB. Route of estrogen administration helps to determine growth hormone (GH) replacement dose in GH-deficient adults. J Clin Endocrinol Metab. 1999;84(11):3956–60.
80. Gibney J, Healy ML, Stolinski M, et al. Effect of growth hormone (GH) on glycerol and free fatty acid metabolism during exhaustive exercise in GH-deficient adults. J Clin Endocrinol Metab. 2003;88(4):1792–7.
81. McMillan CV, Bradley C, Gibney J, Healy ML, Russell-Jones DL, Sonksen PH. Psychological effects of withdrawal of growth hormone therapy from adults with growth hormone deficiency. Clin Endocrinol (Oxf). 2003;59(4):467–75.
82. Kann P, Piepkorn B, Schehler B, et al. Effect of long-term treatment with GH on bone metabolism, bone mineral density and bone elasticity in GH-deficient adults. Clin Endocrinol (Oxf). 1998;48(5):561–8.
83. al-Shoumer KA, Gray R, Anyaoku V, et al. Effects of four years' treatment with biosynthetic human growth hormone (GH) on glucose homeostasis, insulin secretion and lipid metabolism in GH-deficient adults. Clin Endocrinol (Oxf). 1998;48(6):795–802.
84. Valimaki MJ, Salmela PI, Salmi J, et al. Effects of 42 months of GH treatment on bone mineral density and bone turnover in GH-deficient adults. Eur J Endocrinol. 1999;140(6):545–54.
85. Drake WM, Rodriguez-Arnao J, Weaver JU, et al. The influence of gender on the short and long-term effects of growth hormone replacement on bone metabolism and bone mineral density in hypopituitary adults: a 5-year study. Clin Endocrinol (Oxf). 2001;54(4):525–32.
86. Gilchrist FJ et al. The effect of long-term untreated growth hormone deficiency (GHD) and 9 years of GH replacement on the quality of life (QoL) of GH-deficient adults. Clin Endocrinol (Oxf). 2002;57:363–70.

Chapter 12
Quality of Life in Acromegaly and Growth Hormone Deficiency

Susan M. Webb, Eugenia Resmini, Alicia Santos, and Xavier Badia

Abstract Health-related quality of life (HRQoL) is considered an outcome measure in the evaluation of adults with acromegaly and GH deficiency (AGHD), after the unexpected observation of improvement in mood and vitality in initial trials with GH replacement over two decades ago. HRQoL is measured with questionnaires designed to be used for general population or any type of disease (generic questionnaires) or aimed at the specific dimensions affected in a determined condition (disease-generated or specific questionnaires). The latter are more likely to identify impairments related to the underlying disease and benefits of treatment. Examples of disease-generated HRQoL questionnaires are the AcroQoL (for acromegaly), AGHDA (adult GH deficiency assessment), or QLS-H (Questions on Life Satisfaction-Hypopituitarism) questionnaires (for AGHD) and permit further insight into the dimensions most affected in these diseases. These instruments are useful tools for both clinical research and daily practice, since hormone and imaging results do not always correlate with the patients' well-being. Acromegalic patients exhibit severe impairment of HRQoL, which, despite improvement after successful therapy, often remains below normal reference values. The most affected dimension is appearance, suggesting that an earlier diagnosis, before irreversible morphological changes have occurred, would be beneficial. HRQoL is worse in treated acromegaly, which has rendered them GH-deficient, than in those with controlled disease who are GH-sufficient. HRQoL is severely impaired in AGHD also, but improves and may normalize after GH replacement therapy and is maintained over several years; the most affected dimension is vitality. Therapeutic outcomes can now be evaluated by clinicians for HRQoL and may show how adverse effects of the disease on QoL often persist despite successful treatment from an endocrine point of view. Awareness of this is important to prevent

S.M. Webb (✉)
Department of Endocrinology Medicine, Hospital Sant Pau, Universitat Autònoma de Barcelona and CIBERER (Centro de Investigación Biomédica en Red en Enfermedades Raras) Unit 747, Barcelona, Spain
e-mail: swebb@santpau.cat

K. Ho (ed.), *Growth Hormone Related Diseases and Therapy: A Molecular and Physiological Perspective for the Clinician*, Contemporary Endocrinology, DOI 10.1007/978-1-60761-317-6_12,

inappropriate expectations with respect to the long-term results of treatment in patients with pituitary adenomas. HRQoL in these chronic endocrine diseases can be used as an outcome measure for clinical and therapeutic evaluation.

Keywords Health-related quality of life • Acromegaly • GH deficiency • AGHDA • QLS-H • AcroQoL

Introduction

Health-related quality of life (HRQoL) is a concept which refers to how an individual feels, functions, and responds in daily life. Apart from reflecting the overall effects of a disease and treatment, it is influenced by the patient's goals, expectations, standards, concerns, and cultural context, as well as life events. HRQoL can be measured with two types of questionnaires, generic and disease-specific or -generated [1, 2]. Examples of generic questionnaires are the Nottingham Health Profile (NHP) [3], the Psychological General Well-Being Scale (PGWBS) [4], the EuroQol (which includes the EQ-5 Dimensions and EQ-Visual Analog Scale) [5–7], and the Short Form 36 (SF-36) [8]. Among the disease-specific/generated questionnaires for acromegaly, the AcroQoL questionnaire is available [9], while for Adult GH Deficiency both the AGHDA (Adult GH Deficiency Assessment) [10] and QLS-H (Questions on Life Satisfaction-Hypopituitarism) can be used [11, 12]. As for biochemical or any other parameters, all require a validated translation, following standard methodology, when used in different cultural/language settings [1].

Over the last two decades, evidence has shown how HRQoL is impaired in patients with pituitary disease [13–16]. Even though it was thought that HRQoL improved after successful treatment and medical cure, recent findings show how impairment present in active disease tends to improve but may not normalize after biochemical cure [17–19]. Thus, HRQoL assessment is an important outcome measure in clinical practice, as it gives the clinician information from the patient's point of view on issues often not addressed by health professionals, but of importance for the patient's everyday life. This evidence favored the development of specific questionnaires to address the particular impact of a disease on patients' HRQoL and which the clinicians wished to monitor, with the aim of assessing more broadly the impact of these conditions.

Health-Related Quality of Life in Acromegaly

Acromegaly, a chronic disease with physical and psychological limitations (including joint pains, headaches, low energy, and libido) and morphological changes (in body image, excessive sweating, change in the patients voice) often not completely reversible, may go undiagnosed for years despite the presence of signs and symptoms;

Table 12.1 AcroQoL questionnaire scores in different languages from patients in remission (results are from references indicated in brackets)

	Dutch (20) n=118	German (23) n=33	Spanish (14) n=42	French (26) n=36
Global	68±17	68±17	64±20	60±18
Physical	64±21	67±20	60±27	64±16
Psychological	71±17	68±18	66±19	70±16
Appearance	63±22	51±16	56±23	63±20
Personal relations	78±15	73±20	77±19	78±15

thus, the impact of the disease and its treatment on the patients' HRQoL can be conceivably great. Using the SF-36 questionnaire in 36 patients with acromegaly, they were found to have a lower perceived HRQoL than general population in physical function dimensions, but no difference in the mental ones [13]. Successful surgery or response to medical treatment may be followed by marked improvement in the patient's overall health, often, but not always, accompanied by improvement or normalization of biochemical parameters such as GH and IGF-I. Whether HRQoL is also restored to normal levels was unknown until recently the availability of AcroQoL, a questionnaire specifically designed to evaluate the problems typical of acromegaly. AcroQoL has revealed that GH and IGF-I do not always correlate with subjective and clinical improvements experienced by patients and physicians after treatment [19].

The availability of this specific AcroQoL questionnaire has favored its evaluation in randomized clinical trials and clinical practice, contributing to the understanding of the specific impairments which occur in acromegaly [14]. It is currently available in 24 languages (Spanish, English, French, German, Italian, Greek, Dutch, Swedish, Hungarian, Polish, Turkish, Portuguese, Arab, Bulgarian, Czech, Danish, Farsi, Hebrew, Korean, Norwegian, Mandarin Chinese, Russian, Taiwanese and Slovak), with further linguistic adaptations for a further six languages like English for the US, French and Flemish for Belgium, Portuguese for Brazil, and Spanish for Argentina and Mexico. These translations have found similar assessment scores in all languages, thus validating its use in multi-country studies. These evaluations have found lower scores in active disease than in remission, with appearance being the most affected dimension and the personal relations area the least affected [20–25] (Table 12.1).

The AcroQoL questionnaire evaluates physical and psychological dimensions; the latter subdivided into appearance and personal relations domains. The measurement properties of this questionnaire (i.e., validity, reliability, and sensitivity to change) have been demonstrated in different studies and patients. Globally, AcroQoL scored worse in patients with active disease studied basally and 6 months after treatment, than in acromegalic patients with treated, stable disease, studied twice within 1 month. Appearance was the most affected subscale in acromegaly. Longitudinal retesting disclosed no change in the stable group, demonstrating good test-retest

Table 12.2 Features determining HRQoL in acromegaly

Feature	Worsens/Improves	References
Disease activity	Worsens	[14, 22–26]
Disease duration	Worsens	[20, 21]
Older age	Worsens	[9, 20, 21, 62]
Female gender	Worsens	[9]
Musculoskeletal pain	Worsens	[21, 27]
Coexisting diabetes mellitus	Worsens	[62]
Higher BMI	Worsens	[62]
Neurosurgery (as compared to medical therapy only)	Improves	[26]
Adding Pegvisomant to patients controlled on Somatostatin analogs, despite normal IGF-I	Improves	[19]
Radiotherapy	Worsens	[31, 62]
Becoming GH-deficient after treatment	Worsens	[62–64]

reliability, while in the active group after 6 months of treatment, there was improvement in AcroQoL score [25]. No correlation between AcroQoL and GH or IGF-I was observed, although there was a near significant trend between the subscale of appearance (the most affected dimension of AcroQoL) and IGF-I ($p=0.051$). These results support the idea that the patients' perception of his/her HRQoL cannot be inferred directly from hormone values.

Experience using the AcroQoL questionnaire in clinical practice in the UK confirmed severe impairment HRQoL in 80 patients with acromegaly [22], using two generic questionnaires (PGWS, EQ-5D), a symptom and signs score and AcroQoL. AcroQoL scores were correlated with GH and IGF-I. Acromegalic patients scored worse on the PGWBS as compared to the general population and in patients treated for a nonfunctioning pituitary adenoma, mainly in the domains of general health and vitality, and similarly bad or worse than in adults with GH deficiency. The correlations observed between different dimensions of the generic questionnaires and AcroQoL indicate that the dimensions evaluated by AcroQoL are those which impact HRQoL.

HRQoL has been shown to be severely impaired in a large group ($n=118$) of acromegalic patients after an average of 12 years in remission from The Netherlands, using generic questionnaires (SF-36 and NHP), for which normative data from Dutch general population are available. Disease duration, age, treatment with radiotherapy, and presence of joint problems were negatively correlated with the AcroQoL scores [20, 21] (Table 12.2). Endocrine control was not associated with normalization of HRQoL; only the appearance dimension (which was the most impaired) was weakly but significantly correlated with circulating GH levels ($p<0.05$).

A recent double-blind, placebo-controlled, randomized, cross-over clinical study investigated the effect of combining somatostatin analogs and pegvisomant on HRQoL in acromegaly [19]. Twenty acromegalic patients with normal IGF-I receiving somatostatin analog treatment were studied; pegvisomant or placebo was added for two consecutive treatment periods of 16 weeks, separated by a washout period of 4 weeks to 20 acromegalic patients. The AcroQoL questionnaire and the patient-assessed acromegaly symptom questionnaire (PASQ) were used to assess HRQoL. After the addition of pegvisomant, HRQoL improved significantly, despite no significant change of IGF-I. Since evaluation of perceived HRQoL and clinical improvement with these questionnaires seemed more sensitive than IGF-I, the current recommendations on assessment of disease activity in acromegaly using GH and IGF-I are questioned, using this combined somatostatin analog and pegvisomant therapy. Furthermore, the authors highlight the importance of including HRQoL assessment in daily practice, as its improvement may not be correlated with a reduction in IGF-I values.

A recent French study of 93 acromegalic patients comprising 36 with controlled and 57 with active disease [26] showed no statistical difference between groups for the total AcroQoL score. However, the psychological subscale appearance was better in cured (63 ± 20%) than active patients (58 ± 17%, $p = 0.035$). In patients with active disease, the appearance subscale was better in patients after surgery (65 ± 18%) than after medical treatment (54 ± 14%, $p = 0.009$), when IGF-I was also less elevated (410 ± 225 ng/mL vs. 588 ± 353, $p < 0.038$). A weak inverse correlation was found between the psychological subscale appearance score of the AcroQoL questionnaire and IGF-I value ($r = -0.22$, $p = 0.039$, $n = 89$). Together these results suggest that control of GH/IGF-I excess in acromegalic patients mainly influences the psychological subscale "appearance" of AcroQoL; furthermore, neurosurgery despite not always attaining an endocrine cure is associated with a greater improvement in HRQoL when compared to medical therapy alone (Table 12.2).

A cross-sectional study from the UK approached the impact of musculoskeletal problems on HRQoL in long-standing acromegaly [27]. Fifty-eight patients with acromegaly diagnosed at least 5 years before underwent a careful rheumatologic assessment and completed the SF-36, the AIMS-2 (arthritis impact measurement scales 2), and the AcroQoL questionnaires. The SF-36 scores in these acromegalic patients were worse than those of normal population for physical and social functioning, role physical, energy/vitality, pain, and general health; when comparing results to those obtained in patients with nonfunctioning pituitary tumors, physical and social functioning, pain, and general health scored worse than in the acromegalic patients. Ninety percent of the patients reported current pain at evaluation; when comparing those with and without pain, however, only three of the eight domains of SF-36 and 1 of the 12 items evaluated by AIMS2 differed between groups, while all the AcroQoL scores (global, physical, and psychological) were significantly worse in those referring pain. Given the relevance of musculoskeletal pain on QoL, these authors state that patients with acromegaly should be asked about these problems and offered advice on analgesia, physiotherapy, occupational therapy, foot wear, and weight loss. Awareness by the clinician of the strong impact of musculoskeletal symptoms on HRQoL impairment will favor an active approach to allow improvement. Interestingly, only 66% of the patients recalling musculoskeletal symptoms at diagnosis had this information in the medical files, highlighting how often these important issues for perceived HRQoL are overlooked.

Psychiatric morbidity associated with dysfunction and poorer HRQoL present in over one third of acromegalic patients has been described in India. Psychosocial morbidity in 17 patients with acromegaly was compared with 17 matched controls [28]. A complete profile including quality of life, sociodemographic and clinical profile, life events, social support, coping, dysfunction, and social morbidity were assessed. Psychiatric morbidity was found in 6 of the 17 patients and was related to a greater number of life events in the entire lifetime, more number of emotional coping strategies, more dysfunction, and poorer QoL (assessed by the QoL scale Bref WHOQOL-B).

Since personality traits may impact HRQoL, these were investigated in acromegaly in a German study [29]. Seventy acromegalic patients were compared to 58 nonfunctioning pituitary adenomas and 140 controls, using standardized personality questionnaires (the short German version of the Eysenck questionnaire EPQ-RK, and the tridimensional personality questionnaire TPQ). When compared to healthy controls, acromegalic patients showed specific personality features with increased anxiety-related traits. Patients described themselves as more harm avoidant and neurotic and showed a high social conformity. This pattern was also found in nonfunctioning pituitary adenomas; but acromegalic patients also showed a reduction in the novelty-seeking behavior, especially in terms of a decreased impulsiveness, not seen in nonfunctioning pituitary adenomas. These personality features represent a pattern related to increased anxiety, which may affect important areas as patients' QoL, treatment adherence, and patient–doctor contact.

A global comparison of HRQoL in different pituitary tumors has been performed by determining *Z*-scores (or standard deviation scores) for different questionnaires (HADS: the Hospital Anxiety and Depression Scale; MFI-20: Multidimension Fatigue Inventory; NHP: Nottingham Health Profile; SF-36: Short Form Health Survey). In this way, differences in age and gender (two determinants of HRQoL) were accounted for and comparisons with reference populations were possible [30]. Total HRQoL score and all the subscales of the questionnaires were worse in acromegalic patients compared to controls, demonstrating impairment of HRQoL during long-term follow-up after treatment. More impairment for physical ability and functioning and more bodily pain were seen than in patients treated for nonfunctioning pituitary adenomas (NFPT) or prolactinomas, while in Cushing's disease greater anxiety determined more impairment in physical functioning than those with NFPT. Hypopituitarism further impaired multiple aspects of HRQoL. Awareness of these persistent adverse effects of pituitary disease on HRQoL favors discussion with the patient, thus preventing inappropriate expectations with respect to the long-term results of treatment. These same authors have also shown how previous radiotherapy (performed in 33% of patients) was the predominant indicator of progressive impairment in QoL, in a subtle but progressive way during 4 years of follow-up in patients cured from acromegaly [31], using both generic (SF-36, MMFI, HADS, and NHP) and the AcroQoL questionnaire. Specifically, radiotherapy negatively influenced energy, pain and social isolation (NHP), physical fatigue and reduction in activity and motivation (MFI-20), depression and total anxiety and depression scores (HADS), and physical performance (AcroQoL). During follow-up, scores in 5 of

26 HRQoL subscales significantly worsened (namely, physical and social functioning of the SF-36, physical fatigue of the MFI-20, and psychological well-being, and personal relations of the AcroQoL questionnaire).

Thus, the availability of a disease-generated questionnaire as AcroQoL has provided strong evidence confirming that HRQoL is impaired in acromegaly, especially if the disease is active, if medical therapy (not surgery) is provided (with greatest impact on the appearance dimension) and if musculoskeletal symptoms (mainly pain) are present, globally and for each subscale. Patients with acromegaly experience increased anxiety (as the nonfunctioning pituitary adenomas), but specifically also lead to a reduction in impulsiveness, which may impact patients' HRQoL. The physical dimensions of the AcroQoL questionnaire were more sensitive than circulating IGF-I to detect patient's improvement after adding pegvisomant to somatostatin analog treatment in patients who at baseline were considered "controlled," since their IGF-I was within the normal range.

Quality of Life in Adult Patients with Growth Hormone Deficiency

Impaired QoL in adult patients with growth hormone deficiency (AGHD) was observed in the initial randomized, placebo-controlled clinical trials performed two decades ago [32–34], but had been suggested in 1962 by Raben [35]. A number of generic questionnaires designed to cover domains related to perception of health problems (NHP), psychiatric morbidity (PWBS and the General Health Questionnaire, GHQ), and life events in the preceding month which might bias current mood (Life events inventor, LEI) were initially included in clinical trials [33, 36]. AGHD patients were found to have worse scores than matched controls for both the NHP and PGWS at baseline, showing psychological compromise, to an extent that 40% would have justified a psychiatric referral. Improvements in energy and emotional reactions tend to occur within the first 6 months (but may even occur later) and are still present after several years, so that treated patients do not differ from matched controls anymore, while untreated patients show no improvement in HRQoL [12, 37–43]. The majority of adult patients with severe GHD who have experienced the benefits of rhGH choose to continue on daily s.c. injections, which is in sharp contrast with diabetic patients who often ask if they could not change insulin for oral antidiabetic drugs. Together, these findings argue strongly against the idea that improvement of HRQoL constitutes a placebo-effect of rhGH replacement therapy and supports the concepts that GH replacement should be considered standard practice, as with other hormone deficiencies of hypopituitarism [44, 45].

Since AGHD patients do not always score poorly on HRQoL questionnaires, skeptical physicians have expressed doubt on the relevance of HRQoL as an endpoint for the therapeutic benefits of GH replacement. However, whenever AGHD cohorts have been compared with general populations, the former have always scored significantly worse [43], and GH-replaced AGHD patients maintain improved

HRQoL, while untreated subjects show a gradual decline over the subsequent years, especially for vitality [46–48]. The degree of improvement in HRQoL with GH replacement has been shown to be proportional to the baseline deviation from normality. This explains that in some randomized, double-blind placebo-controlled studies unselected at study entry in terms of HRQoL, no changes in HRQoL or cognitive function were observed after 18 months of GH therapy [49]. Notably, only generic questionnaires were used in those studies that showed no improvement [50]. This is not really surprising since patients with normal QoL at outset can hardly be expected to achieve any improvement during substitution therapy. When instruments specifically developed for AGHD patients have been used to assess HRQoL, consistent improvement with GH replacement therapy have been reported, mainly in the dimension of vitality [43].

The validated AGHDA score includes dimensions related to dislike of body image, low energy, poor concentration and memory, and increased irritability. Normative data are available for general population in several countries (including Spain, Sweden, England and Wales, and The Netherlands), where AGHD patients were found to score worse (except for patients aged over 60 years) and to improve after starting substitution therapy with rhGH, which lasted for several years of therapy [46, 51–53]. Its initial definition as "disease-specific" has been criticized, since it was also found to identify impaired QoL in acromegaly [54]. A more appropriate denomination would probably be "disease- generated" or "disease-orientated" questionnaire, even though it will reflect the patients' global situation, i.e., health dimensions (including, symptoms, deficiencies, or treatments), but also professional, hobby, humor, libido, and other related areas. It should also be remembered that extreme situations like GHD and acromegaly can both share common symptoms (i.e., lack of energy, dislike of body image, etc.), so no HRQoL instrument may provide absolute discrimination between different patients or control groups, making individual clinical assessment of paramount importance.

A recent pharmaco-economic approach aimed at obtaining a cost-utility analysis of GH substitution therapy in adults [55]. This was done by converting AGHDA scores into utility scores gathered form the EQ-5D questionnaire and examining the impact of demographic and clinical characteristics of patients with AGHD. These authors also confirm that a greater number of pituitary deficits as well as adult-onset versus childhood-onset GHD determined a lower QoL score.

In all groups of adults with GHD, an improvement from baseline was observed during follow-up, mainly the first year, despite impact of age, primary cause of GHD, disease onset in childhood or adult life, and comorbidities. Furthermore, those dimensions which are least impaired (like problems with socializing and tenseness) tend to normalize earlier than self-confidence and tiredness, while the most affected dimension (memory and concentration) is the last in normalizing [56].

Data from the KIMS database, a postmarketing surveillance study of AGHD patients, have shown improvement of HRQoL using the AGHDA questionnaire in both irradiated and nonirradiated hypopituitary patients due to pituitary tumors or craniopharyngioma, 1 and 2 years after rhGH therapy. Interestingly, those who had received radiotherapy scored significantly worse than those who had not been irradiated, both at baseline and during follow-up [57]. Further results from the KIMS database have shown how improvements in HRQoL correlate with a reduction in the use of healthcare resources (visits to doctor, days in hospital, sick leave, and assistance with daily activities) and improved physical activity and satisfaction with physical activity during leisure time in patients with previously untreated AGHD after 12 months of rhGH [56–58].

The QLS-H questionnaire has the advantage of weighting each item by the individual patient; patients are first asked how important each item is to them and after how satisfied they are with each item included in this QLS-H questionnaire. Translated into several languages, normative data are available in several European countries and the USA [11]. It was used to evaluate 66 patients with severe childhood-onset GHD in the transition phase to adulthood. QLS-H scores for ability to become sexually aroused, ability to tolerate stress, body shape, concentration, initiative/drive, physical stamina and self-confidence were lower than the normal average. However, overall baseline HRQoL was not compromised; this is probably due to the adaptation phenomenon, namely that the patients with a disability only take up activities/challenges that are within their own capabilities. In other words, to maintain a good HRQoL, they approach less demanding aspirations, in accordance with perceived limitations. This adaptation is more common in childhood-onset GHD than in adult-onset patients. Dimensions related to age-specific psychological problems (i.e., sexual arousal and body shape) responded positively to GH replacement [59, 60].

The mechanism by which AGHD determines QoL impairment is probably multifactorial, in parallel to the widespread physiological actions of GH; they include abnormal body composition, decreased exercise capacity and muscle strength, decreased body water which has been related to fatigue, metabolic disturbances, and possible neuroendocrine effects in the central nervous system. This is further supported by the lack of definite correlations between improvements in QoL and biochemical, metabolic, or body composition parameters in treated AGHD.

Of interest is that in adults, no loss of improved QoL or clinical benefit has been observed in recent years despite a reduction of GH dose to around half or a third of that initially prescribed, an adjustment that significantly reduced side effects like edema and arthralgia [61].

HRQoL in Acromegalic Patients who Become GH-Deficient After Treatment

A U-shaped curve of HRQoL in acromegaly was pointed out in a study from Finland, where it was observed to be better in those patients with normal GH (i.e., between 0.3 and 1 μg/L), than if GH was higher reflecting active disease, or lower indicating that these patients had become GHD [62]. A number of groups have used the AGHDA score in patients who are currently GH-deficient due to prior treatment of acromegaly with surgery and/or radiotherapy. In 45 patients with controlled disease (26 GH-deficient and 19 GH-sufficient), HRQoL was measured by the AGHDA score, the SF-36, and a symptom questionnaire. Significant correlations between the GH level after GHRH-arginine stimulation and HRQoL were found, in the sense that a lower GH response was associated with a lower HRQoL score [63]. In a smaller study of 16 patients from the Netherlands, HRQoL and other efficacy parameters (cardiac function, body composition, bone mineral density, lipids, glucose and bone turnover markers) evaluated with the AGHDA, MMFI, HADS, and NHP questionnaires did not change after 1 year of rhGH therapy [64]. However, the mean age of these patients was 56 years, and it is known that improvements of HRQoL are smaller in older patients; furthermore, no attempt to evaluate HRQoL of these patients with an acromegaly-specific questionnaire has been reported. Anecdotally, two women in their late 30s, treated for acromegaly in their early 20s who became GH-deficient, have repeatedly manifested to us that experiencing GH deficiency was much worse than having GH excess and did experience significant improvements in energy and general well-being after starting GH substitution therapy. A recent study has reported benefits of rhGH replacement on the HRQoL in acromegalic patients with a mean age of 46 years, rendered GH deficiency after treatment [65].

Summary

HRQoL is emerging as an important measure to assess the outcome of clinical management and treatment of both acromegaly and adult GH deficiency. Acromegaly severely impairs HRQoL; successful therapy is associated with improvement, but disease control does not normalize scores completely. The availability of a questionnaire specifically designed to reflect the problems experienced by these patients in over 20 languages has broadened knowledge on how acromegaly impacts patients' well-being. Endocrine control parameters like GH and IGF-I are not always correlated with the HRQoL scores. The most affected dimension is appearance. Worse HRQoL is associated with disease activity, musculoskeletal pain, radiotherapy, older age, becoming GH-deficient after treatment for acromegaly, female gender, disease duration, coexisting diabetes mellitus, and a higher BMI, while neurosurgery – compared to medical therapy alone and adding Pegvisomant to patients with a normal IGF-I on somatostatin analogs – improves HRQoL.

Adult patients with GHD experience impaired HRQoL, especially if the deficiency appeared in adult life rather than present since childhood (when they usually adapt to their disability). Any cause of GHD appears to impair HRQoL similarly. Improvement of impaired HRQoL is invariably shown after substitution therapy with rhGH when evaluated with specific questionnaires, which persists over several years of treatment. Despite a reduction in the dose of rhGH used in recent years, associated with a reduction in side-effects, clinical benefit and improved HRQoL persist.

Further investigations in patients who after therapy for acromegaly become GH-deficient deserve attention to optimize the long-term management and prognosis of these patients.

References

1. Patrick DL, Erickson P. Health states and health policy: quality of life in health care evaluation and resource allocation. Oxford: Oxford University Press; 1993.
2. McDowell I, Newell C. Measuring health: a guide to rating scales and questionnaires. 2nd ed. Oxford: Oxford University Press; 1996.
3. Hunt SM, McKenna SP, McEwen J, et al. The Nottingham Health profile: subjective health status and medical consultations. Soc Sci Med. 1981;15A:221–9.
4. Gray LC, Goldsmith HF, Livieratos BB, et al. Individual and contextual social-status contributions to psychological well-being. Sociol Soc Res. 1983;68(1):78–95.
5. Dolan P. Modelling valuations for EuroQol health states. Med Care. 1997;35:1095–108.
6. Badia X, Herdman M, Schiaffino A. A determining correspondence between scores on the EQ-5D "thermometer" and a 5-point categorical rating scale. Med Care. 1999;37:671–7.
7. Brooks R. EuroQol: the current state of play. Health Policy. 1996;37:53–7.
8. Ware JE, Snow KK, Kosinski M, et al. SF-36 Health Survey. Manual and interpretation guide. Boston, MA: The Health Institute, New England Medical Center; 1993.
9. Webb SM, Prieto L, Badia X, et al. Acromegaly quality of life questionnaire (ACROQOL) a new health-related quality of life questionnaire for patients with acromegaly: development and psychometric properties. Clin Endocrinol. 2002;57:251–8.
10. McKenna SP, Doward LC, Alonso J, et al. The QoL.AGHDA: an instrument for the assessment of quality of life in adults with growth hormone deficiency. Qual Life Res. 1999;8:373–83.
11. Blum WF, Shavrikova EP, Edwards DJ, et al. Decreased quality of life in adult patients with growth hormone deficiency compared with general populations using the new, validated, self-weighted questionnaire, questions on life satisfaction hypopituitarism module. J Clin Endocrinol Metab. 2003;88(9):4158–67.
12. Rosilio M, Blum WF, Edwards DJ, et al. Long-term improvement of quality of life during growth hormone (GH) replacement therapy in adults with GH deficiency, as measured by QLS-H©. J Clin Endocrinol Metab. 2004;89:1684–93.
13. Johnson MD, Woodburn CJ, Vance ML. Quality of life in patients with a pituitary adenoma. Pituitary. 2003;6(2):81–7.
14. Webb SM, Badia X, Surinach NL. Validity and clinical applicability of the acromegaly quality of life questionnaire AcroQoL: a 6-month prospective study. Eur J Endocrinol. 2006;155:269–77.
15. Van Aken MO, Pereira AM, Biermasz NR, et al. Quality of life in patients after long-term biochemical cure of Cushing's disease. J Clin Endocrinol Metab. 2002;90:3279–86.
16. Santos A, Resmini E, Martínez MA, et al. Quality of life in patients with pituitary tumors. Curr Opin Endocrinol Diab. 2009;16:299–303.
17. Lindsay JR, Nansel T, Baid S, et al. Long-term impaired quality of life in Cushing's syndrome despite initial improvement after surgical remission. J Clin Endocrinol Metab. 2006;91:447–53.
18. Nielsen EH, Lindholm J, Laurberg P, et al. Nonfunctioning pituitary adenoma: incidence, causes of death and quality of life in relation to pituitary function. Pituitary. 2007;10:67–73.
19. Neggers SJ, van Aken MO, de Herder WW, et al. Quality of life in acromegalic patients during long-term somatostatin analog treatment with and without pegvisomant. J Clin Endocrinol Metab. 2008;93(10):3853–9.
20. Biermasz NK, Van Thiel SW, Pereira AM, et al. Decreased quality of life in patients with acromegaly despite long-term cure of growth hormone excess. J Clin Endocrinol Metab. 2004;89:5369–76.
21. Biermasz NK, Pereira AM, Smit JWA, et al. Morbidity after long-term remission for acromegaly: persisting joint-related complaints cause reduced quality of life. J Clin Endocrinol Metab. 2005;90:2731–9.
22. Rowles SV, Prieto L, Badia X, et al. Quality of life (QOL) in patients with acromegaly is severely impaired: use of a novel measure of QOL: acromegaly quality of life questionnaire. J Clin Endocrinol Metab. 2005;90:3337–41.

23. Trepp R, Everts R, Stettler C, et al. Assessment of quality of life in patients with uncontrolled versus controlled acromegaly using the acromegaly quality of life questionnaire (AcroQoL). Clin Endocrinol. 2005;63:103–10.
24. Deyneli O, Yavuz D, Gozu H, et al. Evaluation of quality of life in Turkish patients with acromegaly (abstract P3-508). In: Programs and abstracts of the 84th Annual Meeting of the Endocrine Society, Philadelphia, 2003.
25. Webb SM, Badia X, Lara-Surinach N, et al. Validity and clinical applicability of the acromegaly quality of life questionnaire AcroQoL: a six months prospective study. Eur J Endocrinol. 2006;155:269–77.
26. Matta MP, Couture E, Cazals L, et al. Impaired quality of life of patients with acromegaly: control of GH/IGF-I excess improves psychological subscale appearance. Eur J Endocrinol. 2008;158(3):305–10.
27. Miller A, Doll H, David J, Wass J. Impact of musculoskeletal disease on quality of life in long-standing acromegaly. Eur J Endocrinol. 2008;158(5):587–93.
28. Mattoo SK, Bhansaku AK, Gupta N, et al. Psychosocial morbidity in acromegaly: a study from India. Endocr. 2008;34(1–3):17–22.
29. Sievers C, Ising M, Pfister H, et al. Personality in patients with pituitary adenomas is characterized by increased anxiety related traits: comparison of 70 acromegalic patients to patients with non-functioning pituitary adenomas and age and gender matched controls. Eur J Endocrinol. 2009;160(3):367–73.
30. Van der Klaauw AA, Kars M, Biermasz NR, et al. Disease-specific impairments in quality of life during long-term follow-up of patients with different pituitary adenomas. Clin Endocrinol. 2008;69:775–84.
31. Van der Klaauw AA, Biermasz NR, Hoftijzer HC, et al. Previous radiotherapy negatively influences quality of life during 4 years of follow-up in patients cured from acromegaly. Clin Endocrinol (Oxf). 2008;69:123–8.
32. Degerblad M, Grunditz R, Hall K, et al. Substitution therapy with recombinant growth hormone (Somatrem) in adults with growth hormone deficiency. Acta Paediatr Scand. 1987;337(Suppl):170–1.
33. Salomon F, Cuneo R, Hesp R, et al. The effects of treatment with recombinant human growth hormone on body composition and metabolism in adults with growth hormone deficiency. N Engl J Med. 1989;321:1797–803.
34. Cuneo R, Salomon F, McGauley G, et al. The growth hormone deficiency syndrome in adults. Clin Endocrinol. 1992;37:387–97.
35. Raben MS. Clinical use of human growth hormone. N Engl J Med. 1962;266:82–6.
36. McGauley GA, Cuneo RC, Salomon F, et al. Psychological well-being before and after growth hormone treatment in adults with growth hormone deficiency. Horm Res. 1990;33 Suppl 4:52–4.
37. Wiren L, Bengtsson BA, Johannsson G. Beneficial effects of long-term GH replacement therapy on quality of life in adults with GH deficiency. Clin Endocrinol. 1998;48(5):613–20.
38. McMillan CV, Bradley C, Gibney J, et al. Psychological effects of withdrawal of growth hormone therapy from adults with growth hormone deficiency. Clin Endocrinol. 2003;59(4):467–75.
39. Malik IA, Foy P, Wallymahmed M, et al. Assessment of quality of life in adults receiving long-term growth hormone replacement compared to control subjects. Clin Endocrinol. 2003;59(1):75–81.
40. Hakkaart-van Roijen L, Beckers A, Stevenaert A, et al. The burden of illness of hypopituitary adults with growth hormone deficiency. Pharmacoeconomics. 1998;14(4):395–403.
41. Sonksen PH, McGauley G. Lies, damn lies and statistics. Growth Horm IGF Res. 2005;15:173–6.
42. Gibney J, Johannsson G. Clinical monitoring of growth hormone replacement in adults. Front Horm Res. 2005;33:86–102.
43. Woodhouse LJ, Mukherjee A, Shalet SM, et al. The influence of growth hormone status on physical impairments, functional limitations, and health-related quality of life in adults. Endocr Rev. 2006;27:287–317.

44. Consensus guidelines for the diagnosis and treatment of adults with growth hormone deficiency. Summary statement of the growth hormone research society workshop on adult GHD. J Clin Endocrinol Metab. 1998;83(2):379–81.
45. Molitch ME, Clemmons DR, Malozowski S, for The Endocrine Society's Clinical Guidelines Subcommittee, et al. Evaluation and treatment of adult growth hormone deficiency: an endocrine society clinical practice guideline. J Clin Endocrinol Metab. 2006;91(5):1621–34.
46. Badia X, Lucas A, Sanmarti A, et al. One-year follow-up of quality of life in adults with untreated growth hormone deficiency. Clin Endocrinol. 1998;49(6):765–71.
47. Gilchrist FJ, Murray RD, Shalet SM. The effect of long-term untreated growth hormone deficiency (GHD) and 9 years of GH replacement on the quality of life (QoL) of GH-deficient adults. Clin Endocrinol. 2002;57(3):363–70.
48. Abs R, Bengtsson BA, Hernberg-Stahl E, et al. GH replacement in 1034 growth hormone deficient hypopituitary adults: demographic and clinical characteristics, dosing and safety. Clin Endocrinol. 1999;50(6):703–13.
49. Baum HB, Katznelson L, Sherman JC, et al. Effects of physiological growth hormone (GH) therapy on cognition and quality of life in patients with adult-onset GH deficiency. J Clin Endocrinol Metab. 1998;83(9):3184–9.
50. Arwert LI, Deijen JB, Drent ML. Effects of growth hormone deficiency and growth hormone treatment on quality of life in growth hormone-deficient adults. Front Horm Res. 2005;33:196–208.
51. Koltowska-Häggström M, Mattsson AF, Monson JP, et al. Does long-term GH replacement therapy in hypopituitary adults with GH deficiency normalise quality of life? Eur J Endocrinol. 2006;155:109–19.
52. Saller B, Mattsson AF, Kann PH, et al. Healthcare utilization, quality of life and patient-reported outcomes during two years of GH replacement therapy in GH-deficient adults – comparison between Sweden, The Netherlands and Germany. Eur J Endocrinol. 2006;154:843–50.
53. Wiren L, Whalley D, McKenna S, et al. Application of a disease-specific, quality-of-life measure (QoL-AGHDA) in growth hormone-deficient adults and a random population sample in Sweden: validation of the measure by rasch analysis. Clin Endocrinol. 2000;52(2):143–52.
54. Barkan AL. The "quality of life-assessment of growth hormone deficiency in adults" questionnaire: can it be used to assess quality of life in hypopituitarism? J Clin Endocrinol Metab. 2001;86(5):1905–7.
55. Koltowska-Häggström M, Kind P, Monson JP, et al. Growth hormone (GH) replacement in hypopituitary adults with GH deficiency evaluated by a utility-weighted quality of life index: a precursor to cost–utility analysis. Clin Endocrinol. 2008;68:122–9.
56. Koltowska-Häggström M, Mattsson AF, Shalet SM. Assessment of quality of life in adult patients with GH deficiency: KIMS contribution to clinical practice and pharmacoeconomic evaluations. Eur J Endocrinol. 2009;161 Suppl 1:S51–64.
57. Maiter D, Abs R, Johannsson G, et al. Baseline and follow-up QoL in KIMS patients with pituitary adenoma or craniopharyngioma, irradiated or not. Eur J Endocrinol. 2006;155:253–60.
58. Hernberg-Ståhl E, Luger A, Abs R, et al. Healthcare consumption decreases in parallel with improvements in quality of life during GH replacement in hypopituitary adults with GH deficiency. J Clin Endocrinol Metab. 2001;86:5277–81.
59. Attanasio AF, Shavrikova EP, Blum WF, et al. Quality of life in childhood onset growth hormone-deficient patients in the transition phase from childhood to adulthood. J Clin Endocrinol Metab. 2005;90(8):4525–9.
60. Murray RD, Shalet SM. The use of self-rating questionnaires as a quantitative measure of quality of life in adult growth hormone deficiency. J Endocrinol Invest. 1999;22(5 Suppl):118–26.
61. Webb SM, Strasburger CJ, Mo M, on behalf of the HypoCCS International Advisory Board, et al. Changing patterns of the adult growth hormone deficiency diagnosis documented in a decade-long global surveillance database. J Clin Endocrinol Metab. 2009;94:392–9.
62. Kauppinen-Mäkelin R, Sane T, Sintonen H, et al. Quality of life in treated patients with acromegaly. J Clin Endocrinol Metab. 2006;91:3891–6.

63. Wexler T, Gunnell L, Omer Z, et al. Growth hormone deficiency is associated with decreased quality of life in patients with prior acromegaly. J Clin Endocrinol Metab. 2009;94:2471–7.
64. Van der Klaauw AA, Bax JJ, Roelfsema F, et al. Limited effects of growth hormone replacement in patients with GH deficiency during long-term cure of acromegaly. Pituitary. 2009;12(4):339–46.
65. Miller KK, Wexler T, Fazeli P, et al. Growth hormone deficiency after treatment of acromegaly: a randomized, placebo-controlled study of growth hormone replacement. J Clin Endocrinol Metab. 2010;95:567–77.

Part IV
Acromegaly

Chapter 13
The Value of GH and IGF-I Measurements in the Management of Acromegaly

Pamela U. Freda

Abstract Measurements of growth hormone (GH) and the GH-dependent peptide insulin-like growth factor-I (IGF-I) are essential for diagnosing and managing acromegaly. The IGF-I level can be easily measured as a single, random sample, and when done properly and compared to a well-characterized, age-adjusted normative database, elevation of the serum IGF-I level is a sensitive and specific indicator for the presence of acromegaly or persistent disease after therapy. The most common approach to GH assessment is to assess the degree of GH suppression after oral glucose administration (oral glucose tolerance test [OGTT]). Failure of GH to fall into the range expected for the healthy population, along with an elevated IGF-I, is confirmatory of active acromegaly. To distinguish active acromegaly from remission, an OGTT nadir GH cut-off of 1 μg/L has been found to be reliable for use with some GH assays, but this cut-off may be as low as 0.3 μg/L with others. Because GH assays are heterogeneous, uniform, clinically relevant GH criteria for acromegaly are difficult to establish. Caveats exist to the testing of GH or IGF-I and therefore the validity of reliance on the measurement of either of these alone remains controversial. In some settings, the acromegaly patient is monitored by the IGF-I level alone, and in others, a combined assessment is preferred, although discrepant results are not uncommon. This chapter reviews our current understanding of the value of GH and IGF-I measurements in the diagnosis and monitoring of acromegaly during therapy.

Keywords Acromegaly • Growth hormone • IGF-I

P.U. Freda (✉)
Department of Medicine, Columbia University, College of Physicians and Surgeons, 650 West 168th Street, 9-905, New York, NY 10032, USA
e-mail: pufl@columbia.edu

K. Ho (ed.), *Growth Hormone Related Diseases and Therapy: A Molecular and Physiological Perspective for the Clinician*, Contemporary Endocrinology,
DOI 10.1007/978-1-60761-317-6_13,

Introduction

Acromegaly is a rare and underrecognized disease that originates in nearly all cases in a growth hormone (GH)-producing tumor of the pituitary gland. Once suspected, which typically occurs only after the development of its multisystem clinical manifestations [1], the diagnosis is pursued by investigating for the presence of GH hypersecretion. A proper biochemical evaluation is therefore essential. The approaches taken to this evaluation vary. In the past, the evaluation rested only on GH values, but the availability of reliable assays for the measurement of circulating levels of the GH-dependent peptide insulin-like growth factor 1 (IGF-I) has radically altered the way in which acromegaly is diagnosed and monitored. In addition, advances in GH assay methodologies have changed our understanding of what constitutes GH excess. In most cases, for diagnosis and monitoring during therapy, a dynamic test of GH secretion, GH suppression after oral glucose, as well as measurement of a serum IGF-I level are performed. However, the results of the two tests can be discrepant and their interpretation may not be straightforward. Controversies still exist about the relative roles for measurements of these two hormones. This chapter reviews our current approaches to the biochemical diagnosis and monitoring of acromegaly.

Biochemical Diagnosis of Acromegaly

Growth Hormone

Measurement of GH is essential for the diagnosis of acromegaly. The normal physiology of GH secretion as well as its pathophysiology in acromegaly needs to be appreciated in order to understand the principles and pitfalls of testing for GH excess. In healthy humans, GH is secreted in a pulsatile pattern that is under neuroregulatory control [2]. In acromegaly, however, GH neuro-regulation is disordered and the pituitary tumor secretes GH in an abnormal pattern with high baseline levels and blunted pulses. In between pulses, GH falls below the detection limits of most GH assays in healthy persons, but in acromegaly GH levels do not, which results in persistent overall GH excess. The presence of dysregulation is exploited during dynamic tests of GH secretion such as the oral glucose tolerance test (OGTT). A number of approaches to assessment for GH excess can be taken. GH is most often measured as single or mean of random circulating GH levels or in the setting of a dynamic test of the GH axis. The pros and cons of each of these are discussed in the sections to follow.

Regardless of the approach taken to GH testing, it is important to keep in mind the limitations of the GH assay itself and that these limitations impact the values reported as well as how we interpret them. GH assays have become increasingly sensitive and specific, evolving in recent years from polyclonal radioimmunoassays to immunoradiometric and then immunofluorescence assays. As a result, current assays can quantitate GH at much lower levels. This has revealed that GH values that were once considered normal can be found in some patients with active acromegaly. With increased assay precision has also come an expectation that we should be able to differentiate GH levels within a tenth of a μg/L around cut-offs. However, whether such precision is truly possible with current GH assay methodologies or clinically relevant are not known. In addition, GH assay methodologies can vary considerably and different methods can yield very different results. This heterogeneity impedes our ability to develop uniform, clinically relevant GH criteria for acromegaly. Current guidelines call for GH values to be interpreted in conjunction with those of IGF-I. The value of serum IGF-I measurements is discussed in a subsequent section.

Random GH Measurements

Random GH levels lack the specificity necessary to secure the diagnosis or to exclude it [3]. Random GH levels are persistently very high in many newly diagnosed patients, but are often between 2 and 10 μg/L and can overlap with the range of pulsatile GH secretion in healthy subjects [3, 4]. Even if the GH level is frankly elevated, an IGF-I level and in most cases also a dynamic test should be performed. Random GH levels can also be elevated in poorly controlled diabetes mellitus, renal failure, malnutrition as well as in the setting of stress or during exercise and sleep [2]. Single random GH levels below a specific level cannot reliably exclude active acromegaly as they can be <1 μg/L in some newly diagnosed patients [5, 6]. It has been proposed that a random GH <0.4 μg/L along with a normal IGF-I level excludes the diagnosis [7]. The validity of a random GH <0.4 mg/L alone has not been established as some newly diagnosed patients have spontaneous GH levels below this value [6]. Rather than a random sample, some centers advocate use of mean values obtained from multiple samples (day series or curve) for the evaluation of acromegaly, but this testing is usually done in monitoring during therapy (see below). Rather than a random GH level, dynamic testing should be used to determine the presence of GH excess.

Oral Glucose Tolerance Test (OGTT)

Measurement of GH suppression after oral glucose (OGTT) administration is the dynamic test most often performed to diagnose acromegaly. Acute plasma glucose elevation suppresses pituitary GH secretion in healthy humans, but impaired GH suppression, and often a paradoxical rise, is a characteristic of active acromegaly [8, 9]. After an overnight fast, GH is measured before and serially over a 2-h period after ingestion of a 100 or 75 g glucose drink and the nadir GH achieved is compared to a specific cut-off. Considerable work has been done in an attempt to determine the specific cut-off above which active acromegaly can be confirmed and below which it is excluded.

The criteria for "normal" GH suppression after oral glucose have evolved in recent years in parallel with the availability of more sensitive and specific GH assays. Fifteen years ago, a nadir GH level <2.0 μg/L, as measured by a polyclonal radio-immunoassay (RIA), was considered normal. However, it is now clear that RIA assays did not detect the full extent of normal GH suppression and that criteria developed with them are no longer valid. With assays utilizing monoclonal antibodies, it became evident that normal levels of GH after glucose are much less than 1 μg/L [10, 11]. In two studies, after 100 g oral glucose, mean nadir GH levels were 0.25 μg/L in young women and 0.029 μg/L in young men [10] and 0.09 in females and 0.08 μg/L in males [12]. In other healthy subjects, after a 75 g OGTT, mean nadir GH levels were 0.15 μg/L [9] and 0.07 μg/L [13]. However, large interindividual variation in nadir GH levels in healthy subjects exists, especially in young woman whose levels may be much higher, making it difficult to develop generally applicable "normal" ranges [12, 14, 15]. One source of this variability may be differences in assay standards, which have varied from older polyclonal to the latest 22K rhGH ones [6, 9–11, 13, 14]. Other assay characteristics are also variable. When OGTT samples from a group of healthy subjects were measured with three different assays calibrated to the rhGH standard (98/574), mean nadir GH concentrations varied from 0.13 μg/L (Immulite 2000) to 0.06 μg/L (Nichols Advantage) and to 0.015 μg/L (DSL Elisa) and ranged from 0.00066 to 0.99 μg/L with these three assays [14]. Thus, it is clear that an optimal cut-off applicable to all laboratories is difficult to establish because GH assays vary.

A variety of cut-offs for nadir GH that can exclude or diagnose acromegaly have been proposed based on the data available from healthy subjects and some newly diagnosed patients. In general, nadir GH values are to be interpreted along with a serum IGF-I level. Some guidelines [7] and research studies propose that nadir GH levels below 1 μg/L can exclude acromegaly [13, 14, 16]. However, other studies demonstrate that when measured by highly sensitive and specific assays, nadir GH levels can be less than 1 μg/L in newly diagnosed patients [6, 12]. With some assays, GH levels below 0.5 μg/L [14] exclude acromegaly. Exactly where below 1 μg/L the true cut-off should be cannot be uniformly agreed upon because the data vary depending upon the GH assay used [9, 12, 13]. The specificity for acromegaly of a GH above one of these cuts-offs is also unclear. An elevated IGF-I value along with failure of GH suppression is consistent with acromegaly, but whether failure of GH suppression alone based on one of these new cut-offs is sufficient evidence of the diagnosis has not yet been established. Although further refinements to these cut-offs are available from data in postoperative patients, it is unclear if dynamic testing criteria developed in the postoperative setting, where GH secretory dynamics may be perturbed and discrepancies between GH and IGF-I levels are not uncommon, can be extrapolated to evaluating untreated patients. Additional work using highly sensitive and specific GH assays in newly diagnosed acromegaly is needed.

The interpretation of GH values after oral glucose should also consider that abnormal suppression can occur in chronic renal insufficiency, liver failure, active hepatitis, hyperthyroidism, diabetes mellitus, anorexia nervosa, and other forms of malnutrition [3, 17, 18] (Table 13.1). Abnormal GH suppression in these settings,

Table 13.1 Conditions other than acromegaly that can influence GH And IGF-I values

Growth hormone
Failure of normal GH suppression
Chronic renal insufficiency
Liver failure
Active hepatitis
Anorexia nervosa
Malnutrition
Hyperthyroidism
Diabetes mellitus
Adolescence
Insulin-like growth factor-I
Lowering of IGF-I level
Nutrient deprivation, malnutrition
Anorexia nervosa
Liver disease
Hypothyroidism
Poorly controlled insulin-dependent diabetes mellitus
Oral estrogen use
Elevation of IGF-I Level
Adolescence
Pregnancy
Hyperthyroidism (mild elevation)
IGF-I assay problems

however, should not be associated with a persistently elevated IGF-I level, and in many of these conditions, IGF-I levels are likely to be low. GH suppression cut-offs may also need to be higher in adolescents than in adults [19, 20]. Interpretation of nadir GH values may also need to consider the patient's age, body mass index (BMI), and gender. As age increases GH secretion decreases and a negative relationship between nadir GH levels and increasing age, in particular in women, has been demonstrated [2, 13, 14, 21, 22]. Recent data also suggest a negative relationship between mass index (BMI) and GH suppression [14, 21, 23]. BMI-specific GH cut-offs have yet to be developed. Gender-specific criteria for glucose-suppressed GH levels, particularly in young women, may also be needed. Nadir GH levels were found to be higher in healthy women than men in some [10, 13, 24], but not in other studies [12, 23]. More detailed gender, age, and BMI-specific GH cut-offs should be developed.

Other Diagnostic Tests

Acromegaly has also been investigated using a number of hormonal stimulation tests that produce paradoxical responses in these patients. However, these lack specificity and do not provide an advantage over OGTT suppression of GH and IGF-I measurements [71]. With TRH stimulation testing, a paradoxical >50% rise of GH after TRH administration is found in only about 50% of patients with acromegaly [3, 4]. Persistent postoperative paradoxical GH stimulation to TRH may be predictive of disease recurrence [25]. In response to GHRH, GH stimulation occurs in about 80% of patients, but the testing is not useful in differentiating patients with acromegaly from normal subjects [3]. Frequent GH measurements over a 24-h period can yield valuable information about GH secretion when done in the research setting, but in most patients with acromegaly this testing is neither practical nor necessary.

Insulin-Like Growth Factor-I (IGF-I)

Measurement of the serum IGF-I level is undertaken in all patients under evaluation for acromegaly. IGF-I, a GH-regulated peptide produced predominantly in the liver, circulates complexed to two GH-regulated carrier proteins, IGFBP-3 and acid-labile subunit (ALS), which play a role in IGF-I regulation and extend its half-life [26, 27]. IGF-I mediates the majority of growth-promoting and anabolic actions of GH [26–29]. Serum concentrations of total IGF-I are relatively stable over a 24-h period in healthy humans [28] and can be measured as a random sample.

Documentation of an elevated IGF-I level is necessary for the diagnosis of acromegaly. IGF-I elevation is a specific indicator of GH excess. In acromegaly, increasing tumoral GH secretion increases IGF-I levels which plateau when the 24-h mean GH level is above ~20 μg/L [30–34]. IGF-I levels are elevated in acromegaly to a range that is distinct from that in healthy subjects [35, 36] and normalize in those who are successfully treated [35–38]. IGF-I elevation has proven to be a sensitive marker for GH excess as IGF-I elevation has revealed mild GH hypersecretion at GH levels once considered "normal" [6, 11].

Accurate interpretation of IGF-I levels should consider important physiological regulators of circulating IGF-I levels including nutritional status, age, gender, and the IGF-binding proteins [39]. Nutrient deprivation of different etiologies, including protein and calorie malnutrition and anorexia nervosa, produces a resistance to GH action, lowering IGF-I production [29, 40, 41] (Table 13.1). Prolonged fasting can lower IGF-I levels in healthy subjects by up to 50% and refeeding returns them to baseline [28, 40, 42]. Circulating IGF-I levels are also determined by age. IGF-I levels peak in the second stage of puberty and decline during adulthood [43]. IGF-I levels can be elevated above the "normal" range in some adolescents without acromegaly [44]. Comparison to a well-characterized, age-adjusted normative database is crucial to the proper interpretation of IGF-I [43, 45]. A rise in IGF-I levels occurs in normal pregnancy, so during and shortly after this, they are not a reliable marker for acromegaly [28]. Woman secrete more GH than men to achieve similar IGF-I levels because of a relative resistance to GH-stimulated IGF-I production [46–49], but the increase in GH secretion in women compensates for this, so distinct IGF-I ranges in female and male adults have generally not been demonstrated [43, 48, 50–53].

IGF-I level interpretation should also consider the potential influence of chronic and critical illness, including thyroid disease, renal failure, and severe liver disease on IGF-I levels [28]. In hypothyroidism serum, IGF-I levels can be somewhat lowered and in hyperthyroidism they can be raised, but should correct to baseline with correction of the thyroid abnormality [28, 54]. Total IGF-I is usually normal [55], but may be low [56] in chronic renal failure. Liver failure lowers serum IGF-I levels [57]. Poorly controlled insulin-dependent diabetes mellitus can be associated with a reduction in IGF-I levels [40, 58], but in patients with newly diagnosed acromegaly and diabetes mellitus, IGF-I levels should still be elevated.

Interpretation of IGF-I levels should also consider IGF-I assay methodology [59, 60]. For proper measurement, IGF-I must be removed from its binding proteins by an extraction process or equivalent blocking procedure [59]. However, IGF-I assays are not equal in the quality of binding protein removal, inter- and intraassay variability, and other characteristics.

Despite these caveats to its use, IGF-I remains a very sensitive and specific marker for diagnosis.

IGF Binding Protein 3 (IGF-BP3) and Acid-Labile Subunit (ALS)

Measurements of the levels of IGFBP-3, the principal IGF-I binding protein, or the glycoprotein ALS, the third member of the IGF-I-IGFBP-3 complex, have been proposed as adjunctive tests for determination of disease status in acromegaly [61, 62]. Production of IGFBP-3 and ALS are GH-dependent [63]. Levels of IGFBP-3 and ALS are affected by nutritional status, liver and renal failure, and age similarly to those of IGF-I, generally remain fairly constant over a day, and typically change in parallel with those of IGF-I [29, 57, 63–66]. IGFBP-3 levels are elevated in most patients with active acromegaly and normalize with successful treatment [61, 64, 67, 68], but can overlap in patients with active acromegaly and healthy subjects [11, 65, 68–70], so in general IGFBP-3 measurements do not provide an advantage over IGF-I levels for the routine monitoring of disease status. ALS levels are elevated in 80–88% of patients with active acromegaly [63, 71–73] and fall with successful therapy [48, 62]. However, ALS assay methodologies and their accuracy vary and more validation of these assays will be required for their clinical use [48, 62, 71].

Biochemical Goals of Acromegaly Management

The biochemical goals of acromegaly management are based on GH and IGF-I normative data as well as on data demonstrating that normalization of these hormones will improve or normalize the signs, symptoms, comorbidities, and excess mortality of acromegaly. Goals for treatment have become increasingly stringent as GH and IGF-I assay methodologies have improved and as increasing data are available linking tight biochemical control to normalization of the morbidities and excess mortality of acromegaly. However, it has also become clear that our efforts to make these goals uniformly applicable are hampered by the heterogeneity that still exists in assay methodologies and thus the variability in the criteria that are derived from different centers. Therefore, what constitutes sufficient "normalization" of these hormones is still debated. Nevertheless, some reasonable guidelines can be provided for the GH and IGF-I levels that should be sought during acromegaly management.

Growth Hormone

Single and Mean GH Measurements

Single GH measurements are not reliable for monitoring disease status in acromegaly. Although a very low random GH level, particular below assay detection, suggests remission, a random GH can be ≤1 μg/L in some patients with persistent active acromegaly after therapy and this alone cannot ensure complete remission [11]. GH may appear to fall markedly after therapy, even into the normal range on a random sample, but because of the logarithmic-linear relationship between GH and IGF-I, respectively, modest elevations of GH can still be associated with high IGF-I levels and persistent acromegaly [11].

In some centers, management of acromegaly includes taking the mean of serial GH samples collected over ~8–10-h period (day series or curve) [3, 74]. Mean GH values can correlate with nadir GH during an OGTT [31] and IGF-I levels [75] and when <2.5 mIU/L with normalization of mortality [76–78]. However, mean GH levels may not predict the integrated peripheral GH effect, as reflected in the IGF-I level, because the pattern of GH secretion is also a determinant of IGF-I production [32, 74]. Day sampling mean GH levels can poorly predict mean 24 h GH levels [79], which can overlap in acromegaly (with elevated IGF-I levels) and healthy subjects [6, 32, 80]. In addition, the mean GH cut-off of 2.5 μg/L, which was developed with polyclonal RIAs [76–78], is no longer valid and a cut-off for use with current assays is unavailable. GH sampling cannot practically be performed outside of a research setting, so it is rarely used clinically for routine disease surveillance.

Oral Glucose Tolerance Test (OGTT)

The traditional and most widely used method for monitoring of GH values during therapy is GH suppression after oral glucose (OGTT). Adequate GH suppression is an important aim of acromegaly treatment [81–83]. The target nadir GH during treatment has become progressively lower with the use of increasingly sensitive and specific GH assays. Most studies do suggest that GH levels should be less than 1 μg/L to establish its remission [7, 11, 13, 14, 16, 84], but whether a more specific cut-off below 1 μg/L is needed is debated. Studies utilizing other assays suggest cut-offs of 0.5 μg/L [14, 85] or 0.25 μg/L [15]. With a 22K GH-specific assay that is calibrated to a 22K rhGH standard, a cut-off of 0.3 μg/L provided the best differentiation (although not complete separation) between active acromegaly (increased IGF-I) and those in remission (normal IGF-I) [11]. When the serum IGF-I level is normal, suppression of the GH into the range seen in most healthy subjects has not been found to further improve the clinical manifestations of acromegaly such as insulin resistance [86].

Agreement on a suitable cut-off for glucose-suppressed GH levels during acromegaly therapy is also difficult to achieve because criteria for monitoring have been developed with GH assays that also varied in methodology including the antibodies and standards used as well as other characteristics [11, 14, 87–89]. The general adoption of a new recombinant human GH reference preparation is considered an important initial step toward establishing uniformity of GH assays [14, 87]. However, even when the same group of acromegaly samples was run in two different assays both calibrated to the rhGH standard (98/574), significant differences were demonstrated [14]; the mean GH nadir in controlled acromegaly patients was 0.98 μg/L (Immulite 2000) assay vs. 0.5 μg/L (Nichols Advantage), and in patients with active disease, these were 7.98 and 4.5 μg/L with these two assays, respectively [14]. Greater uniformity of GH assay methodology will be needed before generalizable criteria can be proposed.

Interpretation of nadir GH levels after oral glucose may also need to consider the acromegaly patient's age, BMI, and gender. Nadir GH levels in patients with acromegaly have been found to be lower with increasing age and BMI in some [13, 14, 21, 90], but not other studies [12, 91]. Gender-specific nadir GH criteria may also be needed as OGTT nadir values may be higher in some young women than men [10, 12, 13, 23, 24].

Insulin-Like Growth Factor-I (IGF-I)

Monitoring of serum IGF-I levels in acromegaly patients is essential. When measured properly and examined relative to a well-characterized, age-adjusted normative database, a high circulating IGF-I level is a sensitive and specific indicator of persistent disease. IGF-I elevation correlates with persistent GH hypersecretion despite treatment [92] and monitoring of IGF-I levels can detect mild GH excess [6, 11].

IGF-I levels normalize in patients with acromegaly after successful surgical [35], somatostatin analog [37] or GH receptor antagonist, pegvisomant [38, 93], therapy. During treatment, IGF-I normalization can also correlate with the reduction of GH excess [30], normalization of pulsatile GH secretion [32], and with suppressed GH levels obtained after oral glucose administration [12, 14, 31]. IGF-I normalization also correlates with the clinical signs and symptoms of the disease, morbidities such as insulin resistance and glucose intolerance, cardiovascular disease, and the excess mortality rate in acromegaly. It is unknown whether achieving an IGF-I level anywhere within the normal range is sufficient for therapy or whether where within the spectrum of normal the patient's IGF-I level lies is also important with respect to optimizing clinical parameters in acromegaly. In one study, IGF-I levels in the upper quartile of the normal range were associated with more insulin resistance than those in the lower three quartiles [86]. However, utilizing current normative data, these upper quartile IGF-I levels would actually be elevated suggesting that the upper limits of older normative data were too high [86]. Although IGF-I levels are not reliable for diagnosing GH deficiency in adults, low normal IGF-I levels during therapy should be avoided as these could be a marker of functional GH deficiency [94]. Further work is needed to determine if clinical disease activity is significantly impacted by where within the IGF-I normal range the IGF-I level lies.

Interpretation of IGF-I levels during treatment should also consider physiological as well as pathologic conditions that can alter IGF-I levels as discussed above as well as gonadal steroid status. Changes in estrogen status could alter disease status during acromegaly treatment because oral estrogen lowers IGF-I levels [28, 95, 96]. By contrast, androgen status and androgen administration do not appear to influence serum IGF-I levels.

Free IGF-I, the major biologically active form of IGF-I, has also been proposed as a marker for treatment of acromegaly [97, 98]. Free IGF-I levels are elevated in active acromegaly and are distinct from levels in healthy subjects in most [36, 69, 99], but not all [100], studies and fall during treatment [38]. However, free IGF-I measurements do not provide an advantage overall total IGF-I for monitoring of acromegaly [101]. It is more difficult to reliably measure free than total IGF-I levels [40, 102] and free IGF-I assays are not available out of a research setting so total IGF-I remains preferred at this time [40].

Combined Testing of GH and IGF-I Levels

Traditionally, a combined assessment of GH and IGF-I levels has been used to monitor patients with acromegaly. These tests provide complimentary information, the OGTT assesses neuro-regulation of GH secretion and the IGF-1 level average GH secretion. A combined assessment is often favored because caveats and pitfalls to each individual test do exist. In patients who have undergone surgical removal of the tumor, both tests are useful. OGTT nadir GH abnormalities can detect persistent GH dysregulation in the postoperative patient [103], and in the first 3 months postoperatively, GH suppression data may more reliably indicate true remission or not than the IGF-I level [85]. During long-term surveillance or during medical therapy, some prefer to use IGF-I alone for monitoring. The OGTT can be performed if the assessment of glucose tolerance is needed [60]. Others would always monitor both GH and IGF-I because of concern that relatively less data are available to support the validity of IGF-I as a marker of disease [104]. Some favor monitoring the efficacy of somatostatin analog therapy only by the IGF-I level because during this therapy discrepancies between IGF-1 and GH are more common and IGF-1 levels could be more consistent [105, 106]. Although traditionally, the OGTT was also monitored periodically during somatostatin analog therapy; in clinical practice, as the OGTT is more cumbersome, we also focus our monitoring on IGF-I levels. As the GH receptor antagonist, pegvisomant, blocks GH action and only lowers IGF-I and not GH, patients receiving this therapy can only be monitored by IGF-1 level. In the patient who has undergone radiotherapy, when GH neurosecretory dysregulation is common, IGF-1 may also be more reliable. In patients with diabetes mellitus, the OGTT is also unreliable and IGF-1 levels are preferred for monitoring.

When a combined testing approach is taken, the results of GH and IGF-I values are often congruent, but discrepancies do occur. In research studies, somewhat arbitrary conclusions can be drawn depending on which test is chosen as the "gold standard." However, in the patient care setting, the presence of these discrepancies can be important clinically. One pattern of discrepant results, abnormal GH suppression with a normal IGF-I level, is reported in 9–39% of patients [11, 13–15, 107]. In these patients, abnormal GH suppression may represent persistent GH neurodysregulation despite overall normal GH secretion [32, 108], as reflected in normal IGF-I levels. If the GH cut-off being used is inaccurate or has not been properly adjusted for the patient's age, BMI and gender GH suppression could also appear to be "abnormal" [13, 14, 21, 90]. Abnormal GH suppression or a lowering of IGF-I levels can also occur in conditions other than acromegaly as discussed above [109]. In some patients, abnormal GH suppression may be a sign of increased risk of disease recurrence as defined by development of a high IGF-I as was found on longitudinal follow-up of 5 of 19 [103] and 1 of 3 [110] such patients. Although the significance of such a discrepant pattern is not clear, so long as the IGF-I remains normal and other clinical factors in acromegaly are under control, these patients can be closely followed and treated only if the IGF-I level becomes consistently abnormal.

The opposite pattern of divergent GH and IGF-I results has also been reported; GH suppression is "normal," but IGF-I levels are elevated. On further examination in many such cases, GH suppression is actually abnormal when assay-appropriate cut-offs are employed. When GH values are related to cut-offs that are too high for the assay, the rate of this discrepancy is high, from 24 to 62% [11, 15, 103, 107, 110]. In particular, using a cut-off of 1 μg/L that is inappropriately high for a highly sensitive and specific assay leads to discrepant GH and IGF-I data in up to 50% of patients [12, 15]. However, using appropriate cut-offs, the rate of this discrepancy is low, generally <5% [11, 14, 16, 111]. One should also consider whether the IGF-I measurement could be inaccurate in patients who are found to have elevated IGF-I levels and normal GH suppression. Early postoperative IGF-I levels may be still falling and may not reliably indicate disease status [85, 105]. At 3 months postoperatively, we typically perform an OGTT, along with the IGF-I level, to establish the outcome of surgery. IGF-I levels are rarely above normal in conditions other than acromegaly, but can be in some adolescents and in pregnancy [109]. IGF-I assay methodology and normative data should also be considered.

When the GH and IGF-I results are discrepant, the testing should first be repeated in a different laboratory, if possible. Investigation into possible causes for the discrepancy should be undertaken. A clear cause of the discrepant pattern, however, usually cannot be identified. The decision then needs to be made as to which biochemical marker should be the one to guide further monitoring and treatment decisions.

In most cases, the IGF-I level is used to guide our therapeutic decisions. IGF-I levels are easier to monitor frequently as they are assessed by a single measurement. Normalization of morbidities such as insulin resistance [86] and excess mortality in acromegaly also correlate with IGF-I normalization. Therefore, so long as the patients' IGF-I levels are persistently normal and no factors potentially confounding IG-I interpretation, it is generally our practice to consider patients with persistently normal IGF-I levels and abnormal GH suppression in remission and follow them or continue treatment as prescribed. We do closely follow patients with abnormal GH suppression as they may be at increased risk of recurrence. No data are available to support the need to suppress GH into the range of healthy subjects when the IGF-I level is normal. It is possible that some patients could benefit symptomatically from reduction of IGF-I into the mid from upper normal range, but this has not been proven and further lowering of IGF-I should be done cautiously as a low normal IGF-I level could be a sign of GH deficiency [94]. Treatment is usually adjusted based on the IGF-I level despite a suppressible GH because some patients with active acromegaly, even newly diagnosed, can have low GH levels [6]. IGFBP-3 measurement may be helpful in some cases with discrepancies [61].

Conclusion

A variety of approaches can be taken to the biochemical assessment of acromegaly. Controversies still exist, however, about the relative roles for GH vs. IGF-I in the diagnostic evaluation and in monitoring during therapy. Some prefer only to monitor by IGF-I level, especially during medical therapy, while others feel that the use of the IGF-I level as the sole marker of disease is not appropriate and a combined assessment is needed [104]. The testing approach may need to be individualized, in particular based on the particular mode of therapy being administered. It is clear that a major hindrance to the development of widely applicable criteria for GH in particular, but also for IGF-I, is the heterogeneity of assays that are in clinical use for the measurement of these hormones. As a result, despite considerable research on this topic, only general guidelines for the use of GH and IGF-I measurements can be made. Additional work needs to be done to develop more widely applicable GH and IGF-I criteria for the diagnosis and monitoring of acromegaly.

Acknowledgments Funded in part by NIH Grants DK064720 and DK073040.

References

1. Reid TJ, Post KD, Bruce JN, Nabi Kanibir M, Reyes-Vidal CM, Freda PU. Features at diagnosis of 324 patients with acromegaly did not change from 1981 to 2006; Acromegaly remains under-recognized and under-diagnosed. Clin Endocrinol (Oxf). 2010;72:203–8.
2. Hartman ML, Veldhuis JD, Thorner MO. Normal control of growth hormone secretion. Horm Res. 1993;40(1–3):37–47.

3. Duncan E, Wass JA. Investigation protocol: acromegaly and its investigation. Clin Endocrinol (Oxf). 1999;50(3):285–93.
4. Chang-DeMoranville BM, Jackson IM. Diagnosis and endocrine testing in acromegaly. Endocrinol Metab Clin North Am. 1992;21(3):649–68.
5. Freda PU, Reyes CM, Nuruzzaman AT, Sundeen RE, Bruce JN. Basal and glucose-suppressed GH levels less than 1 μg/L in newly diagnosed acromegaly. Pituitary. 2003;6:175–80.
6. Dimaraki EV, Jaffe CA, DeMott-Friberg R, Chandler WF, Barkan AL. Acromegaly with apparently normal GH secretion: implications for diagnosis and follow-up. J Clin Endocrinol Metab. 2002;87(8):3537–42.
7. Giustina A, Barkan A, Casanueva FF, et al. Criteria for cure of acromegaly: a consensus statement. J Clin Endocrinol Metab. 2000;85(2):526–9.
8. Earll JM, Sparks LL, Forsham PH. Glucose suppression of serum growth hormone in the diagnosis of acromegaly. JAMA. 1967;201(8):628–30.
9. Hattori N, Shimatsu A, Kato Y, et al. Growth hormone responses to oral glucose loading measured by highly sensitive enzyme immunoassay in normal subjects and patients with glucose intolerance and acromegaly. J Clin Endocrinol Metab. 1990;70(3):771–6.
10. Chapman IM, Hartman ML, Straume M, Johnson ML, Veldhuis JD, Thorner MO. Enhanced sensitivity growth hormone (GH) chemiluminescence assay reveals lower postglucose nadir GH concentrations in men than women. J Clin Endocrinol Metab. 1994;78(6):1312–9.
11. Freda PU, Post KD, Powell JS, Wardlaw SL. Evaluation of disease status with sensitive measures of growth hormone secretion in 60 postoperative patients with acromegaly. J Clin Endocrinol Metab. 1998;83(11):3808–16.
12. Freda PU, Landman RE, Sundeen RE, Post KD. Gender and age in the biochemical assessment of cure of acromegaly. Pituitary. 2001;4:163–71.
13. Costa AC, Rossi A, Martinelli Jr CE, Machado HR, Moreira AC. Assessment of disease activity in treated acromegalic patients using a sensitive GH assay: should we achieve strict normal gh levels for a biochemical cure? J Clin Endocrinol Metab. 2002;87(7):3142–7.
14. Arafat AM, Mohlig M, Weickert MO, et al. Growth hormone response during oral glucose tolerance test: the impact of assay method on the estimation of reference values in patients with acromegaly and in healthy controls, and the role of gender, age, and body mass index. J Clin Endocrinol Metab. 2008;93(4):1254–62.
15. Serri O, Beauregard C, Hardy J. Long-term biochemical status and disease-related morbidity in 53 postoperative patients with acromegaly. J Clin Endocrinol Metab. 2004;89(2):658–61.
16. Mercado M, Espinosa de los Monteros AL, Sosa E, et al. Clinical-biochemical correlations in acromegaly at diagnosis and the real prevalence of biochemically discordant disease. Horm Res. 2004;62(6):293–9.
17. Becker MD, Cook GC, Wright AD. Paradoxical elevation of growth hormone in active chronic hepatitis. Lancet. 1969;2(7629):1035–9.
18. Vinik A, Pimstone B, Buchanan-Lee B. Impairment of hyperglycemic induced growth hormone suppression in hyperthyroidism. J Clin Endocrinol Metab. 1968;28(11):1534–8.
19. Holl RW, Bucher P, Sorgo W, Heinze E, Homoki J, Debatin KM. Suppression of growth hormone by oral glucose in the evaluation of tall stature. Horm Res. 1999;51(1):20–4.
20. Pieters GF, Smals AG, Kloppenborg PW. Defective suppression of growth hormone after oral glucose loading in adolescence. J Clin Endocrinol Metab. 1980;51(2):265–70.
21. Vierhapper H, Heinze G, Gessl A, Exner M, Bieglmayr C. Use of the oral glucose tolerance test to define remission in acromegaly. Metabolism. 2003;52(2):181–5.
22. Zadik Z, Chalew SA, McCarter Jr RJ, Meistas M, Kowarski AA. The influence of age on the 24-hour integrated concentration of growth hormone in normal individuals. J Clin Endocrinol Metab. 1985;60(3):513–6.
23. Ronchi CL, Varca V, Giavoli C, et al. Long-term evaluation of postoperative acromegalic patients in remission with previous and newly proposed criteria. J Clin Endocrinol Metab. 2005;90(3):1377–82.
24. Markkanen H, Pekkarinen T, Valimaki MJ, et al. Effect of sex and assay method on serum concentrations of growth hormone in patients with acromegaly and in healthy controls. Clin Chem. 2006;52(3):468–73.

25. Biermasz NR, Smit JW, van Dulken H, Roelfsema F. Postoperative persistent thyrotrophin releasing hormone-induced growth hormone release predicts recurrence in patients with acromegaly. Clin Endocrinol (Oxf). 2002;56(3):313–9.
26. Jones JI, Clemmons DR. Insulin-like growth factors and their binding proteins: biological actions. Endocr Rev. 1995;16(1):3–34.
27. Thissen JP, Ketelslegers JM, Underwood LE. Nutritional regulation of the insulin-like growth factors. Endocr Rev. 1994;15(1):80–101.
28. Clemmons DR, Van Wyk JJ. Factors controlling blood concentration of somatomedin C. Clin Endocrinol Metab. 1984;13(1):113–43.
29. Clemmons DR, Underwood LE. Nutritional regulation of IGF-I and IGF binding proteins. Annu Rev Nutr. 1991;11:393–412.
30. Barkan AL, Beitins IZ, Kelch RP. Plasma insulin-like growth factor-I/somatomedin-C in acromegaly: correlation with the degree of growth hormone hypersecretion. J Clin Endocrinol Metab. 1988;67(1):69–73.
31. Dobrashian RD, O'Halloran DJ, Hunt A, Beardwell CG, Shalet SM. Relationships between insulin-like growth factor-1 levels and growth hormone concentrations during diurnal profiles and following oral glucose in acromegaly. Clin Endocrinol (Oxf). 1993;38(6):589–93.
32. Ho KY, Weissberger AJ. Characterization of 24-hour growth hormone secretion in acromegaly: implications for diagnosis and therapy. Clin Endocrinol (Oxf). 1994;41(1):75–83.
33. Holly JM, Cotterill AM, Jemmott RC, et al. Inter-relations between growth hormone, insulin, insulin-like growth factor-I (IGF-I), IGF-binding protein-1 (IGFBP-1) and sex hormone-binding globulin in acromegaly. Clin Endocrinol (Oxf). 1991;34(4):275–80.
34. Rieu M, Girard F, Bricaire H, Binoux M. The importance of insulin-like growth factor (somatomedin) measurements in the diagnosis and surveillance of acromegaly. J Clin Endocrinol Metab. 1982;55(1):147–53.
35. Roelfsema F, Frolich M, Van Dulken H. Somatomedin-C levels in treated and untreated patients with acromegaly. Clin Endocrinol (Oxf). 1987;26(2):137–44.
36. van der Lely AJ, de Herder WW, Janssen JA, Lamberts SW. Acromegaly: the significance of serum total and free IGF-I and IGF-binding protein-3 in diagnosis. J Endocrinol. 1997;155 Suppl 1:S9–13; discussion S15–6.
37. Lamberts SW, Uitterlinden P, Schuijff PC, Klijn JG. Therapy of acromegaly with sandostatin: the predictive value of an acute test, the value of serum somatomedin-C measurements in dose adjustment and the definition of a biochemical "cure". Clin Endocrinol (Oxf). 1988;29(4): 411–20.
38. Trainer PJ, Drake WM, Katznelson L, et al. Treatment of acromegaly with the growth hormone-receptor antagonist pegvisomant [see comments]. N Engl J Med. 2000;342(16):1171–7.
39. LeRoith D, Clemmons D, Nissley P, Rechler MM. NIH conference. Insulin-like growth factors in health and disease [see comments]. Ann Intern Med. 1992;116(10):854–62.
40. Hall K, Hilding A, Thoren M. Determinants of circulating insulin-like growth factor-I. J Endocrinol Invest. 1999;22(5 Suppl):48–57.
41. Smith WJ, Underwood LE, Clemmons DR. Effects of caloric or protein restriction on insulin-like growth factor-I (IGF-I) and IGF-binding proteins in children and adults. J Clin Endocrinol Metab. 1995;80(2):443–9.
42. Merimee TJ, Zapf J, Froesch ER. Insulin-like growth factors in the fed and fasted states. J Clin Endocrinol Metab. 1982;55(5):999–1002.
43. Brabant G, von zur Muhlen A, Wuster C, et al. Serum insulin-like growth factor I reference values for an automated chemiluminescence immunoassay system: results from a multicenter study. Horm Res. 2003;60(2):53–60.
44. Evain-Brion D, Garnier P, Schimpff RM, Chaussain JL, Job JC. Growth hormone response to thyrotropin-releasing hormone and oral glucose-loading tests in tall children and adolescents. J Clin Endocrinol Metab. 1983;56(3):429–32.
45. Brabant G. Insulin-like growth factor-I: marker for diagnosis of acromegaly and monitoring the efficacy of treatment. Eur J Endocrinol. 2003;148 Suppl 2:S15–20.

46. Dall R, Longobardi S, Ehrnborg C, et al. The effect of four weeks of supraphysiological growth hormone administration on the insulin-like growth factor axis in women and men. GH-2000 Study Group. J Clin Endocrinol Metab. 2000;85(11):4193–200.
47. Ho KY, Evans WS, Blizzard RM, et al. Effects of sex and age on the 24-hour profile of growth hormone secretion in man: importance of endogenous estradiol concentrations. J Clin Endocrinol Metab. 1987;64(1):51–8.
48. Strasburger CJ, Bidlingmaier M, Wu Z, Morrison KM. Normal values of insulin-like growth factor I and their clinical utility in adults. Horm Res. 2001;55 Suppl 2:100–5.
49. van den Berg G, Veldhuis JD, Frolich M, Roelfsema F. An amplitude-specific divergence in the pulsatile mode of growth hormone (GH) secretion underlies the gender difference in mean GH concentrations in men and premenopausal women. J Clin Endocrinol Metab. 1996;81(7): 2460–7.
50. Ghigo E, Aimaretti G, Gianotti L, Bellone J, Arvat E, Camanni F. New approach to the diagnosis of growth hormone deficiency in adults. Eur J Endocrinol. 1996;134(3):352–6.
51. Gomez JM, Maravall FJ, Gomez N, Navarro MA, Casamitjana R, Soler J. Interactions between serum leptin, the insulin-like growth factor-I system, and sex, age, anthropometric and body composition variables in a healthy population randomly selected. Clin Endocrinol (Oxf). 2003;58(2):213–9.
52. Hilding A, Hall K, Wivall-Helleryd IL, Saaf M, Melin AL, Thoren M. Serum levels of insulin-like growth factor I in 152 patients with growth hormone deficiency, aged 19-82 years, in relation to those in healthy subjects. J Clin Endocrinol Metab. 1999;84(6):2013–9.
53. Landin-Wilhelmsen K, Wilhelmsen L, Lappas G, et al. Serum insulin-like growth factor I in a random population sample of men and women: relation to age, sex, smoking habits, coffee consumption and physical activity, blood pressure and concentrations of plasma lipids, fibrinogen, parathyroid hormone and osteocalcin. Clin Endocrinol (Oxf). 1994;41(3):351–7.
54. Miell JP, Taylor AM, Zini M, Maheshwari HG, Ross RJ, Valcavi R. Effects of hypothyroidism and hyperthyroidism on insulin-like growth factors (IGFs) and growth hormone- and IGF-binding proteins. J Clin Endocrinol Metab. 1993;76(4):950–5.
55. Feld S, Hirschberg R. Growth hormone, the insulin-like growth factor system, and the kidney. Endocr Rev. 1996;17(5):423–80.
56. Goldberg AC, Trivedi B, Delmez JA, Harter HR, Daughaday WH. Uremia reduces serum insulin-like growth factor I, increases insulin-like growth factor II, and modifies their serum protein binding. J Clin Endocrinol Metab. 1982;55(6):1040–5.
57. Schalch DS, Kalayoglu M, Pirsch JD, Yang H, Raslich M, Rajpal S. Serum insulin-like growth factors and their binding proteins in patients with hepatic failure and after liver transplantation. Metabolism. 1998;47(2):200–6.
58. Herlihy OM, Perros P. Elevated serum growth hormone in a patient with Type 1 diabetes: a diagnostic dilemma. Diabetes Metab Res Rev. 2000;16(3):211–6.
59. Clemmons DR. Commercial assays available for insulin-like growth factor I and their use in diagnosing growth hormone deficiency. Horm Res. 2001;55 Suppl 2:73–9.
60. Clemmons DR. IGF-I assays: current assay methodologies and their limitations. Pituitary. 2007;10(2):121–8.
61. Grinspoon S, Clemmons D, Swearingen B, Klibanski A. Serum insulin-like growth factor-binding protein-3 levels in the diagnosis of acromegaly. J Clin Endocrinol Metab. 1995;80(3)927–32.
62. Morrison KM, Wu Z, Bidlingmaier M, Strasburger CJ. Findings and theoretical considerations on the usefulness of the acid-labile subunit in the monitoring of acromegaly. Growth Horm IGF Res. 2001;11(Suppl A):S61–3.
63. Baxter RC. The binding protein's binding protein – clinical applications of acid-labile subunit (ALS) measurement. J Clin Endocrinol Metab. 1997;82(12):3941–3.
64. Blum WF, Ranke MB. Use of insulin-like growth factor-binding protein 3 for the evaluation of growth disorders. Horm Res. 1990;33 Suppl 4:31–7.
65. Juul A, Main K, Blum WF, Lindholm J, Ranke MB, Skakkebaek NE. The ratio between serum levels of insulin-like growth factor (IGF)-I and the IGF binding proteins (IGFBP-1, 2 and 3)

decreases with age in healthy adults and is increased in acromegalic patients. Clin Endocrinol (Oxf). 1994;41(1):85–93.
66. Tonshoff B, Blum WF, Mehls O. Derangements of the somatotropic hormone axis in chronic renal failure. Kidney Int Suppl. 1997;58:S106–13.
67. Jorgensen JO, Moller N, Moller J, Weeke J, Blum WF. Insulin-like growth factors (IGF)-I and -II and IGF binding protein-1, -2, and -3 in patients with acromegaly before and after adenomectomy. Metabolism. 1994;43(5):579–83.
68. Thissen JP, Ketelslegers JM, Maiter D. Use of insulin-like growth factor-I (IGF-I) and IGF-binding protein-3 in the diagnosis of acromegaly and growth hormone deficiency in adults. Growth Regul. 1996;6(4):222–9.
69. Marzullo P, Di Somma C, Pratt KL, et al. Usefulness of different biochemical markers of the insulin-like growth factor (IGF) family in diagnosing growth hormone excess and deficiency in adults. J Clin Endocrinol Metab. 2001;86(7):3001–8.
70. de Herder WW, van der Lely AJ, Janssen JA, Uitterlinden P, Hofland LJ, Lamberts SW. IGFBP-3 is a poor parameter for assessment of clinical activity in acromegaly. Clin Endocrinol (Oxf). 1995;43(4):501–5.
71. Arosio M, Garrone S, Bruzzi P, Faglia G, Minuto F, Barreca A. Diagnostic value of the acid-labile subunit in acromegaly: evaluation in comparison with insulin-like growth factor (IGF) I, and IGF-binding protein-1, -2, and -3. J Clin Endocrinol Metab. 2001;86(3):1091–8.
72. Baxter RC. Circulating levels and molecular distribution of the acid-labile (alpha) subunit of the high molecular weight insulin-like growth factor-binding protein complex. J Clin Endocrinol Metab. 1990;70(5):1347–53.
73. Khosravi MJ, Diamandi A, Mistry J, Krishna RG, Khare A. Acid-labile subunit of human insulin-like growth factor-binding protein complex: measurement, molecular, and clinical evaluation. J Clin Endocrinol Metab. 1997;82(12):3944–51.
74. Peacey SR, Toogood AA, Veldhuis JD, Thorner MO, Shalet SM. The relationship between 24-hour growth hormone secretion and insulin-like growth factor I in patients with successfully treated acromegaly: impact of surgery or radiotherapy. J Clin Endocrinol Metab. 2001;86(1):259–66.
75. Bates AS, Evans AJ, Jones P, Clayton RN. Assessment of GH status in acromegaly using serum growth hormone, serum insulin-like growth factor-1 and urinary growth hormone excretion. Clin Endocrinol (Oxf). 1995;42(4):417–23.
76. Rajasoorya C, Holdaway IM, Wrightson P, Scott DJ, Ibbertson HK. Determinants of clinical outcome and survival in acromegaly. Clin Endocrinol (Oxf). 1994;41(1):95–102.
77. Kaltsas GA, Isidori AM, Florakis D, et al. Predictors of the outcome of surgical treatment in acromegaly and the value of the mean growth hormone day curve in assessing postoperative disease activity. J Clin Endocrinol Metab. 2001;86(4):1645–52.
78. Bates AS, Vanthoff W, Jones JM, Clayton RN. Does treatment of acromegaly affect life expectancy. Metabolism. 1995;44(1):1–5.
79. Bajuk Studen K, Barkan A. Assessment of the magnitude of growth hormone hypersecretion in active acromegaly: reliability of different sampling models. J Clin Endocrinol Metab. 2008;93(2):491–6.
80. Hartman ML, Veldhuis JD, Vance ML, Faria AC, Furlanetto RW, Thorner MO. Somatotropin pulse frequency and basal concentrations are increased in acromegaly and are reduced by successful therapy. J Clin Endocrinol Metab. 1990;70(5):1375–84.
81. Melmed S, Jackson I, Kleinberg D, Klibanski A. Current treatment guidelines for acromegaly. J Clin Endocrinol Metab. 1998;83(8):2646–52.
82. Jaffe CA, Barkan AL. Acromegaly. Recognition and treatment. Drugs. 1994;47(3):425–45.
83. Camacho-Hubner C. Assessment of growth hormone status in acromegaly: what biochemical markers to measure and how? Growth Horm IGF Res. 2000;10(Suppl B):S125–9.
84. De Marinis L, Mancini A, Bianchi A, et al. Preoperative growth hormone response to thyrotropin-releasing hormone and oral glucose tolerance test in acromegaly: a retrospective evaluation of 50 patients. Metabolism. 2002;51(5):616–21.

85. Feelders RA, Bidlingmaier M, Strasburger CJ, et al. Postoperative evaluation of patients with acromegaly: clinical significance and timing of oral glucose tolerance testing and measurement of (free) insulin-like growth factor I, acid-labile subunit, and growth hormone-binding protein levels. J Clin Endocrinol Metab. 2005;90(12):6480–9.
86. Puder JJ, Nilavar S, Post KD, Freda PU. Relationship between disease-related morbidity and biochemical markers of activity in patients with acromegaly. J Clin Endocrinol Metab. 2005;90(4):1972–8.
87. Pokrajac A, Wark G, Ellis AR, Wear J, Wieringa GE, Trainer PJ. Variation in GH and IGF-I assays limits the applicability of international consensus criteria to local practice. Clin Endocrinol (Oxf). 2007;67(1):65–70.
88. Seth J, Ellis A, Al-Sadie R. Serum growth hormone measurements in clinical practice: an audit of performance from the UK National External Quality Assessment scheme. Horm Res. 1999;51 Suppl 1:13–9.
89. Ebdrup L, Fisker S, Sorensen HH, Ranke MB, Orskov H. Variety in growth hormone determinations due to use of different immunoassays and to the interference of growth hormone-binding protein. Horm Res. 1999;51 Suppl 1:20–6.
90. Colao A, Pivonello R, Cavallo LM, et al. Age changes the diagnostic accuracy of mean profile and nadir growth hormone levels after oral glucose in postoperative patients with acromegaly. Clin Endocrinol (Oxf). 2006;65(2):250–6.
91. Rosario PW, Furtado MS. Growth hormone after oral glucose overload: revision of reference values in normal subjects. Arq Bras Endocrinol Metabol. 2008;52(7):1139–44.
92. Barreca A, Ciccarelli E, Minuto F, Bruzzi P, Giordano G, Camanni F. Insulin-like growth factor I and daily growth hormone profile in the assessment of active acromegaly. Acta Endocrinol (Copenh). 1989;120(5):629–35.
93. van der Lely AJ, Hutson RK, Trainer PJ, et al. Long-term treatment of acromegaly with pegvisomant, a growth hormone receptor antagonist. Lancet. 2001;358(9295):1754–9.
94. Mukherjee A, Monson JP, Jonsson PJ, Trainer PJ, Shalet SM. Seeking the optimal target range for insulin-like growth factor I during the treatment of adult growth hormone disorders. J Clin Endocrinol Metab. 2003;88(12):5865–70.
95. Weissberger AJ, Ho KK, Lazarus L. Contrasting effects of oral and transdermal routes of estrogen replacement therapy on 24-hour growth hormone (GH) secretion, insulin-like growth factor I, and GH-binding protein in postmenopausal women. J Clin Endocrinol Metab. 1991;72(2):374–81.
96. Cardim HJ, Lopes CM, Giannella-Neto D, da Fonseca AM, Pinotti JA. The insulin-like growth factor-I system and hormone replacement therapy. Fertil Steril. 2001;75(2):282–7.
97. Janssen JA, Lamberts SW. Is the measurement of free IGF-I more indicative than that of total IGF-I in the evaluation of the biological activity of the GH/IGF-I axis? J Endocrinol Invest. 1999;22(4):313–5.
98. Frystyk J, Ivarsen P, Stoving RK, et al. Determination of free insulin-like growth factor-I in human serum: comparison of ultrafiltration and direct immunoradiometric assay. Growth Horm IGF Res. 2001;11(2):117–27.
99. Frystyk J, Skjaerbaek C, Dinesen B, Orskov H. Free insulin-like growth factors (IGF-I and IGF-II) in human serum. FEBS Lett. 1994;348(2):185–91.
100. Stoffel-Wagner B, Springer W, Bidlingmaier F, Klingmuller D. A comparison of different methods for diagnosing acromegaly. Clin Endocrinol (Oxf). 1997;46(5):531–7.
101. Sneppen SB, Lange M, Pedersen LM, et al. Total and free insulin-like growth factor I, insulin-like growth factor binding protein 3 and acid-labile subunit reflect clinical activity in acromegaly. Growth Horm IGF Res. 2001;11(6):384–91.
102. Frystyk J. Utility of free IGF-I measurements. Pituitary. 2007;10(2):181–7.
103. Freda PU, Nuruzzaman AT, Reyes CM, Sundeen RE, Post KD. Significance of "abnormal" nadir growth hormone levels after oral glucose in postoperative patients with acromegaly in remission with normal insulin-like growth factor-I levels. J Clin Endocrinol Metab. 2004;89(2):495–500.

104. Brooke AM, Drake WM. Serum IGF-I levels in the diagnosis and monitoring of acromegaly. Pituitary. 2007;10(2):173–9.
105. Espinosa-de-los-Monteros AL, Mercado M, Sosa E, et al. Changing patterns of insulin-like growth factor-I and glucose-suppressed growth hormone levels after pituitary surgery in patients with acromegaly. J Neurosurg. 2002;97(2):287–92.
106. Carmichael JD, Bonert VS, Mirocha JM, Melmed S. The utility of oral glucose tolerance testing for diagnosis and assessment of treatment outcomes in 166 patients with acromegaly. J Clin Endocrinol Metab. 2009;94(2):523–7.
107. Alexopoulou O, Bex M, Abs R, T'Sjoen G, Velkeniers B, Maiter D. Divergence between growth hormone and insulin-like growth factor-I concentrations in the follow-up of acromegaly. J Clin Endocrinol Metab. 2008;93(4):1324–30.
108. Ho PJ, Jaffe CA, Friberg RD, Chandler WF, Barkan AL. Persistence of rapid growth hormone (GH) pulsatility after successful removal of GH-producing pituitary tumors. J Clin Endocrinol Metab. 1994;78(6):1403–10.
109. Freda PU. Current concepts in the biochemical assessment of the patient with acromegaly. Growth Horm IGF Res. 2003;13(4):171–84.
110. Minniti G, Jaffrain-Rea ML, Esposito V, Santoro A, Tamburrano G, Cantore G. Evolving criteria for post-operative biochemical remission of acromegaly: can we achieve a definitive cure? An audit of surgical results on a large series and a review of the literature. Endocr Relat Cancer. 2003;10(4):611–9.
111. Machado EO, Taboada GF, Neto LV, et al. Prevalence of discordant GH and IGF-I levels in acromegalics at diagnosis, after surgical treatment and during treatment with octreotide LAR. Growth Horm IGF Res. 2008;18(5):389–93.

Chapter 14
The Role of Somatostatin Analogues in Treatment of Acromegaly

Haliza Haniff and Robert D. Murray

Abstract Somatostatin receptor ligand (SSRL) therapy is the mainstay of medical management of patients with acromegaly, either as adjuvant or primary therapy. Octreotide long-acting release (LAR) and lanreotide autogel (ATG) account for almost all clinical use of these analogues in acromegaly. SSRL acts via stimulation of one or more somatostatin receptor subtypes (SSTR1–5). Inhibition of GH secretion occurs primarily through SSTR2; however, SSTR1 and -5 also play a role. Anti-proliferative effects encompass cell cycle arrest and apoptosis for which SSTR3 is the predominant receptor subtype implicated.

Octreotide LAR and lanreotide ATG improve symptoms in almost all patients, control GH and IGF-I secretion in around 60 and 50%, respectively, and inhibit tumoural growth in at least 98% of patients. Significant tumour shrinkage is observed in 40–70%. A greater incidence and magnitude of tumour shrinkage is seen in primary therapy compared with adjuvant therapy. Tolerability is generally good, though gastrointestinal side effects are frequent, but rapid tachyphlaxis occurs. Inhibition of insulin secretion leads to problems with carbohydrate handling and chronic use leads to formation of gallstones in up to 30% of patients. Recent advances support the use of both octreotide LAR and lanreotide ATG 6 weekly for some patients, and lanreotide ATG can be self-injected to improve convenience for the patient. Recent data suggest prolonged remission can occur following SSRL withdrawal in patients who achieve long-term control on treatment. Whether use of SSRL pre-operatively to induce tumour shrinkage improves the outcome of surgery remains controversial; however, there are undoubtedly benefits to reducing anaesthetic risk.

Future improvements in SSRL take advantage of the synergy between SSTR subtypes, or the SSTR and dopaminergic systems. Two molecules are currently in clinical

R.D. Murray (✉)
Department of Endocrinology, Leeds Teaching Hospitals NHS Trust, Leeds, UK
e-mail: robert.murray@leedsth.nhs.uk

K. Ho (ed.), *Growth Hormone Related Diseases and Therapy: A Molecular and Physiological Perspective for the Clinician*, Contemporary Endocrinology,
DOI 10.1007/978-1-60761-317-6_14, © Springer Science+Business Media, LLC 2011

trials, pasireotide (SOM230) which has high affinity for SSTR1, 2, 3, and 5; and BIM23A760 which is a "dopastatin" and binds DR2 and SSTR2 and 5. In addition, the prospect of prolonged acting formulations of octreotide to improve patient convenience is also in development.

Keywords Octreotide • Lanreotide SR • Octreotide LAR • Lanreotide autogel • Role of SSRL in the treatment of acromegaly • Pre-operative use of SRIF analogues • Interaction of the dopaminergic and somatostatin pathways

Introduction

In 1968, Krulich et al., while studying the distribution of growth hormone-releasing factor in the hypothalamus of rats, made a surprise discovery of growth hormone inhibiting factor [1]. Further isolation and characterization of this somatotropin-release inhibiting factor (SRIF, somatostatin) was undertaken by Brazeau et al. in the early 1970 [2]. Somatostatin has two predominant active isoforms comprising of 14 and 28 amino acids, cleaved from a large pre-prohormone. SRIF-14 is predominant in the brain and SRIF-28 in the peripheral organs. Somatostatin modulates neurotransmission in the brain and acts as a neurohormone in the pituitary gland to inhibit the secretion of GH and TSH. It also exerts inhibitory effects on the gastrointestinal tract and the pancreas. The biological effects of SRIF are imparted through a family of specific seven transmembrane G-protein-coupled receptors (SSTR) that are differentially expressed in a tissue-specific pattern, thereby conferring functional specificity of SRIF action [3]. The five known SSTR subtypes (SSTR1–5) were cloned and characterized between 1992 and 1995 [4]. Once bound to its receptor, somatostatin inhibits the activation of adenyl cyclase with consequent reduction of intracellular cyclic AMP levels and inhibition of calcium channels preventing intracellular calcium flux. In the endocrine system, the end result is the inhibition of secretion of various hormones including GH, TSH, insulin, and glucagon. Somatostatin also modulates the activity of phosphotyrosine phosphatase and mitogen-activated protein kinase (MAPK) and can thereby induce cell cycle arrest [5] or apoptosis [6, 7].

Insights into the physiological regulation of GH secretion by GHRH and somatostatin (SRIF) have allowed development of pharmacological interventions capable of suppression of somatotrophinoma hormonal hypersecretion, and in the majority of cases, shrinkage of tumour size. Somatotrophinomas retain many of the characteristics derived from their pituitary cells of origin, including the SSTR system. Of these, SSTR2 is expressed ubiquitously in GH-secreting adenoma, and SSTR5 is reported to be co-expressed in the majority of tumours [6, 8–11]. SSTR1 and SSTR3 are expressed in around 50% of tumours [6, 8, 11–13]. SSTR4 expression within somatotrophinomas, when reported to be present, has been at a very low prevalence [9].

SSTR2 is the pivotal receptor subtype through which GH secretion is regulated, supported by the finding that in vivo suppression of GH secretion from somatotrophinomas correlates with SSTR2 expression [10, 11, 13]. Further evidence for the importance of SSTR2 is derived from studies using selective SSTR2 antagonists (BIM-23627 and BIM-23454) [14], which are able to reverse the inhibitory effect of SRIF-14 on GH secretion from primary cultures of rat pituitary cells. In addition to the direct action of SRIF on the pituitary, an inhibitory effect is also observed on the peripheral actions of GH [15].

Clinically Available SRIF Analogue Formulations

The therapeutic utility of native SRIF was precluded by its short half-life of 2–3 min, and concurrent physiological effects. These problems have been largely superseded by development of long-acting somatostatin receptor ligands (SSRL) which have been extensively used since the mid-1980 for the medical treatment of acromegaly.

Octreotide

The development of octreotide (Sandostatin), an octapeptide SRIF analogue by Bauer et al. in 1982, represented a real therapeutic advance to the treatment of acromegaly [16]. Its relative resistance to enzymatic degradation prolongs its half-life to 113 min after subcutaneous injection [17]. It does, however, remain the shortest acting SSRL formulation currently available. Octreotide has high affinity for SSTR2, but 12-fold lower affinity for SSTR5, and 25 times lower affinity at SSTR3 compared with SRIF-14 [8]. Octreotide shows minimal affinity for either SSTR1 or SSTR4.

Octreotide inhibits GH secretion with potency 45 times greater than native SRIF. After a single subcutaneous injection of 50 or 100 μg, octreotide levels peak at 30–60 min and GH secretion is suppressed for up to 5 h. Octreotide is capable of inhibiting spontaneous and GHRH- or TRH-stimulated GH release from somatotrophinomas [18, 19]. To maintain GH suppression, subcutaneous octreotide needs to be administered 2–4 times daily. Despite this, loss of biochemical control is often observed towards the end of an injection with GH levels rising immediately before the next dose [20]. The initial dose is 100–250 μg 3 times daily and total dose of up to 1,500 μg in 24 h can be given safely. Notably, subcutaneous octreotide has an immediate, dramatic analgesic effect on headaches in acromegalic patients.

Lanreotide SR

To improve convenience for the patient and avoid the undesirable biochemical "escape" frequently seen at the end of the dosing interval with subcutaneous octreotide, long-acting SSRL formulations were developed. Lanreotide SR (Somatuline LP) was the first long-acting SSRL to market. The half-life of this drug is considerably prolonged by incorporating the active peptide into microspheres of biodegradable polymer. Lanreotide has high affinity for SSTR2, but 18-fold lower affinity for SSTR5 compared with SRIF-14, and negligible affinity for SSTR1, 3, or 4 [8, 21]. After injection, lanreotide is released in a biphasic pattern characterized by an initial release phase peaking at 2 days followed by prolonged release for 10–14 days [22]. Lanreotide SR is available in Europe as a 30 mg vial which is injected intramuscularly every 7–14 days, the dosing interval dependent on clinical and biochemical responses [22].

Octreotide LAR

Octreotide has also been formulated in a LAR preparation, octreotide LAR (Sandostatin LAR), in which the active drug is encapsulated in microspheres of biodegradable polymer permitting a prolonged release. Similar to lanreotide SR, it has a biphasic release pattern. The initial rise in octreotide levels reflects immediate release of drug from the surface of the microspheres. A second peak occurs over the following 7–10 days with drug levels thereafter remaining elevated for 20–30 days [23]. A typical starting dose of octreotide LAR is 20 mg injected intramuscularly every 4 weeks. Steady-state conditions are achieved after the third injection without further drug accumulation [24]. The dose of octreotide LAR can be optimized on the basis of GH and IGF-1 levels, with doses of 40 mg 4 weekly being considered maximal. In patients who achieve good control, the dose can be reduced to 10 mg 4 weekly, and in subset of patients, the dosing interval can be prolonged beyond 4 weeks [25].

Lanreotide Autogel

Acknowledgement of the limitations of lanreotide SR which frequently required patients to have weekly injections administered by healthcare professionals led to the development of lanreotide autogel (ATG) (Somatuline Autogel), a prolonged acting formulation of lanreotide. The combination of hydrophobic and hydrophilic residues, together with the disulphide bridge, means lanreotide acetate naturally self-aggregates when mixed with water forming a homogenous semisolid gel. This preparation is formulated as a prefilled syringe containing 60, 90, or 120 mg lanreotide and is administered as a deep subcutaneous injection every 4 weeks. Alternately, a higher dosage can be used and the injection frequency prolonged up to 8 weeks without loss of biochemical control [26, 27]. Maximal serum concentrations are reached after 1–2 days in healthy subjects and 3.8–7.7 days in patients with acromegaly, depending on the dose administered [28]. The half-life is 23–29 days. Lanreotide ATG exhibits linear pharmacokinetics and steady-state conditions are achieved after 3–4 injections, similar to octreotide LAR. In the long term, higher steady-state levels are achieved with lanreotide ATG compared to lanreotide SR [28, 29].

Role of SSRL in the Treatment of Acromegaly

Medical therapy plays a vital role in a large proportion of patients with acromegaly; not only those who failed to be cured by surgery, but also when the probability of a surgical cure is low, for example, in large extrasellar tumours with no compressive effects. In recently published updated consensus guidelines for acromegaly management, SSRLs were recommended as first-line therapy in the latter scenario. SSRLs are also used as first-line therapy in patients who are not suitable candidates for anaesthesia and in those refusing surgery. In patients post-radiotherapy, while awaiting the suppressive effects of this treatment to manifest, the use of SSRL has been employed with good biochemical control. SSRLs are also used pre-operatively to optimize co-morbidities to improve surgical outcome [30]. Efficacy of treatment of acromegaly depends on attainment of several goals:

1. Achievement of biochemical control defined by GH levels <2.5 μg/L and normalization of IGF-1 levels to within gender- and age-matched limits.
2. Control of tumour growth.
3. Improvement in clinical signs and symptoms.
4. Tolerability of treatment.

Biochemical Control

Overall, somatostatin analogues successfully control GH hypersecretion (GH <2.5 μg/L) and normalize IGF-1 in approximately 60 and 50% of patients with acromegaly, respectively. Evidence to support the efficacy of sc octreotide in the treatment of acromegaly is prevalent in the literature. The two largest studies of sc octreotide undertaken in the early 1990 recruited 189 and 115 patients, respectively [31, 32]. These studies demonstrated that biochemical and clinical improvement in acromegaly could be achieved in around 45–60% of patients using thrice-daily octreotide. The use of sc octreotide has, however, been eclipsed by the longer-acting depot SSRLs (lanreotide SR, octreotide LAR, and Lanreotide ATG), which provide greater convenience and similar efficacy to their short-acting counterpart.

Baldelli et al. [33] and Verhelst et al. [34] studied the efficacy of lanreotide SR in 118 and 66 patients with acromegaly, respectively. In both studies, patients were switched from subcutaneous octreotide to lanreotide SR, and the dose of the latter optimized on the basis of GH levels. Overall, lanreotide SR was found to be more efficacious in controlling GH and IGF-1 levels than sc octreotide, with 44–77% of patients achieving biochemical targets. A large meta-analysis in 2005 which included 914 patients from 19 studies showed a normalized GH rate of 48% and IGF-1 of 47% after a mean of 15.5 months of treatment with lanreotide SR [35]. The use of octreotide LAR is similarly supported by two notably large studies [36, 37]. The first was a European multicentre open-label 12-month study using octreotide LAR at a 4 weekly interval which recruited 151 patients with acromegaly responsive to sc octreotide [37].

Approximately, 70 and 66% patients achieved GH ≤2.5 μg/L and normalized IGF-1 values, respectively. In the second study, 110 patients proven to be responsive to octreotide LAR in the first 6 months of treatment were followed up for a median of 30 months [36]. Individually tailored doses of octreotide LAR induced control of GH to ≤2.5 μg/L in 72% and GH <1.0 μg/L in 27% of patients with 75% achieving normalization of IGF-1. In both patient cohorts, octreotide LAR was equally effective whether used as primary therapy or as adjunctive therapy following previous pituitary surgery [36, 38]. The aforementioned meta-analysis of efficacy of SSRL included 12 studies of octreotide LAR in a total of 612 patients with acromegaly after a mean follow-up period of 15 months [35]. Safe GH and normalized IGF-1 levels were reported in 57 and 67%, respectively. Of note, however, is that the overall percentage of patients controlled by lanreotide SR and octreotide LAR was not unduly different to that observed in previous studies strictly optimizing the dose of subcutaneous octreotide.

There has been controversy over the relative efficacy of the long-acting somatostatin analogues, lanreotide SR and octreotide LAR, in the management of acromegaly. Although the proportion of patients achieving biochemical control of GH (48 vs. 54%) and IGF-I (42 vs. 63%) was reportedly different when comparing efficacy of lanreotide SR and octreotide LAR in the 2005 meta-analysis [35], this was only significant in individuals not selected for responsiveness to SSRLs. Differences in the degree of responsiveness in this meta-analysis likely related, at least in part, to methodological variances in the included studies. Confounding variables included lack of standardization of assays, the robustness of the IGF-I reference range, and variable pre-treatment GH and IGF-I values. A number of studies have directly compared efficacy of these two analogues on biochemical markers of the GH axis [39–43], the cumulative data from which suggest octreotide LAR to be modestly more efficacious compared with lanreotide SR [44]. Given that both octreotide and lanreotide exhibit similar affinities for SSTR2 and SSTR5 in vitro, the apparent differences in clinical efficacy, therefore, likely relate to differences in pharmacodynamics and pharmacokinetics of the various formulations.

The emergence of lanreotide ATG into the market in 2002 provided endocrinologist with a more prolonged acting formulation of lanreotide that was easy to administer and which made this drug much more desirable. Lanreotide ATG has now replaced lanreotide SR in clinical practice. The first large multicentre crossover trial enrolled 144 patients responsive to lanreotide SR, as defined by a GH <10 μg/L [45]. Patients were switched directly from lanreotide SR 30 mg every 7–14 days to the equivalent monthly dose of lanreotide ATG 60, 90, or 120 mg every 28 days. After 3 months of lanreotide ATG, comparable mean GH and IGF-1 values were demonstrated in the lanreotide ATG (GH: 2.87 ± 0.22 μg/L; IGF-1: 317 ± 15 ng/mL) compared with lanreotide SR (GH: 2.82 ± 0.19 μg/L; IGF-1: 323 ± 16 ng/mL) phase. One hundred and thirty patients entered the extension phase of the study for a further year [46]. Mean GH (2.4 ± 0.2 μg/L) and IGF-I (287 ± 12 ng/mL) levels after this period were significantly lower compared with lanreotide SR. Control of GH and IGF-1 improved from 49 and 44 to 68 and 50%, respectively. Fourteen of the study patients were followed for a period of up to 3 years during which GH control improved from 36 to 77% and IGF-1 normalization increased from 36 to 54% at the last follow-up with no evidence of tachyphylaxis [47]. A further large study of 93 patients confirmed lanreotide ATG to be at least as effective as lanreotide SR in controlling the hormonal end-points in acromegaly [48].

A recent multicentre study set out to look at long-term efficacy and safety of lanreotide ATG in 99 unselected patients who completed 52 weeks' treatment [49]. Patients on prior medical treatment for acromegaly underwent a washout period and were then randomized to 60, 90, or 120 mg of lanreotide ATG or placebo in a double-blinded construct. Following a single injection, none of the patients on placebo compared with 63% of patients on lanreotide ATG achieved >50% decrease from baseline in serum GH levels, demonstrating a rapid onset of action. The placebo group was re-allocated to lanreotide ATG at fixed doses for four injections. After four injections, 72% of patients had a greater than 50% reduction in ambient GH values, 49% had a GH less than 2.5 μg/L, 54% normalized their IGF-1, and 38% achieved the combined criteria of normalizing GH and IGF-1. The last phase of the study was open-labelled with dose adjustment according to response for a further eight injections. By week 52, 54 and 59% had normalized their GH and IGF-1 respectively, with 43% achieving the combined criteria of hormonal control [49].

Five relatively small studies have compared the efficacy of lanreotide ATG to that of octreotide LAR in patients with acromegaly [50–54]. Collectively, these studies show non-inferiority of lanreotide ATG compared with octreotide LAR [44] (Table 14.1). Of note is that three of these five studies assessed the efficacy of lanreotide ATG after only three injections following either initiation of therapy or dose change, and thus before steady-state levels had been obtained. Secondary end-points of these studies suggest lanreotide ATG leads to fewer technical problems with injections and possibly a lower incidence of local injection site reactions, though this is not borne out in all studies [54]. Because of the ease of administration of lanreotide ATG, under study conditions, patients were given the option of continuing to receive their injections from healthcare professionals or receiving their injections at home either by self-administration or by their partner [55]. Local injection tolerability was good for both groups and safety profiles were similar. Patients/partners administered lanreotide ATG with no detrimental effect on efficacy [55].

Use of SSRL as Primary Therapy

The pioneering study by Newman et al. [56] demonstrated that long-term subcutaneous octreotide of up to 5 years in treatment-naive acromegalic patients was as effective as when used as adjuvant therapy post-operatively or post-pituitary irradiation in normalizing hormonal end-points. Unsurprisingly, further studies have confirmed this finding to also be true when using octreotide LAR [57–62]. In a prospective study of 67 consecutive treatment-naïve acromegalic patients, treated with octreotide LAR

Table 14.1 Summary of biochemical end-points of studies comparing efficacy of octreotide LAR and Lanreotide ATG

	SRIF analogue	(*n*)	Baseline GH (μg/L)	Baseline IGF-I (ng/mL)	Final GH (μg/L)	Final IGF-I (ng/mL)	GH <2.5 μg/L (*n*)	Normal IGF-I (*n*)	Normal IGF-I GH <2.5 μg/L	Mean dose (mg/4 weeks)	Study design
Alexopoulou et al. [51]	OCT	25	30.2±25.2[a]	917±431[b]	2.4±1.8	337±201	16	13	9	25.2±5.9	Open, OCT to LAN, no WO, LAN titrated
	LAN	25	–	–	2.9±2.4	332±193	12	13	8	108±21.2	
Ashwell et al. [52]	OCT	10	8.0±3.7[f]	485±105	3.0±1.7[f]	212±70	9[e]	6	6[e]	20±0.0	Open, GH<10 μg/L on OCT20, OCT to LAN, no WO, LAN titrated
	LAN	10	–	–	3.3±1.6[f]	154±61	9[e]	8	8[e]	93.0±22.1	
van Thiel et al. [50]	OCT	7	–	59.5±32.7[c]	3.0±0.8	40±4[d]	4	3	3[e]	24.3±5.3	Open. Pts OCT responsive.. OCT to LAN. WO x 10 wks. LAN titrated
	LAN	7	–	–	5.3±2.7	55±8[d]	3	3	3[e]	111±22.7	
Andries et al. [54]	OCT	10	1.7±2.0	266±104	0.9±0.8	265±133	10	5	5	–	Open, randomized crossover, no WO, on optimal OCT at baseline, "equivalent" LAN dose
	LAN	10	0.9±0.7	271±136	1.4±1.6	261±155	7	6	5	–	
Ronchi et al. [53]	OCT	23	15.2±17.4	519±275	4.0±2.5	333±177	10	8	4	23.9±6.6	Open, OCT to LAN120, 3 months WO, LAN titrated
	LAN	22	9.9±11.3	544±312	3.8±5.7	356±187	13	9	7	96.4±27.4	
OCT LAR							49/75 (65.3%)	35/75 (46.7%)	27/75 (36.0%)		
LAN ATG							44/74 (59.5%)	39/74 (52.7%)	31/74 (41.9%)		

From Murray and Melmed [44]

SRIF somatostatin; *OCT* octreotide LAR; *LAN* lanreotide autogel; "OCT to LAN" initial treatment with octreotide LAR followed by lanreotide autogel; *WO* wash-out

[a]Data available on n=25
[b]Data available on *n*=19
[c]Data available on *n*=6 in nmol/L
[d]Data in nmol/L
[e]GH target <5.0 mU/L
[f]Data in mU/L

for a median of 48 months, 68.7 and 70.1% normalized GH (<2.5 μg/L) and IGF-1 levels, respectively [59]. This study demonstrated that the best predictor of final hormonal values were the respective values obtained by 6 months. The majority of the decrease in GH and IGF-1 levels occurred within the first 6–12 months. Drug efficacy improved with prolonged treatment with GH and IGF1 levels continuing to decline, even after 9 years of sustained treatment. No tachyphylaxis was noted. No difference was observed in achievement of hormonal end-points between macroadenomas and microadenomas [38, 59]; however, most other studies have found that patient with microadenomas tended to have better suppression of GH hypersecretion compared with macroadenomas [36, 60–62]. Colao et al. reported octreotide LAR to successfully control GH and IGF-1 in 100% patients harbouring microadenomas and 50% with macroadenomas [60]. It has repeatedly been shown that the lower the basal GH value, the better the chances of attaining GH and IGF-1 suppression with octreotide LAR [35, 58, 60, 61], though this has not been a universal finding [36, 38, 57, 59]. A further large study examining primary therapy with octreotide LAR enrolled 98 patients over a 48-week period [61]. At the end of the study, 44 and 34% of patients, respectively, achieved a mean GH <2.5 μg/L and normalized IGF-1 level. Both criteria of biochemical control were realized in 25% of patients. The proportion of patients achieving treatment target were lower in this study compared to the two previous large studies [59, 60]. This discrepancy may reflect selection bias, differences in sample size and ethnic background, or study design.

The only study of lanreotide ATG as primary therapy in newly diagnosed acromegalic patients was recently published [26]. Most of the patients had macroadeomas. Twenty-six patients were treated with a fixed dose of lanreotide ATG 120 mg every 4–8 weeks, with the dosing interval tailored according to response. After 12 months, 58 and 58% of patients attained GH <1.9 μg/L and IGF-1 normalization, respectively. Approximately, one third of patients each ended up on having their injections every 4-, 6-, and 8-week intervals, with patients with extended interval maintaining hormonal control.

Interestingly, after medical treatment of prolactinomas for around 4 years, dopamine agonist withdrawal can be undertaken in those whose serum prolactin level has been well controlled and who have shown significant resolution of the tumour on imaging. A significant proportion of patients thereafter retain remission for a prolonged period of time [63]. Ronchi et al. followed a similar protocol for 27 patients with GH-secreting adenoma treated with SSRL therapy for a mean of 48 months and who had achieved GH levels less than 2.5 μg/L and normalized IGF-I levels [64]. At first assessment at 12–14 weeks, 15 patients had undergone rapid disease relapse and were recommenced on SSRL therapy. At time of publication, only 10 of the 12 patients in remission had follow-up in excess of 6 months, one of whom had shown disease recurrence. Five of the nine patients in remission at 6 months showed 12 months follow-up, all of whom remained in remission [64]. Attainment of prolonged remission could not be predicted by age, previous treatment, duration of SSRL treatment, or pre-treatment GH/IGF-I levels. These data are preliminary, but suggest a proportion of patients (at least 20%) treated long-term with SSRL may enter prolonged remission and that SSRL therapy may not be a lifelong requirement.

Tumour Shrinkage

Data from a number of distinct cell culture and animal models support an anti-proliferative role for somatostatin, and its analogues [65–72], through a myriad of putative mechanisms. These mechanisms are either direct effects on the tissue including cycle arrest and induction of apoptosis [73, 74], or indirect mechanisms including inhibition of growth factor production and angiogenesis [75–78]. Induction of the anti-proliferative effects of SSRLs is initiated through the SSTRs, with the tissue effects being dictated by the receptor subtype expressed. Activation of SSTR2 or SSTR3 can induce apoptosis [79–81], whereas activation of SSTR1, 2, 4, and 5 leads to cell cycle arrest [74, 80, 82–85]. Studies of the effect of SSRLs on primary cultures of GH-secreting adenoma have confirmed the ability of these molecules to inhibit somatotroph cell proliferation [6, 86]. Immunohistochemical comparison of tumours that received and did not receive pre-treatment with SSRLs prior to operative intervention reveals evidence of reduced proliferation (Ki-67), a greater proportion of cells in G_1 and M phase of the cell cycle, interstitial fibrosis, and moderate reduction in cytoplasmic, but not nuclear volume in those tumours which received SSRLs [87–89].

Clinical studies using octreotide and lanreotide result in tumour shrinkage of GH-secreting adenomas in 30–70% of patients [73]. The significant variability of success between studies can be attributed to variable study design, heterogenous populations, different definitions of significant tumour shrinkage, and non-blinded and non-standardized interpretation of pituitary imaging which could suffer from observer subjectivity. Moreover, some studies report on changes in tumour diameter and others on tumour volume. With these caveats in mind, results from individual studies should be interpreted carefully.

A review of 37 studies examining tumour size in 920 patients who were treated with SSRLs either as adjunctive therapy (post-pituitary surgery and post- radiotherapy) or as primary therapy reported a significant decrease in tumour size in 382 (42%) [3]. Further analysis of this dataset was performed after stratification by whether SSRL treatment was primary therapy or was used as adjuvant therapy. Fifty-two percent of patients receiving SSRL as primary therapy showed tumour shrinkage, in contrast to just 21% when used adjuvant therapy [73]. Differences in the proportion experiencing tumour shrinkage were observed between subcutaneous octreotide, lanreotide SR, and octreotide LAR (Table 14.2). The lower proportion achieving significant adenoma shrinkage with lanreotide SR likely relates to prior treatment with octreotide and only a short washout period in many of the studies. Prior treatment with surgery or radiotherapy leads to inherent difficulty in assessing pituitary adenoma size accurately as a consequence of alterations in sella anatomy, tumour morphology, and as a result of other histological changes such as fibrosis that may follow these interventions. In a further review, Melmed et al. performed a critical analysis on 14 studies using SSRL as primary therapy in treatment of acromegaly [90]. They revealed significant tumour shrinkage (defined as between 10 and 45% in individual studies) to occur in 36.6% of patients. In those patients in whom a

Table 14.2 Proportion of patients with GH-secreting adenoma shrinkage observed with primary somatostatin receptor ligand (SSRL) or following prior adjuvant therapy with surgery or radiotherapy according to the formulation used

	Subcutaneous octreotide	Lanreotide SR	Octreotide LAR	All treatment
Studies (*n* = %)	22	8	7	37
Patients (*n* = %)	478	263	180	921
Tumour shrinkage (*n* = %)	217/478 (45%)	62/263 (24%)	103/180 (57%)	382/921 (42%)
Primary therapy tumour shrinkage (*n* = %)	110/217 (51%)	40/130 (31%)	81/101 (80%)	231/448 (52%)
Adjuvant therapy tumour shrinkage (*n* = %)	22/82 (27%)	8/97 (9%)	22/79 (28%)	52/248 (21%)
Mixed studies tumour shrinkage (*n* = %)	85/179 (47%)	14/46 (30%)		

Adapted from Bevan et al. [73]

significant reduction in tumour size was observed, the mean tumour volume reduction with SSRL primary therapy was in the region of 50% [90]. In a third review, the authors undertook a meta-analysis of tumour shrinkage with SSRL incorporating 26 primary therapy and eight secondary therapy studies [35]. They concluded that tumour shrinkage was more prevalent when SRIF analogues were used as primary compared with adjuvant therapy. It was also concluded that both subcutaneous octreotide and octreotide LAR were more efficacious in inducing tumour shrinkage compared with lanreotide SR [35]. No data were available for lanreotide ATG in any of the three reviews performed.

Considering the individual larger studies with a more robust construct, Cozzi et al. treated 67 patients using octreotide LAR as primary therapy for a median of 48 months, of whom 82% achieved tumour shrinkage of >25% [59]. Seventy percent of patients showed >50% tumour shrinkage and 44.1% achieved >75% tumour shrinkage. Pituitary adenoma disappeared in three patients, with progressive shrinkage to empty sella in five patients, and in three others, apparent cavernous sinus invasion on imaging disappeared. Mercado et al. published the largest study to employ blinded analysis and centralized tumour volume measurement [61]. Sixty-eight newly diagnosed acromegalic patients were treated for 12 months with octreotide LAR 10–30 mg 4 weekly. Significant tumour volume reduction, defined as greater than 20%, was noted in 63 and 75% of patients at weeks 24 and 48, respectively [61]. This is in keeping with a previous prospective multicentre study which also systematically evaluated tumour volume changes during primary therapy with octreotide and found a greater than 30% reduction in tumour volume in 73% of patients [58]. Most studies of lanreotide ATG have not reported on tumour shrinkage as most studies switched patients from either octreotide LAR or lanreotide SR or had previously undergone pituitary surgery or irradiation. A recent study on primary therapy using lanreotide ATG 120 mg every 4, 6, or 8 weeks revealed 77% of the 26 patients studied attained >25% tumour shrinkage [26].

Tumour volume reduction can be detected as early as 12 weeks [58, 61] with progressive volumetric reduction over time with long-term SSRL treatment [59, 61, 91]. The majority of shrinkage does, however, occur within the first 6 months of treatment [35]. With this knowledge, a role for SSRL has been proposed as treatment for patients with GH-secreting tumours presenting with visual field defects. Amato et al. reported improvement in visual fields of three patients treated with primary SSRL therapy [62], whereas Colao et al. reported 4 of 21 patients previously treated by surgery and with visual field defects not to show visual improvement with octreotide LAR [57].

In addition to pre-treatment, a number of additional factors are proposed to influence the degree of tumour shrinkage. Data concerning differences in tumour shrinkage between octreotide LAR and lanreotide ATG remain unclear at present due to limited studies with the latter preparation. The dose of subcutaneous octreotide has been purported to be important in tumour shrinkage [32]; however, this relationship has not been observed with the long-acting formulations [36]. It has been suggested that a higher proportion of patients harbouring macroadenomas achieve tumour shrinkage compared with microadenomas [32, 36, 57, 59, 60, 62, 92], though this has not been a universal finding [58, 61]. The absolute reduction in size of macroadenomas is, however, far greater than in microadenomas [57–59]. A weak correlation is observed between tumour shrinkage and the degree of GH/IGF-1 suppression [59, 93, 94] in some studies, but not others [36, 57, 58, 62, 92]. Interestingly, tumour shrinkage is observed to occur before that of normalization of the GH/IGF-I axis [59]. Notably, for an individual tumour, the degree of suppression of the GH axis does not predict tumour shrinkage; however, tumour shrinkage is unlikely in the absence of GH/IGF-I reductions [32].

Although a large proportion of patients do not show significant reductions in tumour size with SSRL treatment, it is important to note that growth of GH-secreting adenomas while on therapy is relatively unusual and equates to less than 2% [35, 73]. This is an important benefit from therapy as local symptoms from tumour mass can be avoided. Where increases in adenoma size are reported in patients receiving SSRL therapy, this has been in invasive tumours that also fail to show a reduction in biochemical markers [93], cystic components of the tumour [95], or in early in treatment before optimal dosing had been achieved [94]. Interestingly, tumour shrinkage has been recorded even in patients who failed to reach safe GH levels [59, 61, 96]. This discrepancy between the anti-secretory and anti-proliferative activity of SSRLs is hypothesized to be due to diverse molecular mechanisms and prevalence of somatostatin receptor subtypes involved [96, 97].

Clinical Response and Tolerability

SSRL treatment promotes amelioration of the troublesome symptoms of acromegaly. Significant clinical improvement has been reported in patients experiencing headaches (21–66%), hyperhidrosis (21–42%), fatigue (26–65%), joint pains (21–61%), paraesthesia (38–77%), carpal tunnel syndrome (15–90%), and soft tissue swelling (up to 100%) [37, 57, 60, 61]. Improvements in severity of sleep apnoea were reported in 14 patients treated with octreotide LAR for 6 months which positively correlated with the improvement of GH/IGF-1 levels [98].

Cardiomyopathy is a recognized consequence of acromegaly conferring increased morbidity and mortality in this population. GH and IGF-1 act both directly and indirectly via its receptors on the cardiac myocytes to exert its effect. By tackling hypersecretion of GH and IGF-1, SRIF treatment improves morbidity and mortality. A meta-analysis conducted through to June 2006, which included 290 patients from 18 studies, found that SRIF treatment was associated with significant improvement in the structural and functional haemodynamic cardiac parameters in patients with acromegaly [99]. In particular, there were significant reductions of heart rate and left ventricular hypertrophy via improvements in left ventricular mass index, interventricular septum thickness, and left ventricular posterior wall thickness [99]. Additionally, SRIF also improved exercise tolerance with trend towards improvement in left ventricular end-diastolic dimension and left ventricular ejection fraction [99]. These beneficial effects correlated with the degree of improvement in IGF-1 and GH levels [99]. No significant effect was found on blood pressure. Pathological prolonged QT interval on electrocardiography in acromegalic patient with active disease has been shown to normalize with SRIF treatment [100].

Data on the effect of SRIF on glucose metabolism are rather controversial with heterogeneous results reported. Excess GH level in acromegaly has a pro-diabetogenic effect leading to increased rates of glucose intolerance and diabetes mellitus. GH hypersecretion impairs glucose utilization by peripheral tissues and inhibits insulin-induced suppression of hepatic gluconeogenesis [101]. SRIF treatment induces the inhibition of GH and glucagon secretion thus improving insulin sensitivity. It also reduces gastrointestinal absorption rates of glucose and fructose [102] and potentiates the ability of insulin to suppress hepatic glucose production [102]. However, SRIF also directly decreases pancreatic beta cell secretion of insulin, which could in turn worsen glucose tolerance. Thus, the balance between the two divergent effects of SRIF and the baseline glucose tolerance in an individual with acromegaly determines the outcome of glycoregulation. This may be why there is such a huge variability of effect of SRIF on glucose metabolism ranging from no change to deterioration/improvement of glucose tolerance or even hypoglycaemia.

A meta-analysis conducted on 619 patients from 31 studies from 1987 to 2008 assessed the clinical impact of SRIF on glucose metabolism [103]. Studies included patients treated for 3 weeks to 96 months with SRIF. This meta-analysis showed that SRIF significantly reduced fasting plasma insulin levels, although there were high levels of inconsistency between studies. There were no significant change of fasting plasma glucose and HbA1c [103]. There was a statistically significant variation over time of serum glucose response to OGTT, but again with a high level of heterogeneity among trials. The authors concluded that although SRIF decreased fasting serum insulin levels, there were no consistent effect on glucose metabolism [103]. Of note, in studies that reported an increase in HbA1c, changes did not exceed 0.5%. On an individual basis, given the variability of the effect of SRIF on glucose metabolism, blood glucose levels should be monitored in these patients.

Treatments with octreotide LAR and lanreotide ATG are generally well tolerated and the incidence of major adverse events is low. SSRLs are associated with low rates of treatment withdrawal as a result of treatment-related adverse events [37, 49]. Side effects with SSRL treatment are usually of mild-to-moderate intensity and mainly of gastrointestinal in origin [45, 48, 51]. The most frequently reported adverse events are diarrhoea, abdominal discomfort, flatulence, nausea, and steatorrhoea. In most patients, gastrointestinal side effects become less frequent during long-term treatment and do not lead to cessation of therapy. The most potentially important adverse event is increased tendency for gallstone formation as SSRL attenuates gallbladder contractility and delays emptying. New gallstone or biliary sludge formation has been reported in 8–30%, with a variable geographical incidence [37, 49]. Most cholelithiasis are asymptomatic and occur within the early years of therapy [91]. Furthermore, it has been suggested that acute cholelithiasis and biliary colic may be more common on discontinuation of SSRLs with the restoration of gallbladder contractility [104]. Frank cholecystitis is very rarely reported in these patients. Injection site pain and swelling have been reported in 5–20% of patients. Subcutaneous granulomas at injection sites are rare, although few cases have been reported to occur with lanreotide ATG [47] and octreotide LAR [105]. As discussed previously, disturbance of glycoregulation can lead to hypo- or hyperglycaemia. No changes to haematological or biochemical parameters were noted [45, 47, 49, 60]. Other reported side effects include hair loss and memory loss [106].

Pre-Operative Use of SRIF Analogues

The beneficial effects of SRIF analogues on tumour shrinkage, biochemical markers of GH secretion, and symptoms and signs of acromegaly led to several groups proposing the use of SRIF analogues pre-operatively as early as the 1980s [107, 108]. Surgical outcomes are highly dependent on adenoma size and degree of invasiveness [109]. Significantly more favourable results occur for microadenomas than for invasive macroadenomas. It is intuitive to think that, by induction of tumour shrinkage in large and invasive tumours, more of the tumour will be brought in to the operative field aiding surgical removal of previously inaccessible parts of the tumour. This would hopefully be reflected in lower ambient GH and IGF-I levels post-operatively.

There has been general agreement from studies of pre-operative use of SRIF analogues in acromegaly that their use results in symptomatic benefit, abrogated hormonal hypersecretion, tumour shrinkage, and improved imaging in the pre-operative period [92, 107, 109–112]. It has also been suggested that the use of SRIF analogues pre-operatively results in a change in consistency of the tumour, with treated tumour being subjectively softer and more clearly delineated from the remaining normal tissue [108, 111, 113]. This finding, however, has not been universal [109, 110], with one study reporting pre-treatment to result in a greater proportion of firm tumours [30]. Histologically, a greater incidence of cellular atypia, characterized by reduced cytoplasmic and nuclear area, is observed in adenoma that has been treated pre-operatively with SRIF analogues [109–113]. Perivascular fibrosis is reported to occur at increased prevalence [107, 111–113].

What remains more contentious is whether pre-operative use of SRIF analogues improves surgical outcomes, specifically relating to achieving biochemical control. Despite documented tumour shrinkage being observed in a significant proportion of patients, improvements in tumour invasiveness and ease of removal are often not observed [110].

Early studies, inclusive of only small numbers (≤10 patients), compared surgical outcome of patients pre-treated with SRIF analogues with historical reference data and reported improved surgical cure rates in invasive macroadenomas of 60–80% [92, 107, 112]. These data were promising and led to larger studies of a similar design. Stevenaert et al. in a series of studies [109, 111, 113] compared outcome of a cohort of 64 patients who received pre-operative subcutaneous octreotide therapy (300–1,500 μg/day) for between 3 weeks and 39 months with historical control data. Overall remission rates showed no significant improvement [109, 111, 113]; however, an improved remission rate, defined by GH values of <2 μg/L and normal IGF-I levels, was observed for enclosed adenomas but not invasive adenomas [109, 111, 113]. Colao et al. analysed the surgical outcomes of 22 patients with acromegaly treated pre-operatively for 3–6 months with subcutaneous octreotide (150–600 μg/day) and compared this with 37 patients who remained untreated pre-operatively [110]. A greater proportion of the octreotide pre-treated patients achieved target GH and IGF-I values in the early post-operative period; however, mean GH and IGF-I levels were not significantly different to patients who received no intervention pre-operatively [110]. Biermasz et al. in a prospective analysis attempted to improve upon the quality of previous data by individually matching patients who received pre-operative octreotide with those who did not [114]. Patients were matched for tumour class, tumour grade, and pre-intervention GH levels. The pre-operative octreotide dose was titrated according to GH levels (mean dose 530 μg/day) and administered for a mean of 4.3 months. Post-operative GH levels and the proportion with GH <2 μg/L (5 mU/L) were not different between the groups [114]. IGF-I values were lower in the untreated group, with a trend towards a greater proportion with normal IGF-I values in the direct surgery group [114].

Although these latter studies have included a control arm, they have failed to randomize individuals to pre-operative treatment with SRIF analogues or direct surgical intervention, relying on retrospective analyses and the inherent biases in patient selection. Unfortunately, despite all the studies described above, the effect of pre-operative use of SRIF analogues on surgical outcome remains unclear. To try and effectively answer this question, Carlsen et al. undertook a prospective multicentre trial in patients with newly diagnosed acromegaly and randomized volunteers to direct surgery or 6-month pre-operative therapy with octreotide LAR 20 mg 4 weekly for 6 months [30]. Thirty-one patients were randomized to each arm of the study. At baseline, the pre-treatment group showed a greater proportion of males and lower IGF-I values. When assessed 3 months post-operatively, a normal IGF-I was observed in 14 pre-treated compared with seven untreated patients (45 vs. 23%,

$P = 0.11$) [30]. This became significant when the analysis was confined to only those individuals displaying macroadenomas (50 vs. 16%, $P = 0.02$). Notably, only one of ten patients with an invasive macrodenoma achieved a normal IGF-I levels post-operatively; this individual having been randomized to direct surgical intervention. Using the combined criteria of normal IGF-I and GH nadir <1 μg/L during the OGTT, 38% of macroadenomas that were pre-treated and 16% of those who were not achieved target ($P = 0.12$) [30].

In addition to the putative benefits on surgical cure rates, it could be proposed that SRIF analogue pre-treatment would reduce the incidence of hypopituitarism, surgical complications, and duration of hospital stay. Reductions in duration of hospitalization are reported [110], as are low rates of complications [112] and hypopituitarism [107]. No difference between SRIF analogue pre-treated and direct surgery groups with regard to these secondary end-points, however, is reported equally as frequently [30, 114].

Despite the current literature, the value of pre-operative SRIF analogues remains unclear. The recently published prospective randomized study [30] has provided compelling insights, but the surgical cure rates of only 16% in macroadenomas which went straight to surgery are below that of good surgical series in the literature [115]. It is therefore questionable whether SRIF analogue pre-treatment would have an effect in neurosurgical units already achieving cure rates in macroadenomas of 40–50%. The putative beneficial effects on tumour consistency are not consistently reported in all studies and improvements in other secondary end-point have also been variably reported. As we move further towards endoscopic transphenoidal resection of pituitary adenoma in preference to microscopic, we will need to consider whether the value of pre-operative use of SRIF analogues on biochemical outcomes needs to be investigated again. There is, however, no question that further studies need to be performed in a prospective multicentre randomized structure.

Leaving aside the direct effects of SRIF analogues on the tumour, the beneficial effects of pre-operative treatment on glucose tolerance, blood pressure, cardiac abnormalities, and obstructive nasopharyngeal soft tissues [110] may be hypothesized to affect the safety of anaesthetic procedures. On the basis of these benefits alone, pre-operative use of SRIF analogues should be considered where there is evidence of metabolic or cardiac complications, as well as to alleviate the patient's symptoms prior to surgery.

Optimizing SRIF Analogue Therapy

The Somatostatin System

As discussed earlier, SSTR2 is the pivotal receptor subtype through which GH secretion is regulated. Expression of SSTR2 by GH-secreting adenoma is, however, variable. The resultant variation in the density of SSTR2 on the cell surface of GH-secreting adenoma may, at least in part, explain resistance to currently available SRIF analogues [4, 116]. Somatotrophinomas that respond to octreotide show a high density of SSTR2, whereas non-responders show minimal SSTR2 expression, but frequently show significantly greater expression of SSTR5 [117]. SSTR expression of non-responders is reportedly up to eightfold greater than in responders.

Development of agonists specific for one, or two, SSTRs has allowed characterization of the role of each receptor subtype in control of GH secretion. In vitro SRIF analogues with preference for SSTR5 inhibit GH secretion from primary cultures of rat pituitary cell, human foetal pituitaries, and GH-secreting adenoma [6, 117–119], to a similar degree to SSTR2 analogues [118]. Analogues with enhanced specificity for either SSTR2 (BIM-23197, BIM-23190, NC-4-28B) or SSTR5 (BIM-23268) in primary cultures of GH-secreting adenomas [6, 10, 13, 117, 118, 120] show more potent inhibition of GH release compared with octreotide or lanreotide [119]. The GH inhibition by the differing analogues between individual tumours is heterogeneous [6, 13] with around half the tumours showing inhibition of GH release when exposed to either SSTR2 or SSTR5-directed analogues [13, 120]. The other tumours show inhibition of GH with the SSTR2 analogue but not the SSTR5-specific analogue, or conversely are responsive to the SSTR5 analogue but not the SSTR2 analogue [13, 120]. GH-secreting adenomas responsive to octreotide in vivo respond well to SSTR2-directed analogues in vitro, with little response to SSTR5-specific analogues [117]. In octreotide-non-responsive tumours, SSTR5 analogues result in a more potent inhibition of GH secretion than SSTR2-specific analogues [117]. In these latter tumours, the inhibitory response to SSTR5-specific analogues correlates with the density of SSTR5 expression of the tumour [10].

The SSTR1-selective analogue BIM-23926 and SRIF-14 reduced GH in primary cultures of GH-secreting adenomas [9] by 32 versus 45%, respectively. Similarly, prolactin was inhibited by 20 and 16%, respectively [9], and cell viability was reduced by both drugs. In vitro, the effect of the selective SSTR1 agonist BIM-23745 on GH-secreting adenomas expressing SSTR1, shown to be partial or non-responders to octreotide or lanreotide in vivo, was significantly greater inhibition of GH compared with octreotide (59 vs. 32%) [12]. These data, although complex, support a putative role for both SSTR1- and SSTR5-selective agonists in GH-secreting adenomas resistant to currently available SRIF analogues.

In primary cultures of GH-secreting adenomas, analogue combinations containing elements with affinity for both SSTR2 and SSTR5 were more potent than either SSTR2 and SSTR5 analogues alone, or combinations of compounds specific for the same receptor subtype [119]. Thus, analogues with high affinity to both SSTR2 and SSTR5 may be of therapeutic advantage. In primary cultures of octreotide-responsive and partially responsive tumours [117], a greater inhibition of GH was observed with BIM-23244 (SSTR2 and SSTR5 affinity) in octreotide-responsive tumours compared with octreotide; likely the result of higher affinity of BIM-23244 for SSTR2 compared with octreotide. In partial responders, BIM-23244 produced greater inhibition of GH than octreotide (44 vs. 36%) [117], suggesting greater efficacy of biselective analogues. In the partial responders, the combination of BIM-23268 (SSTR5) and BIM-23197 (SSTR2) was equally as effective as the biselective analogue BIM-23244 [117]. In GH-secreting tumours partially resistant to octreotide in vivo, BIM-23244 resulted in greater maximal GH suppression at a lower EC50 than octreotide [121].

The enhanced efficacy of analogues with affinity for both SSTR2 and SSTR5 may result simply from signalling through both receptors, though it is likely to be contributed to by more complex receptor interactions and enhanced trafficking [122–124]. Heterodimerization of different G-protein-coupled receptors leads to modification of ligand binding and synergy of receptor activation. SSTR2 and SSTR5 form constitutive homodimers at the cell surface, whereas SSTR1 is present in the monomeric state [122–124]. Heterodimers between SSTR1 and SSTR5 result in crossover activation [124] and between SSTR2 and SSTR3 is characterized by a predominance of SSTR2 signalling, SSTR3 silencing, and resistance to agonist-induced desensitization [123]. Heterodimers between SSTR2 and SSTR5 are not yet reported, and cooperation between SSTR2 and SSTR5 may thus occur through a different mechanism. If SSTR5 and SSTR2 are co-transfected and co-stimulated with an SSTR2 analogue then internalization of SSTR2 is modified [125]. Formation of heterodimers in various combinations modifies the effect of the receptor [122] and is likely directed, at least in part, by the specifics of the ligand. This putative agonist-specific trafficking of receptor signalling is complex and results in agonists of equivalent affinity resulting in different quantitative effects on GH secretion [8].

Molecules with high-affinity binding to SSTR1, 2, 3, and 5 [21, 126] in addition to direct agonist activity at each receptor subtype may have a synergistic effect through receptor homo- and heterodimerization. Given the multiple potential modes of action, one could expect that a greater proportion of patients may respond to treatment with these moieties. The most thoroughly investigated multiligand SRIF analogue is SOM230 (pasireotide), which is currently in phase 3 clinical studies.

Interaction of the Dopaminergic and Somatostatin Pathways

Dopamine signals through five G-protein-coupled receptors, D1R–D5R. The SSTR and dopamine receptor families show ~30% sequence homology [127]. The dopaminergic D2R is expressed almost universally in GH-secreting adenomas [128]; however, there is significant variability in the cell surface receptor density [8]. Higher levels of D2R occur in mammosomatotroph tumours compared with pure GH-secreting adenomas [128]. Dopaminergic agonists as single therapy have limited efficacy in achieving biochemical control of somatotrophinomas [129–131]. Nonetheless, clinical studies have suggest the combination of a SRIF and D2R analogue is more effective than either therapy alone [132, 133]. In patients with acromegaly on stable therapy with SRIF analogues, but who fail to obtain target GH and IGF-I normalization, cabergoline resulted in GH and IGF-I control in 21 and 42% of patients, respectively [132].

Heterodimerization of SSTR5 and the D2R in transfected cells resulted in greater affinity for selective SSTR5 and D2 agonists, with consequent enhanced coupling with adenylyl cyclase, a synergistic effect on activation of transduction pathways, and increased receptor trafficking. Heterodimerization of SSTR2 and D2R is associated with reciprocal modification of ligand binding, improved EC50 for cyclic adenosine monophosphate (cAMP) inhibition, and increased internalization of SSTR2 [134]. The SSTR and D2R share a number of signal transduction pathways that may be augmented by heterodimerization of receptors.

Chimeric molecules able to stimulate the SSTR and D2R, "dopastatins", have been formulated to optimize the interaction between these receptor subtypes. BIM-23A387 is characterized by high affinity for SSTR2 and D2R. In primary cultures of GH-secreting adenomas, BIM-23A387 produced similar maximal GH suppression to the SSTR2-selective analogue (BIM-23023), dopaminergic D2 agonist (BIM-53097), and combination of BIM-23023 and BIM-53097 [128]. The mean EC50 for GH suppression was, however, 50 times lower for BIM-23A387 [128]. BIM-23A387 displays similar or greater efficacy in inhibiting GH secretion from cultures of rat pituicytes, human foetal pituitary cells, and GH-secreting adenoma compared with combination of agonists to SSTR2 and D2R (BIM-53097 and BIM-23023) [135]. In 18 GH-secreting adenomas, partially responsive to octreotide in vivo, BIM-23A387 resulted in greater maximal GH suppression at a lower EC50 in 13 tumours and similar responses to octreotide in the remaining five tumours [121].

Evidence suggests incorporation of D2R and SSTR moieties within the same molecule is more effective than individual D2 and SSTR analogues. Several dopastatin molecules with affinity for D2R, SSTR2, and SSTR5 have been developed (BIM-23A760 and BIM-23A761). In cultures of somatotrophinomas, BIM-23A387 showed improved efficacy compared with octreotide [121]. Maximal GH secretion with octreotide, BIM-23A387, and BIM-23A761 was 26, 32, and 41%, respectively [121]. Comparison of BIM-23A761 with a combination of monospecific analogues to SSTR2, SSTR5, and D2R showed inhibition of GH by 49 and 38%, respectively [121]. Progression into phase 2 clinical trials of BIM-23A760 is currently underway.

Drug Pipeline for Acromegaly

Of the multiple drugs examined in preclinical studies, only pasireotide (SOM230) and BIM-23A760 are actively being pursued in clinical studies. There is clearly potential to optimize control of GH and IGF-I and improve convenience for patients with the additional moieties in development.

Pasireotide (SOM230)

Pasireotide (Novartis Pharma AG, Basel, Switzerland) is a stable, synthetic, cyclohexapeptide moiety with high-binding affinity to SSTR1, 2, 3, and 5 (Table 14.3). The essential structural elements of SRIF-14 were identified and transposed into a cyclopeptide template containing both natural and modified amino acids [126]. Binding affinity for the individual SSTR shows a 20–30, 5-, and 40–100-fold greater affinity to SSTR1, SSTR3, and SSTR5 compared with octreotide, respectively [21, 126].

Table 14.3 Human somatostatin receptor and dopamine receptor subtype- binding affinities from transfected CHO cells

Compound	SSTR1	SSTR2	SSTR3	SSTR4	SSTR5	D2R
SRIF-14	2.3	0.2	1.4	1.8	1.4	–
Octreotide	>1,000	0.6	34	>1,000	7.0	–
Lanreotide	>1,000	0.7	98	>1,000	12.7	–
Pasireotide	9.3	1.0	1.5	>100	0.2	–
Cabergoline	–	–	–	–	–	3.0
BIM-23023	>1,000	0.4	87	>1,000	4.2	>1,000
BIM-23120	>1,000	0.3	412	>1,000	214	–
BIM-23190	>1,000	0.3	217	>1,000	11.1	–
BIM-23197	>1,000	0.2	27	897	9.8	>1,000
BIM-23206	>1,000	166	>1,000	>1,000	2.4	–
BIM-23244	>1,000	0.3	133	>1,000	0.7	–
BIM-23268	12	28	5.5	36	0.4	–
BIM-23745	42	>1,000	>1,000	>1,000	>1,000	–
BIM-23926	3.6	>1,000	>1,000	833	788	–
BIM-53097	>1,000	>1,000	>1,000	>1,000	>1,000	22
BIM-23A387	293	0.2	77	>1,000	26	22
BIM-23A760	622	0.03	160	>1,000	42	15
BIM23A761	462	0.06	52	>1,000	3.7	27
BIM-23A779	2.5	0.3	0.6	20	0.6	–
NC-4-28B	–	0.002	–	–	>5	–

Adapted from Shimon et al. [118], Saveanu et al. [117], Tulipano et al. [120], Bruns et al. [21], Ren et al. [135], Jaquet et al. [121], Zatelli et al. [10], Fusco et al. [152]
SRIF-14 somatostatin; *SSTR* somatostatin receptor subtype; *D2R* dopamine receptor subtype 2; *CHO* Chinese hamster ovary

The molecule additionally has high affinity for SSTR2, but two- to threefold less than octreotide and lanreotide. In in vitro studies using primary cultures of rat pituitary and human foetal pituitary cells, pasireotide showed equivalent or greater potency compared with SRIF-14 and octreotide in inhibiting GH secretion [21, 136].

In in vivo rat studies, a single octreotide dose showed a lower ED50 for inhibition of GH secretion at 1 h post dose, indicating greater potency than pasireotide [21, 126]. However, at 6 h post dose, pasireotide showed the lower ED50 conversant with its longer biological half-life (2 vs. 23 h) [21, 126]. With continuous pasireotide, octreotide, or placebo for 4 months, suppression of IGF-I levels was more pronounced and long-lasting during pasireotide therapy [21, 126] and was reflected in reduced growth and weight gain [126]. Single-dose experiments in rats show pasireotide inhibits insulin secretion at lower doses than octreotide, suggesting this molecule may be potentially diabetogenic [21]. In chronically treated rats, no adverse effect of pasireotide on glucose levels was observed [21]. In vivo studies in monkeys and dogs confirmed the superiority of pasireotide to octreotide in inhibiting the GH axis [126]. Plasma glucose, insulin, and glucagon levels remained unchanged during 7 days of pasireotide treatment in monkeys [126].

In primary cultures of GH-secreting adenomas, pasireotide induced similar maximum GH secretion to octreotide and SRIF-14 [136], but inhibited GH secretion in a greater proportion of individual cultures studied [11, 136], and induced greater suppression of prolactin secretion from mammosomatotroph tumours. The IC50 value for GH inhibition with pasireotide is higher in both rat pituicyte and GH-secreting adenoma cell cultures compared with octreotide and SRIF-14 [11].

A proof-of-concept study in patients with acromegaly using modelling analysis showed the EC50 for pasireotide to be significantly greater than for octreotide; however, pasireotide had lower clearance and a longer duration of action [137]. The reduced potency but superior pharmacokinetics/dynamics of pasireotide results in a dose of 250 μg being equivalent to ~100 μg of octreotide [137]. In a single-dose comparison study of pasireotide (100 and 250 μg) and octreotide (100 μg) in patients with active acromegaly, similar decreases in mean GH levels were shown [138]. A similar magnitude of GH inhibition was observed with pasireotide and octreotide in eight tumours; whereas a greater degree of inhibition was observed with pasireotide and octreotide in three and one patients, respectively [138]. In single-dose studies, octreotide led to inhibition of insulin secretion not observed with pasireotide [138, 139], and although no differences in glucose levels between treatments were observed following lunch, an early increase in glucose 1 h after injection was observed only with pasireotide [139]. No significant side effects or local injection site reactions were observed with pasireotide.

A 16-week, phase 2 randomized multicentre open-label crossover study of pasireotide (200–600 μg b.d.) was performed in 60 patients with active acromegaly (GH >5 μg/L and elevated IGF-I), either treatment-naive or following surgery [140]. Patients initially received subcutaneous octreotide 100 mg 3 times daily and were then randomized to pasireotide 200, 400, or 600 μg twice daily in random order, each phase for 4 weeks. Following 4 weeks of octreotide, 9% showed full control (GH levels less than 2.5 μg/L and normal IGF-I) and 26% GH levels less than 2.5 μg/L. After 3 months of pasireotide, 27% showed a full response and 49% GH values less than 2.5 μg/L [140]. Tumour volume reduced by a mean of 14% with pasireotide, with 39% achieving more than a 20% decrease. As expected, symptoms and signs of acromegaly improved. Side effects were generally mild to moderate and consisted primarily of gastrointestinal upset [140]. At baseline, 38 patients showed normal fasting glucose, 17 impaired fasting glucose (5.6–7.0 mmol/L) and four diabetes (fasting glucose >7.0 mmol/L). After 3 months of pasireotide, 30 patients retained normal, 13 showed impaired, fasting glucose, and 16 showed diabetes [140], suggestive of a pro-diabetic tendency of this drug.

Pasireotide has since been incorporated in a LAR formulation similar to that of octreotide LAR. In a single-dose comparison of pasireotide LAR (4 and 8 mg/kg) with octreotide LAR in rats and mice over 35 days, octreotide LAR reduced IGF-I by 31% on day 1 with tachyphylaxis occurring thereafter [141]. In contrast, pasireotide LAR reduced IGF-I by 40% on day 1 and maintained suppression to ~50% of control values until day 35 [84]. Plasma glucose levels tended to be slightly higher in the pasireotide LAR-treated rats. Insulin was inhibited to a similar extent by both drugs [141]. The interim results of an ongoing phase 1 study of pasireotide

LAR were presented at the 90th Meeting of the American Endocrine Society [142]. Thirty-five patients with de-novo, persistent, or recurrent acromegaly, and baseline GH >5 μg/L and elevated IGF-I levels were randomized to 20, 40, or 60 mg of intramuscular pasireotide LAR. Assessment at 3 months showed a dose-dependent reduction in IGF-I levels, with normalized values obtained with 20, 40, and 60 mg in 50, 64, and 62% of patients, respectively. Epidemiologically "safe" GH levels of <2.5 μg/L were obtained in 50, 50, and 54%, respectively. GH <2.5 μg/L and normalization of IGF-I levels were obtained in 40, 36, and 54% of patients receiving 20, 40, and 60 mg, respectively [142].

BIM-23A760 Chimeric/Dopastatin

BIM-23A760 has high affinity for SSTR2, SSTR5, and D2R (Table 14.3) [121] and has been chosen as the moiety to undergo future development in clinical studies.

In rat pituitary cell cultures incubated with BIM-23A760, BIM-23120 (SSTR2), and octreotide for 24 h, GH secretion was inhibited in 50, 36, and 34% of cultures, respectively [143]. No effect of D2R or SSTR5-selective analogues was observed. In the rat mammosomatotroph cell line (GH3), which expresses only SSTR2, BIM-23A760 and the SSTR2-selective ligands were equally effective in inhibiting GH release, whereas SSTR5 and D2R selective analogues had no effect [143]. In primary cultures of tumours derived from acromegalic patients who were partially responsive to octreotide [121], BIM-23A760 resulted in greater GH suppression compared with octreotide [121]. To examine the importance of receptor expression on efficacy of SRIF analogues, SSTR2, SSTR5, and D2R levels were quantitated in individual GH-secreting tumours [144]. Analogues with affinity for only one receptor subtype showed GH suppression with octreotide (SSTR2) in 61%, BIM-23268 (SSTR5) in 19%, and cabergoline (D2R) in 21% of tumours. In those tumours non-responsive to octreotide, a significantly greater inhibition of GH secretion was observed with BIM-23A760 (52 vs. 28%) [144].

Octreotide Implants

Endo Pharmaceutical Solutions Inc. (Indevus pharmaceuticals; Lexington, MA, USA) has been involved in developing a formulation of octreotide using its Hydron polymer technology. In this process, octreotide is incorporated into non-biodegradable hydro-gel reservoirs resulting in sustained release of the drug at a constant and predetermined rate for 6 months. The implant is inserted subcutaneously in the inner aspect of the upper arm. A phase I/II proof-of-concept study was completed in 11 patients with acromegaly in 2004. The results of a 6-month, open-label, phase 2 trial were reported in November 2007. The trial enrolled 34 patients with acromegaly who previously showed complete or partial responsiveness to octreotide. Half of the patients had baseline GH levels of <5 μg/L on enrolment to the study [145]. The octreotide implant maintained GH at this level in 94% and normalized IGF-I in around 60% of patients, similar to the effect achieved with octreotide LAR. The patients with baseline GH levels of >5 μg/L showed the octreotide implant to suppress GH to <5 and <2.5 μg/L in 59 and 35%, respectively, similar to the effect of octreotide LAR in partial responders [145]. Endo pharmaceutical solutions initiated a phase 3 study to test the efficacy, safety, and tolerability of their octreotide implant in 34 sites across six countries in early 2008, with the aim of recruiting 140 subjects across the USA and Europe [146]. Results are awaited.

Octreotide C2L

Octreotide C2L 30 mg is a prolonged release preparation administered intramuscularly at 6 weekly intervals being developed by Ambrilia Biopharma Inc. (QC, Canada) [147]. In May 2008, Ambrilia announced the results of a 24-week, phase 3 clinical trial evaluating safety and efficacy in 65 patients with acromegaly. Individuals switched from octreotide LAR to C2L at 12 weeks and reassessed at 24 weeks. At 24 weeks, GH levels were controlled in 42% of patients on octreotide C2L compared with 33% receiving octreotide LAR during the initial 12 weeks. Concurrent values for IGF-I normalization were 42% at week 24 and 45% at week 12. Improvement in clinical symptoms of acromegaly was equivalent with both treatments. Side effects were mild and transient. Octreotide C2L therefore shows non-inferiority to octreotide LAR, but with less-frequent injections in an easier product to reconstitute [148, 149]. Following this success, the company initiated a phase 3 extension trial to evaluate the long-term safety of C2L over 96 weeks. The company is simultaneously conducting phase 3 trials designed to evaluate C2L octreotide 10- and 20-mg formulations [150].

Octreotide Atrigel

QLT Inc. (Vancouver, Canada) has incorporated octreotide within the Atrigel drug delivery system, aiming to provide steady-state levels over a period of 30 days. The Atrigel matrix consists of biodegradable polymers dissolved in a biocompatible carrier. A phase 2a study aimed at enrolling 16 patients with acromegaly, and due to commence in 2006, has been delayed following the finding of adverse events in a primate toxicology study [151]. Further toxicology studies prior to human studies were a requisite. QLT Inc. has decided not to proceed with the octreotide Atrigel program at the current time.

References

1. Krulich L, Dhariwal AP, McCann SM. Stimulatory and inhibitory effects of purified hypothalamic extracts on growth hormone release from rat pituitary in vitro. Endocrinology. 1968;83(4):783–90.
2. Brazeau P, Vale W, Burgus R, Ling N, Butcher M, Rivier J, et al. Hypothalamic polypeptide that inhibits the secretion of immunoreactive pituitary growth hormone. Science. 1973;179(68):77–9.
3. Schonbrunn A. Somatostatin receptors present knowledge and future directions. Ann Oncol. 1999;10 Suppl 2:S17–21.
4. Hofland LJ, Lamberts SW. The pathophysiological consequences of somatostatin receptor internalization and resistance. Endocr Rev. 2003;24(1):28–47.
5. Lania A, Mantovani G, Spada A. Genetic abnormalities of somatostatin receptors in pituitary tumors. Mol Cell Endocrinol. 2008;286(1–2):180–6.
6. Danila DC, Haidar JN, Zhang X, Katznelson L, Culler MD, Klibanski A. Somatostatin receptor-specific analogs: effects on cell proliferation and growth hormone secretion in human somatotroph tumors. J Clin Endocrinol Metab. 2001;86(7):2976–81.
7. Hu C, Yi C, Hao Z, Cao S, Li H, Shao X, et al. The effect of somatostatin and SSTR3 on proliferation and apoptosis of gastric cancer cells. Cancer Biol Ther. 2004;3(8):726–30.
8. Saveanu A, Jaquet P, Brue T, Barlier A. Relevance of coexpression of somatostatin and dopamine D2 receptors in pituitary adenomas. Mol Cell Endocrinol. 2008;286(1–2):206–13.
9. Zatelli MC, Piccin D, Tagliati F, Ambrosio MR, Margutti A, Padovani R, et al. Somatostatin receptor subtype 1 selective activation in human growth hormone (GH)- and prolactin (PRL)-secreting pituitary adenomas: effects on cell viability, GH, and PRL secretion. J Clin Endocrinol Metab. 2003;88(6):2797–802.
10. Zatelli MC, Piccin D, Tagliati F, Bottoni A, Ambrosio MR, Margutti A, et al. Dopamine receptor subtype 2 and somatostatin receptor subtype 5 expression influences somatostatin analogs effects on human somatotroph pituitary adenomas in vitro. J Mol Endocrinol. 2005;35(2):333–41.
11. Hofland LJ, van der Hoek J, van Koetsveld PM, de Herder WW, Waaijers M, Sprij-Mooij D, et al. The novel somatostatin analog SOM230 is a potent inhibitor of hormone release by growth hormone- and prolactin-secreting pituitary adenomas in vitro. J Clin Endocrinol Metab. 2004;89(4):1577–85.
12. Matrone C, Pivonello R, Colao A, Cappabianca P, Cavallo LM. Del Basso De Caro ML, et al. Expression and function of somatostatin receptor subtype 1 in human growth hormone secreting pituitary tumors deriving from patients partially responsive or resistant to long-term treatment with somatostatin analogs. Neuroendocrinology. 2004;79(3):142–8.
13. Jaquet P, Saveanu A, Gunz G, Fina F, Zamora AJ, Grino M, et al. Human somatostatin receptor subtypes in acromegaly: distinct patterns of messenger ribonucleic acid expression and hormone suppression identify different tumoral phenotypes. J Clin Endocrinol Metab. 2000;85(2):781–92.
14. Tulipano G, Soldi D, Bagnasco M, Culler MD, Taylor JE, Cocchi D, et al. Characterization of new selective somatostatin receptor subtype-2 (sst2) antagonists, BIM-23627 and BIM-23454. Effects of BIM-23627 on GH release in anesthetized male rats after short-term high-dose dexamethasone treatment. Endocrinology. 2002;143(4):1218–24.
15. Murray RD, Kim K, Ren SG, Chelly M, Umehara Y, Melmed S. Central and peripheral actions of somatostatin on the growth hormone-IGF-I axis. J Clin Invest. 2004;114(3):349–56.
16. Bauer W, Briner U, Doepfner W, Haller R, Huguenin R, Marbach P, et al. SMS 201–995: a very potent and selective octapeptide analogue of somatostatin with prolonged action. Life Sci. 1982;31(11):1133–40.
17. del Pozo E, Neufeld M, Schluter K, Tortosa F, Clarenbach P, Bieder E, et al. Endocrine profile of a long-acting somatostatin derivative SMS 201–995. Study in normal volunteers following subcutaneous administration. Acta Endocrinol (Copenh). 1986;111(4):433–9.

18. Barkan AL, Kelch RP, Hopwood NJ, Beitins IZ. Treatment of acromegaly with the long-acting somatostatin analog SMS 201–995. J Clin Endocrinol Metab. 1988;66(1):16–23.
19. Pieters GF, Smals AE, Smals AG, von Gennep JA, Kloppenborg PW. The effect of minisomatostatin on anomalous growth hormone responses in acromegaly. Acta Endocrinol (Copenh). 1987;114(4):537–42.
20. Battershill PE, Clissold SP. Octreotide. A review of its pharmacodynamic and pharmacokinetic properties, and therapeutic potential in conditions associated with excessive peptide secretion. Drugs. 1989;38(5):658–702.
21. Bruns C, Lewis I, Briner U, Meno-Tetang G, Weckbecker G. SOM230: a novel somatostatin peptidomimetic with broad somatotropin release inhibiting factor (SRIF) receptor binding and a unique antisecretory profile. Eur J Endocrinol. 2002;146(5):707–16.
22. Anthony LB. Long-acting formulations of somatostatin analogues. Ital J Gastroenterol Hepatol. 1999;31 Suppl 2:S216–8.
23. Lancranjan I, Bruns C, Grass P, Jaquet P, Jervell J, Kendall-Taylor P, et al. Sandostatin LAR: a promising therapeutic tool in the management of acromegalic patients. Metabolism. 1996;45(8 Suppl 1):67–71.
24. Chen T, Miller TF, Prasad P, Lee J, Krauss J, Miscik K, et al. Pharmacokinetics, pharmacodynamics, and safety of microencapsulated octreotide acetate in healthy subjects. J Clin Pharmacol. 2000;40(5):475–81.
25. Jenkins PJ, Akker S, Chew SL, Besser GM, Monson JP, Grossman AB. Optimal dosage interval for depot somatostatin analogue therapy in acromegaly requires individual titration. Clin Endocrinol (Oxf). 2000;53(6):719–24.
26. Colao A, Auriemma RS, Rebora A, Galdiero M, Resmini E, Minuto F, et al. Significant tumour shrinkage after 12 months of lanreotide Autogel-120 mg treatment given first-line in acromegaly. Clin Endocrinol (Oxf). 2009;71(2):237–45.
27. Lombardi G, Minuto F, Tamburrano G, Ambrosio MR, Arnaldi G, Arosio M, et al. Efficacy of the new long-acting formulation of lanreotide (lanreotide Autogel) in somatostatin analogue-naive patients with acromegaly. J Endocrinol Invest. 2009;32(3):202–9.
28. Bronstein M, Musolino N, Jallad R, Cendros JM, Ramis J, Obach R, et al. Pharmacokinetic profile of lanreotide Autogel in patients with acromegaly after four deep subcutaneous injections of 60, 90 or 120 mg every 28 days. Clin Endocrinol (Oxf). 2005;63(5):514–9.
29. Cendros JM, Peraire C, Troconiz IF, Obach R. Pharmacokinetics and population pharmacodynamic analysis of lanreotide Autogel. Metabolism. 2005;54(10):1276–81.
30. Carlsen SM, Lund-Johansen M, Schreiner T, Aanderud S, Johannesen O, Svartberg J, et al. Preoperative octreotide treatment in newly diagnosed acromegalic patients with macroadenomas increases cure short-term postoperative rates: a prospective, randomized trial. J Clin Endocrinol Metab. 2008;93(8):2984–90.
31. Vance ML, Harris AG. Long-term treatment of 189 acromegalic patients with the somatostatin analog octreotide. Results of the International Multicenter Acromegaly Study Group Arch Intern Med. 1991;151(8):1573–8.
32. Ezzat S, Snyder PJ, Young WF, Boyajy LD, Newman C, Klibanski A, et al. Octreotide treatment of acromegaly. A randomized, multicenter study. Ann Intern Med. 1992;117(9):711–8.
33. Baldelli R, Colao A, Razzore P, Jaffrain-Rea ML, Marzullo P, Ciccarelli E, et al. Two-year follow-up of acromegalic patients treated with slow release lanreotide (30 mg). J Clin Endocrinol Metab. 2000;85(11):4099–103.
34. Verhelst JA, Pedroncelli AM, Abs R, Montini M, Vandeweghe MV, Albani G, et al. Slow-release lanreotide in the treatment of acromegaly: a study in 66 patients. Eur J Endocrinol. 2000;143(5):577–84.
35. Freda PU, Katznelson L, van der Lely AJ, Reyes CM, Zhao S, Rabinowitz D. Long-acting somatostatin analog therapy of acromegaly: a meta-analysis. J Clin Endocrinol Metab. 2005;90(8):4465–73.
36. Cozzi R, Attanasio R, Montini M, Pagani G, Lasio G, Lodrini S, et al. Four-year treatment with octreotide-long-acting repeatable in 110 acromegalic patients: predictive value of short-term results? J Clin Endocrinol Metab. 2003;88(7):3090–8.

37. Lancranjan I, Atkinson AB. Results of a European multicentre study with Sandostatin LAR in acromegalic patients. Sandostatin LAR Group. Pituitary. 1999;1(2):105–14.
38. Ayuk J, Stewart SE, Stewart PM, Sheppard MC. Efficacy of Sandostatin LAR (long-acting somatostatin analogue) is similar in patients with untreated acromegaly and in those previously treated with surgery and/or radiotherapy. Clin Endocrinol (Oxf). 2004;60(3):375–81.
39. Chanson P, Boerlin V, Ajzenberg C, Bachelot Y, Benito P, Bringer J, et al. Comparison of octreotide acetate LAR and lanreotide SR in patients with acromegaly. Clin Endocrinol (Oxf). 2000;53(5):577–86.
40. Turner HE, Vadivale A, Keenan J, Wass JA. A comparison of lanreotide and octreotide LAR for treatment of acromegaly. Clin Endocrinol (Oxf). 1999;51(3):275–80.
41. Cozzi R, Dallabonzana D, Attanasio R, Barausse M, Oppizzi G. A comparison between octreotide-LAR and lanreotide-SR in the chronic treatment of acromegaly. Eur J Endocrinol. 1999;141(3):267–71.
42. Kendall-Taylor P, Miller M, Gebbie J, Turner S. al-Maskari M. Long-acting octreotide LAR compared with lanreotide SR in the treatment of acromegaly Pituitary. 2000;3(2):61–5.
43. Ronchi CL, Orsi E, Giavoli C, Cappiello V, Epaminonda P, Beck-Peccoz P, et al. Evaluation of insulin resistance in acromegalic patients before and after treatment with somatostatin analogues. J Endocrinol Invest. 2003;26(6):533–8.
44. Murray RD, Melmed S. A critical analysis of clinically available somatostatin analog formulations for therapy of acromegaly. J Clin Endocrinol Metab. 2008;93(8):2957–68.
45. Caron P, Beckers A, Cullen DR, Goth MI, Gutt B, Laurberg P, et al. Efficacy of the new long-acting formulation of lanreotide (lanreotide Autogel) in the management of acromegaly. J Clin Endocrinol Metab. 2002;87(1):99–104.
46. Caron P, Bex M, Cullen DR, Feldt-Rasmussen U. Pico Alfonso AM, Pynka S, et al. One-year follow-up of patients with acromegaly treated with fixed or titrated doses of lanreotide Autogel. Clin Endocrinol (Oxf). 2004;60(6):734–40.
47. Caron P, Cogne M, Raingeard I, Bex-Bachellerie V, Kuhn JM. Effectiveness and tolerability of 3-year lanreotide Autogel treatment in patients with acromegaly. Clin Endocrinol (Oxf). 2006;64(2):209–14.
48. Lucas T, Astorga R. Efficacy of lanreotide Autogel administered every 4–8 weeks in patients with acromegaly previously responsive to lanreotide microparticles 30 mg: a phase III trial. Clin Endocrinol (Oxf). 2006;65(3):320–6.
49. Melmed S, Cook D, Schopohl J, Goth MI, Lam KS, Marek J. Rapid and sustained reduction of serum growth hormone and insulin-like growth factor-1 in patients with acromegaly receiving lanreotide Autogel therapy: a randomized, placebo-controlled, multicenter study with a 52 week open extension. Pituitary. 2010;13(1):18–28.
50. van Thiel SW, Romijn JA, Biermasz NR, Ballieux BE, Frolich M, Smit JW, et al. Octreotide long-acting repeatable and lanreotide Autogel are equally effective in controlling growth hormone secretion in acromegalic patients. Eur J Endocrinol. 2004;150(4):489–95.
51. Alexopoulou O, Abrams P, Verhelst J, Poppe K, Velkeniers B, Abs R, et al. Efficacy and tolerability of lanreotide Autogel therapy in acromegalic patients previously treated with octreotide LAR. Eur J Endocrinol. 2004;151(3):317–24.
52. Ashwell SG, Bevan JS, Edwards OM, Harris MM, Holmes C, Middleton MA, et al. The efficacy and safety of lanreotide Autogel in patients with acromegaly previously treated with octreotide LAR. Eur J Endocrinol. 2004;150(4):473–80.
53. Ronchi CL, Boschetti M. Degli Uberti EC, Mariotti S, Grottoli S, Loli P, et al. Efficacy of a slow-release formulation of lanreotide (Autogel) 120 mg in patients with acromegaly previously treated with octreotide long acting release (LAR): an open, multicentre longitudinal study. Clin Endocrinol (Oxf). 2007;67(4):512–9.
54. Andries M, Glintborg D, Kvistborg A, Hagen C, Andersen M. A 12-month randomized crossover study on the effects of lanreotide autogel and octreotide long-acting repeatable on GH and IGF-l in patients with acromegaly. Clin Endocrinol (Oxf). 2007;68(3):473–80.

55. Bevan JS, Newell-Price J, Wass JA, Atkin SL, Bouloux PM, Chapman J, et al. Home administration of lanreotide Autogel® by patients with acromegaly, or their partners, is safe and effective. Clin Endocrinol (Oxf). 2008;68(3):343–9.
56. Newman CB, Melmed S, George A, Torigian D, Duhaney M, Snyder P, et al. Octreotide as primary therapy for acromegaly. J Clin Endocrinol Metab. 1998;83(9):3034–40.
57. Colao A, Ferone D, Marzullo P, Cappabianca P, Cirillo S, Boerlin V, et al. Long-term effects of depot long-acting somatostatin analog octreotide on hormone levels and tumor mass in acromegaly. J Clin Endocrinol Metab. 2001;86(6):2779–86.
58. Bevan JS, Atkin SL, Atkinson AB, Bouloux PM, Hanna F, Harris PE, et al. Primary medical therapy for acromegaly: an open, prospective, multicenter study of the effects of subcutaneous and intramuscular slow-release octreotide on growth hormone, insulin-like growth factor-I, and tumor size. J Clin Endocrinol Metab. 2002;87(10):4554–63.
59. Cozzi R, Montini M, Attanasio R, Albizzi M, Lasio G, Lodrini S, et al. Primary treatment of acromegaly with octreotide LAR: a long-term (up to nine years) prospective study of its efficacy in the control of disease activity and tumor shrinkage. J Clin Endocrinol Metab. 2006;91(4):1397–403.
60. Colao A, Pivonello R, Rosato F, Tita P, De Menis E, Barreca A, et al. First-line octreotide-LAR therapy induces tumour shrinkage and controls hormone excess in patients with acromegaly: results from an open, prospective, multicentre trial. Clin Endocrinol (Oxf). 2006;64(3):342–51.
61. Mercado M, Borges F, Bouterfa H, Chang TC, Chervin A, Farrall AJ, et al. A prospective, multicentre study to investigate the efficacy, safety and tolerability of octreotide LAR (long-acting repeatable octreotide) in the primary therapy of patients with acromegaly. Clin Endocrinol (Oxf). 2007;66(6):859–68.
62. Amato G, Mazziotti G, Rotondi M, Iorio S, Doga M, Sorvillo F, et al. Long-term effects of lanreotide SR and octreotide LAR on tumour shrinkage and GH hypersecretion in patients with previously untreated acromegaly. Clin Endocrinol (Oxf). 2002;56(1):65–71.
63. Colao A, Di Sarno A, Cappabianca P, Di Somma C, Pivonello R, Lombardi G. Withdrawal of long-term cabergoline therapy for tumoral and nontumoral hyperprolactinemia. N Engl J Med. 2003;349(21):2023–33.
64. Ronchi CL, Rizzo E, Lania AG, Pivonello R, Grottoli S, Colao A, et al. Preliminary data on biochemical remission of acromegaly after somatostatin analogs withdrawal. Eur J Endocrinol. 2008;158(1):19–25.
65. Kokudo N, Kothary PC, Eckhauser FE, Raper SE. Inhibitory effects of somatostatin on rat hepatocyte proliferation are mediated by cyclic AMP. J Surg Res. 1991;51(2):113–8.
66. Ferjoux G, Bousquet C, Cordelier P, Benali N, Lopez F, Rochaix P, et al. Signal transduction of somatostatin receptors negatively controlling cell proliferation. J Physiol Paris. 2000;94 (3–4):205–10.
67. Scambia G, Panici PB, Baiocchi G, Perrone L, Iacobelli S, Mancuso S. Antiproliferative effects of somatostatin and the somatostatin analog SMS 201–995 on three human breast cancer cell lines. J Cancer Res Clin Oncol. 1988;114(3):306–8.
68. Pagliacci MC, Tognellini R, Grignani F, Nicoletti I. Inhibition of human breast cancer cell (MCF-7) growth in vitro by the somatostatin analog SMS 201–995: effects on cell cycle parameters and apoptotic cell death. Endocrinology. 1991;129(5):2555–62.
69. Hofland LJ, van Koetsveld PM, Wouters N, Waaijers M, Reubi JC, Lamberts SW. Dissociation of antiproliferative and antihormonal effects of the somatostatin analog octreotide on 7315b pituitary tumor cells. Endocrinology. 1992;131(2):571–7.
70. Weckbecker G, Liu R, Tolcsvai L, Bruns C. Antiproliferative effects of the somatostatin analogue octreotide (SMS 201–995) on ZR-75-1 human breast cancer cells in vivo and in vitro. Cancer Res. 1992;52(18):4973–8.
71. Santini V, Lamberts SW, Krenning EP, Backx B, Lowenberg B. Somatostatin and its cyclic octapeptide analog SMS 201–995 as inhibitors of proliferation of human acute lymphoblastic and acute myeloid leukemia. Leuk Res. 1995;19(10):707–12.

72. Florio T, Thellung S, Arena S, Corsaro A, Spaziante R, Gussoni G, et al. Somatostatin and its analog lanreotide inhibit the proliferation of dispersed human non-functioning pituitary adenoma cells in vitro. Eur J Endocrinol. 1999;141(4):396–408.
73. Bevan JS. Clinical review: The antitumoral effects of somatostatin analog therapy in acromegaly. J Clin Endocrinol Metab. 2005;90(3):1856–63.
74. Patel YC. Somatostatin and its receptor family. Front Neuroendocrinol. 1999;20(3):157–98.
75. Weckbecker G, Raulf F, Bodmer D, Bruns C. Indirect antiproliferative effect of the somatostatin analog octreotide on MIA PaCa-2 human pancreatic carcinoma in nude mice. Yale J Biol Med. 1997;70(5–6):549–54.
76. Garcia de la Torre N, Wass JA, Turner HE. Antiangiogenic effects of somatostatin analogues. Clin Endocrinol (Oxf) 2002;57(4):425–41.
77. Florio T, Morini M, Villa V, Arena S, Corsaro A, Thellung S, et al. Somatostatin inhibits tumor angiogenesis and growth via somatostatin receptor-3-mediated regulation of endothelial nitric oxide synthase and mitogen-activated protein kinase activities. Endocrinology. 2003;144(4):1574–84.
78. Kumar M, Liu ZR, Thapa L, Chang Q, Wang DY, Qin RY. Antiangiogenic effect of somatostatin receptor subtype 2 on pancreatic cancer cell line: Inhibition of vascular endothelial growth factor and matrix metalloproteinase-2 expression in vitro. World J Gastroenterol. 2004;10(3):393–9.
79. Sharma K, Patel YC, Srikant CB. Subtype-selective induction of wild-type p53 and apoptosis, but not cell cycle arrest, by human somatostatin receptor 3. Mol Endocrinol. 1996;10(12):1688–96.
80. He Y, Yuan XM, Lei P, Wu S, Xing W, Lan XL, et al. The antiproliferative effects of somatostatin receptor subtype 2 in breast cancer cells. Acta Pharmacol Sin. 2009;30(7):1053–9.
81. Teijeiro R, Rios R, Costoya JA, Castro R, Bello JL, Devesa J, et al. Activation of human somatostatin receptor 2 promotes apoptosis through a mechanism that is independent from induction of p53. Cell Physiol Biochem. 2002;12(1):31–8.
82. Sharma K, Patel YC, Srikant CB. C-terminal region of human somatostatin receptor 5 is required for induction of Rb and G1 cell cycle arrest. Mol Endocrinol. 1999;13(1):82–90.
83. Buscail L, Esteve JP, Saint-Laurent N, Bertrand V, Reisine T, O'Carroll AM, et al. Inhibition of cell proliferation by the somatostatin analogue RC-160 is mediated by somatostatin receptor subtypes SSTR2 and SSTR5 through different mechanisms. Proc Natl Acad Sci U S A. 1995;92(5):1580–4.
84. Cordelier P, Esteve JP, Bousquet C, Delesque N, O'Carroll AM, Schally AV, et al. Characterization of the antiproliferative signal mediated by the somatostatin receptor subtype sst5. Proc Natl Acad Sci U S A. 1997;94(17):9343–8.
85. Li M, Wang X, Li W, Li F, Yang H, Wang H, et al. Somatostatin receptor-1 induces cell cycle arrest and inhibits tumor growth in pancreatic cancer. Cancer Sci. 2008;99(11):2218–23.
86. Florio T, Thellung S, Corsaro A, Bocca L, Arena S, Pattarozzi A, et al. Characterization of the intracellular mechanisms mediating somatostatin and lanreotide inhibition of DNA synthesis and growth hormone release from dispersed human GH-secreting pituitary adenoma cells in vitro. Clin Endocrinol (Oxf). 2003;59(1):115–28.
87. Losa M, Ciccarelli E, Mortini P, Barzaghi R, Gaia D, Faccani G, et al. Effects of octreotide treatment on the proliferation and apoptotic index of GH-secreting pituitary adenomas. J Clin Endocrinol Metab. 2001;86(11):5194–200.
88. Thapar K, Kovacs KT, Stefaneanu L, Scheithauer BW, Horvath E, Lloyd RV, et al. Antiproliferative effect of the somatostatin analogue octreotide on growth hormone-producing pituitary tumors: results of a multicenter randomized trial. Mayo Clin Proc. 1997;72(10):893–900.
89. Ezzat S, Horvath E, Harris AG, Kovacs K. Morphological effects of octreotide on growth hormone-producing pituitary adenomas. J Clin Endocrinol Metab. 1994;79(1):113–8.
90. Melmed S, Sternberg R, Cook D, Klibanski A, Chanson P, Bonert V, et al. A critical analysis of pituitary tumor shrinkage during primary medical therapy in acromegaly. J Clin Endocrinol Metab. 2005;90(7):4405–10.
91. Maiza JC, Vezzosi D, Matta M, Donadille F, Loubes-Lacroix F, Cournot M, et al. Long-term (up to 18 years) effects on GH/IGF-1 hypersecretion and tumour size of primary somatostatin

analogue (SSTa) therapy in patients with GH-secreting pituitary adenoma responsive to SSTa. Clin Endocrinol (Oxf). 2007;67(2):282–9.
92. Plockinger U, Reichel M, Fett U, Saeger W, Quabbe HJ. Preoperative octreotide treatment of growth hormone-secreting and clinically nonfunctioning pituitary macroadenomas: effect on tumor volume and lack of correlation with immunohistochemistry and somatostatin receptor scintigraphy. J Clin Endocrinol Metab. 1994;79(5):1416–23.
93. Abe T, Ludecke DK. Effects of preoperative octreotide treatment on different subtypes of 90 GH-secreting pituitary adenomas and outcome in one surgical centre. Eur J Endocrinol. 2001;145(2):137–45.
94. Lucas T, Astorga R, Catala M. Preoperative lanreotide treatment for GH-secreting pituitary adenomas: effect on tumour volume and predictive factors of significant tumour shrinkage. Clin Endocrinol (Oxf). 2003;58(4):471–81.
95. Attanasio R, Barausse M, Cozzi R. GH/IGF-I normalization and tumor shrinkage during long-term treatment of acromegaly by lanreotide. J Endocrinol Invest. 2001;24(4):209–16.
96. Casarini AP, Pinto EM, Jallad RS, Giorgi RR, Giannella-Neto D, Bronstein MD. Dissociation between tumor shrinkage and hormonal response during somatostatin analog treatment in an acromegalic patient: preferential expression of somatostatin receptor subtype 3. J Endocrinol Invest. 2006;29(9):826–30.
97. Resmini E, Dadati P, Ravetti JL, Zona G, Spaziante R, Saveanu A, et al. Rapid pituitary tumor shrinkage with dissociation between antiproliferative and antisecretory effects of a long-acting octreotide in an acromegalic patient. J Clin Endocrinol Metab. 2007;92(5):1592–9.
98. Ip MS, Tan KC, Peh WC, Lam KS. Effect of Sandostatin LAR on sleep apnoea in acromegaly: correlation with computerized tomographic cephalometry and hormonal activity. Clin Endocrinol (Oxf). 2001;55(4):477–83.
99. Maison P, Tropeano AI, Macquin-Mavier I, Giustina A, Chanson P. Impact of somatostatin analogs on the heart in acromegaly: a metaanalysis. J Clin Endocrinol Metab. 2007;92(5):1743–7.
100. Fatti LM, Scacchi M, Lavezzi E. Pecori Giraldi F, De Martin M, Toja P, et al. Effects of treatment with somatostatin analogues on QT interval duration in acromegalic patients. Clin Endocrinol (Oxf). 2006;65(5):626–30.
101. Hansen I, Tsalikian E, Beaufrere B, Gerich J, Haymond M, Rizza R. Insulin resistance in acromegaly: defects in both hepatic and extrahepatic insulin action. Am J Physiol. 1986;250(3 Pt 1):E269–73.
102. Moller N, Petrany G, Cassidy D, Sheldon WL, Johnston DG, Laker MF. Effects of the somatostatin analogue SMS 201–995 (sandostatin) on mouth-to-caecum transit time and absorption of fat and carbohydrates in normal man. Clin Sci (Lond). 1988;75(4):345–50.
103. Mazziotti G, Floriani I, Bonadonna S, Torri V, Chanson P, Giustina A. Effects of somatostatin analogs on glucose homeostasis: a metaanalysis of acromegaly studies. J Clin Endocrinol Metab. 2009;94(5):1500–8.
104. Paisley AN, Roberts ME, Trainer PJ. Withdrawal of somatostatin analogue therapy in patients with acromegaly is associated with an increased risk of acute biliary problems. Clin Endocrinol (Oxf). 2007;66(5):723–6.
105. Rideout DJ, Graham MM. Buttock granulomas: a consequence of intramuscular injection of Sandostatin detected by In-111 octreoscan. Clin Nucl Med. 2001;26(7):650.
106. Jallad RS, Musolino NR, Salgado LR, Bronstein MD. Treatment of acromegaly with octreotide-LAR: extensive experience in a Brazilian institution. Clin Endocrinol (Oxf). 2005;63(2):168–75.
107. Barkan AL, Lloyd RV, Chandler WF, Hatfield MK, Gebarski SS, Kelch RP, et al. Preoperative treatment of acromegaly with long-acting somatostatin analog SMS 201–995: shrinkage of invasive pituitary macroadenomas and improved surgical remission rate. J Clin Endocrinol Metab. 1988;67(5):1040–8.
108. Spinas GA, Zapf J, Landolt AM, Stuckmann G, Froesch ER. Pre-operative treatment of 5 acromegalics with a somatostatin analogue: endocrine and clinical observations. Acta Endocrinol (Copenh). 1987;114(2):249–56.

109. Stevenaert A, Harris AG, Kovacs K, Beckers A. Presurgical octreotide treatment in acromegaly. Metabolism. 1992;41(9 Suppl 2):51–8.
110. Colao A, Ferone D, Cappabianca P. del Basso De Caro ML, Marzullo P, Monticelli A, et al. Effect of octreotide pretreatment on surgical outcome in acromegaly J Clin Endocrinol Metab. 1997;82(10):3308–14.
111. Stevenaert A, Beckers A. Presurgical octreotide: treatment in acromegaly. Metabolism. 1996;45(8 Suppl 1):72–4.
112. Lucas-Morante T, Garcia-Uria J, Estrada J, Saucedo G, Cabello A, Alcaniz J, et al. Treatment of invasive growth hormone pituitary adenomas with long-acting somatostatin analog SMS 201–995 before transsphenoidal surgery. J Neurosurg. 1994;81(1):10–4.
113. Stevenaert A, Beckers A. Presurgical octreotide treatment in acromegaly. Acta Endocrinol (Copenh). 1993;129 Suppl 1:18–20.
114. Biermasz NR, van Dulken H, Roelfsema F. Direct postoperative and follow-up results of transsphenoidal surgery in 19 acromegalic patients pretreated with octreotide compared to those in untreated matched controls. J Clin Endocrinol Metab. 1999;84(10):3551–5.
115. Lissett CA, Peacey SR, Laing I, Tetlow L, Davis JR, Shalet SM. The outcome of surgery for acromegaly: the need for a specialist pituitary surgeon for all types of growth hormone (GH) secreting adenoma. Clin Endocrinol (Oxf). 1998;49(5):653–7.
116. Reubi JC, Landolt AM. The growth hormone responses to octreotide in acromegaly correlate with adenoma somatostatin receptor status. J Clin Endocrinol Metab. 1989;68(4):844–50.
117. Saveanu A, Gunz G, Dufour H, Caron P, Fina F, Ouafik L, et al. Bim-23244, a somatostatin receptor subtype 2- and 5-selective analog with enhanced efficacy in suppressing growth hormone (GH) from octreotide-resistant human GH-secreting adenomas. J Clin Endocrinol Metab. 2001;86(1):140–5.
118. Shimon I, Taylor JE, Dong JZ, Bitonte RA, Kim S, Morgan B, et al. Somatostatin receptor subtype specificity in human fetal pituitary cultures. Differential role of SSTR2 and SSTR5 for growth hormone, thyroid-stimulating hormone, and prolactin regulation. J Clin Invest. 1997;99(4):789–98.
119. Shimon I, Yan X, Taylor JE, Weiss MH, Culler MD, Melmed S. Somatostatin receptor (SSTR) subtype-selective analogues differentially suppress in vitro growth hormone and prolactin in human pituitary adenomas. Novel potential therapy for functional pituitary tumors J Clin Invest. 1997;100(9):2386–92.
120. Tulipano G, Bonfanti C, Milani G, Billeci B, Bollati A, Cozzi R, et al. Differential inhibition of growth hormone secretion by analogs selective for somatostatin receptor subtypes 2 and 5 in human growth-hormone-secreting adenoma cells in vitro. Neuroendocrinology. 2001;73(5):344–51.
121. Jaquet P, Gunz G, Saveanu A, Dufour H, Taylor J, Dong J, et al. Efficacy of chimeric molecules directed towards multiple somatostatin and dopamine receptors on inhibition of GH and prolactin secretion from GH-secreting pituitary adenomas classified as partially responsive to somatostatin analog therapy. Eur J Endocrinol. 2005;153(1):135–41.
122. Grant M, Patel RC, Kumar U. The role of subtype-specific ligand binding and the C-tail domain in dimer formation of human somatostatin receptors. J Biol Chem. 2004;279(37):38636–43.
123. Pfeiffer M, Koch T, Schroder H, Klutzny M, Kirscht S, Kreienkamp HJ, et al. Homo- and heterodimerization of somatostatin receptor subtypes. Inactivation of sst(3) receptor function by heterodimerization with sst(2A). J Biol Chem. 2001;276(17):14027–36.
124. Rocheville M, Lange DC, Kumar U, Sasi R, Patel RC, Patel YC. Subtypes of the somatostatin receptor assemble as functional homo- and heterodimers. J Biol Chem. 2000;275(11):7862–9.
125. Sharif N, Gendron L, Wowchuk J, Sarret P, Mazella J, Beaudet A, et al. Coexpression of somatostatin receptor subtype 5 affects internalization and trafficking of somatostatin receptor subtype 2. Endocrinology. 2007;148(5):2095–105.
126. Weckbecker G, Briner U, Lewis I, Bruns C. SOM230: a new somatostatin peptidomimetic with potent inhibitory effects on the growth hormone/insulin-like growth factor-I axis in rats, primates, and dogs. Endocrinology. 2002;143(10):4123–30.

127. Rocheville M, Lange DC, Kumar U, Patel SC, Patel RC, Patel YC. Receptors for dopamine and somatostatin: formation of hetero-oligomers with enhanced functional activity. Science. 2000;288(5463):154–7.
128. Saveanu A, Lavaque E, Gunz G, Barlier A, Kim S, Taylor JE, et al. Demonstration of enhanced potency of a chimeric somatostatin-dopamine molecule, BIM-23A387, in suppressing growth hormone and prolactin secretion from human pituitary somatotroph adenoma cells. J Clin Endocrinol Metab. 2002;87(12):5545–52.
129. Melmed S, Ho K, Klibanski A, Reichlin S, Thorner M. Clinical review 75: Recent advances in pathogenesis, diagnosis, and management of acromegaly. J Clin Endocrinol Metab. 1995;80(12):3395–402.
130. Abs R, Verhelst J, Maiter D, Van Acker K, Nobels F, Coolens JL, et al. Cabergoline in the treatment of acromegaly: a study in 64 patients. J Clin Endocrinol Metab. 1998;83(2):374–8.
131. Colao A, Ferone D, Marzullo P, Di Sarno A, Cerbone G, Sarnacchiaro F, et al. Effect of different dopaminergic agents in the treatment of acromegaly. J Clin Endocrinol Metab. 1997;82(2):518–23.
132. Cozzi R, Attanasio R, Lodrini S, Lasio G. Cabergoline addition to depot somatostatin analogues in resistant acromegalic patients: efficacy and lack of predictive value of prolactin status. Clin Endocrinol (Oxf). 2004;61(2):209–15.
133. Lamberts SW, Zweens M, Verschoor L, del Pozo E. A comparison among the growth hormone-lowering effects in acromegaly of the somatostatin analog SMS 201–995, bromocriptine, and the combination of both drugs. J Clin Endocrinol Metab. 1986;63(1):16–9.
134. Baragli A, Alturaihi H, Watt HL, Abdallah A, Kumar U. Heterooligomerization of human dopamine receptor 2 and somatostatin receptor 2 Co-immunoprecipitation and fluorescence resonance energy transfer analysis. Cell Signal. 2007;19(11):2304–16.
135. Ren SG, Kim S, Taylor J, Dong J, Moreau JP, Culler MD, et al. Suppression of rat and human growth hormone and prolactin secretion by a novel somatostatin/dopaminergic chimeric ligand. J Clin Endocrinol Metab. 2003;88(11):5414–21.
136. Murray RD, Kim K, Ren SG, Lewis I, Weckbecker G, Bruns C, et al. The novel somatostatin ligand (SOM230) regulates human and rat anterior pituitary hormone secretion. J Clin Endocrinol Metab. 2004;89(6):3027–32.
137. Ma P, Wang Y, van der Hoek J, Nedelman J, Schran H, Tran LL, et al. Pharmacokinetic-pharmacodynamic comparison of a novel multiligand somatostatin analog, SOM230, with octreotide in patients with acromegaly. Clin Pharmacol Ther. 2005;78(1):69–80.
138. van der Hoek J, de Herder WW, Feelders RA, van der Lely AJ, Uitterlinden P, Boerlin V, et al. A single-dose comparison of the acute effects between the new somatostatin analog SOM230 and octreotide in acromegalic patients. J Clin Endocrinol Metab. 2004;89(2):638–45.
139. van der Hoek J, van der Lelij AJ, Feelders RA, de Herder WW, Uitterlinden P, Poon KW, et al. The somatostatin analogue SOM230, compared with octreotide, induces differential effects in several metabolic pathways in acromegalic patients. Clin Endocrinol (Oxf). 2005;63(2):176–84.
140. Petersenn S, Schopohl J, Barkan A, Mohideen P, Colao A, Abs R, et al. Pasireotide (SOM230) demonstrates efficacy and safety in patients with acromegaly: a randomized, multicenter, phase II trial. J Clin Endocrinol Metab. 2010;95(6):2781–9.
141. Schmid HA, Brueggen J, Guitard P. Effects of a long-acting release formulation of pasireotide (SOM230) on hormone secretion in rats. In: 89th Annual Meeting of the American Endocrine Society; 2007; Toronto. Canada; 2007 June;2007:P3–337.
142. Petersenn S, Bollerslev J, Arafat AM, Glusman JE, Serri O, Hu M, et al. Pasireotide LAR shows efficacy in patients with acromegaly: interim results from a randomised, multi-centre, phartmacokinetic/pharmacodynamic, phase I study. In: 90th Annual Meeting of the American Endocrine Society, San Francisco; June 2008. p. OR41–5.
143. Gruszka A, Ren SG, Dong J, Culler MD, Melmed S. Regulation of growth hormone and prolactin gene expression and secretion by chimeric somatostatin-dopamine molecules. Endocrinology. 2007;148(12):6107–14.

144. Saveanu A, Gunz G, Guillen S, Dufour H, Culler MD, Jaquet P. Somatostatin and dopamine-somatostatin multiple ligands directed towards somatostatin and dopamine receptors in pituitary adenomas. Neuroendocrinology. 2006;83(3–4):258–63.
145. Open label study of octreotide implant in patients with acromegaly. http://clinicaltrials.gov/ct2/show/NCT00913055. Accessed 17 Oct 2010.
146. Efficacy and safety study of octreotide implant in patients with acromegaly. http://clinicaltrials.gov/ct2/show/NCT00765323. Accessed 17 Oct 2010.
147. Ambrillia product pipeline: acromegaly – octreotide (C2L) http://www.ambrilia.com/en/products/acromegaly-octreotide.php. Accessed 17 Oct 2010.
148. Ambrilla biopharma announces positive phase III results for octreotide C2L. http://www.drugs.com/clinical_trials/ambrilla-biopharma-announces-positive-phase-iii-results-octreotide-c2l-4260.html. Accessed 17 Oct 2010.
149. Efficacy and safety of C2L-OCT-01 PR in acromegalic patients http://clinicaltrials.gov/ct2/show/NCT00616551. Accessed 17 Oct 2010.
150. Safety and biological activity of C2L-OCT-01 PR in acromegalic patients. http://clinicaltrials.gov/ct2/show/NCT00642421. Accessed 17 Oct 2010.
151. QLT delays initiation of phase IIa atrigel/octreotide program. http://qltinc.com/newsCenter/2006/060518.htm. Accessed 17 Oct 2010.
152. Fusco A, Gunz G, Jaquet P, Dufour H, Germanetti AL, Culler MD, et al. Somatostatinergic ligands in dopamine-sensitive and -resistant prolactinomas. Eur J Endocrinol. 2008;158(5):595–603.

Chapter 15
The Role of External Beam Radiation Therapy and Stereotactic Radiosurgery in Acromegaly

Bruce E. Pollock

Abstract The treatment goals for patients with growth hormone (GH)-secreting pituitary adenomas are tumor control, biochemical remission, preservation of anterior pituitary function, and minimal treatment-related complications. Surgical resection is the preferred treatment for most patients with acromegaly because tumor removal provides rapid biochemical remission thereby eliminating the need for lifelong medical therapy. However, patients with pituitary macroadenomas and patients with tumors that extend into the cavernous sinuses generally cannot be cured with surgery alone and some form of adjuvant radiation is indicated. Both external beam radiation therapy (EBRT) and stereotactic radiosurgery (SRS) provide high rates of tumor control and significantly reduce serum GH and insulin-like growth factor-I. The primary risk of both EBRT and SRS for acromegalic patients is new pituitary hormone deficiencies. The time to biochemical remission is shorter after SRS compared to EBRT, and the risk of radiation-induced tumors is significantly lower with SRS. These factors combined with the practical benefits of SRS (1–5 days vs. 5–6 weeks of treatment time) have made SRS the preferred radiation technique for properly selected acromegalic patients who fail or cannot tolerate surgical resection or medical therapy.

Keywords Acromegaly • Growth hormone • Pituitary adenoma • Radiation therapy • Radiosurgery

B.E. Pollock (✉)
Department of Neurological Surgery, Mayo Clinic College of Medicine, Rochester, MN 55905, USA

Department of Radiation Oncology, Mayo Clinic College of Medicine, Rochester, MN 55905, USA
e-mail: pollock.bruce@mayo.edu

K. Ho (ed.), *Growth Hormone Related Diseases and Therapy: A Molecular and Physiological Perspective for the Clinician*, Contemporary Endocrinology, DOI 10.1007/978-1-60761-317-6_15,

Introduction

The preferred management for patients with growth hormone (GH)-secreting pituitary adenomas is tumor resection, most commonly performed via the transsphenoidal approach. In the majority of patients, serum GH and insulin-like growth factor-I (IGF-I) levels are rapidly normalized after surgery reducing the long-term risks of GH oversecretion including diabetes mellitus and cardiovascular disease. Unfortunately, patients with larger tumors and those with tumors that extend into the cavernous sinus much less frequently achieve biochemical remission postoperatively, necessitating either medical therapy or radiation treatment. Medical therapy with long-acting somatostatin analogs (SSA) or GH receptor agonists (pegvisomant) is often initiated to minimize the metabolic consequences of GH oversecretion. However, not every patient responds or tolerates medical therapy and the expense of either SSA or GH receptor therapy is significant, especially if one considers that lifelong treatment is likely necessary. Consequently, radiation treatment is frequently utilized as a salvage technique to stop tumor growth and produce biochemical remission. External beam radiation therapy (EBRT) has long been used as an adjuvant to surgical treatment or as primary treatment for patients with inoperable pituitary tumors [1–9]. Despite tumor control rates greater than 90% after radiation therapy of GH-producing tumors, the time required to achieve biochemical remission is often 10 years or more. In addition, radiotherapy can cause hypothalamo-pituitary dysfunction, cognitive decline, and has a risk of radiation-induced neoplasms [10–14]. Over the past 30 years, stereotactic radiosurgery (SRS) has emerged as an effective alternative to EBRT for many patients with pituitary adenomas [15–25]. In this chapter, the role of EBRT and SRS will be reviewed for patients with GH-secreting pituitary adenomas.

Biology of Radiation Treatments

Ionizing radiation is composed of particulate radiation and electromagnetic radiation. Although some radiation treatments use electrons, protons, or neutrons (particulate radiation), the vast majority of radiation treatments are delivered via photons (electromagnetic radiation). When photon radiation hits a cell, it will most likely interact with water since cells are primarily composed of water. This creates unstable molecules called free radicals that can interact with a cell's DNA, potentially resulting in cell injury or death (indirect effect of radiation). Less frequently, photon radiation may interact with directly with DNA or other cellular structures leading to injury and possibly cell death (direct effect of radiation). Heavy particle radiation such as proton beam therapy causes far more damage to cells from the direct effect of radiation than photon-based techniques, which exert their therapeutic effect almost exclusively from the indirect effect of radiation.

Radiation of cells leads to DNA injury, but not all of the damage is equally capable of producing cellular death. Disruption of DNA-protein cross-links, direct base damage, and single-strand breaks play a relatively small role in the lethality of ionizing radiation. Provided the opposite strand is intact, cells are capable of repairing single-strand breaks with great efficiency. Therefore, the critical lesions caused by radiation are double-strand DNA breaks. The most common types of cell death after radiotherapy are mitotic cell death and apoptosis. If a cell accumulates sufficient radiation damage, its later progeny will be unable to go divide successfully. Several cell divisions may occur before the lethal effects of radiation are expressed. Consequently, rapidly dividing normal tissues or malignant tumors express radiation damage sooner than slower dividing normal tissues or benign tumors such as pituitary adenomas. Not all cells have to go through mitosis in order to succumb to radiation, but instead die from programmed cell death or apoptosis.

Radiation can cause damage to normal tissues in addition to the desired effect on tumors. To be clinically useful, radiation must differentially damage a tumor more than the adjacent normal tissues. This differential effect is referred to as the therapeutic ratio. The therapeutic ratio can be improved by designing radiation treatments that deliver high doses of radiation to the tumor with less radiation going to nearby structures (conformality), or through dose fractionation. In general, normal tissues are better able to repair DNA damage than tumors. However, a single, large dose of radiation is able to overcome these repair mechanisms and damage normal tissues along with tumor. Therefore, unless the radiation dose delivery is highly conformal, it is advantageous to deliver multiple small doses of radiation to maximize its cytotoxic effects while reducing the damage to the adjacent normal tissues.

External Beam Radiation Therapy

Immobilization and accurate reproducibility of the treatment position are essential for patients undergoing radiotherapy. Most radiation therapy centers fabricate a thermoplastic mask for each patient to prevent head movement during radiation delivery. Fiducial markers are then placed on the mask to provide consistent references points. These markers are aligned with lasers in the linear accelerator (LINAC) treatment room each day prior to radiation delivery. Additional head fixation can be achieved with a relocatable head frame stabilized by a mouth-guard type bite block for patients undergoing stereotactic radiation therapy (SRT). A day-to-day accuracy of 2 mm or less is possible with such "pin-less" stereotactic systems.

After the immobilization device has been constructed, the patient will undergo a treatment simulation in the treatment position. Modern simulators use computed tomography (CT) to image the region of interest; image fusion with magnetic resonance imaging (MRI) is typically performed to facilitate tumor volume definition and identify adjacent neuro-vascular structures. The CT simulators incorporate imaging workstations allowing target contouring, image manipulation, and simulation of beam arrangements. Once the planning CT images are in the treatment planning computer, the radiation oncologist defines the gross tumor volume (GTV), the clinical target volume (CTV), and critical normal structures. A planning target volume (PTV) is then generated by enlarging the CTV with a margin of approximately 5 mm to account for patient set-up error and inaccuracies in imaging. A treatment plan is then constructed to deliver a homogenous radiation dose to cover the PTV, while minimizing dose to critical structures outside the PTV. A homogenous dose distribution avoids regions of increased dose within the treatment volume and is essential during EBRT to decreasing the risk of radiation damage outside the PTV.

Contemporary computerized planning systems and utilization of MRI permit EBRT to deliver high dose of radiation to the sellar region while minimizing the radiation dose to the surrounding normal structures. Different methods to accomplish this dosimetric goal include three-dimensional conformal radiotherapy (use of multiple, nonoverlapping, conformal radiation beams shaped by custom-made blocks to the configuration of the tumor volume) and use of multileaf collimators (individual leaves that shift to conforming the radiation beam to the target). The planning and designing of beams in this process are done manually and guided by experience and comparison of few different sets of beam arrangements to achieve the optimal treatment plan.

Intensity modulated radiation therapy (IMRT) provides a completely different approach to treatment planning [6]. Rather than clinicians defining beam directions, beam weights, margins, and then displaying dose distributions to assess whether the treatment plan will lead to an acceptable outcome, IMRT is based on the concept of inverse treatment planning. With inverse treatment planning, the radiation oncologist specifies the prescribed target dose and the dose constraints for normal structures. Algorithms of the inverse treatment planning system allow the evaluation a vast array of possible beam arrangements, and then determine the clinically optimized treatment plan. Hundreds to thousands of small, modulated radiation beams strike a tumor site with varying intensities and from many angles to cover the tumor in a three-dimensional manner.

Radiation therapy is typically performed with the patient receiving one treatment a day, five treatments a week, Monday through Friday. Most conventional fractionation schedules are delivered with 1.8–2.0 Gy fractions each day. Smaller doses per fraction allow more healing of sublethal damage between treatments of the adjacent normal tissues. Hypofractionation using fewer fractions but larger doses per fraction (3–5 Gy) to achieve a lower total dose (20–30 Gy) over has also been employed by some centers. Despite the theoretical radio-biological equivalence of such hypofractionated schedules with either conventional dose fractionation or high-dose, single-fraction techniques, long-term studies on tumor control or toxicities still do not exist.

Stereotactic Radiosurgery

SRS is a surgical procedure that combines the principles of stereotactic localization with the precise delivery of radiation to an imaging-defined target. Over the past three decades, SRS has gained acceptance and is now an integral part of both neurosurgery and radiation oncology. Advances in neuroimaging and improved computer software have made SRS safer and more effective in the management of a wide range of disorders affecting the nervous system. Today, radiosurgery is available to patients

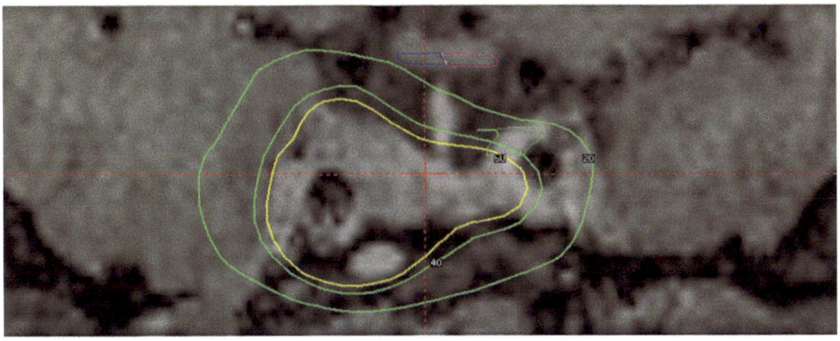

Fig. 15.1 Dose plan for a 38-year-old woman with a residual GH-secreting macroadenoma extending into the right cavernous sinus. Ten isocenters of radiation were used to cover a volume of 6.9 cm^3. The tumor margin dose was 25 Gy, and the maximum radiation dose was 50 Gy. The adjacent right optic nerve received a maximum dose of 10.1 Gy

at the majority of large medical centers. In contrast to EBRT, SRS does not rely on the differential radiation sensitivity of the target compared to the normal brain for its effectiveness. Instead, radiosurgery derives its safety from conformal dose plans that deliver large radiation doses into the target with a rapid fall-off of radiation at the edges of the dose plan (Fig. 15.1). Several different radiosurgical systems satisfy these essential criteria by using various technical solutions.

The Leksell Gamma Knife® (Elekta Instruments, Norcross, GA) has been used to manage patients with a wide variety of intracranial disease for more than 40 years. The most recent version of this system, the Leksell Gamma Knife® PERFEXION™, contains 192 individual cobalt-60 sources which naturally decay to nickel (^{60}Ni). During this process, photons are released and precisely directed through circular channels drilled into a high-density metal helmet. Secondary collimation is accomplished through eight sectors containing 4, 8, and 16 mm openings as well as a closed position. Dose planning is achieved by arranging a number of radiation isocenters, often with different sector openings or blockage of sectors reduce radiation from that direction. The Gamma Knife® employs fixed radiation sources with a constant relation to the collimator helmet and stereotactic frame, thus isocenter verification films or port films are unnecessary. Patient positioning is robotically controlled, repositioning the patient at each isocenter by digitally transferred information from the operator's console. LINAC-based radiosurgical systems have been employed since the 1980s. These systems generally consist of a stereotactic head frame, floor-stand, and a 6-MV LINAC. Multiple sagittally oriented radiation arcs are delivered using a wide range of circular collimators. Detection of mechanical inaccuracy is performed by a phantom-target film technique for every treatment arc. The majority of targets are treated at the 70–80% isodose line. John Adler and colleagues at Stanford University developed a new radiosurgical device called the CyberKnife® (Accuray, Sunnyvale, CA). The CyberKnife® has a compact, lightweight LINAC mounted on a robotic arm. For intracranial targets, patients are fitted with a noninvasive flexible mesh mask to limit large movements. Skeletal landmarks are used to define stereotactic space, therefore fixation using a stereotactic head frame is not required. The system has dynamic tracking software that can follow lesions in six dimensions (three translational and three rotational axes) to account and compensate for any patient movement. If any movement of the target during treatment is detected, the LINAC is repositioned to the new target location. The majority of targets are treated at the 70–80% isodose line. Hypofractionated treatments over 3–5 sessions are frequently delivered with the CyberKnife®.

Patient Selection

A complete patient evaluation is essential before initiating EBRT or SRS for patients with pituitary adenomas. Information on a patient's surgical procedures, pathological review, previous radiation exposure, current endocrine status, and visual function are reviewed to determine the feasibility of radiation treatments for these patients. Critical in the decision-making process is the size and location of the tumor, especially its relationship to the optic nerves and chiasm, as determined by high-quality MRI. Patients with a clearly defined tumor that does not directly involve the optic apparatus are generally considered good candidates for SRS, whereas patients with poorly defined tumors or those with compression of the optic nerves or chiasm are considered better patients for EBRT. Patients with new or progressive visual field deficits in the setting of an enlarging tumor are referred for surgical resection to reduce mass effect and improve neurologic function.

EBRT for GH-Secreting Pituitary Adenomas

Radiation therapy has been used for over 30 years to manage patients with GH-secreting pituitary adenomas. Most centers utilize conventional dose fractionation schedules of 1.8–2.0 Gy per fraction and a total dose of 45–50 Gy. Tumor growth control exceeds 90% with such dosing regimens. However, despite relatively uniform treatment parameters, a wide range of biochemical remission rates have been noted following EBRT (Table 15.1). The discrepancy in hormonal outcomes relates to a number of factors including the method of IGF-I determination, definition of biochemical remission, and duration of follow-up. Most recent studies have shown a clear time-dependent decrease in serum GH and IGF-I levels following EBRT. Minniti et al. reported 47 patients having EBRT from 1982 until 1994 (median follow-up, 12 years) [7]. Age-corrected IGF-I levels were normalized off medications in 8% at 2 years, 23% at 5 years, 42% at 10 years, and 61% at 15 years. In the largest series on EBRT for patients with acromegaly, Jenkins et al. reported outcomes for 656 patients on behalf of the UK National Acromegaly Register Study Group [5]. The median follow-up was 7 years, with 20% of patients having follow-up more than 10 years. Using a criteria of GH less than 2.5 ng/mL and a normal age-corrected IGF-I level off medications, the proportion of patients

Table 15.1 Results of EBRT for patients with GH-secreting pituitary adenomas

References	No. of patients	Radiation dose	Follow-up	Biochemical remission	Hypopituitarism
Barkan et al. [1]	38	Median, 46 Gy	Mean, 6.8 years	5%[a]	Not stated
Powell et al. [9]	32	Mean, 47.4 Gy	Mean, 5.6 years	44%[a]	32%
Cozzi et al. [2]	49	Mean, 45.1 Gy	Median, 14 years	16% at 10 years[a]	8%
Epaminoda et al. [3]	67	Mean, 53.6 Gy	Median, 10 years	47% at 10 years[a]	60%
Gutt et al. [4]	41	Median, 50 Gy	Median, 12.8 years	17% at 7 years[b]	55% at 5 years
Minniti et al. [7, 12]	47	Median, 45 Gy	Median, 12 years	42% at 10 years[c]	78% at 10 years
Jenkins et al. [5]	656	Median, 45 Gy	Median, 7 years	63% at 10 years[b]	18–27% at 10 years
Mullan et al. [8]	57	Median 45 Gy	Mean, 13.9 years	30% at 10 years[d]	33%

[a]Normal IGF-I off medications
[b]GH < 2.5 ng/mL and normal age-corrected IGF-I off medications
[c]Oral glucose-suppressed GH level < 1 μg/L and normal age-corrected IGF-I off medications
[d]Normal IGF-I off medications for at least 1 year

reaching biochemical remission was 38% at 2 years, 50% at 5 years, and 63% at 10 years. Of note, both this study and the study by Minniti et al. found that the chance of GH/IGF-I normalization after EBRT correlated with the preirradiation hormonal levels. Consequently, in reviewing the results of EBRT, one must be mindful of selection bias skewing the proportion of patients achieving biochemical remission down simply based on the fact that patients having EBRT likely have difficult tumors that are not controlled with either medical therapy or surgical resection.

A number of endocrinologists are reluctant to recommend EBRT for acromegalic patients due to the risk of radiation-related complications. Although noted in earlier papers, the risk of visual decline or brain necrosis is extremely uncommon using current techniques and dose prescription guidelines. New anterior pituitary deficits clearly do occur after EBRT, but similar to the chance of biochemical remission, a wide range (8–78%) of new hormonal deficits has been reported. It is critical that preexisting hypopituitarism and the variable radio-sensitivity of the different hypothalamic-pituitary axes be considered when examining long-term needs for hormonal replacement therapy. Although most series have reported rates of hypopituitarism of 60–80% at last follow-up, many patients have altered pituitary function before EBRT related either to the tumor itself or previous surgery. In addition, although gonadal function and GH levels are frequently depressed, thyroid (20–25%) or steroid (15–20%) replacement therapy is generally required by a minority of patients. The most significant concern of EBRT is the risk of radiation-induced tumors. Several large series have noted approximately a 2% risk of delayed tumor formation [10, 12], but no secondary tumors were seen in the large study of Jenkins et al. with more than 4,000 years of patient follow-up [5]. Finally, cognitive dysfunction is a frequently discussed though poorly documented phenomenon after EBRT [11]. A recent paper by van der Klaauw et al. used four health-related and one disease-specific acromegaly quality of life (QoL) questionnaires to assess the longitudinal changes in QoL in a cohort of patients with controlled acromegaly [14]. Eighty-two patients (27 treated with EBRT) with biochemical control (GH less than 1.9 μg/L and normal age- and sex-adjusted IGF-I levels) were compared at baseline then again 4 years later. The physical and social functioning of the SF-36 decreased, physical fatigue increased in the MFI-20, and psychological well-being and personal relations were progressively affected in the ACRO-QoL. Increasing patient age and previous radiotherapy were the primary factors associated with impairment of QoL despite good biochemical control of their acromegaly.

SRS for GH-Secreting Pituitary Adenomas

SRS has emerged as an effective alternative to EBRT for many acromegalic patients who fail surgical or medical therapy. Figure 15.2 shows an example of marked reduction in tumor size in a patient 10 years after SRS. Similar to EBRT, SRS has a tumor growth control rate of approximately 95%. Still, direct comparison of the reported series is difficult because of small patient numbers, limited follow-up intervals, and varied criteria of biochemical remission (Table 15.2). For example, numerous studies have noted biochemical remission rates from 50 to 60% using normal age- and sex-adjusted IGF-I levels [18, 19, 22, 23], whereas Vik-Mo et al. applied a more rigid criteria requiring oral glucose suppression testing and noted hormonal normalization in only 17% of patients [25]. In our most recent review of this topic, we followed 46 patients with GH-secreting pituitary adenomas for a median of 63 months [23]. We defined biochemical remission as a fasting GH level less than 2 ng/mL and a normal age- and sex-adjusted IGF-I level off all pituitary suppressive medications. By these criteria, 60% of patients achieved biochemical remission 5 years after SRS. Multivariate analysis found that biochemical remission correlated with lower IGF-I levels before SRS and the absence of pituitary suppressive medications at the time of the procedure. Biochemical remission rates exceeded 80% for patients with lower IGF-I levels before radiosurgery (less than 2.25 times the upper limit of normal) and for those not taking pituitary suppressive medications at the time of radiosurgery. Other centers have also noted that the degree of hypersecretion relates GH and IGF-I normalization after SRS [17, 19, 22]. Castinetti et al. reported that the mean GH and IGF-I levels of patients in remission after SRS were 7.1 and 495 ng/mL before the procedure compared to 25.3 and 673 ng/mL for patients who continued to have active acromegaly after SRS, respectively [17].

The concept that treatment with pituitary suppressive medications at the time of SRS is a negative predictor of biochemical remission is controversial. Landolt et al. found that octreotide treatment reduced the likelihood of biochemical remission from 60 to 11% in acromegalic patients having SRS [21]. They hypothesized that octreotide therapy reduced the metabolic activity in the adenoma cells thereby

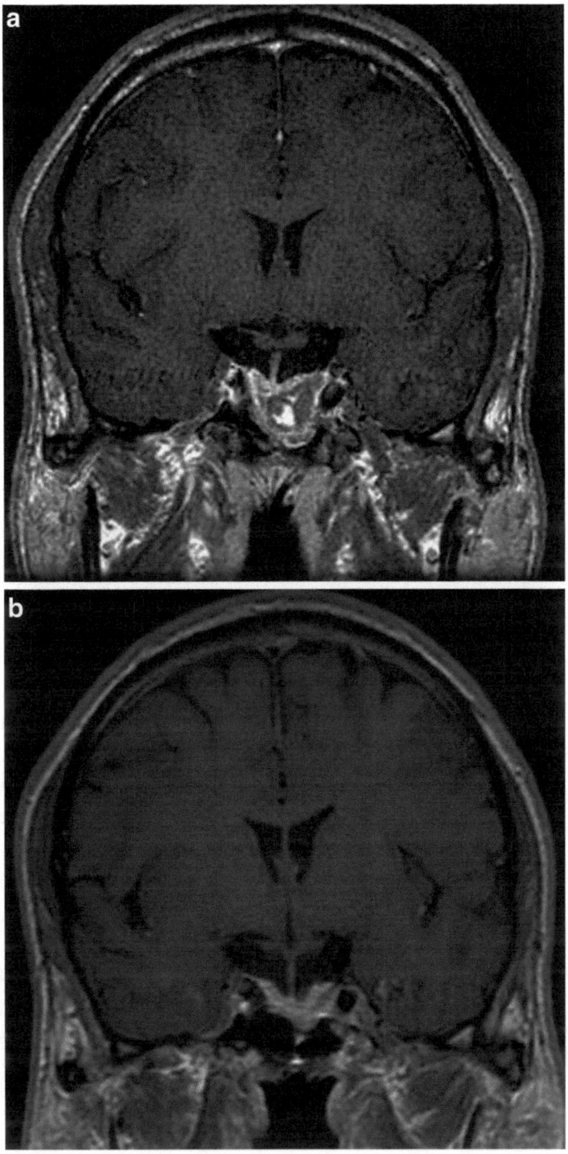

Fig. 15.2 Coronal postgadolinium MRIs of a patient with a residual GH-secreting macroadenoma extending into the left cavernous sinus after prior transsphenoidal surgery. (**a**) MRI before radiosurgery. The tumor margin dose at radiosurgery was 20 Gy. (**b**) MRI performed 10 years after radiosurgery shows the tumor to be decreased in size

providing a radio-protective effect. In addition to our center, the University of Virginia also noted reduced efficacy in patients using pituitary suppressive medications at the time of SRS [18]. Of the 27 patients on a SSA at the time of SRS, 10 (37%) were able to achieve successful remission compared to 40 of 68 patients (59%) off SSA therapy. Conversely, several studies have found no difference in the endocrine outcome between patients who received somatostatin agonists at the time of radiosurgery and those who did not [15, 17, 22]. One possible explanation

Table 15.2 Results of SRS for patients with GH-secreting pituitary adenomas

References	No. of patients	Radiation dose	Follow-up	Biochemical remission	Hypo-pituitarism
Attanasio et al. [15]	30	Mean, 20 Gy	Median, 46 months	30% at 5 years[a]	7%
Castinetti et al. [17]	82	Median, 25 Gy	Mean, 49.5 months	17%[b]	17%
Jezkova et al. [19]	96	Median, 35 Gy	Mean, 53.7 months	56% at 5 years[a]	27%
Pollock et al. [32]	46	Median, 20 Gy	Median, 63 months	60% at 5 years[b]	33% at 5 years
Vik-Mo et al. [25]	53	Mean, 26.5 Gy	Mean, 66 months	17%[c]	15%
Jagannathan et al. [18]	95	Mean, 22 Gy	Mean, 57 months	53%[a]	34%
Losa et al. [22]	83	Median, 21.5 Gy	Median, 69 months	53% at 5 years[d]	9%

[a]Normal IGF-I off medications

[b]GH<2.0 ng/mL and normal age-corrected IGF-I off medications

[c]Oral glucose-suppressed GH level<2.6 mIU/L and normal age-corrected IGF-I off medications

[d]GH<2.5 ng/mL and normal age-corrected IGF-I off medications

for this discrepancy may relate to the completeness of tumor coverage during the radiosurgical procedure. The median treatment volume in our series was 3.3 cm^3 compared to 1.4 cm^3 in the paper by Attanasio et al. [15] and 1.1 cm^3 in the study of Losa et al. [22]. Despite the difference in these studies on the potential radio-protective effect of pituitary suppressive medications, even centers that have not found a negative effect recommend stopping SSA and dopamine agonist prior to SRS [17, 22]. A randomized trial will be needed to accurately determine whether pituitary suppressive medications have any impact on biochemical remission of GH-secreting tumors after SRS.

Complications after pituitary adenoma radiosurgery can include radiation necrosis of the temporal lobe, stenosis of the internal carotid artery, or diplopia from radiation injury to the third, fourth, or sixth cranial nerves [17, 24]. However, the risk of these complications is very low (1% or less) and is most frequent in patients who have undergone prior EBRT. Radiation-induced damage to the anterior visual pathways has been well described, but the risk is between 1 and 2% if the maximum radiation dose to the optic nerves or chiasm does not exceed 12 Gy [26, 27]. The most frequent complication after SRS of pituitary adenomas are new anterior pituitary deficits. Recent studies have documented the incidence of new hormonal insufficiency after pituitary adenoma SRS from 7 to 41% [16, 17, 22–25, 28–30]. The different rates of hypopituitarism reported likely relate to variation in the patient characteristics including history of prior surgery or EBRT, radiation dose prescribed, treatment volume, follow-up intervals, and the completeness of the patients' endocrine evaluation. Attanasio et al. reported that only two of 30 acromegalic patients (7%) having SRS developed new pituitary deficits [15], whereas Jagannathan et al. noted new endocrine deficiencies in 34% of patients in their series [2]. Jezkova et al. [19] and Vladyka et al. [30] performed detailed dosimetric analysis of pituitary adenoma patients having SRS. They determined the safe mean pituitary dose was 15 Gy to maintain gonadotropic and thyrotropic function and 18 Gy to maintain adrenocorticotropic function. Our center recently analyzed 82 patients (secreting tumors, $n=53$; nonsecreting tumors, $n=29$) using similar dose-volume analysis to determine factors related to pituitary insufficiency after SRS [29]. The median endocrinological follow-up was 63 months. Thirty-four patients (41%) developed new anterior pituitary deficits at a median of 32 months after SRS. The risk of developing new anterior pituitary deficits was 16 and 45% at 2 and 5 years, respectively. Multivariate analysis showed that poor visualization of the pituitary gland and increasing mean pituitary gland radiation dose correlated with new anterior pituitary deficits. New anterior pituitary deficits are stratified by mean pituitary gland radiation dose: ≤7.5 Gy, 0%; 7.6–13.2 Gy, 29%; 13.3–19.1 Gy, 39%; >19.1 Gy, 83%. We now try to limit the radiation dose to pituitary gland during SRS to increase the probability of preserving pituitary function.

Comparison of EBRT and SRS for Acromegalic Patients

Despite lacking long-term follow-up studies of patients having SRS for GH-secreting pituitary adenomas, most centers favor SRS over EBRT for three primary reasons. First, SRS which is performed over 1–5 days is simpler and more convenient for patients and their families than EBRT that is typically performed over 5–6 weeks. Second, the time required for biochemical remission appears less after SRS for patients with hormone-secreting tumors. Landolt et al. compared 16 patients having SRS to 50 patients undergoing EBRT for persistent acromegaly [20]. The mean interval to remission was significantly shorter for the radiosurgery group (1.4 vs. 7.1 years). Kong et al. compared the efficacy of EBRT (64 patients, mean dose, 50.4 Gy) and single-fraction SRS (61 patients, mean dose, 25.1 Gy) for patients with both secreting and nonfunctional pituitary adenomas [31]. At a mean follow-up of 37 months, no difference was noted in either tumor growth control or the chance of biochemical remission. However, the time to biochemical remission was less after SRS compared to EBRT (median, 26 vs. 63 months). They concluded that SRS should be recommended over EBRT for suitable pituitary adenoma patients. Third, SRS has a lower chance of radiation-induced tumors when compared to EBRT. The incidence of radiation-induced has been documented between 2 and 3% after pituitary adenoma radiotherapy [10, 12, 13]. By comparison, the estimated risk of radiation-induced tumors after SRS is approximately 0.01%. Rowe et al. compared the incidence of new central nervous system malignancies in their radiosurgical practice to the national incidence in the United Kingdom over several decades [32]. Based on over 30,000 patient-years of follow-up, they did not note an increased incidence in their radiosurgical patients compared to the age- and sex-adjusted national cohort. The primary weakness of this study is the relative short mean follow-up interval (6.1 years) after SRS.

References

1. Barkan AL, Halasz I, Dornfeld KJ, et al. Pituitary irradiation is ineffective in normalizing plasma insulin-like growth factor I in patients with acromegaly. J Clin Endocrinol Metab. 1997;82:3187–91.
2. Cozzi R, Barausse M, Asnaghi D, et al. Failure of radiotherapy in acromegaly. Eur J Endocrinol. 2001;145:717–26.
3. Epaminoda P, Porretti S, Cappiello V, et al. Efficacy of radiotherapy in normalizing serum IGF-I, acid-labile subunit (ALS), and IGFBP-3 levels in acromegaly. Clin Endocrinol (Oxf). 2001;55:183–9.
4. Gutt B, Hatzack C, Morrison K, et al. Conventional pituitary irradiation is effective in normalizing plasma IGF-I in patients with acromegaly. Eur J Endocrinol. 2001;144:109–16.
5. Jenkins PJ, Bates P, Carson N, et al. Conventional pituitary irradiation is effective in lowering serum growth hormone and insulin-like growth factor-I in patients with acromegaly. J Clin Endocrinol Metab. 2006;91:1239–45.
6. Mackley HB, Reddy CA, Lee SY, et al. Intensity-modulated radiotherapy for pituitary adenomas: the preliminary report of the Cleveland Clinic experience. Int J Radiat Oncol Biol Phys. 2007;67:232–9.
7. Minniti G, Jaffrain-Rea ML, Osti M, et al. The long-term efficacy of conventional radiotherapy in patients with GH-secreting pituitary adenomas. Clin Endocrinol. 2005;62:210–6.
8. Mullan K, Sanabria C, Abram WP, et al. Long term effect external pituitary irradiation on IGF 1 levels in patients with acromegaly free of adjunctive treatment. Eur J Endocrinol. 2009;161:547–51.
9. Powell JS, Wardlaw SL, Post KD, et al. Outcome of radiotherapy for acromegaly using normalization of insulin-like growth factor I to define cure. J Clin Endocrinol Metab. 2000;85: 2068–71.
10. Breen P, Flickinger JC, Kondziolka D, et al. Radiotherapy for nonfunctional pituitary adenoma: analysis of long-term tumor control. J Neurosurg. 1998;89:933–8.
11. Constine LS, Woolf PD, Cann D, et al. Hypothalamic-pituitary dysfunction after radiation for brain tumors. N Engl J Med. 1993;328:87–94.
12. Minniti G, Traish D, Ashley S, et al. Risk of second brain tumor after conservative surgery and radiotherapy for pituitary adenoma: update after an additional 10 years. J Clin Endocrinol Metab. 2005;90:800–4.
13. Simmons NE, Laws Jr ER. Glioma occurrence after sellar irradiation: case report and review. Neurosurgery. 1998;42:172–8.
14. van der Klaauw AA, Biermasz NR, Hoftijzer HC, et al. Previous radiotherapy negatively influences quality of life during 4 years of follow-up in patients cured from acromegaly. Clin Endocrinol. 2008;69:123–8.
15. Attanasio R, Epaminonda P, Motti E, et al. Gamma-knife radiosurgery in acromegaly: a 4-year follow-up study. J Clin Endocrinol Metab. 2003;88:3105–12.
16. Castinetti F, Nagai M, Morange I, et al. Long-term results of stereotactic radiosurgery in secretory pituitary adenomas. J Clin Endocrinol Metab. 2009;94:3400–7.
17. Castinetti F, Taieb D, Kuhn JM, et al. Outcome of gamma knife radiosurgery in 82 patients with acromegaly: correlation with initial hypersecretion. J Clin Endocrinol Metab. 2005;90:4483–8.
18. Jagannathan J, Sheehan JP, Pouratian N, et al. Gamma knife radiosurgery for acromegaly: outcomes after failed transsphenoidal surgery. Neurosurgery. 2008;62:1262–70.
19. Jezkova J, Marek J, Vaclav H, et al. Gamma knife radiosurgery for acromegaly-long-term experience. Clin Endocrinol. 2006;64:588–95.
20. Landolt AM, Haller D, Lomax N, et al. Stereotactic radiosurgery for recurrent surgically treated acromegaly: comparison with fractionated radiotherapy. J Neurosurg. 1998;88:1002–8.
21. Landolt AM, Haller D, Lomax N, et al. Octreotide may act as a radioprotective agent in acromegaly. J Clin Endocrinol Metab. 2000;85:1287–9.

22. Losa M, Gioia L, Picozzi P, et al. The role of stereotactic radiotherapy in patients with growth hormone-secreting pituitary adenoma. J Clin Endocrinol Metab. 2008;93:2546–52.
23. Pollock BE, Jacob JT, Brown PD, et al. Radiosurgery of growth hormone producing pituitary adenomas: factors associated with endocrine cure. J Neurosurg. 2007;106:833–8.
24. Pollock BE, Nippoldt TB, Stafford SL, et al. Results of stereotactic radiosurgery in patients with hormone-producing pituitary adenomas: factors associated with endocrine normalization. J Neurosurg. 2002;97:525–30.
25. Vik-Mo EO, Oksnes M, Pedersen PH, et al. Gamma knife stereotactic radiosurgery for acromegaly. Eur J Endocrinol. 2007;157:255–63.
26. Leber KA, Bergloff J, Pendl G. Dose-response of the visual pathways and cranial nerves of the cavernous sinus to stereotactic radiosurgery. J Neurosurg. 1998;88:43–50.
27. Stafford SL, Pollock BE, Leavitt JA, et al. A study on the radiation tolerance of the optic nerves and chiasm after stereotactic radiosurgery. Int J Radiat Oncol Biol Phys. 2003;55: 1177–81.
28. Feigel GC, Bonelli M, Berghold A, et al. Effects of gamma knife radiosurgery of pituitary adenomas on pituitary function. J Neurosurg. 2002;97(Suppl):415–21.
29. Leenstra J, Tanaka S, Kline R, et al. Factors associated with endocrine deficits after stereotactic radiosurgery of pituitary adenomas. Neurosurgery. 2010;67:27–32.
30. Vladyka V, Liscak R, Novotny Jr J, et al. Radiation tolerance of functioning pituitary tissue in gamma knife surgery for pituitary adenomas. Neurosurgery. 2003;52:309–17.
31. Kong D, Lee J, Lim DH, et al. The efficacy of fractionated radiotherapy and stereotactic radiosurgery for pituitary adenomas. Long term results of 125 consecutive patients treated in a single institution. Cancer. 2007;110:854–60.
32. Rowe J, Grainger A, Walton L, et al. Risk of malignancy after gamma knife stereotactic radiosurgery. Neurosurgery. 2007;60:60–6.

Chapter 16
Mortality and Morbidity in Acromegaly: Impact of Disease Control

Ian M. Holdaway

Abstract The excess production of growth hormone (GH) and insulin-like growth factor-I (IGF-I) in patients with acromegaly is associated with a range of complications which have important effects on quality of life and which lead to reduced life expectancy. The increased mortality of the disorder has been recognized in 18 studies, totaling 4,806 individuals with acromegaly and including 1,116 deaths during follow-up, with meta-analyses indicating a standardized mortality estimate of 1.7 (95% CI 1.5–2). Univariate and multivariate analyses within the individual studies have identified the presence of hypertension, diabetes, and cardiac disease as important predictors of survival, as well as estimated duration from onset of the condition to date of effective treatment. However, the most important impact on survival comes from reduction of circulating GH and IGF-I concentrations, with serum GH measured by radioimmunoassay of <2.5 μg/L and a serum IGF-I measurement within the normal range for age providing mortality estimates indistinguishable from the general population. The major comorbidities of acromegaly, including cardiovascular disease, stroke, diabetes, hypertension, arthropathy, and sleep apnea syndrome, also benefit from biochemical remission following treatment, and management of these complications of the disorder appears to be important in maintaining quality of life as well as assisting with reduction in mortality.

Keywords Acromegaly morbidity • Acromegaly mortality • Radiotherapy and acromegaly • Meta-analysis • Biochemical remission of acromegaly

I.M. Holdaway (✉)
Department of Endocrinology, Greenlane Clinical Centre and Auckland Hospital, Greenlane Road West, Private Bag, 92189, Auckland, New Zealand
e-mail: Ian@adhb.govt.nz

K. Ho (ed.), *Growth Hormone Related Diseases and Therapy: A Molecular and Physiological Perspective for the Clinician*, Contemporary Endocrinology,
DOI 10.1007/978-1-60761-317-6_16, © Springer Science+Business Media, LLC 2011

Introduction

For many years, clinicians have recognized that excess production of growth hormone (GH) in acromegaly causes serious metabolic problems and often interferes markedly with quality of life. Despite its relative rarity, acromegaly is an extensively studied disorder, and a consensus is emerging which highlights the important morbidity and increased mortality of the disorder and the place of various treatments, together with the goals of therapy particularly in terms of remission and cure [1, 2]. This chapter reviews current knowledge concerning the morbidity and mortality of acromegaly.

Morbidity of Acromegaly

The complications of acromegaly are listed in Table 16.1. Some have been shown in univariate or multivariate analysis to directly contribute to increased mortality, whereas others are very likely to lead to increased morbidity and mortality but specific evidence is currently lacking, probably due to the relatively small numbers of patients studied to date. Other complications such as joint disease are unlikely to influence mortality to any significant extent, but can seriously impair quality of life.

Cardiovascular Complications

Acromegaly has been associated with a range of cardiovascular abnormalities (Table 16.1), and cardiovascular disease is the principal cause of death in the disorder [3]. Hypertension is a frequent complication and appears to be an independent risk factor for mortality [4]. Other cardiovascular abnormalities include cardiomyopathy, cardiac rhythm disturbance, valvular heart disease, and IHD (reviewed by Clayton [5] and Colao et al. [6]). There have been a wide range of studies indicating a state of generalized vasculopathy in acromegaly, including impaired vascular reactivity [7], capillary abnormalities [8], coagulopathy, and dyslipidemia [9].

Hypertension: Approximately 30–40% of patients presenting with untreated acromegaly have an elevated blood pressure, especially those with a family history of hypertension. The mechanisms leading to hypertension remain uncertain, with likely contributions from sodium retention, blunted diurnal rhythm of catecholamine release, the effect of associated insulin resistance on the vasculature, increased cardiac output in the initial phase of the disorder, and increased peripheral vascular resistance. Remission of acromegaly reduces the overall prevalence of hypertension [10, 11] and often makes elevated blood pressure more easily controlled in those with persistent hypertension.

Cardiomyopathy: A number of studies dating from the 1970s have reported abnormalities of cardiac function in acromegaly. Receptors for GH and insulin-like growth factor-I (IGF-I) are present in cardiac myocytes and coronary vessels, as

Table 16.1 Complications of acromegaly

Complications shown to contribute to increased mortality in univariate or multivariate analysis
Hypertension
Cardiac disease, with contributions from
Cardiomyopathy
Ischaemic heart disease
Valvular heart disease
Arrhythmia
Diabetes
Cerebrovascular disease
Colon cancer
Treatment-related complications
Cerebrovascular disease due to external pituitary radiotherapy
Hypopituitarism (ACTH deficiency and excess cortisol replacement)
Complications which may possibly contribute to excess mortality, but not as yet demonstrated in formal studies
Sleep apnea syndrome
Respiratory disease
Dyslipidemia
Insulin resistance
Abnormal vascular reactivity, prothrombotic changes, capillary abnormalities, abnormal endothelial function
Cancer other than colon cancer
Other complications which may affect quality of life
Arthropathy, including spinal disease
Osteopenia
Carpal tunnel syndrome
Integumentary changes including acne, hirsutism, hyperhydrosis, skin tags, cosmetic appearance
Peripheral neuropathy and myopathy
Dental abnormalities
Headache
Colonic polyps and megacolon
Direct complications of pituitary adenoma (visual compromise and other cranial nerve damage)

well as the peripheral vasculature. Echocardiographic studies show that a progressive cardiomyopathy complicates acromegaly in most individuals, commencing with initially increased heart rate and systolic output accompanied by the development of concentric biventricular hypertrophy, progressing to diastolic dysfunction and impaired systolic function on exercise, and finally systolic dysfunction at rest with ventricular dilatation and clinical cardiac failure [6]. Newer techniques such as cardiac MRI indicate myocardial edema in some patients, which can reverse during treatment [12]. With modern treatment of both the GH excess and the cardiac disorder, progression to terminal cardiac failure is now uncommon [10]. Treatment of acromegaly with pituitary surgery or somatostatin receptor analogs (SSRAs) can reduce LV mass and improve myocardial function [13], and echocardiographic abnormalities may be more readily reversed in younger subjects. Cardiac disease can progress if there is delayed biochemical correction of acromegaly as may occur with treatment by external pituitary radiotherapy. Hypopituitarism induced by treatment may also affect myocardial function.

Ischemic heart disease: The direct cardiomyopathic effects of excess GH and IGF-I on the heart may be amplified by coexisting coronary artery disease. There is impaired endothelium-dependent vasodilatation in acromegaly, and many individuals have abnormal lipid profiles and prothrombotic metabolic changes with or without diabetes, all of which increase coronary vascular risk. Framingham risk scores are higher in those with active compared with controlled acromegaly [14], and Agaston coronary calcification scores appear to be increased in both active and controlled GH excess [15]. Estimates of the prevalence of coronary artery disease in unselected patients range from 17 to 40% in various studies [16], and abnormal myocardial perfusion scans are detected in up to 20% of acromegalic individuals, with improvement during treatment.

Valvular heart disease: There is an increased frequency of regurgitant cardiac valvular disease in acromegaly associated with increased cytokine and metalloproteinase accumulation in valve tissue leading to myxomatous degeneration [17, 18]. Up to 80% of individuals have some echocardiographic evidence of aortic or mitral valvular incompetence secondary to these changes. In a longitudinal study of mitral regurgitation, 32% of acromegalic subjects had MR at initial assessment progressing to 60% over 2 years of follow-up [18]. Short-term treatment with SSRAs did not seem to lead to improved valve function [19], but although long-term effects of treatment have not been reported, it seems likely that therapy would be beneficial. Tricuspid and other valvular insufficiencies with high-dose cabergoline therapy of acromegaly might also occur in occasional patients.

Cardiac arrhythmias: The electrocardiographic QTc interval is often prolonged in acromegaly [20] and can be normalized with SSRA therapy. There appears to be an increased prevalence of arrhythmias such as paroxysmal atrial fibrillation, supraventricular and ventricular tachycardia, sick sinus syndrome, and bundle branch block which can develop in up to 40% of acromegalic individuals, with a tendency to an increased arrhythmogenic effect of exercise (reviewed by Colao et al. [6]). ECG monitoring has shown that one third of patients have >50 ventricular ectopic beats per 24 h. It is, however, unclear whether death from cardiac arrhythmia is increased in acromegaly.

Cerebrovascular and Other Vascular Complications

There is evidence that the prevalence of death from stroke is increased in acromegaly [21]. This is likely to be due to both the increased rate of hypertension and also the generalized vasculopathy of the disorder, and there are also increasing data to suggest a causal relationship between the development and mortality of cerebrovascular disease and previous treatment by external beam pituitary radiotherapy [21, 22]. A number of acromegalics have increased aortic root diameter, although ascending aortic aneurysm appears rare. It seems likely that the rate of peripheral vascular disease and renovascular disease is also increased in active acromegaly, but formal studies of these complications are awaited.

Diabetes

GH exerts significant effects upon body fuel economy, and elevated levels are associated with impaired fasting glycemia, impaired glucose tolerance, and overt diabetes, with some impairment of glucose tolerance found in 16–59% of acromegalics in various series, and clinical diabetes in up to 23% of subjects [23]. Clamp studies indicate that this mainly stems from insulin resistance at the level of the liver and muscle, with resulting poorly controlled hepatic glucose production. By contrast, IGF-I appears to increase muscle sensitivity to glucose, but this effect is swamped by the GH-induced insulin resistance. The likelihood of insulin resistance increases steadily as the nadir serum GH following oral glucose testing (oGTT) rises above 2 μg/L [24]. Clamp studies also indicate prominent utilization of fatty acids as an energy source in acromegaly. The prevalence of diabetes in acromegaly appears to increase with age, higher plasma GH, family history of diabetes, and duration of GH excess [10], and some studies have found correlations between the presence of diabetes and either hypertension or cardiac disease [25]. A recent study has found the prevalence of diabetes to be 50% increased in carriers of the exon 3-deleted form of the GH receptor, presumably due to amplified GH signaling through this receptor isoform [26].The presence of diabetes appears to predict increased mortality in acromegaly in univariate analysis [27]. Successful treatment of the disorder significantly reduces the prevalence of diabetes [11, 28]. Formal testing of insulin resistance has revealed measurements similar to controls in those with normal IGF-I after treatment, and this improvement appears to correlate better with IGF-I normalization rather than restoration of normal GH secretion, since borderline nonsuppressibility of GH was not itself associated with increased insulin resistance [29]. There has been concern that use of SSRAs for treatment of acromegaly might suppress pancreatic insulin secretion and hence worsen diabetes, but formal studies indicate similar rates of glucose intolerance in those treated with SSRAs compared with treatment by pituitary surgery. However, studies of glucose intolerance in patients treated with the GH receptor blocker, Pegvisomant, which should not intrinsically affect glucose tolerance, have found better metabolic profiles than for those treated with SSRAs, especially in patients with preexisting diabetes [30].

Lipid abnormalities: Early studies of acromegaly identified a range of serum lipid abnormalities [9], and hyperlipidemia is present in up to 68% of cases [31]. GH influences lipoprotein lipase and has variable effects upon Lp(a) and LDL subfractions, as well as elevating serum levels of small dense LDL particles. Lipoprotein abnormalities may persist after successful therapy of GH excess in some individuals, but in most cases the lipid profile improves with remission following either successful pituitary surgery or SSRA therapy. Improvement in HDL and Lp(a) concentrations has also been observed with Pegvisomant therapy [30]. Although it seems likely that lipoprotein abnormalities contribute to the morbidity and mortality of acromegaly, their precise role and the effect of lipid-lowering treatment remain to be determined in long-term studies.

Acromegalic Arthropathy

GH and IGF-I act on chondrocytes in articular cartilage to promote DNA synthesis, cell replication, and production of proteoglycan and glycosaminoglycan (reviewed in Colao et al. [6]), as well as inducing collagen synthesis and enhancing osteoblast turnover in bone. These actions lead to cartilage thickening and widening of the joint space, with resulting alteration of joint geometry, increasing synovial hypertrophy, periarticular softening, and proliferation of regenerative fibrocartilage, leading to joint osteoarthritis. Cartilage expansion can be observed clinically in the costal cartilages (the "acromegalic rosary"). Arthritis is one of the most frequent complications of acromegaly, with self-reported joint pain in up to 77% of individuals and overt osteoarthritis in as many as 70% of patients, resulting in a significant deterioration in their quality of life [32]. Joint pain was present at diagnosis in 17% [33] and 41% of patients [3] in different series. Radiology of the hips and knees can be abnormal in more than 50% of cases, and shoulder ultrasound reveals abnormalities in >80% of acromegalic individuals [34]. Spinal pain and reduced spinal mobility are common, with DISH, spinal stenosis, and facet joint arthritis in some individuals. Besides osteoarthritis of the major joints, there may be temporal-mandibular joint problems, in part related to prognathism. Joint problems appear unrelated to age or serum GH and IGF-I levels, but may be more prominent in females [32]. There are varying reports of a relationship to disease duration. Bone density is variably reported as normal, increased, or decreased in acromegaly. The frequency of vertebral fractures has been found to be increased in a cross-sectional study of men treated for acromegaly (57 vs. 22% in matched controls [35]), particularly in those with hypogonadism or long duration of active acromegaly. Control of acromegaly does not appear to influence the course of established arthropathy, although joint thickness as assessed by ultrasound may improve [36].

Cancer

IGF-I can influence cell growth in vitro at the level of the cell cycle, with promotion of the growth of transformed cells. GH can also act independently of IGF-I to promote replication in some cell lines in vitro. Epidemiological studies in the general nonacromegalic population have linked serum levels of IGF-I in the upper part of the population normal range with subsequent development of breast, colon, and prostate carcinoma. A recent meta-analysis of the epidemiologic data suggests a relative risk (RR) of 1.38 for development of prostate cancer (comparing highest vs. lowest quintile of serum IGF-I [37]), and a further meta-analysis calculated a RR of 1.49 for prostate cancer and 1.65 for premenopausal breast cancer [38]. However, data linking elevation of GH and IGF-I in acromegaly to increased rates of tumor development and elevated mortality from cancer are not compelling. Cancer prevalence in various series of acromegalic patients ranges from 6 to 25% (mean 11%), and cancer as a cause of death ranges from 9 to 50% with a mean of 26%. The most powerful correlation with mortality was seen with death from bowel cancer in the large UK study reported by Orme et al. [39]. The RR of death from colon cancer was significantly elevated by 2.5-fold, although the overall death rate from malignant disease was not increased. However, a further study could not replicate this finding [40]. Nabarro noted an increase in female breast cancer in his series [10], and an increase in cancer mortality in females was noted by Bengtsson et al. [41]. Although some reports suggest increased mortality from breast and prostate cancer in acromegaly, there is no consistent evidence to support these findings. A special topic of relevance in acromegaly is the potential for second cerebral tumor development after radiotherapy treatment of GH-secreting pituitary adenomas. In a series of reported cases of second tumors following radiotherapy of pituitary neoplasms, 38% developed in acromegalic patients [42].

Colonic polyps and screening for cancer: Despite initial controversy, there is a growing consensus that the prevalence of noncancerous and precancerous colonic polyps is increased in acromegaly. A recent meta-analysis suggests a RR of 3.6 for noncancerous hyperplastic polyps, an RR of 2.5 for potentially cancerous adenomatous polyps, and an RR of 4.3 for colon cancer (all $p<0.001$ [43]). A family history of colon carcinoma appears to increase the risk of colonic neoplasia in acromegaly. It appears prudent to recommend regular colonoscopic examination as well as mammography and prostate cancer screening in acromegalic subjects. Consensus guidelines have suggested that colonoscopy should start at age 50 years, and thereafter as per the recommendations for the general population [1, 2], although a recent study suggests that, following initial colonoscopy screening, further screening should be at 5-year intervals for those with a previous adenoma (or if there is a high ongoing GH level), but only every 10 years for those in remission. Acromegaly-associated redundant colon and megacolon may cause problems when performing colonoscopy in some individuals.

Sleep Apnea Syndrome

Although snoring, daytime somnolence, and sleep disturbance have been recognized in acromegalic individuals for many years, it is only recently that sleep apnea has been identified as a serious and highly prevalent complication of GH excess. A number of small studies have found 60–88% of unselected patients show evidence of sleep apnea syndrome (SAS) [44], dominantly of obstructive type associated with hypertrophy of soft tissues in the upper airway. However, central apnea can also occur. Some but not all studies have found associations of SAS with age, gender, and biochemical activity of acromegaly. In nonacromegalic subjects, SAS has been associated with an increased risk of vascular disease, metabolic abnormalities, and weight gain. The extent to which these problems are linked with SAS in acromegalic subjects remains to be determined, but a number of studies have found significantly higher rates of SAS, particularly higher apnea-hypopnea scores, in acromegalic patients with hypertension and diabetes compared with controls. Importantly, treatment of acromegaly has been associated with reduction in the prevalence and severity of SAS [45], and therapy of SAS with cPAP can often reduce the prevalence of hypertension in these patients.

Respiratory Disease

Early studies of mortality in acromegaly indicated significant increases in death from respiratory disorders [39, 46, 47]. GH & IGF-I excess induces changes in the thoracic cage, narrowing of small airways, lung perfusion defects, subclinical hypoxemia, and a range of alterations in respiratory function. However, recent studies have failed to show an increase in death from respiratory causes in acromegaly [4, 21]. Smoking history has not been recorded in most mortality studies, but a history of cigarette smoking did not seem higher in deceased compared with surviving patients in a New Zealand study [4] and was not an independent predictor of survival (although data were only available for 64% of the patient group at final follow-up).

Headache

Headache is a common symptom in acromegaly, sometimes due to dural stretching by the GH-secreting pituitary adenoma, but more commonly due to chronic or episodic migraine or primary stabbing headache [48]. Curiously, headache often responds well to treatment with SSRAs independent of achieving biochemical cure, and cases of acromegaly dependent on octreotide therapy for headache control occur, whereas headache is unpredictably responsive to dopamine agonist therapy [48]. The prevalence of headache is significantly reduced by treatment of acromegaly [27].

Other Complications

A range of additional complications have been described in acromegaly, including skin abnormalities such as skin tags and acanthosis nigricans, carpal tunnel syndrome, acne, hirsuitism, peripheral neuropathy, thyroid disorder [10], and hyperparathyroidism. Dental problems are reasonably common, in part related to malocclusion and prognathism. Symptoms such as excess sweating and carpal tunnel syndrome are significantly improved by treatment [10, 27].

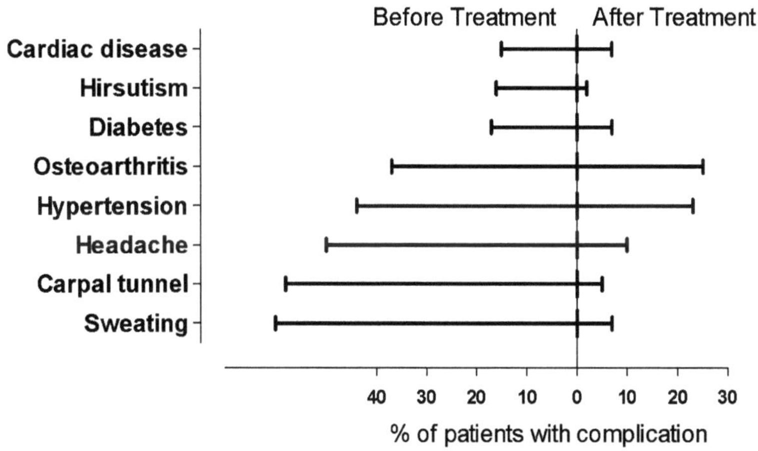

Fig. 16.1 Prevalence of various complications of acromegaly in New Zealand patients before and after biochemical remission of the disorder (serum GH <2.5 μg/L). Data from refs. [3, 27]

Morbidity According to GH and IGF-I Levels Following Treatment

Although there is now a considerable literature concerning the effect of lowering serum levels of GH and IGF-I on mortality in acromegaly (see below), there is less information available about the effect of biochemical remission on the prevalence and severity of complications of the disorder. Where available, information concerning the individual complications has been recorded above. The prevalence of various complications of acromegaly at last follow-up in the Auckland (New Zealand) population is outlined in Fig. 16.1 according to whether patients had achieved a posttreatment random GH above or below 2.5 μg/L as measured by radioimmunoassay [3, 27]. Reduction of GH to <2.5 μg/L significantly reduced the prevalence of the major complications of the disorder. Others have detected a similar order of improvement with treatment. The advent of more sensitive assays for GH has allowed a more precise identification of normal compared with abnormal GH production as judged from GH suppressibility during oGTT or by 24 h GH estimations, but it remains uncertain whether such assays and assessments will provide a clearer cut-off for GH below which complications will be minimal compared with earlier assays. Elevated serum IGF-I following treatment appears to be associated with higher complication rates [11, 29].

Mortality of Acromegaly

Age at Death

The average age of death in those with acromegaly from the major series where data are available [4, 21, 28, 41, 46–47, 49–53] was 63 years. Although it is not possible to calculate an average reduction in life expectancy for acromegalic individuals from these data, it is likely that survival was reduced by at least 10 years compared with the nonacromegalic population in these series. However, with effective treatment, it is likely that life expectancy will be considerably improved or normalized (see below). In support of this, the mean age at death in New Zealand acromegalic subjects has improved decade by decade from 49 ± 9 years in 1960–1970 to 66 ± 15 years in 1990–2000 ($p = 0.02$).

Causes of Death in Acromegaly

The cause of death in acromegaly as determined from a number of studies is shown in Fig. 16.2. The spread of diagnoses is broadly similar to the causes of death in the general population, and the frequency of death from most causes is not greatly different from that expected for Western populations. Earlier studies from the UK [39, 46] found an apparent increase in the proportion of deaths from respiratory illness, which has not been observed in more recent reports, and which might reflect relatively high rates of occupational lung disease [47] and possibly cigarette

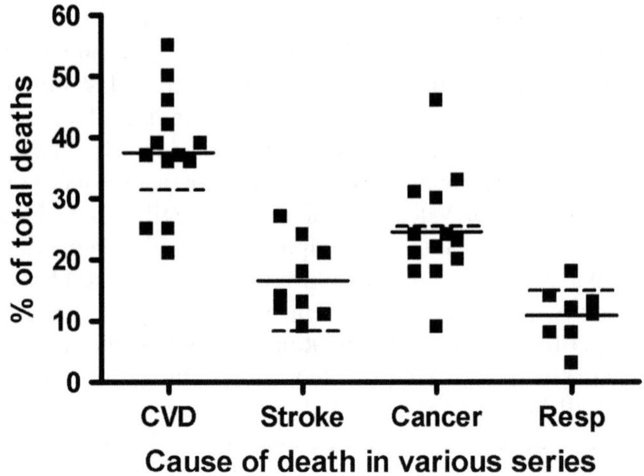

Fig. 16.2 Cause of death in acromegaly (data from refs. [3, 11]). *Solid line* = mean, *dotted line* = expected mortality in general New Zealand population 1996 (NZ Yearbook)

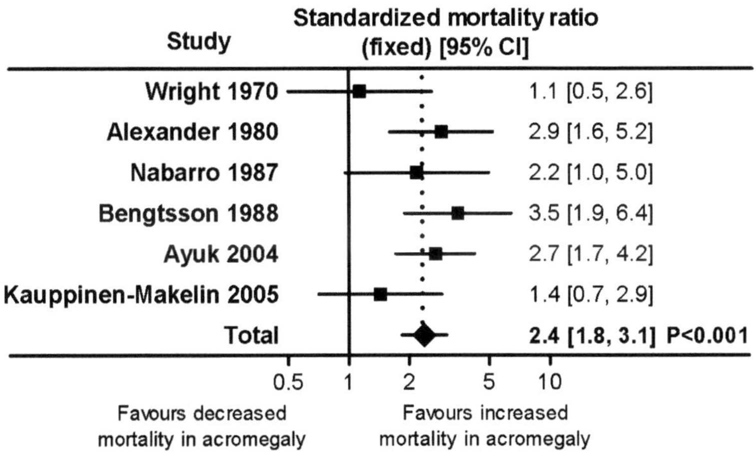

Fig. 16.3 Meta-analysis of mortality from stroke in patients with acromegaly. Pooled standardized mortality ratios (SMRs) with 95% confidence intervals from refs. [10, 21, 41, 46, 52, 53]

smoking in the UK over the duration of these studies, which may have had a disproportionate effect on the acromegalic population. It seems likely that there is a slight excess of observed compared with expected deaths from cardiac disease [10] (women only), [39, 41, 46] (men only), [53] (men only), [56]. An analysis of recorded death from stroke (Fig. 16.3) also indicates a significantly increased mortality from this complication. It does not appear that death from cancer is overrepresented in the acromegalic population (Fig. 16.2), although several studies have noted a significant increase in deaths from all forms of cancer [41, 53] (men), [57] or specifically from colon cancer [39].

Death from cerebrovascular disease: The observed increase in death from stroke in the acromegalic population may relate to the high prevalence of hypertension in these individuals, but recent studies also implicate the late effects of radiotherapy as a contributing factor [21, 22, 49]. Until recently, pituitary radiotherapy was widely used as either primary or adjuvant treatment for acromegaly, mainly administered as external beam fractionated radiotherapy or sometimes delivered by radionuclide implantation into the pituitary gland. There is a borderline significant positive correlation between the extent of use of radiotherapy and overall mortality in major studies ($r=0.52$, $p=0.04$), and a limited meta-analysis confirms increased mortality in those treated with external pituitary radiotherapy (Fig. 16.4). Unpublished observations from the New Zealand series, where 38 of 208 patients were treated by yttrium implantation of the pituitary [27], indicate a significantly worse late mortality rate in these individuals than seen in those treated by pituitary surgery alone, despite similar rates of biochemical remission. However, the implant group had a high rate of rapidly developing hypopituitarism following treatment which may have influenced the mortality findings.

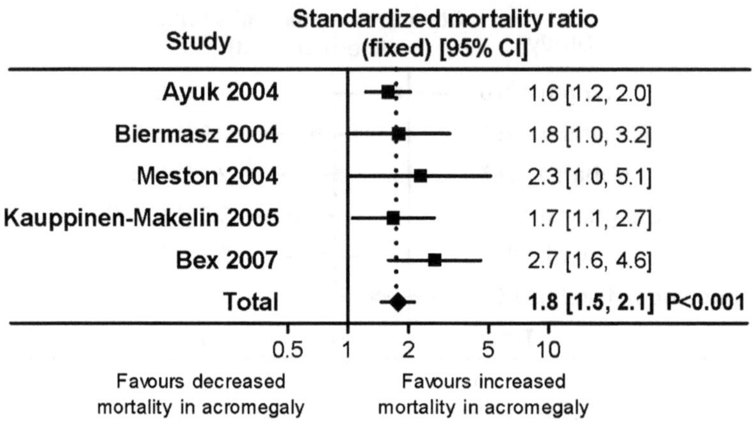

Fig. 16.4 Meta-analysis of mortality in patients with acromegaly treated with pituitary radiotherapy as whole or part management of the disorder. Pooled SMRs with 95% confidence intervals from refs. [21, 31, 49, 52, 59]

Overall Mortality in Acromegaly

Two recent meta-analyses have confirmed an increase in mortality in acromegalic subjects [54, 55]. Eighteen studies have provided standardized mortality ratios (SMRs) for the acromegalic population [4, 10, 21, 28, 31, 39, 46, 47, 50–53, 56–61]. It is possible that there may be some replication of data between the early UK studies in this series, but omission of these reports does not alter the overall results [54]. The most recent meta-analysis [55] provides data from 4,806 patients including 1,116 deaths and indicates an SMR for acromegaly of 1.7 (95% CI 1.5–2) compared with the general population (Fig. 16.5). This increase in mortality was apparent despite treatment of the disorder, suggesting either an intrinsic increase in mortality from the condition, or more likely, relatively inefficient treatment of the disorder and its complications. In support of this latter possibility, the more recent studies in Fig. 16.5 have, in general, found lesser increases in mortality [62], presumably reflecting adoption of stricter targets for biochemical remission in terms of achieved GH and IGF-I levels following treatment, together with a wider range of treatments able to achieve these goals than was earlier available, as well as restricted treatment with external pituitary radiotherapy. These possibilities are supported by the demonstration of improved mortality figures in studies where SSRAs were used to achieve remission where surgery had been unsuccessful (Fig. 16.6), and in studies where high rates of biochemical remission (>70%) were achieved (Fig. 16.7). Mortality is also likely to have been influenced by the introduction of national and international therapeutic guidelines for managing the major comorbidities of acromegaly, including targets for effective treatment of hypertension and diabetes and screening guidelines for colon cancer. Effective therapies for these complications of

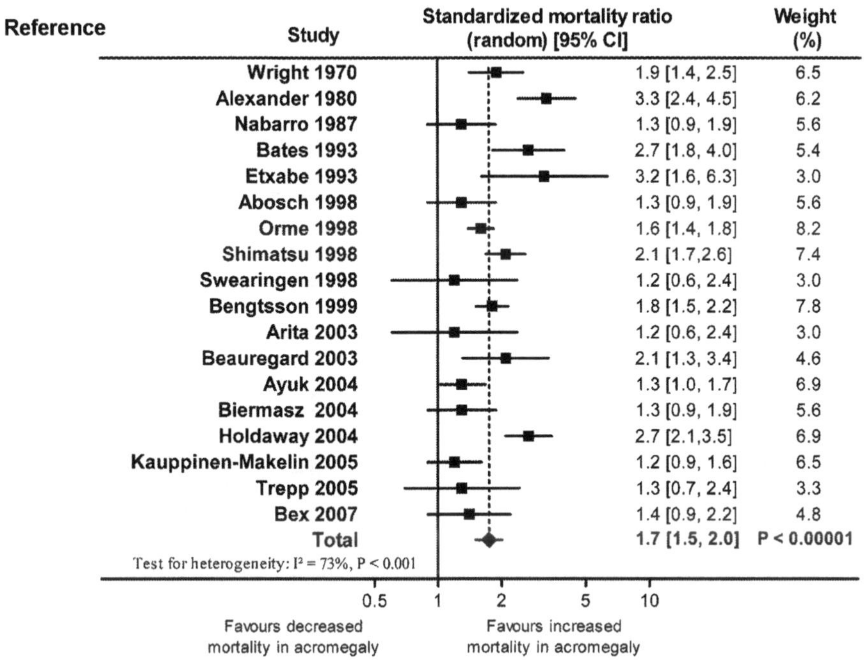

Fig. 16.5 Overall mortality in individuals with acromegaly. Pooled SMRs with 95% confidence intervals from published studies (from ref. [55], with permission). Individual references as listed in text

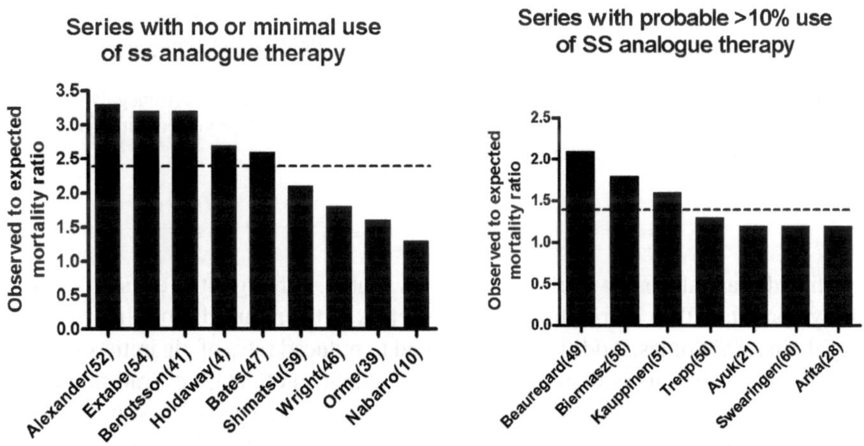

Fig. 16.6 Mortality in studies of acromegaly grouped according to greater or lesser likely use of SSRAs to achieve remission. Reproduced from ref. [62] with permission. *Dashed line* = mean

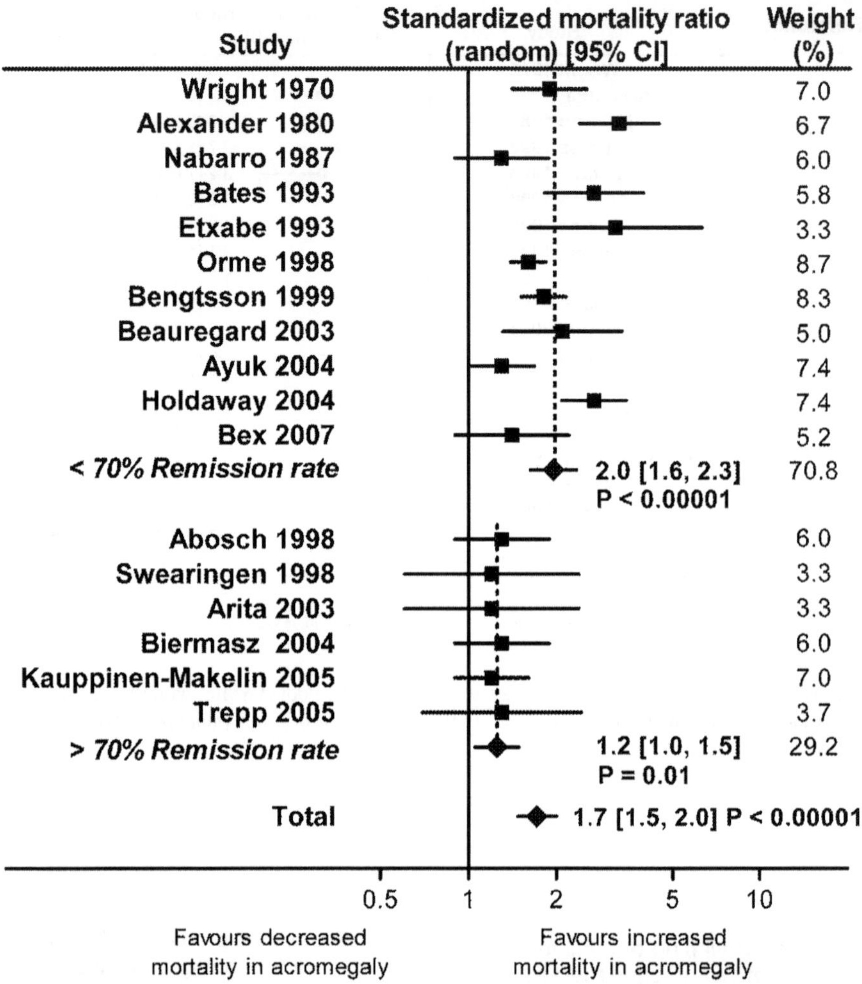

Fig. 16.7 Pooled SMRs with 95% confidence intervals from studies of acromegaly grouped according to the actual or likely remission rate following treatment. Reproduced with permission from ref. [55]. Data are from references as in text for Fig. 16.5

the disorder are also now more commonly available, although it is difficult to precisely quantitate the contribution of these factors to the reduced mortality of acromegaly in recent times. Additionally, the trend to reduced rates of cigarette smoking in the last 25–30 years is likely to have had a beneficial effect on overall mortality. Of course, such improvements should have influenced mortality in both the acromegalic and general populations, and so might not influence the SMR values, but it is likely that they would have a proportionally greater effect in acromegaly given the increased prevalence of vascular and metabolic problems in this group.

Factors Influencing Mortality in Acromegaly

By means of retrospective analysis of death rates in acromegaly, and application of regression analysis, a number of factors have been found to influence mortality in univariate analyses. These include age [4, 21, 59], gender [10, 52, 53, 57], the presence of hypertension [4, 46] or diabetes [27, 46], cardiac disease [4, 41, 46], pre- and posttreatment GH and/or IGF-I levels [4, 21, 39, 47, 50, 52, 58, 61], the estimated duration of the disorder prior to treatment [4, 59], and the use of external pituitary radiotherapy [21, 22, 31]. It should be noted that there are no large studies that have assessed the effect of smoking or the presence and/or treatment of sleep apnea on mortality, but it seems likely that these would adversely affect death rates in the disorder. When these data are subject to multivariate analysis, there are, as would be expected, fewer variables which influence mortality in acromegaly. The dominant factor found in most studies is the serum GH and/or IGF-I achieved by treatment [4, 21, 52, 59], with age, hypertension, estimated duration of disease prior to therapy [4], and use of pituitary radiotherapy [22] also independently affecting mortality in various studies. Surprisingly, given the increase in mortality noted for nonacromegalic hypopituitary subjects (average SMR 1.76 from five studies), only one study to date has identified hypopituitarism (specifically ACTH deficiency) as a significant predictor of mortality in acromegaly [22].

Biochemical Predictors of Mortality

GH: Early studies of the treatment of acromegaly indicated that reduction of GH levels appeared to improve long-term outcome. However, a series of reports in the 1990s identified specific GH concentrations following treatment which were associated with mortality figures similar to the general population. A meta-analysis of mortality in acromegaly according to posttreatment GH levels above or below a cut-off value of 2.5 μg/L (mostly measured in randomly collected samples and estimated by radioimmunoassay) is shown in Fig. 16.8. These data include some estimates from patients where GH was only one of a group of measurements (usually including glucose-suppressed GH and serum IGF-I values), or where GH was either measured as a random clinic sample or as a mean of a series of measurements. However, despite these limitations, there is clear evidence of an elevated mortality if final posttreatment serum GH is >2.5 μg/L (measured by radioimmunoassay), whereas mortality estimates overlap the expected value for the general population when GH is below 2.5 μg/L. There are fewer data for glucose-suppressed GH, but this measurement does not appear to give a superior mortality estimate compared with basal serum GH alone [55]. Deaths from cancer fell sequentially with lowering of GH of levels in the UK study of Orme et al. [39], and death from cardiovascular disease was reduced to expected levels when posttreatment GH was <2.5 μg/L measured by RIA [4]. Ayuk et al. found a sharp increase in death rate per 1,000 individuals as posttreatment GH rose above 2.5 μg/L [21].

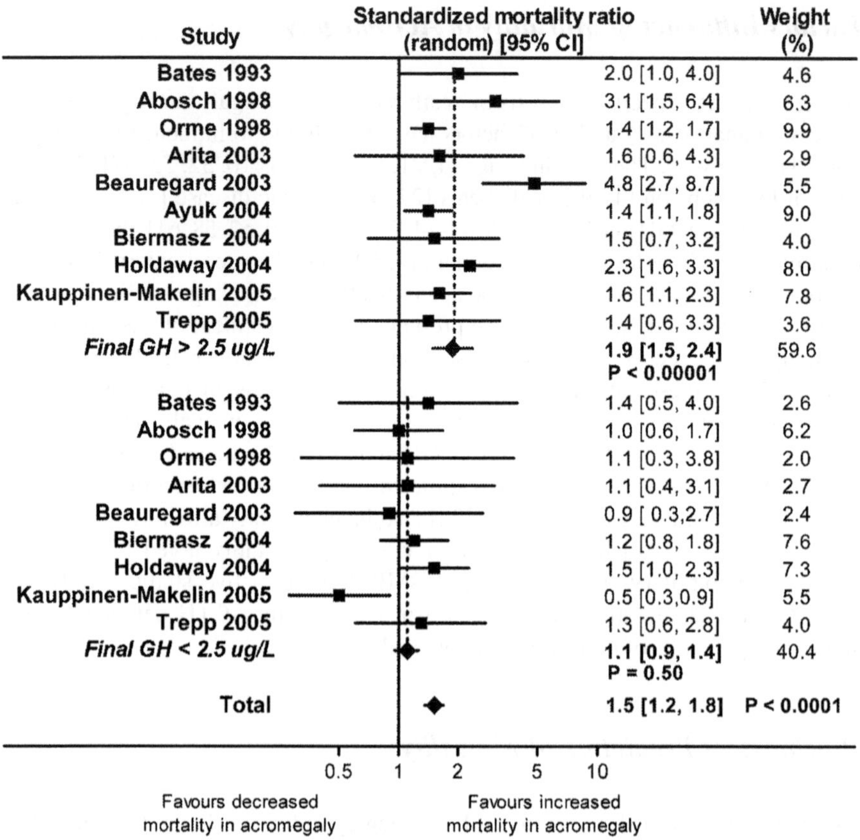

Fig. 16.8 Pooled SMRs with 95% confidence intervals for studies of acromegaly grouped according to serum GH level (measured mainly by relatively insensitive radioimmunoassay) at final follow-up. Reproduced with permission from ref. [55]. Data from references as in text for Fig. 16.5

IGF-I: Serum IGF-I is tightly correlated with GH levels and provides an integrated estimate of GH production over time. Although there are fewer studies linking IGF-I measurements to mortality compared with GH, serum IGF-I following treatment also appears on meta-analysis (Fig. 16.9) to significantly predict mortality, with SMR values indistinguishable from unity when IGF-I levels are in the age-adjusted normal range. Several studies, however, have not found IGF-I to be a significant predictor of mortality compared with GH [21, 52].

Discordant remission data comparing serum GH with IGF-I: As many as 46% of patients treated for acromegaly have discordant remission estimates using both GH and IGF-I measurements [63]. At present, there are no published mortality data to indicate the correct approach to such patients. Unpublished analysis from the New Zealand series shows patients with remission values for GH, but elevated IGF-I measurements have worse mortality outcomes than seen in those with normal IGF-I but elevated GH, but the confidence limit for these estimates was wide. In one study,

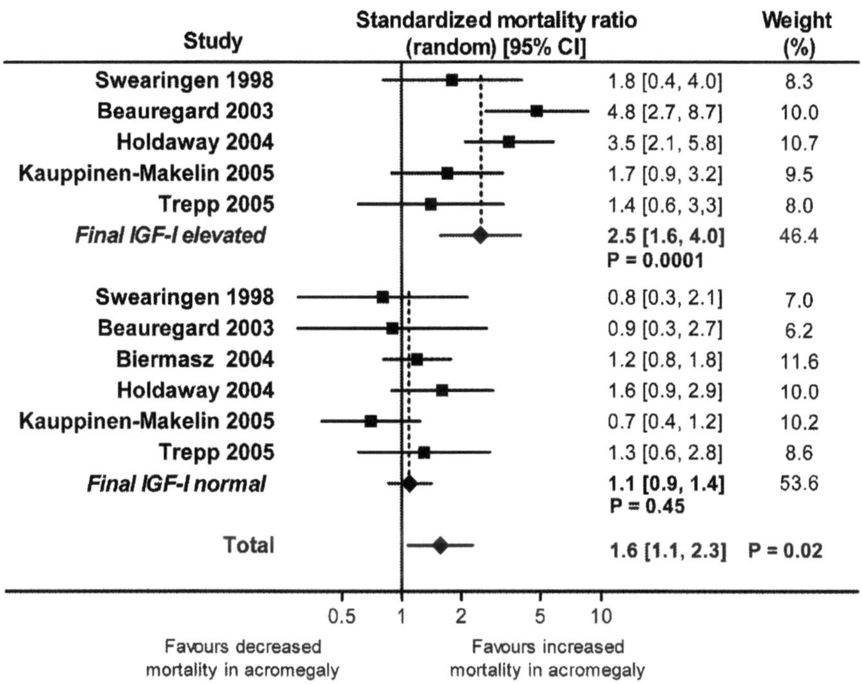

Fig. 16.9 Pooled SMRs with 95% confidence intervals for studies of acromegaly grouped according to serum IGF-I level at final follow-up. Reproduced with permission from ref. [55]. Data from references as in text for Fig. 16.5

a normal serum IGF-I was a stronger predictor of normal metabolic parameters than GH, since modest GH elevations did not worsen metabolic measurements if IGF-I was normal [29]. At present, consensus guidelines suggest using clinical judgment if GH and IGF-I results do not agree when determining remission status. Others have suggested utilizing IGF-I levels for decision making when GH values are elevated but close to the remission cut point [64].

Measurement of Serum GH and IGF-I Using Modern Assays, and the Effect on the Criteria for Remission of Acromegaly

GH: Modern two-site immunoradiometric and chemiluminescent GH assays are more sensitive than earlier radioimmunoassays and now use recombinant 22kd GH for assay standardization instead of an extracted pituitary preparation. As a result, basal and 24-h GH levels and, in particular, nadir GH values during oGTT are lower in normal subjects than seen with earlier assays. Unfortunately, modern GH assays have not been used to any significant extent in the mortality series reported to date,

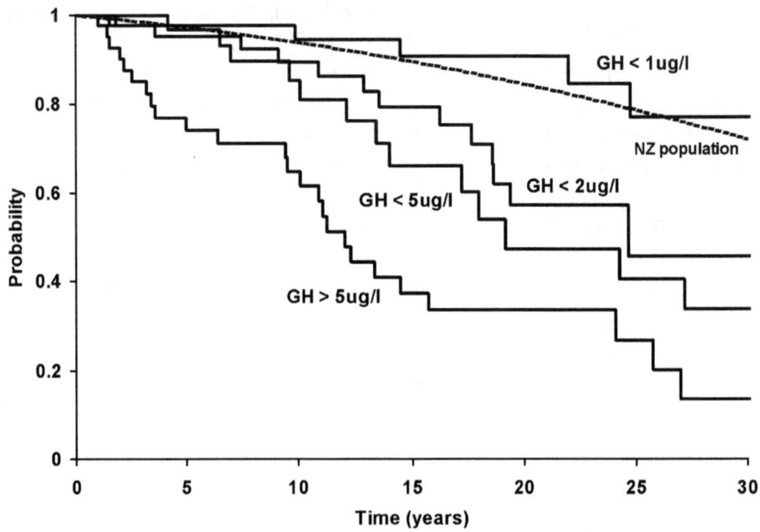

Fig. 16.10 Probability of survival of New Zealand patients with acromegaly according to serum GH at last follow-up. Reproduced with permission from ref. [4]. The *dotted line* represents the probability of survival for the general New Zealand population. $p<0.0001$ by log rank

so the target GH level for remission of acromegaly using current highly sensitive assays remains uncertain. In the New Zealand series [4], approximately half of the final GH values were determined using a two-site immunoradiometric assay, and the lowest mortality (exactly the same as the general population) was seen when random serum GH was <1 μg/L (Fig. 16.10). Other studies have shown that elevated GH >0.14 μg/L measured by sensitive immunoradiometric assay after oGTT was not associated with impaired insulin sensitivity if IGF-I levels were normal [29]. It thus appears sensible to aim for a random serum GH of <1 μg/L and glucose-suppressed GH of <0.4 μg/L following treatment of acromegaly when measured with current sensitive assays (possibly down to 0.14 μg/L with ultrasensitive assays). However, formal studies with mortality as the endpoint are needed to clarify the GH remission criteria using modern GH measurement systems. At present, consensus guidelines suggest a post-oGTT GH value less than 0.4 μg/L [1] or 0.3 μg/L [2] as an indicator of remission when using sensitive GH assays, but have not recommended a value for random GH measurements.

IGF-I: The meta-analyses indicate that achieving a serum IGF-I in the normal range for age and gender following treatment of acromegaly lowers mortality to levels expected for the nonacromegalic population. Since IGF-I and GH levels are tightly correlated, it is not possible to state whether either one of the two measurements provides superior estimates of improvement in mortality [4]. Several groups, however, have found serum IGF-I to be less predictive of mortality than GH measurements [21, 52], with a suggestion that the GH/IGF-I relationship may be distorted in those previously treated with pituitary radiotherapy [63], making serum IGF-I less reliable for therapy decisions in such patients. Circulating IGF-I concentrations are strongly influenced by serum IGF-binding proteins and acid-labile subunit values, and although these latter variables do not seem to provide superior predictive data for mortality compared with IGF-I itself, abnormal IGFBP-3 levels in certain situations may make IGF-I less reliable as a measure of remission. As with GH assays, it is important that IGF-I levels are standardized uniformly with a recombinant IGF-I assay standard.

Conclusions

An individual who developed acromegaly before the advent of modern treatments often had a reduced quality of life and limited life expectancy. The recognition and treatment of the comorbidities of the disorder, the establishment of appropriate biochemical criteria for assessing remission, and refinements in techniques of treatment have all helped to reduce the impact of the condition. However, acromegaly remains a challenging disorder, requiring the interaction of multiple groups for effective management, including clinical chemists, endocrinologists, surgeons, and radiotherapists, as well as subspecialty input for assessment and treatment of specific comorbidities such as cardiac disease, arthropathy, and SAS. Recognition of the increased morbidity and mortality of the disorder has been pivotal in advancing the management of acromegaly.

References

1. Melmed S, Colao A, Barkan A, et al. Guidelines for acromegaly management: an update. J Clin Endocrinol Metab. 2009;94:1509–17.
2. Strasburger CJ. Consensus statement. Biochemical assessment and long-term monitoring in patients with acromegaly: statement from a joint consensus conference of the Growth Hormone Research Society and The Pituitary Society. J Clin Endocrinol Metab. 2004;89:3099–102.
3. Holdaway IM, Rajasoorya C. Epidemiology of acromegaly. Pituitary. 1999;2:29–41.
4. Holdaway IM, Rajasoorya RC, Gamble GD. Factors influencing mortality in acromegaly. J Clin Endocrinol Metab. 2004;89:667–74.
5. Clayton RN. Cardiovascular function in acromegaly. Endocr Rev. 2003;24:272–7.
6. Colao A, Ferone D, Marzullo P, et al. Systemic complications of acromegaly:epidemiology, pathogenesis, and management. Endocr Rev. 2004;25:102–52.
7. Paisley AN, Izzard AS, Gemmell I, et al. Small vessel remodelling and impaired endothelial-dependent dilatation in subcutaneous resistance arteries from patients with acromegaly. J Clin Endocrinol Metab. 2009;94:1111–7.
8. Schiavon F, Maffei P, Martini C, et al. Morphologic study of microcirculation in acromegaly by capillaroscopy. J Clin Endocrinol Metab. 1999;84:3151–5.
9. Nikkila EA, Pelkonen R. Serum lipids in acromegaly. Metabolism. 1975;24:829–38.
10. Nabarro JD. Acromegaly. Clin Endocrinol. 1987;26:481–512.
11. Holdaway IM, Rajasoorya CR, Gamble GD, et al. Long-term treatment outcome in acromegaly. Growth Horm IGF Res. 2003;13:185–92.

12. Gouya H, Vignaux O, Le Roux P, et al. Rapidly reversible myocardial edema in patients with acromegaly: assessment with ultrafast T2 mapping in a single-breath-hold MRI sequence. Am J Roentgenol. 2008;190:1576–82.
13. Pivonello P, Galderisi M, Auriemma RS, et al. Treatment with growth hormone receptor antagonist in acromegaly: effect on cardiac structure and function. J Clin Endocrinol Metab. 2007;92:476–82.
14. Bogazzi F, Battolla L, Spinelli C, et al. Risk factors for development of coronary heart disease in patients with acromegaly: a 5-year prospective study. J Clin Endocrinol Metab. 2007;92:4271–7.
15. Cannavo S, Almoto B, Cavalli G, et al. Acromegaly and coronary disease: an integrated evaluation of conventional coronary risk factors and coronary calcifications detected by computed tomography. J Clin Endocrinol Metab. 2006;91:3766–72.
16. Sacca L, Cittadini A, Fazio S. Growth hormone and the heart. Endocr Rev. 1994;15:555–73.
17. Pereira AM, Van Thiel SW, Lindner JR, et al. Increased prevalence of regurgitant valvular heart disease in acromegaly. J Clin Endocrinol Metab. 2004;89:71–5.
18. Van der Klaauw AA, Bax JJ, Roelfsema F, et al. Uncontrolled acromegaly is associated with progressive mitral valvular regurgitation. Growth Horm IGF Res. 2006;16:101–7.
19. Colao A, Marek J, Goth MI, et al. No greater incidence or worsening of cardiac valve regurgitation with somatostatin analog treatment of acromegaly. J Clin Endocrinol Metab. 2008;93: 2243–8.
20. Fatti LM, Scacchi M, Lavezzi E, et al. Effects of treatment with somatostatin analogues on QT interval duration in acromegalic patients. Clin Endocrinol. 2006;65:626–30.
21. Ayuk J, Clayton RN, Holder G, et al. Growth hormone and pituitary radiotherapy, but not serum insulin-like growth factor-I concentrations, predict excess mortality in patients with acromegaly. J Clin Endocrinol Metab. 2004;89:1613–7.
22. Sherlock M, Reulen RC, Aragon Alonso A, et al. ACTH deficiency, higher doses of hydrocortisone replacement, and radiotherapy are independent predictors of mortality in patients with acromegaly. J Clin Endocrinol Metab. 2009;94:4216–23.
23. Resmini E, Minuto F, Colao A, et al. Secondary diabetes associated with principal endocrinopathies: the impact of new treatment modalities. Acta Diabetes. 2009;46:85–95.
24. Coculescu M, Niculescu D, Lichiardopol R, et al. Insulin resistance and insulin secretion in non-diabetic acromegalic patients. Exp Clin Endocrinol Metab. 2007;115:308–16.
25. Jaffrain-Rea ML, Moroni C, Baldelli R, et al. Relationship between blood pressure and glucose tolerance in acromegaly. Clin Endocrinol. 2001;54:189–95.
26. Mercado M, Gonzalez B, Sandoval C, et al. Clinical and biochemical impact of the d3 growth hormone receptor genotype in acromegaly. J Clin Endocrinol Metab. 2008;93:3411–5.
27. Rajasoorya C, Holdaway IM, Wrightson P, et al. Determinants of clinical outcome and survival in acromegaly. Clin Endocrinol. 1994;41:95–102.
28. Arita K, Kurisu K, Tominaga A, et al. Mortality in 154 surgically treated patients with acromegaly – a 10-year follow-up survey. Endocr J. 2003;50:163–72.
29. Puder JJ, Nilavar S, Post KD, et al. Relationship between disease-related morbidity and biochemical markers of activity in patients with acromegaly. J Clin Endocrinol Metab. 2005;90:1972–8.
30. Hodish I, Barkan A. Long-term effects of pegvisomant in patients with acromegaly. Nat Clin Pract Endocrinol Metab. 2008;4:324–32.
31. Bex M, Abs R, T'Sjoen G, et al. AcroBel – the Belgian registry on acromegaly: a survey of the "real-life" outcome on 418 acromegalic subjects. Eur J Endocrinol. 2007;157:399–409.
32. Biermasz NR, Pereira AM, Smit JWA, et al. Morbidity after long-term remission for acromegaly: persisting joint-related complaints cause reduced quality of life. J Clin Endocrinol Metab. 2005;90:2731–9.
33. Dons RF, Rosselet P, Pastakia B, et al. Arthropathy in acromegalic patients before and after treatment: a long-term follow-up study. Clin Endocrinol. 1988;28:515–24.
34. Maffei P, Schiavon F, Ragazzi R, et al. Shoulder arthropathy in acromegaly. Abstracts of the 85th annual meeting US endocrinol society. 2003. p. P1–639.

35. Mazziotti G, Bianchi A, Bonadonna S, et al. Prevalence of vertebral fractures in men with acromegaly. J Clin Endocrinol Metab. 2008;93:4649–55.
36. Colao A, Cannavo S, Marzullo P, et al. Twelve months of treatment with octreotide-LAR reduces joint thickness in acromegaly. Eur J Endocrinol. 2003;148:31–8.
37. Roddam AW, Allen NE, Appleby P, et al. Insulin-like growth factors, their binding proteins, and prostate cancer risk: an analysis of individual patient data from 12 prospective studies. Ann Int Med. 2008;149:461–71.
38. Renehan AG, Zwahlen M, Minder C, et al. Insulin-like growth factor (IGF)-I, IGF binding protein-3, and cancer risk: systematic review and meta-regression analysis. Lancet. 2004;363:1346–53.
39. Orme SM, McNally RJ, Cartwright RA, et al. Mortality and cancer incidence in acromegaly: a retrospective cohort study. United Kingdom Acromegaly Study Group. J Clin Endocrinol Metab. 1998;83:2730–4.
40. Renehan A, Bhaskar P, Painter JE, et al. The prevalence and characteristics of colorectal neoplasia in acromegaly. J Clin Endocrinol Metab. 2000;85:3417–24.
41. Bengtsson B-A, Eden S, Ernest I, et al. Epidemiology and long-term survival in acromegaly. Acta Med Scand. 1988;223:327–35.
42. Jones A. Radiation oncogenesis in relation to the treatment of pituitary tumours. Clin Endocrinol. 1991;35:379–97.
43. Rokkas T, Pistiolas D, Sechopoulos P, et al. Risk of colorectal neoplasm in patients with acromegaly: a meta-analysis. World J Gastroenterol. 2008;14:3484–9.
44. Van Haute FRB, Taboada GF, Correa LL, et al. Prevalence of sleep apnea and metabolic abnormalities in patients with acromegaly and analysis of cephalometric parameters by magnetic resonance imaging. Eur J Endocrinol. 2008;158:459–65.
45. Sze L, Schmid C, Bloch KE, et al. Effect of transsphenoidal surgery on sleep apnoea in acromegaly. Eur J Endocrinol. 2007;156:321–9.
46. Wright AD, Hill DM, Lowy C, et al. Mortality in acromegaly. Q J Med. 1970;39:1–16.
47. Bates AS, Van't Hoff W, Jones JM, et al. An audit of outcome of treatment in acromegaly. Q J Med. 1993;86:293–9.
48. Levy MJ, Matharu MS, Meeran K, et al. The clinical characteristics of headache in patients with pituitary tumours. Brain. 2005;128:1921–30.
49. Mestron A, Webb SM, Astorga R, et al. Epidemiology, clinical characteristics, outcome, morbidity and mortality in acromegaly based on the Spanish Acromegaly Registry (Registro Espanol de Acromegalia, REA). Eur J Endocrinol. 2004;151:439–46.
50. Beauregard C, Truong U, Hardy J, et al. Long-term outcome and mortality after transsphenoidal adenomectomy for acromegaly. Clin Endocrinol. 2003;58:86–91.
51. Trepp R, Stettler C, Zwahlen M, et al. Treatment outcomes and mortality of 94 patients with acromegaly. Acta Neurochir (Wien). 2005;147:243–51.
52. Kauppinen-Makelin R, Sane T, Reunanen A, et al. A nationwide survey of mortality in acromegaly. J Clin Endocrinol Metab. 2005;90:4081–6.
53. Alexander L, Appleton D, Hall R, et al. Epidemiology of acromegaly in the Newcastle region. Clin Endocrinol. 1980;12:71–9.
54. Dekkers OM, Biermasz NR, Pereira AM, et al. Mortality in acromegaly: a meta-analysis. J Clin Endocrinol Metab. 2008;93:61–7.
55. Holdaway IM, Bolland MJ, Gamble GD. A meta-analysis of the effect of lowering serum levels of GH and IGF-I on mortality in acromegaly. Eur J Endocrinol. 2008;159:89–95.
56. Bengtsson BA, Jonsson B, Nilsson B. Increased mortality in acromegaly is mainly caused by cardiovascular diseases. Abstracts of the 81st annual scientific meeting US endocrine society. 1999. p. P3–648.
57. Etxabe J, Gaztambide S, Latorre P, et al. Acromegaly: an epidemiological study. J Endocrinol Invest. 1993;16:181–7.
58. Abosch A, Tyrrell JB, Lamborn KR, et al. Transsphenoidal microsurgery for growth hormone-secreting pituitary adenomas: initial outcome and long-term results. J Clin Endocrinol Metab. 1998;83:3411–8.

59. Biermasz NR, Dekker FW, Pereira AM, et al. Determinants of survival in treated acromegaly in a single center: predictive value of serial insulin-like growth factor I measurements. J Clin Endocrinol Metab. 2004;89:2789–96.
60. Shimatsu A, Yokogoshi Y, Saito S, et al. Long-term survival and cardiovascular complications in patients with acromegaly and pituitary gigantism. J Endocrinol Invest. 1998;21:55–7.
61. Swearingen B, Barker FG, Katznelson L, et al. Long-term mortality after transsphenoidal surgery and adjunctive therapy for acromegaly. J Clin Endocrinol Metab. 1998;83:3419–26.
62. Holdaway IM. Excess mortality in acromegaly. Horm Res. 2007;68 Suppl 5:166–72.
63. Sherlock M, Aragon Alonso A, Reulen RC, et al. Monitoring disease activity using GH and IGF-I in the follow-up of 501 patients with acromegaly. Clin Endocrinol. 2009;71:74–81.
64. Freda PU. Monitoring of acromegaly: what should be performed when GH and IGF-I levels are discrepant? Clin Endocrinol. 2009;71:166–70.

Chapter 17
GHR Antagonist: Efficacy and Safety

Claire E. Higham and Peter J. Trainer

Abstract Pegvisomant is a GH receptor antagonist (GHR) licensed for the treatment of acromegaly. The story of the development of pegvisomant began with the recognition of the therapeutic implications of GH gene transgenic mice having a dwarf, rather than giant, phenotype and culminated in studies demonstrating its ability to normalise IGF-I in up to 97% of patients. The defining studies used pegvisomant as a daily monotherapy but more recent data suggest pegvisomant can be as effective when administered as a weekly preparation or in combination with somatostatin analogues or dopamine agonists. It is well tolerated and available data do not provide any evidence of potentiation of tumour growth. Approximately 3% of patients experience elevation in liver transaminases, typically within weeks of commencing treatment. The aetiology remains uncertain and available data suggest that it resolves on discontinuation of the drug. Important long term safety and efficacy data will continue to be provided by the ACROSTUDY registry that is open to all patients receiving pegvisomant.

Keywords Acromegaly • GHR antagonists • Animal studies • Pegvisomant • Dopamine agonists • Gigantism and McCune • Albright Syndrome • Metabolic markers • Lipohypertrophy

Acromegaly

Acromegaly is the consequence of excess GH secretion after epiphyseal fusion. It has a prevalence of 55–59 patients per million, an incidence of 3–4 per million and is almost exclusively caused by excess production of GH from a benign somatotroph adenoma [1, 2].

P.J. Trainer (✉)
Department of Endocrinology, Christie Hospital, Manchester M20 4BX, UK
e-mail: peter.trainer@manchester.ac.uk

K. Ho (ed.), *Growth Hormone Related Diseases and Therapy: A Molecular and Physiological Perspective for the Clinician*, Contemporary Endocrinology,
DOI 10.1007/978-1-60761-317-6_17, © Springer Science+Business Media, LLC 2011

The onset of acromegaly is insidious and patients typically have the disease for 4–10 years prior to diagnosis. Once thought of, the diagnosis is usually easily confirmed by a raised serum IGF-I level and failure to suppress serum GH during a 75 g oral glucose tolerance test (OGTT). In contrast to healthy individuals, serum GH is always detectable in patients with GH-producing adenomas with increased frequency and amplitude of GH pulses associated with a significantly greater degree of disorder [3].

Active disease is associated with a 2–3-fold increase in mortality, mainly as a result of cardiovascular disease. The excess mortality associated with acromegaly can be reduced to that of the background population with effective treatment to reduce GH and IGF-I levels to within normal limits [1, 2, 4–6]. However, GH and IGF-I results are not always concordant and there is controversy as to which is the more reliable marker of remission and for follow-up [7, 8], although the most recent consensus statement defined biochemical remission as an age- and sex-matched IGF-I value in addition to a GH of less than 0.4 μg/L during a 75 g OGTT [9].

The first line of treatment for acromegaly is trans-sphenoidal surgery, ideally carried out by an experienced surgeon. The role of conventional three-field, multi-fractional radiotherapy as an effective second-line treatment is falling out of favour as a consequence of concerns about hypopituitarism and increased risk of cerebrovascular disease. There may be a specific niche for stereotactic radiotherapy in patients with well-circumscribed lesions away from the optic chiasm, but despite surgery and radiotherapy the majority of patients require medical treatment at some stage. The mainstay of medical treatment is somatostatin analogues (SSAs), which reduce GH secretion from the pituitary and normalise GH and IGF-I levels in up to 60% of patients [10]. Dopamine agonists also reduce GH secretion from somatotroph adenomas by around 30% and therefore have a role in mild disease or combination therapies [11, 12]. However, despite all the above treatments, around 30% of patients with acromegaly remain uncontrolled and in need of additional therapy.

The Development of GHR Antagonists

The beginning of the growth hormone antagonist story was a paradoxical finding during the study of GH analogues designed to have increased activity at the growth hormone receptor (GHR). One analogue had physiological effects that were contrary to expected with the transgenic mouse model having a dwarf phenotype with body weight being negatively correlated to serum levels of the altered protein [13]. The potential therapeutic implications of a GHR antagonist in the treatment of acromegaly or conditions associated with unwanted GH excess, e.g. diabetic complications such as retinopathy, were quickly appreciated and these early experiments led the way to the production and testing of further mutated bovine and human GH DNA constructs.

Systematic site-specific mutations revealed the glycine at position 119 of the bovine GH protein (bGH-G119R) to be vital in the development of the dwarf phenotype [14] and the same result was achieved with a mutation in the equivalent position in the human GH gene (G120R) [15]. In order to make the molecule viable as a potential therapeutic agent, modifications were required; the addition of four to five 5,000 kDa polyethylene glycol (PEG) molecules increased the molecular weight of the protein from 22 kDa to around 50 kDa, thereby reducing renal clearance and prolonging the half-life from 15 min to 70 h. This also had the effect of reducing the immunogenicity of the foreign protein. The addition of these PEG molecules necessitated nine further amino acid changes to prevent PEG binding to sites vital to the interaction with the GHR. This new molecule was named pegvisomant.

Recent advances in the understanding of GH/GHR interactions and downstream GH-signalling pathways suggest that pegvisomant works as an effective GHR antagonist by binding to preformed membrane-bound GHR dimers, preventing conformational changes of a rotational nature within the receptor/peg complex normally induced by GH binding, in turn preventing downstream intracellular signalling and so inhibiting GH action [16].

Efficacy

Blockade of the GHR by pegvisomant leads to an increase in serum GH, and although this is probably of no clinical consequence, it means that in contrast to all other treatment modalities, serum GH levels cannot be used to monitor the efficacy of treatment. Furthermore, pegvisomant interferes in the majority of commercial GH assays [17] as a result of the 1,000-fold excess of pegvisomant compared to GH required to achieve adequate antagonist activity. This leads to a "Hook effect" resulting in spuriously low GH results. Measurement of GH during pegvisomant treatment therefore requires analysis on specific assays which eliminate cross-reactivity, but these are not yet routinely available [18]. As a result, the normalisation of IGF-I to within age- and gender-related reference ranges is the primary endpoint in pegvisomant treatment.

This reliance on IGF-I measurements is not without difficulty. Titration of pegvisomant and determination of normal IGF-I levels rely on dependable serum IGF-I assays and robust reference ranges which are not always in place [19]. As mentioned previously, there remains controversy surrounding the correlation of IGF-I with morbidity and mortality in acromegaly [7]. These caveats aside the studies described below demonstrate that pegvisomant has excellent efficacy in the treatment of acromegaly, and indeed, it is probable that all patients can reach a normal IGF-I during pegvisomant therapy, provided adequate dose titration can be achieved.

Animal Studies

Efficacy studies were first carried out in animals prior to phase I studies in man. Unlike the original mutant bovine (bGHG119R) protein, the pegvisomant molecule has a low affinity for non-primate GHR and is relatively ineffective in rodents [20] and, therefore, animal studies have been largely limited to primates. Administration

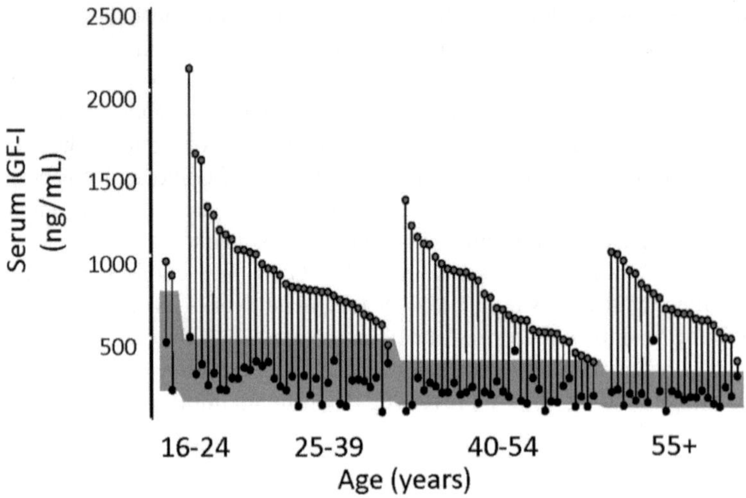

Fig. 17.1 IGF-I results in 90 patients at baseline (*grey dots*) and following 18 months of pegvisomant therapy (*black dots*). *Grey shaded area* represents age-adjusted reference ranges for IGF-I. Adapted from van der Lely et al. [24]

of pegvisomant to ovariectomised female rhesus monkeys reduced plasma IGF-I, IGFBP-3, and ALS levels by 60% within 3 days following a single subcutaneous injection. The fall in IGF-I was mirrored by an increase in serum GH levels [21].

Human Studies

The first studies in healthy male volunteers using a single subcutaneous pegvisomant injection of either 0.3 or 1 mg/kg (approximating to a single 80 mg injection) led to a maximum 50% reduction in circulating IGF-I levels at day 5 with the higher dose [18]. Following this, an initial "proof of concept" placebo-controlled, double-blind study in active acromegaly (baseline IGF-I >50% above the upper limit of normal [ULN]) demonstrated a rather disappointing decrease of only 16% and 31% in serum IGF-I levels following 30 or 80 mg (i.e. the approximate equivalent of 1 mg/kg), respectively, of pegvisomant once weekly for 6 weeks. This corresponded to poor IGF-I normalisation rates [22].

The disappointing results with weekly dosing led the defining phase III, multicentre, double-blind, placebo-controlled trial to use daily dosing to increase serum pegvisomant concentrations and potentially enhance the rate of IGF-I normalisation. Pegvisomant at doses 10, 15, and 20 mg daily for 6 weeks led to 54, 81, and 89% of patients, respectively, achieving an IGF-I within the reference range. IGFBP-3, ALS, and free IGF-I fell in parallel to total serum IGF-I [23]. Eighteen months of treatment in an open-label escalating dose regime extension study showed that using a maximum dose of 40 mg daily, 97% of patients achieved an IGF-I within the reference range (Fig. 17.1). This was an efficacy unequalled by any other medical treatment in acromegaly [24], opening the way for pegvisomant becoming licenced.

Treatment-Resistant Disease

These initial trials were conducted in patients with acromegaly who had an IGF-I above the ULN following washout from other medications, and while these were mainly patients who were uncontrolled on other treatments, this was not a requisite for trial inclusion. However, the role of pegvisomant within the acromegaly treatment algorithm is usually considered as limited to those with treatment-resistant disease as a result of the cost, relative inconvenience of daily injections and paucity of long-term safety data. Subsequent studies have confirmed that pegvisomant is also efficacious in those patients where other treatments have previously failed [25–27]. In these studies, patients were converted to pegvisomant after failing conventional treatments with surgery, radiotherapy and SSAs. IGF-I was normalised in 78% [27] to 100% of patients [25, 26].

Other Modes of Treatment

Weekly Therapy

The phase III trial led to the belief that daily pegvisomant is more efficacious than weekly dosing as IGF-I normalisation was achieved in 54% of patients taking 10 mg once daily compared to only 20% of those taking the approximately equivalent 80 mg weekly dose. However, a closer inspection of the two studies reveals that the inclusion criteria differed (phase II; IGF-I > 150% of ULN vs. phase III; IGF-I > 130% of ULN), and when the decrease in IGF-I achieved is expressed as a percentage of baseline, it is approximately equivalent between the weekly and daily dose regimens (27 and 31%, respectively). The difference in normalisation rates was therefore probably a result of the different baseline IGF-I values and poor normalisation with the initial weekly regimen was most likely due to inadequate dosing. There is now good evidence for the successful use of weekly pegvisomant, both as pegvisomant monotherapy [28] and in combination with SSA therapy [29]. We successfully converted five patients from daily administration (median dose 15 mg, range 10–20 mg) to the equivalent dose as a twice weekly followed by weekly injection with no change in IGF-I (mean IGF-I 145 ng/mL daily dosing vs. 127 ng/mL weekly dosing), safety parameters or quality of life over 32 weeks [28].

Weekly dosing is an attractive strategy as a means of reducing injection frequency and potentially improving compliance. At present, this is limited by the lack of pegvisomant vials containing greater than 20 mg. Reconstitution of each vial with 1 mL of water results in a large volume of injection for the average weekly requirement.

Combination Treatment

Combination treatment with pegvisomant and SSAs is effective at controlling IGF-I levels; the addition of weekly pegvisomant (up to 160 mg) in patients uncontrolled

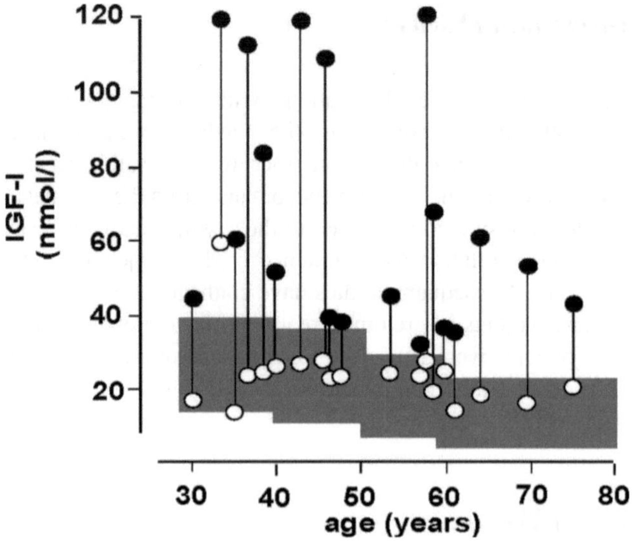

Fig. 17.2 Serum insulin-like growth factor (IGF-I) concentration in 19 patients with acromegaly before (*filled dots*) and after (*open dots*) 6 weeks of combined somatostatin analogue (SSA) and pegvisomant therapy. *Shaded area* represents age-adjusted IGF-I reference ranges. Adapted from Feenstra et al. [29]

on maximum dose of SSAs normalised IGF-I in all 32 patients reported by Neggers et al. [30] (Fig. 17.2). More recently, a prospective trial randomised patients uncontrolled on maximum dose of octreotide to either stop octreotide and start pegvisomant as monotherapy, or to continue octreotide and have pegvisomant added. Overall, the rates of normalisation of IGF-I were the same in both limbs, with patients on pegvisomant monotherapy requiring on average 5 mg more pegvisomant per day than those on combination therapy [31]. As far as control of IGF-I is concerned, for patients who are uncontrolled on maximum dose SSAs, there appears to be little difference between switching to pegvisomant monotherapy or initiating combination therapy. Choice of treatment regime should be based on individual need and choice, for example a patient who has experienced tumour shrinkage with SSAs would be best treated with combination therapy, whereas a patient with diabetes may benefit more from pegvisomant monotherapy.

The original study of combination treatment with pegvisomant and SSA suggested a pegvisomant dose sparing and therefore cost saving benefit. This dose sparing is probably a result of reduced serum GH levels and therefore reduced competition at the GHR, although there is also a suggestion that serum pegvisomant levels may be increased in the presence of SSA [29]. However, as described above, the only study to randomise patients to monotherapy (pegvisomant 10 mg od) or combination therapy (oct 30 mg and pegvisomant 10 mg) showed on average only a 5 mg difference in dose between monotherapy and combination treatment which suggests the two options to be of approximately equivalent cost [31].

Combination therapy of pegvisomant and dopamine agonists is an attractive proposition considering the relative inexpense and oral nature of the DA. Furthermore, cabergoline has a similar half-life as pegvisomant and both drugs could potentially be given on a weekly basis. We have recently concluded a prospective open-label trial of low-dose pegvisomant therapy and cabergoline that suggests cabergoline could have a pegvisomant dose sparing effect [32].

Long-Term Treatment

There are accumulating data that the efficacy of pegvisomant is sustained outside clinical trials. Our retrospective analysis of 57 patients (many of whom had participated in the original clinical trials) treated for a median of 18 months but up to 10 years in two tertiary referral centres in the UK demonstrated that 95% of patients normalised IGF-I with a median dose of 15 mg daily. In the great majority of patients, the maintenance dose remained stable; however, two patients in this cohort required increasing doses of pegvisomant with time raising the possibility of tachyphylaxis [33].

The original clinical trials involved around only 200 patients and it was recognised that there was a need for ongoing assessment of efficacy and safety in a prospective, observational manner. ACROSTUDY is a Pfizer-sponsored, international observational multi-centre surveillance registry accumulating non-interventional data to monitor the long-term safety and efficacy of pegvisomant. At the time of the most recent analysis (Feb 2009), 792 patients were registered as having been treated with pegvisomant from 300 centres in ten countries.

From this database mean IGF-I at baseline was 518 ± 296 ng/mL, falling to 277 ± 180 ng/mL 1 year following pegvisomant and thereafter remaining stable. This equated to an IGF-I normalisation rate of only 62%, which did not change over 4 years. The average dose of pegvisomant for those patients with a normal IGF-I was 106 mg/week (15 mg/day). Surprisingly, the average dose for those patients with an elevated IGF-I was not much higher at 113 mg/week (16 mg/day), suggesting failure of adequate dose titration [34]. This lack of dose titration did not occur in our UK study where patients were followed in a tertiary referral setting and may reflect the diversity of follow-up in the ACROSTUDY. The incremental cost of using more than one ampoule of pegvisomant, significant relief of symptoms without achieving normalisation of IGF-I and the opinion that IGF-I normalisation is not absolutely necessary particularly if a patient has levels close to the upper limit of the reference range likely contributed to a failure to dose titrate.

Gigantism and McCune Albright Syndrome

Case reports have supported the efficacy of pegvisomant in gigantism [35–38] and a recent sub-analysis of ACROSTUDY revealed that pegvisomant had arrested linear growth and resulted in a normal IGF-I in all 11 giants in whom other medical therapies had failed [39]. There is also evidence of effectiveness in patients with McCune Albright associated with acromegaly [40, 41].

Factors Affecting Dose Requirement

The mean dose of pegvisomant required to normalise IGF-I is 16.6 mg/day with the spectrum being from 5 mg alternate day to 60 mg/day. The main influences on the dose required are the pre-treatment levels of serum GH and IGF-I [42], but many additional factors contribute.

Women require larger doses of pegvisomant than men to achieve the same plasma concentrations and require higher plasma concentrations to achieve a similar decrease in IGF-I [42–44]. Weight and previous radiotherapy are also determinants of pegvisomant dose requirements, dose required increasing by approximately 0.24 mg for each kg increase in body weight [42]. Previous radiotherapy reduces dose requirements: 15.2 mg/day average dose with prior radiotherapy compared to 18 mg/day with no radiotherapy despite equivalent baseline GH/IGF-I levels.

There has been recent increasing interest in the role of exon-3 deletions of the GHR in GH action, with evidence that a lack of exon-3 of the GHR increases the biological activity of GH. Two studies have recently shown that the presence of an exon-3-deleted GHR is associated with a lower dose requirement for pegvisomant [44, 45], with patients homozygous or heterozygous for the mutation requiring on average a 20% lower dose of pegvisomant per kg body weight [44].

Quality of Life

Much of the focus of clinicians when treating acromegaly revolves around biochemical measures of disease activity which contrasts with patients, who are understandably more concerned with symptom control and quality of life. Initial short-term randomised control trial did demonstrate significant improvements in the symptoms and signs of disease with pegvisomant treatment [23, 24] and a more recent retrospective subgroup analysis of 130 patients within the German ACROSTUDY showed significant improvements in perspiration, soft tissue swelling and perceived health at 12 months of treatment, particularly in those patients with the highest IGF-I levels at baseline [46]. While it is well documented that active acromegaly is associated with a reduced quality of life, there is only one study addressing the role of pegvisomant in improving this and it is an area that needs to be more rigorously studied.

An interesting, prospective, randomised, placebo-controlled, cross-over trial recently demonstrated that the addition of low-dose pegvisomant to SSAs in patients with an IGF-I within the reference range at baseline resulted in a significant improvement in quality of life with no change in serum total IGF-I levels [47]. The mechanisms of this remain elusive, but merit further investigation.

Metabolic Markers

The excess mortality associated with active acromegaly is attributed mainly to increased cardiovascular disease and diabetes and therefore biomarkers of these conditions may provide additional information to guide treatment.

The majority of studies have demonstrated that pegvisomant leads to an improvement in glucose metabolism; with a decrease in fasting glucose, decreased HbA1c and improved insulin sensitivity following pegvisomant [24, 48]. More detailed glucose stable isotope studies suggest the improvements are the result of increased hepatic and peripheral insulin sensitivity [49, 50].

The results from studies analysing cardiovascular risk factors are less consistent with small numbers of patients included and changes in many markers not reaching statistical significance. However, Parkinson et al. demonstrated a significant increase in total and LDL cholesterol [51], from levels below the normal reference range to within normal with pegvisomant treatment, where others showed non-significant trends in this direction [52, 53]. Sesmilo showed a small but significant increase in triglycerides [52], whereas others did not [51, 53, 54]. It should be noted however that cardiovascular risk in acromegaly is the subject of some controversy. Cardiovascular disease in active disease is probably related to hypertrophic remodelling of smooth and cardiac muscle leading to hypertension and cardiomyopathy [55–57] rather than traditional risk factors for atherosclerosis, although a recent, well-conducted retrospective analysis of 133 patients from the German ACROSTUDY showed that female patients with active disease had a significantly higher Framingham Score than matched controls despite significantly lower total and LDL cholesterol levels. This risk was improved with pegvisomant treatment, particularly when IGF-I reduced to within the reference range [58]. However, further understanding of the mechanisms of cardiovascular disease are clearly needed. Defects in cortisol and bone metabolism characteristic of acromegaly are also corrected with pegvisomant therapy [59–61].

Safety

As with any new drug with a novel mode of action, safety monitoring was paramount throughout the sponsored trials with pegvisomant. Data collected from these initial studies were focused on concerns with regard to tumour growth and tachyphylaxis. The data in this regard have been very reassuring. However, an unexpected side effect has been the development of abnormal liver function following initiation of pegvisomant in approximately 3% of patients.

The most recent analysis of ACROSTUDY data showed that from 792 patients there were 56 serious AEs reported during the cumulative 2,625 patient year experience, with 13 being attributed to the drug. Twelve were related to abnormalities in liver enzymes or pituitary tumour growth which are discussed below. There have

Table 17.1 A comprehensive list (in alphabetic order) of adverse events regarded by the reporting investigator as not being pegvisomant related

Abdominal complaint (3)
Allergic reaction
Alopecia (2)
Asthenia (2)
Bone pain
Constipation
Diarrhoea
Dizziness (3)
Eczema (2)
Elevated liver enzymes (8)
Fatigue (2)
Headache (3)
Hypertension
Hypotension
Increased appetite
Increased sweating (3)
Injection site reaction (9)[a]
Insomnia (2)
Intestinal mycosis
Mood problems (2)
Muscle pain
Paraesthesia
Shortness of breath
Skin dries
Stomach bloating
Tingling
Weight gain

Number of patients affected in brackets
Adapted from Trainer [34]
[a] Including three cases of lipohypertrophy

been two deaths, neither of which, from the information available, were attributable to pegvisomant. Table 17.1 contains an overview of the non-serious adverse events reported (Table 17.1); 56 were attributed to pegvisomant, the majority being liver enzyme-related or injection site reactions [34].

Tumour Growth

One of the main concerns with pegvisomant was that the reduction in IGF-I and consequent increase in serum GH levels might induce an increase in tumour size, analogous to Nelson's syndrome following bilateral adrenalectomy for Cushing's disease. Consequently, tumour size has been closely monitored during all clinical studies and annual scans are recommended during routine care.

During the phase III and subsequent 18-month open-label trial, tumour volumes were monitored by MRI scanning and analysed by a single radiologist at baseline and following treatment. In the 12-week study, mean tumour volume was unchanged, and in the 18-month extension trial, overall mean tumour volume again remained unchanged (mean 11.5 months); however, two patients did demonstrate an increase in tumour volume [24].

Analysis of all the data from sponsored clinical trials using pegvisomant has shown that 9 out of 304 patients had tumour growth in the first 12 months of pegvisomant treatment, i.e. an incidence of around 3% with a cumulative experience of 452 patient years [24, 53, 62, 63].

These nine patients have been studied in more detail [64] to ascertain the reasons for tumour growth and are shown in Table 17.2. Importantly, none of these patients had received prior radiotherapy. Six of the nine had documented increased tumour growth prior to starting pegvisomant either on no treatment or SSAs and therefore an aggressive underlying tumour probably accounts for the increased size. Four of these patients went on to have surgery and radiotherapy for the tumour and one continues on pegvisomant in combination with SSA. The remaining two had SSA added to the pegvisomant with good effect.

Three of the nine had previously been treated with SSAs, and when switched to pegvisomant, experienced a rebound increase in tumour size at 6 months as a result of SSA cessation. All three had tumours that remained stable subsequent to this and they continue on pegvisomant monotherapy with no further change in size.

In addition to studying the changes in the first year, 43 of the 403 patients had been treated for more than 18 months (mean duration of 29 months (mean 18–45)) with no significant increase in any patient. The subgroup that had been irradiated showed a significant decrease in tumour size. These data lend support to the increasing confidence that prolonged high serum GH levels in association with pegvisomant treatment do not drive an increase in tumour size.

Similar rates of increased tumour size have been documented outside the clinical trial setting. Within the German ACROSTUDY, patients were recommended to have baseline scans prior to pegvisomant initiation and annually thereafter, although this was left at the discretion of the local physician. In 2009, 245 patients from 371 patients registered had undergone at least two MRI scans. Initially, these scans were reported by local radiologists and re-evaluated centrally by a single, blinded radiologist if tumour growth was suspected. Twenty patients (approx. 6%) were highlighted as having tumour size increase, but only nine (approx. 3%) were confirmed on re-analysis. The reasons for increased size again reflected that seen in the trial patients, with one third having documented tumour growth prior to pegvisomant, two patients apparently having rebound growth following SSA therapy and three further having clinically insignificant change. Only one patient had new significant growth on pegvisomant that required further treatment [65].

We find the data to be reassuring with regard to tumour size in pegvisomant treatment; mild clinically insignificant increases in size may result from rebound growth after cessation of SSA therapy, but do not require termination of pegvisomant therapy, although regular monitoring is required. A small minority of patients have

Table 17.2 Table of characteristics of the nine patients with tumour growth while on pegvisomant

Patient	Sex/age	Tumour growth prior to pegv	Treated with octreotide LAR prior to pegv	Tumour at start of pegv	Time to increase in tumour size after start of pegv	Tumour on continued treatment with pegv	Treatment for tumour progression	Ongoing therapy
1	F 26	Yes[a]	No	Macro	7	NA	Radiotherapy	SSA
2	M 32	No data	Yes[b]	Macro	6	Growth	Surgery, radiotherapy	PegV
3	F 33	Yes[c]	Yes	Macro	6	NA	Surgery	Unknown
4	F 33	Yes[d]	Yes	Macro	6	NA	SSA, surgery	SSA
5	F 61	Yes[c]	Yes	Macro	6	Growth	SSA,[e] surgery, Radiotherapy	Pegv + SSA
6	M 53	No	Yes	Macro	6	Stable	None	PegV
7	M 64	No	Yes	Micro	8	Stable	None	PegV
8	M 43	Yes[a]	Yes	Macro	6	Growth	+SSA	PegV + SSA
9	M 32	Yes[a]	Yes	Macro	6	Growth	+SSA	PegV + SSA

Adapted from Jimenez et al. [64]
[a]While on no medication
[b]Octreotide sc
[c]While on long-acting SSA
[d]During pregnancy while on no treatment
[e]The tumour continued to grow resulting in visual defects while on pasireotide for 6.5 months

significant tumour growth while on pegvisomant, usually those with aggressive tumours, and such patients require very careful monitoring and may require combination treatment or radiotherapy. Of the small series represented in the literature, it appears that younger patients are more likely to have aggressive tumours, particularly in association with gigantism or McCune Albright and these patients require close monitoring and more safety data are required.

Liver Function Disturbance

During the initial trials, one patient out of the 80 randomised to pegvisomant developed elevated liver enzymes. ALT (alanine transaminase) rose to 904 U/L (0–47) and AST (aspartate transaminase) to 389 U/L (0–37). Bilirubin and alkaline phosphatase (ALP) remained within the normal limits and the patient was relatively asymptomatic, only complaining of mild fatigue. Viral serology was negative and liver USS was normal. The patient discontinued pegvisomant and LFTs returned to normal within 8 weeks. However, the abnormality recurred when the patient was re-challenged with pegvisomant [24].

There have now been multiple reports of abnormalities in liver enzymes during treatment and the incidence of the idiosyncratic hepatitis is probably <3% [34].The reaction usually occurs within the first 3 months of therapy (although has occurred later), is not related to the dose of pegvisomant, and in each case, has resolved if pegvisomant is discontinued. If pegvisomant is discontinued, the elevated transaminases may recur on re-challenge [23], but this is not always the case [66]. There have also been reports of resolution of LFTs despite continuation of the drug [66]. Little is known about the hepatic metabolism of pegvisomant and the mechanism underlying the LFT abnormalities is not well understood, although recent studies suggest that male patients and those with the UGT1A1*28 polymorphism associated with Gilbert's syndrome and indicative of reduced hepatic glucuronidation activity are more susceptible [67]. Liver biopsies have shown an active hepatitis picture, possible chronic active type or drug induced [66].

It is important to consider that there are other possible causes of raised liver enzymes in pegvisomant therapy. Gallstone disease is a common feature of acromegaly and incidence is increased by SSA therapy; however, it is rarely symptomatic during SSA treatment as the drug inhibits gallbladder motility. Discontinuation of SSA therapy, a common feature prior to pegvisomant treatment, is associated with an increased incidence of acute biliary problems [68]. In 2006, a study of 12 patients with raised liver enzymes on pegvisomant therapy found that five had gallstone disease demonstrated on ultrasound scan [66]. Furthermore, some patients recorded as having abnormal LFTs on pegvisomant in the ACROSTUDY had abnormal results prior to initiation of pegvisomant [34].

LFTs should be carefully monitored, particularly during the first 6 months of treatment. A complete investigation should be undertaken to differentiate between a predominant transaminase or cholestatic picture and treatment administered accordingly. Relation to pegvisomant can be confirmed if the abnormalities resolve upon discontinuation of the drug and this should be considered, although as patients are generally asymptomatic and there is no evidence of long-term harm, this is not always necessary if elevations are mild.

Lipohypertrophy

Three publications have reported eight patients with pegvisomant-induced lipohypertrophy [69–71]. The mechanism is unknown, but it appears to be more frequent in women and may reflect inhibition of the local action of GH at the adipocyte, leading to a reduction in lipolysis and an accumulation of lipid. USS and MRI imaging of the areas demonstrate focal accumulations of subcutaneous adipose tissue and histology confirms normal subcutaneous and adipose tissue with no evidence of inflammatory infiltrate. Rotation of injection sites was sufficient to allow resolution in some patients, but there were a number who appeared particularly sensitive to the complication and suffered from the deposition at all injection sites that only resolved on discontinuation. Patients must therefore be taught to rotate the sites of pegvisomant administration.

Pregnancy

The average age at diagnosis of acromegaly is 45 years, but it can present at any age. Active acromegaly commonly results in infertility as a result of suppressed gonadotrophins, but many reports exist of successful pregnancy in patients treated with SSAs. There have been no teratogenic effects of pegvisomant reported from animal studies; all the same women with the potential for pregnancy should be using contraception while on pegvisomant. There have been two pregnancies reported with pegvisomant treatment. One patient had a normal pregnancy and caesarean delivery, having been treated with 15–25 mg once daily throughout the pregnancy. At delivery, the foetus and placenta had minimal detectable pegvisomant levels and normal levels of foetal GH and IGF-I were recorded. By 6 months, the child had reached normal developmental milestones [72]. The second lady conceived with her first cycle of IVF and ICSI while on pegvisomant therapy at 20 mg once daily that was discontinued on confirmation of pregnancy. An elective caesarean was performed at 32 weeks that resulted in the delivery of a healthy boy who was developmentally normal at 1 year [73].

Conclusion

Pegvisomant is a very effective drug for treatment-resistant acromegaly with placebo-controlled trials demonstrating normalisation of IGF-I in up to 97% of patients. Providing adequate dose titration is performed and the drug is tolerated, a reduction in IGF-I to within the reference range should be possible in virtually all patients. It can be used successfully in combination with SSA and DA and its efficacy is sustained as a weekly preparation. Side effects are rare and accumulating data are reassuring with regard to tumour growth, but close monitoring is still required with regular MR scanning and checking of LFTs, particularly during the initial 6-month dose titration phase.

We predict future development of pegvisomant will focus on increased use of combination therapies with SSA and possibly dopamine agonists and the use of pegvisomant on a weekly basis, although this will require new preparations of the drug. The potential role of pegvisomant in treatment of various malignancies is an exciting but unexplored area. ACROSTUDY should continue to provide invaluable long-term safety and efficacy data.

References

1. Alexander L, Appleton D, Hall R, Ross WM, Wilkinson R. Epidemiology of acromegaly in the Newcastle region. Clin Endocrinol. 1980;12:71–9.
2. Bengtsson BA, Eden S, Ernest L, Oden A, Sjorgren B. Epidemiology and long-term survival in acromegaly. A study of 166 cases diagnosed between 1955 and 1984. Acta Med Scand. 1988;223:327–35.
3. Hartman ML, Pincus SM, Johnson ML, Matthews DH, Faunt LM, Vance ML, et al. Enhanced basal and disorderly growth hormone secretion distinguish acromegalic from normal pulsatile growth hormone release. J Clin Invest. 1994;94:1277–88.
4. Bates AS, Van't Hoff W, Jones JM, Clayton RN. An audit of outcome of treatment in acromegaly. Q J Med. 1993;86:293–9.
5. Orme SM, McNally RJ, Cartwright RA, Belchetz PE. Mortality and cancer incidence in acromegaly: a retrospective cohort study. United Kingdom Acromegaly Study Group. J Clin Endocrinol Metab. 1998;83:2730–4.
6. Swearingen B, Barker II FG, Katznelson L, Biller BM, Grinspoon S, Klibanski A, et al. Long-term mortality after transsphenoidal surgery and adjunctive therapy for acromegaly. J Clin Endocrinol Metab. 1998;83(10):3419–26.
7. Ayuk J, Clayton RN, Holder G, Sheppard MC, Stewart PM, Bates AS. Growth hormone and pituitary radiotherapy, but not serum insulin-like growth factor-I concentrations, predict excess mortality in patients with acromegaly. J Clin Endocrinol Metab. 2004;89:1613–7.
8. Clemmons DR, Strasburger C. Monitoring the response to treatment in acromegaly. J Clin Endocrinol Metab. 2004;11:5289–91.
9. Giustina A, Chanson P, Bronstein MD, Klibanski A, Lamberts S, Casanueva FF, et al. A consensus on criteria for cure of acromegaly. J Clin Endocrinol Metab. 2010;95:3141–8.
10. Murray RD, Melmed S. A critical analysis of clinically available somatostatin analog formulations for therapy of acromegaly. J Clin Endocrinol Metab. 2008;93:2957–68.
11. Jackson SN, Fowler J, Howlett TA. Cabergoline treatment of acromegaly: a preliminary dose finding study. Clin Endocrinol. 1997;46:745–9.
12. Abs R, Verhelst J, Maiter D, Van Acker K, Nobels F, Coolens JL, et al. Cabergoline in the treatment of acromegaly: a study in 64 patients. J Clin Endocrinol Metab. 1998;83:374–6.
13. Chen WY, White ME, Wagner TE, Kopchick JJ. Functional antagonism between endogenous mouse growth hormone (GH) and a GH analog results in dwarf transgenic mice. Endocrinology. 1991;129:1402–8.
14. Chen WY, Wight DC, Mehta BV, Wagner TE, Kopchick JJ. Glycine 119 of bovine growth hormone is critical for growth-promoting activity. Mol Endocrinol. 1991;5:1845–52.

15. Chen WY, Chen NY, Yun J, Wagner TE, Kopchick JJ. In vitro and in vivo studies of antagonistic effects of human growth hormone analogs. J Biol Chem. 1992;269:15892–7.
16. Brown RJ, Adams JJ, Pelekanos RA, Wan Y, McKinstry WJ, Palethorpe K, et al. Model for growth hormone receptor activation based on subunit rotation within a receptor dimmer. Nat Struct Mol Biol. 2005;12:814–21.
17. Paisley AN, Hayden K, Ellis A, Anderson J, Wieringa G, Trainer PJ. Pegvisomant interference in GH assays results in underestimation of GH levels. Eur J Endocrinol. 2007;156(3):315–9.
18. Thorner MO, Strasburger CJ, Wu Z, Straume M, Bidlingmaier M, Pezzoli SS, et al. Growth hormone (GH) receptor blockade with a PEG-modified GH (B2036-PEG) lowers serum insulin-like growth factor-I but does not acutely stimulate serum GH. J Clin Endocrinol Metab. 1994;84:2098–103.
19. Pokrajac A, Wark G, Ellis AR, Wear J, Wieringa GE, Trainer PJ. Variation in GH and IGF-I assays limits the applicability of international consensus criteria to local practice. Clin Endocrinol (Oxf). 2007;67(1):65–70.
20. Mode A, Tollet P, Wells T, Carmignac DF, Clark RG, Chen WY, et al. The human growth hormone (hGH) antagonist G120RhGH does not antagonize GH in the rat, but has paradoxical agonist activity, probably via the prolactin receptor. Endocrinology. 1996;137:447–54.
21. Wilson ME. Insulin-like growth factor I (IGF-I) replacement during growth hormone receptor antagonism normalizes serum IGF-binding protein-3 and markers of bone formation in ovariectomized rhesus monkeys. J Clin Endocrinol Metab. 2000;85:1557–62.
22. Van der Lely AJ, Lamberts SWJ, Barkan A, Panadya N, Besser GM, Trainer PJ, et al. A six week, double blind, placebo controlled study of a growth hormone antagonist, B2036-PEG (Trovert) in acromegalic patients. Program of the 80th annual meeting of the endocrine society. New Orleans, LA; 1998. p. 57 (Abstract OR4-1).
23. Trainer PJ, Drake WM, Katznelson L, Freda PU, Herman-Bonert V, van der Lely AJ, et al. Treatment of acromegaly with the growth hormone-receptor antagonist pegvisomant. N Engl J Med. 2000;20:1171–7.
24. van der Lely AJ, Hutson RK, Trainer PJ, Besser GM, Barkan AL, Katznelson L, et al. Long-term treatment of acromegaly with pegvisomant, a growth hormone receptor antagonist. Lancet. 2001;358:1754–9.
25. Herman-Bonert VS, Zib K, Scarlett JA, Melmed S. Growth hormone receptor antagonist therapy in acromegalic patients resistant to somatostatin analogs. J Clin Endocrinol Metab. 2000;85:2958–61.
26. Drake WM, Parkinson C, Akker SA, Monson JP, Besser GM, Trainer PJ. Successful treatment of resistant acromegaly with a growth hormone receptor antagonist. Eur J Endocrinol. 2001;145:451–6.
27. Colao A, Pivonello R, Cappabianca P, Auriemma RS, De Martino MC, Ciccarelli A, et al. The use of a GH receptor antagonist in patients with acromegaly resistant to somatostatin analogs. J Endocrinol Invest. 2003;26:53–6.
28. Higham CE, Thomas JD, Bidlingmaier M, Drake WM. Trainer PJ Successful use of weekly pegvisomant administration in patients with acromegaly. Eur J Endocrinol. 2009;161(1):21–5.
29. Feenstra J, de Herder WW, ten Have SM, van den Beld AW, Feelders RA, Janssen JA, et al. Combined therapy with somatostatin analogues and weekly pegvisomant in active acromegaly. Lancet. 2005;365(9471):1644–6. Erratum in: Lancet. 2005;365(9471):1620.
30. Neggers SJ, van Aken MO, Janssen JA, Feelders RA, de Herder WW, van der Lely AJ. Long-term efficacy and safety of combined treatment of somatostatin analogs and pegvisomant in acromegaly. J Clin Endocrinol Metab. 2007;92(12):4598–601.
31. Trainer PJ, Ezzat S, D'Souza GA, Layton G, Strasburger CJ. A randomized, controlled, multicentre trial comparing pegvisomant alone with combination therapy of pegvisomant and long-acting octreotide in patients with acromegaly. Clin Endocrinol (Oxf). 2009;71(4):549–57.
32. Higham CE, Atkinson AB, Aylwin S, Martin NM, Moyes V, Newell-Price J, et al. A prospective clinical trial of combined cabergoline (C) and pegvisomant (pegV) treatment in patients with active acromegaly. Program of the annual meeting of the endocrine society. Washington, DC; 2009 (Abstract P1744).

33. Higham CE, Chung TT, Lawrance J, Drake WM, Trainer PJ. Long-term experience of pegvisomant therapy as a treatment for acromegaly. Clin Endocrinol (Oxf). 2009;71:86–91.
34. Trainer PJ. ACROSTUDY: the first 5 years. Eur J Endocrinol. 2009;161 Suppl 1:S19–24.
35. Rix M, Laurberg P, Hoejberg AS, Brock-Jacobsen B. Pegvisomant therapy in pituitary gigantism: successful treatment in a 12-year-old girl. Eur J Endocrinol. 2005;153(2):195–201.
36. Müssig K, Gallwitz B, Honegger J, Strasburger CJ, Bidlingmaier M, Machicao F, et al. Pegvisomant treatment in gigantism caused by a growth hormone-secreting giant pituitary adenoma. Exp Clin Endocrinol Diabetes. 2007;115(3):198–202.
37. Goldenberg N, Racine MS, Thomas P, Degnan B, Chandler W, Barkan A. Treatment of pituitary gigantism with the growth hormone receptor antagonist pegvisomant. J Clin Endocrinol Metab. 2008;93(8):2953–6.
38. Main KM, Sehested A, Feldt-Rasmussen U. Pegvisomant treatment in a 4-year-old girl with neurofibromatosis type 1. Horm Res. 2006;65(1):1–5.
39. Higham CE, Emy P, Ferone D, Finke R, Laurberg P, Main K, et al. Treatment experience in 11 patients with gigantism. Endocrine. 2010;21:P280 (Abstracts).
40. Akintoye SO, Kelly MH, Brillante B, Cherman N, Turner S, Butman JA, et al. Pegvisomant for the treatment of gsp-mediated growth hormone excess in patients with McCune-Albright syndrome. J Clin Endocrinol Metab. 2006;91(8):2960–6.
41. Galland F, Kamenicky P, Affres H, Reznik Y, Pontvert D, Le Bouc Y, et al. McCune-Albright syndrome and acromegaly: effects of hypothalamopituitary radiotherapy and/or pegvisomant in somatostatin analog-resistant patients. J Clin Endocrinol Metab. 2006;91(12):-4957–61.
42. Parkinson C, Burman P, Messig M, Trainer PJ. Gender, body weight, disease activity, and previous radiotherapy influence the response to pegvisomant. J Clin Endocrinol Metab. 2007;92(1):190–5.
43. Marazuela M, Lucas T, Alvarez-Escolá C, Puig-Domingo M, de la Torre NG, de Miguel-Novoa P, et al. Long-term treatment of acromegalic patients resistant to somatostatin analogues with the GH receptor antagonist pegvisomant: its efficacy in relation to gender and previous radiotherapy. Eur J Endocrinol. 2009;160(4):535–42.
44. Bernabeu I, Alvarez-Escolá C, Quinteiro C, Lucas T, Puig-Domingo M, Luque-Ramírez M, et al. The exon 3-deleted growth hormone receptor is associated with better response to pegvisomant therapy in acromegaly. J Clin Endocrinol Metab. 2010;95(1):222–9.
45. Bianchi A, Mazziotti G, Tilaro L, Cimino V, Veltri F, Gaetani E, et al. Growth hormone receptor polymorphism and the effects of pegvisomant in acromegaly. Pituitary. 2009; 12(3):196–9.
46. Sievers C, Brübach K, Saller B, Schneider HJ, Buchfelder M, Droste M, et al.; German Pegvisomant Investigators. Change of symptoms and perceived health in acromegalic patients on pegvisomant therapy: a retrospective cohort study within the German Pegvisomant Observational Study (GPOS). Clin Endocrinol (Oxf). 2010;73:89–94.
47. Neggers SJ, van Aken MO, de Herder WW, Feelders RA, Janssen JA, Badia X, et al. Quality of life in acromegalic patients during long-term somatostatin analog treatment with and without pegvisomant. J Clin Endocrinol Metab. 2008;93(10):3853–9.
48. Barkan AL, Burman P, Clemmons DR, Drake WM, Gagel RF, Harris PE, et al. Glucose homeostasis and safety in patients with acromegaly converted from long-acting octreotide to pegvisomant. J Clin Endocrinol Metab. 2005;90:5684–91.
49. Lindberg-Larsen R, Moller N, Schmitz O, Nielsen S, Andersen M, Orskov H, et al. The impact of pegvisomant treatment on substrate metabolism and insulin sensitivity in patients with acromegaly. J Clin Endocrinol Metab. 2007;92:1724–8.
50. Higham CE, Rowles S, Russell-Jones D, Umpleby AM, Trainer PJ. Pegvisomant improves insulin sensitivity and reduces overnight free fatty acid concentrations in patients with acromegaly. J Clin Endocrinol Metab. 2009;94(7):2459–63.
51. Parkinson C, Drake WM, Wieringa G, Yates AP, Besser GM, Trainer PJ. Serum lipoprotein changes following IGF-I normalization using a growth hormone receptor antagonist in acromegaly. Clin Endocrinol (Oxf). 2002;56:303–11.

52. Sesmilo G, Fairfield WP, Katznelson L, Pulaski K, Freda PU, Bonert V, et al. Cardiovascular risk factors in acromegaly before and after normalization of serum IGF-I levels with the GH antagonist pegvisomant. J Clin Endocrinol Metab. 2002;87(4):1692–9.
53. Colao A, Pivonello R, Auriemma RS, De Martino MC, Bidlingmaier M, Briganti F, et al. Efficacy of 12-month treatment with the GH receptor antagonist pegvisomant in patients with acromegaly resistant to long-term, high-dose somatostatin analog treatment: effect on IGF-I levels, tumor mass, hypertension and glucose tolerance. Eur J Endocrinol. 2006;154(3):467–77.
54. Paisley AN, O'Callaghan CJ, Lewandowski KC, Parkinson C, Roberts ME, Drake WM, et al. Reductions of circulating matrix metalloproteinase 2 and vascular endothelial growth factor levels after treatment with pegvisomant in subjects with acromegaly. J Clin Endocrinol Metab. 2006;91(11):4635–40.
55. Fazio S, Cittadini A, Sabatini D, Merola B, Colao AM, Biondi B, et al. Evidence for biventricular involvement in acromegaly: a Doppler echocardiographic study. Eur Heart J. 1993;14:26–33.
56. Lie JT. Pathology of the heart in acromegaly: anatomic findings in 27 autopsied patients. Am Heart J. 1980;100:41–52.
57. Paisley AN, Izzard AS, Gemmell I, Cruickshank K, Trainer PJ, Heagerty AM. Small vessel remodeling and impaired endothelial-dependent dilatation in subcutaneous resistance arteries from patients with acromegaly. J Clin Endocrinol Metab. 2009;94(4):1111–7.
58. Berg C, Petersenn S, Lahner H, Herrmann BL, Buchfelder M, Droste M, et al.; Investigative Group of the Heinz Nixdorf Recall Study and the German Pegvisomant Observational Study Board and Investigators. Cardiovascular risk factors in patients with uncontrolled and long-term acromegaly: comparison with matched data from the general population and the effect of disease control. J Clin Endocrinol Metab. 2010;95:3648–56.
59. Trainer PJ, Drake WM, Perry LA, Taylor NF, Besser GM, Monson JP. Modulation of cortisol metabolism by the growth hormone receptor antagonist pegvisomant in patients with acromegaly. J Clin Endocrinol Metab. 2001;86:2989–92.
60. Parkinson C, Kassem M, Heickendorff L, Flyvbjerg A, Trainer PJ. Pegvisomant-induced serum insulin-like growth factor-I normalization in patients with acromegaly returns elevated markers of bone turnover to normal. J Clin Endocrinol Metab. 2003;88(12):5650–5.
61. Fairfield WP, Sesmilo G, Katznelson L, Pulaski K, Freda PU, Stavrou S, et al. Effects of a growth hormone receptor antagonist on bone markers in acromegaly. Clin Endocrinol (Oxf). 2002;57(3):385–90.
62. Frohman LA, Bonert V. Pituitary tumor enlargement in two patients with acromegaly during pegvisomant therapy. Pituitary. 2007;10(3):283–9.
63. Melmed S, Sternberg R, Cook D, Klibanski A, Chanson P, Bonert V, et al. A critical analysis of pituitary tumor shrinkage during primary medical therapy in acromegaly. J Clin Endocrinol Metab. 2005;90(7):4405–10.
64. Jimenez C, Burman P, Abs R, Clemmons DR, Drake WM, Hutson KR, et al. Follow-up of pituitary tumor volume in patients with acromegaly treated with pegvisomant in clinical trials. Eur J Endocrinol. 2008;159(5):517–23.
65. Buchfelder M, Weigel D, Droste M, Mann K, Saller B, Brübach K, et al.; Investigators of German Pegvisomant Observational Study. Pituitary tumor size in acromegaly during pegvisomant treatment: experience from MR re-evaluations of the German Pegvisomant Observational Study. Eur J Endocrinol. 2009;161(1):27–35.
66. Biering H, Saller B, Bauditz J, Pirlich M, Rudolph B, Johne A, et al.; German Pegvisomant Investigators. Elevated transaminases during medical treatment of acromegaly: a review of the German pegvisomant surveillance experience and a report of a patient with histologically proven chronic mild active hepatitis. Eur J Endocrinol. 2006;154(2):213–20.
67. Bernabeu I, Marazuela M, Lucas T, Loidi L, Alvarez-Escolá C, Luque-Ramírez M, et al. Pegvisomant-induced liver injury is related to the UGT1A1*28 polymorphism of Gilbert's syndrome. J Clin Endocrinol Metab. 2010;95(5):2147–54.

68. Paisley AN, Roberts ME, Trainer PJ. Withdrawal of somatostatin analogue therapy in patients with acromegaly is associated with an increased risk of acute biliary problems. Clin Endocrinol (Oxf). 2007;66(5):723–6.
69. Maffei P, Martini C, Pagano C, Sicolo N, Corbetti F. Lipohypertrophy in acromegaly induced by the new growth hormone receptor antagonist pegvisomant. Ann Intern Med. 2006;145(4):310–2.
70. Marazuela M, Daudén E, Ocón E, Moure D, Nattero L. Pegvisomant-induced lipohypertrophy: report of a case with histopathology. Ann Intern Med. 2007;147(10):741–3.
71. Bonert VS, Kennedy L, Petersenn S, Barkan A, Carmichael J, Melmed S. Lipodystrophy in patients with acromegaly receiving pegvisomant. J Clin Endocrinol Metab. 2008;93(9):3515–8.
72. Brian SR, Bidlingmaier M, Wajnrajch MP, Weinzimer SA, Inzucchi SE. Treatment of acromegaly with pegvisomant during pregnancy: maternal and fetal effects. J Clin Endocrinol Metab. 2007;92(9):3374–7.
73. Qureshi A, Kalu E, Ramanathan G, Bano G, Croucher C, Panahloo A. IVF/ICSI in a woman with active acromegaly: successful outcome following treatment with pegvisomant. J Assist Reprod Genet. 2006;23:439–42.

Part V
Use of Growth Hormone

Chapter 18
Long-Acting Growth Hormone Analogues

Alice Thorpe, Helen Freeman, Sarbendra L. Pradhananga, Ian R. Wilkinson, and Richard J.M. Ross

Abstract Growth hormone (GH) is an anabolic cytokine hormone regulating linear growth in childhood and normal body composition in adults. The current therapeutic regimen for GH replacement requires once-daily subcutaneous injections which is inconvenient and expensive. A number of approaches have been taken to create long-acting preparations, including depot preparations and sustained-release formulations. However, although of proven efficacy, such GH preparations are characterised by a dominant early-release profile, causing supraphysiological GH levels, manufacture is expensive and injections may be painful. There is a need for cytokine formulations that minimise manufacturing costs, have good pharmacokinetic profiles, are easy to administer, and are acceptable to patients. Recent focus has been on the generation of long-acting GH analogues by post-translational modification or protein fusions. In this article, we review the new technologies for generating long-acting GH agonists and GH antagonists including pegylation, albumin fusions and ligand-receptor fusions.

Keywords Pegylation • Pegylation of GH • GH antagonist Somavert • Albumin-fusion proteins • Albumin and GH fusions • Ligand-receptor fusions

---

* Disclosure Statement
RR & IW have equity interests in Asterion Ltd.

R.J.M. Ross (✉)
Academic Unit of Diabetes, Endocrinology & Metabolism, Department of Human Metabolism, University of Sheffield, Royal Hallamshire Hospital, Glossop Road, Sheffield, UK
e-mail: r.j.ross@sheffield.ac.uk

K. Ho (ed.), *Growth Hormone Related Diseases and Therapy: A Molecular and Physiological Perspective for the Clinician*, Contemporary Endocrinology,
DOI 10.1007/978-1-60761-317-6_18, © Springer Science+Business Media, LLC 2011

Introduction

Optimal hormone replacement therapy should mimic physiological hormone levels; this is a basic tenet of endocrinology. However, for most endocrine replacement, this is not possible as the secretion profile of endogenous hormones is complex and the pharmacokinetics of most synthetic hormones do not match the endogenous rhythm. Growth hormone (GH) is secreted from the anterior pituitary in pulses every 3–4 h with maximal secretion occurring during slow-wave sleep. GH has a short serum half-life (Table 18.1), and although it would be possible to simulate GH pulses through frequent intravenous injections, this is not practical and current therapy is usually given as a single subcutaneous bolus at night. GH pulses are generated by a dynamic equilibrium between GH-releasing hormone (GHRH) and somatostatin. Therapy with GHRH can induce physiological endogenous GH pulses and even promote normal growth in children whose defect is in the hypothalamus, but therapy requires frequent injections and most patients with GH deficiency have some pituitary deficit [1]. An alternative approach has been the use of long-acting GH preparations, although concern remains that continuous GH exposure may provide a non-physiological response.

GH is a cytokine hormone and most cytokines have a short serum half-life. There have been a number of approaches to generating long-acting biologicals and these include depot preparations or generation of analogues by modification including: pegylation and fusions to other proteins. Depot or sustained-release formulations and continuous subcutaneous infusions of GH in GH-deficient children and adults have proven efficacious [2–5]. However, sustained-release GH preparations are characterised by a dominant early-release profile, resulting in supraphysiological GH levels [6], and manufacture is expensive and injections may be painful [4]. In this chapter, we review some new technologies for generating long-acting GH agonists and GH antagonists including pegylation, albumin fusions and ligand-receptor fusions.

Pegylation

Pegylation was initially described in the 1970s [7, 8], and subsequent extensive research has led to the development of site-specific polyethylene glycol (PEG) of varying molecular weights [9]. Pegylation involves covalent linkage of PEG molecules to the protein of interest and is used to delay protein clearance [10, 11]. Generally, pegylated proteins retain the structure of the PEG-linker-protein and the functional groups used within the protein for pegylation are the primary NH_2 groups at the N-terminus and on lysine residues [11, 12]. Use of lysine residues results in a heterogeneous product due to random pegylation at multiple sites; however; this heterogeneity can be reduced by using site-specific PEG-reagents, such as PEG-malemide,

Table 18.1 Comparative data for long-acting GH analogues in rats, monkeys and humans

	Rat		Monkey		Man	
	t ½ (h)	Clearance ($ml\ h^{-1}\ kg^{-1}$)	t ½ (h)	Clearance ($ml\ h^{-1}\ kg^{-1}$)	t ½ (h)	Clearance ($ml\ h^{-1}\ kg^{-1}$)
Native GH						
hGH (s.c.) [36, 52]	0.66±0.1	820±94	1.7±0.4	154±4.4	1.9–2.5	202–261
Pegylated GH						
Bolder [20]	9 (s.c.)	Detectable up to 72 h (i.v.)	–	–	–	–
Pfizer [21]	–	–	–	–	42	1.3
Pegylated GH antag						
Pegvisomant (s.c.) [53]					75	0.3
Albumin fusion						
Albutropin [36]	6 (s.c.) 3 (i.v.)	30.6 (i.v.)	13–15 (s.c. and i.v.)	7 (i.v.)	–	–
LR-fusion [45]						
GH-LRv2 (s.c.)	25±6.4	2.8±0.2	54±6.8	1.7±0.2	–	–
GH-LRv3 (s.c.)	26±6.8	2.7±0.2	76±4.9	0.96±0.2	–	–

which reacts with the functional group present on free cysteine residues [11]. It is possible to generate site-specific binding by introducing cysteine residues into the protein, as has been used for recombinant granulocyte-macrophage colony-stimulating factor [13]. It is also possible to reduce non-specific conjugation by decreasing the number of primary amine groups present by replacing lysine residues with non-reactive amino acids [12]. Sulfhydryl groups may also be used as the reactive functional group; this is typically used in the site-specific pegylation of antibodies [11]. The linker within pegylated proteins can be varied to either create a permanent linkage to the peptide such as succinimiddyl succinate PEG or releasable, e.g. r-PEG or PEG-bicin, which are hydrolyzed within the body [14]. This releasable technology allows the manufacture of pro-drugs [14]. In addition, the linker can be branched or linear, and PEG itself is variable in size and shape allowing generation of forked/multi-armed PEGs.

Pegylation delays protein clearance through two mechanisms: reduced renal clearance and prevention of proteolysis. Pegylation generates an increase in molecular weight and an apparent increase in size through PEG's ability to bind water, thereby increasing the hydrodynamic radius of the molecule [15]. The hydrodynamic radius of the protein may be 5–10 times greater than that of the un-pegylated molecule [14]. For example, a single-chain variable fragment antibody conjugated with a 40 kDa branched PEG had an expected MW of 65 kDa, but an apparent mass of 670 kDa [9]. The water trapping effect of PEG creates a shell-like structure around the protein preventing proteolytic recognition and degradation within the circulation [9]. For peptides that are cleared via endocytosis and cellular degradation, pegylation does not generally alter the route of clearance, but has been noted to slow the process [14].

Another potential benefit of PEG is its apparent ability to decrease immunogenicity of some proteins [8, 9]. A more significant decrease in immunogenicity is observed with the addition of a higher mass PEG [15]. However, the reduction in immunogenicity is not a universal observation, and a number of patients have developed an IgM immune response to PEG itself [10].

The addition of large PEG moieties results in decreased receptor-binding affinity through steric hindrance, although the increased circulatory residence time may compensate for the decreased affinity [14]. In the development of pegylated pharmaceuticals, it is essential to achieve the optimum balance between decreased receptor binding and increased circulatory residence to achieve maximal potency [16].

One of the disadvantages to pegylation is heterogeneity of product. Pharmaceutical products are subject to stringent analytical controls and the manufacture of a homogenous, well-defined product is paramount. By their nature, PEGs are polydisperse and variation in site and number of attached PEGs is frequently observed and worse as the molecular weight of the PEG increases [10]. Heterogeneity of product is generally high when the functional group selected is on the lysine residues, due to random pegylation at multiple sites. Heterogeneity can be reduced by selecting other functional groups, such as those present on free cysteines [11]. However, the utilisation of cysteine residues for site-specific conjugation may be limited, as native cysteine residues are generally buried within the peptide and their introduction via genetic engineering may result in protein misfolding [10].

PEG is classified as a non-biodegradable substance, although degradation has been observed in the presence of substances such as alcohol dehydrogenase and elimination of PEG from the body is very much dependent on PEG size. PEGs with a lower molecular weight are cleared via the kidney, while larger PEGs exhibit a slower rate of clearance, predominantly via the liver. This may be an important consideration in the long-term use of pegylated peptides, as accumulation and resultant consequences are yet to be established. Pegvisomant, the pegylated GH antagonist, is internalised into cells which may explain why some patients develop a hepatitis as the liver has a high number of GH receptors [17]. Cellular internalisation of pegylated recombinant peptides has also been noted in renal tubular epithelial cells. As a result of this internalisation, the development of vacuoles has been seen due to the simultaneous internalisation of water molecules as a function of the hydrophilic nature of PEG, although these vacuoles were not related to any clinically significant pathology [18]. In rhesus monkeys treated with pegylated haemoglobin, vacuolated macrophages were seen in the spleen and bone marrow [19].

Pegylation of GH

Initial studies with pegylated GH were undertaken by Clark *et al.* [16] who demonstrated that conjugation of GH with PEG results in decreased GH clearance, but that increasing the number of PEG moieties resulted in reduced receptor binding. The optimal balance between decreased receptor-binding affinity and increased circulatory residence was achieved with the addition of 5×5 kDa PEGs [16]. The constructs developed by Clark utilised an amine-specific PEG, resulting in a heterogeneous and non-site-specific pegylated product due to binding at any of the eight lysine residues present in GH or on the N-terminus. Bolder Pharmaceuticals have subsequently developed a homogeneous mono-pegylated GH analogue [20]. This analogue utilises a 20 kDa vinylsulfone PEG, a cysteine-specific PEG that binds to an introduced cysteine residue via substitution of a Threonine3; this construct is known as PEG-TC3. This mono-pegylated GH analogue has a 3–4 times reduced receptor-binding affinity, but an eightfold increase in plasma residency time following subcutaneous injection into rats (Table 18.1), resulting in increased drug potency [20]. Pfizer developed a branched pegylated GH with a 40 kDa PEG that in humans resulted in 10–20-fold increase in plasma half-life (Table 18.1) [21]. This pegylated analogue was administered in doses of between 1 and 4 mg a week for 6 weeks to GH-deficient adults. The pegylated GH increased IGF-I levels both over time and with increasing dose; however, lipoatrophy occurred at the site of injection in 13% of patients [22]. This lipoatrophy seems likely to be a direct consequence of high levels of GH remaining at the site of injection, although the exact mechanism has not been established.

Pegylation of the GH Antagonist Somavert

A GH antagonist Somavert (Pegvisomant) is now marketed for the treatment of acromegaly [23, 24]. The core structure of Pegvisomant is a mutated GH molecule, B2036, which contains eight mutations in the first binding site (H18D, H21N, R167N, K168A, D171S, K172R, E174S, I179T) and a G120K mutation in binding site two [25]. The mutations in site 1 increase affinity for the GH receptor and the mutation in site 2 blocks the conformational change in the receptor dimer required for cell signalling [26]. The mutation in the second binding site also introduces a lysine residue, which is then pegylated, further helping to reduce receptor-binding affinity and increase the antagonistic effect of the drug while delaying clearance. Pegvisomant contains 4–6 PEG molecules covalently linked to lysine residues and the N-terminus within B2036. Genetic manipulation was required to remove a lysine residue within binding site 1 of the GH mutant to prevent pegylation, which would have decreased receptor-binding affinity. Pegvisomant exhibits 20-fold decreased receptor-binding affinity than wild-type GH; however, this is compensated by increased circulatory residence time [25]. Un-pegylated B2036 exhibits a similar circulatory residence time to that of native GH (16 min) due to rapid renal ultrafiltration [27, 28]. However, when conjugated with multiple 5 kDa PEGs, Pegvisomant exhibits a plasma half-life of 72 h [28]. As renal clearance accounts for only 50% of GH clearance and receptor internalisation is not considered an important component of GH clearance, the prolonged circulating half-life for Pegvisomant is presumed to be due to reduced proteolysis. Despite its prolonged half-life (Table 18.1), Pegvisomant is still administered via daily injections to maintain a high circulating level [28]. Pegvisomant at doses of between 10 and 20 mg daily can control acromegaly in the majority of patients [24].

Albumin-Fusion Proteins

Albumin is the most abundant plasma protein and important physiologically for maintaining plasma pH, osmotic pressure and as a carrier for insoluble and hydrophobic endogenous and exogenous compounds such as: fatty acids, metal ions and drugs [29, 30]. Albumin is a multi-binding transporter protein with a large binding capacity made up of three sub-domains each of which are composed of two subunits [29]. Recombinant human serum albumin (rHSA) is safe for human use and demonstrates comparable pharmokinetics and pharmodynamics to native HSA [31]. Fusions with HSA have been used to extend the plasma half-life of biological therapeutics. The technology is based upon the knowledge that albumin is approximately 67 kDa and resultant fusion proteins are of a molecular size too large to be filtered by the kidneys [30]. In addition, albumin has a long plasma half-life, 19 days in humans [30], because it utilises the neonatal Fc receptor (FcRn) [29] which recycles the protein. As with pegylation, albumin-fusion technology has been noted to mask the therapeutic protein, therefore stabilising it against peptidases and reducing immunogenicity [31].

Albumin fusions are generally expressed in mammalian cell lines, although yeast production has also been used [31]. An example of an albumin-fused therapeutic substance is recombinant human factor VIIa (rVIIa), developed for the treatment of bleeding episodes in haemophiliacs unable to tolerate human factor VIII due to antibodies. The rVIIa fusion is linked to the 585-amino acid albumin via a 31-amino acid glycine-serine linker. The glycine-serine linker was necessary to maintain potency of the clotting factor [32]. The albumin-fused FVIIa has a molecular weight of 120 kDa compared to 59 kDa for the wild-type FVIIa and exhibits a sixfold increase in half-life in rats as well as maintaining 60–70% biological activity [33]. Another example is Albulin, an insulin-albumin fusion containing both the A- and B-chains of human insulin, linked together by a dodecapeptide and fused to the NH_2 terminus of human serum albumin. Administered to diabetic mice, Albulin normalises glucose levels over a sustained 24-h period in a smooth fashion, as opposed to other recombinant long-acting insulins, which result in peaks and troughs in glucose concentration [34]. ConjuChem Inc. has developed an albumin-protein fusion technology referred to as drug affinity complex (DAC) technology. In these constructs, therapeutically active peptides are conjugated to albumin via a connector; the site of attachment to the drug and length of which are optimised for maximal drug potency. Most DAC peptides contain the Malemide reactive group, which specifically reacts with cysteine 34 in HSA (present in subdomain 1A). This technology is unique as the reactive group can react with exogenous or endogenous albumin, as cysteine residues are not located in any other plasma proteins. An example of a DAC-conjugate is CJC-1134-PC, an albumin conjugate of Exendin-4 used in the treatment of type 2 diabetes. Other albumin-fusion proteins include interferon α-2b, Pro-urokinase and Hirudin which is linked to albumin via its C-terminus [35].

Albumin and GH Fusions

Albutropin, a GH albumin fusion, has HSA linked at its C-terminus to the N-terminus of GH. The molecule has a MW of 89 kDa and achieved an improved pharmacokinetic profile compared to GH when administered to rats and monkeys (Table 18.1) [36]. In rats, the increase in residency time is due to a slower rate of absorption rather than a decrease in clearance rate. However, in monkeys, a significantly extended terminal half-life can be observed on both s.c. and i.v. administration. This can in part be attributed to a slower rate of absorption following s.c. administration; however, a marked decrease in clearance rate can also be observed, particularly at higher concentrations of 1.5 and 4 mg/kg. Albutropin is biologically active in hypophysectomized rats receiving either daily, every other day, or every 4 days doses where cumulative weight gain and enlargement of tibial epiphyseal growth plate width were observed [36]. Although initial clinical studies were undertaken with Albutropin in healthy volunteers and patients, there have been no reports of the outcome and clinical studies have not progressed.

Ligand-Receptor Fusions and GH

GH acts through a cell-surface type 1 cytokine receptor (GHR). In common with other cytokine receptors, the extracellular domain of the GHR is proteolytically cleaved and circulates as a binding protein (GHBP) [37]. Under physiological conditions, GH is in part bound in the circulation in a 1:1 molar ratio by GHBP and this complex appears to be biologically inactive but is protected from clearance and degradation [38, 39]. A cross-linked complex of GH with GHBP has delayed clearance, but exhibits no biological activity as cross-linking presumably obscures the active GH-binding site [40]. However, co-administration of separately purified GHBP with GH in a 1:1 ratio can augment the anabolic actions of GH through generating a depot [41]. Thus, like many hormonal systems, binding in the circulation provides an inactive circulating reservoir in equilibrium with active free hormone [42].

We have been examining the impact of making fusions between ligand and their cognate extracellular domain receptor or, as in the case of GH, the GHBP (Fig. 18.1a). Initially, we sought to make a GH antagonist and hypothesised that a ligand-receptor fusion (LR-fusion) of cytokine to its extracellular domain receptor would interfere with receptor conformation and block signalling. This was based on our observation that a truncated extracellular domain GHR was a dominant negative inhibitor of the GHR [43, 44]. However, to our surprise, the first LR-fusion of GH proved to be an

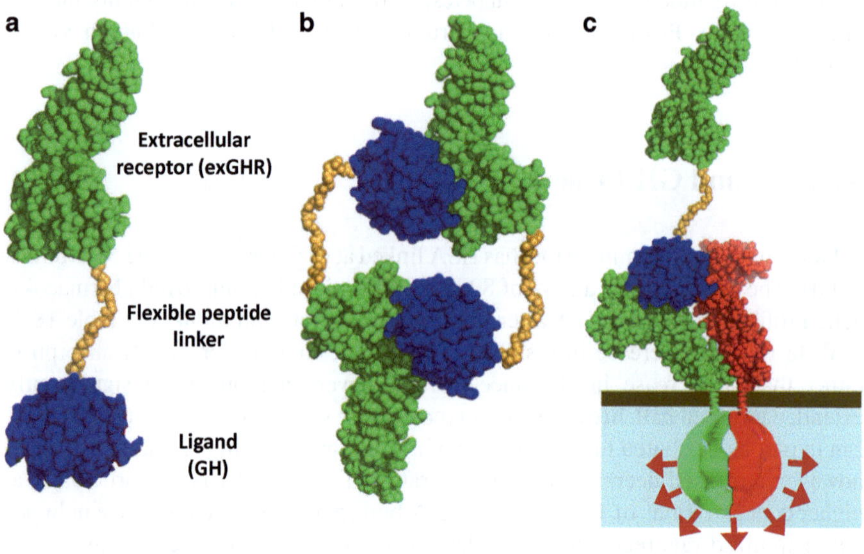

Fig. 18.1 (**a**) The LR-fusion molecule in monomeric form with GH (*blue*) linked (*gold*) to exGHR (*green*). (**b**) Model for the LR-fusion forming a reciprocal head-to-tail dimer where GH (*blue*) in one molecule binds to exGHR (*green*) in the other molecule. Finally, in (**c**) the LR-fusion in monomeric form is capable of binding and activating the GH receptor

agonist and subsequent work has demonstrated that LR-fusions provide a new technology for generating long-acting biologicals with exceptional pharmacokinetic properties (Table 18.1) [45].

The design of the LR-fusion was based on the known crystal structure of the GHR [46]. We used a flexible Gly_4Ser linker with four repeats (predicted length of 80 Å). This long linker was chosen as a relatively flexible tether between GH and the GHR such that the GH moiety could still interact with the cell-surface GHR. Similar Gly_4Ser linkers have been used in recombinant single-chain Fv antibody production because of stability and lack of immunogenicity [47]. The LR-fusion was appropriately folded, appearing on both native PAGE gels and in gel filtration as two distinct species, i.e. potentially monomer and dimer. The presence of dimers was confirmed by analytical ultracentrifugation. We propose that the LR-fusion forms a reciprocal head-to-tail dimer through intermolecular binding of the GH moiety in each LR-fusion molecule to the receptor moiety in the other (Fig. 18.1b). The LR-fusion appeared as two bands on SDS-PAGE, with a molecular weight difference of 5 kDa, presumably due to glycosylation [48, 49]. The LR-fusion was more potent *in vivo* compared to GH, but *in vivo* bioactivity was 10-times less. This discrepancy can be attributed to dimerisation of the LR-fusion. In a static *in vivo* bioassay, the dimer would be biologically inactive as seen with the native GH/GHBP complex [50, 51]. However *in vivo*, the dimer provides a reservoir of inactive hormone in equilibrium with biologically active monomer (Fig. 18.1c).

After i.v. administration to rats, the LR-fusion had a 300-times reduced clearance compared to GH and a 10–30-times reduced clearance compared to that previously reported for a GH/GHBP complex or conjugate [16, 38]. We propose that the greatly reduced clearance of our LR-fusion is attributable to both reduced renal clearance and a conformation that prevents proteolysis. In hypophysectomised rats, our LR-fusion given only once during 10 days produced a similar increase in weight to that seen with daily injections of GH (Fig. 18.2a, b). It has previously been shown that GHBP co-administered as 1:1 molar complex with GH augments growth [41]. Using the same protocol, the LR-fusion protein promoted growth over 10 days after a single injection, whereas the GH/GHBP complex required daily injections and the LR-fusion generated a higher IGF-I level than that seen after GH/GHBP co-administration. GH is biologically inactive when conjugated to GHBP and the non-covalently linked complex lacks the stability of the LR-fusion [40, 41]. The greater biological action of the LR-fusion may relate to its increased stability and its ability to activate the GHR in monomeric form. LR-fusion administration resulted in clearly elevated IGF-I levels compared to GH injection (Fig. 18.2c). We suggest that the dose-response to GH of growth and IGF-I differs in hypophysectomised rats. Thus, the dose of LR-fusion used in our study was in excess of that required to promote a maximal growth response, but still capable of stimulating IGF-I generation. Rats display more rapid renal clearance than humans making it difficult to predict the dosing regimen that will be required in man. One might expect that the LR-fusion could be used at lower doses and much less frequently than GH. The attraction of the LR-fusion concept is its relative simplicity for manufacture and its native configuration which might be anticipated to be less immunogenic.

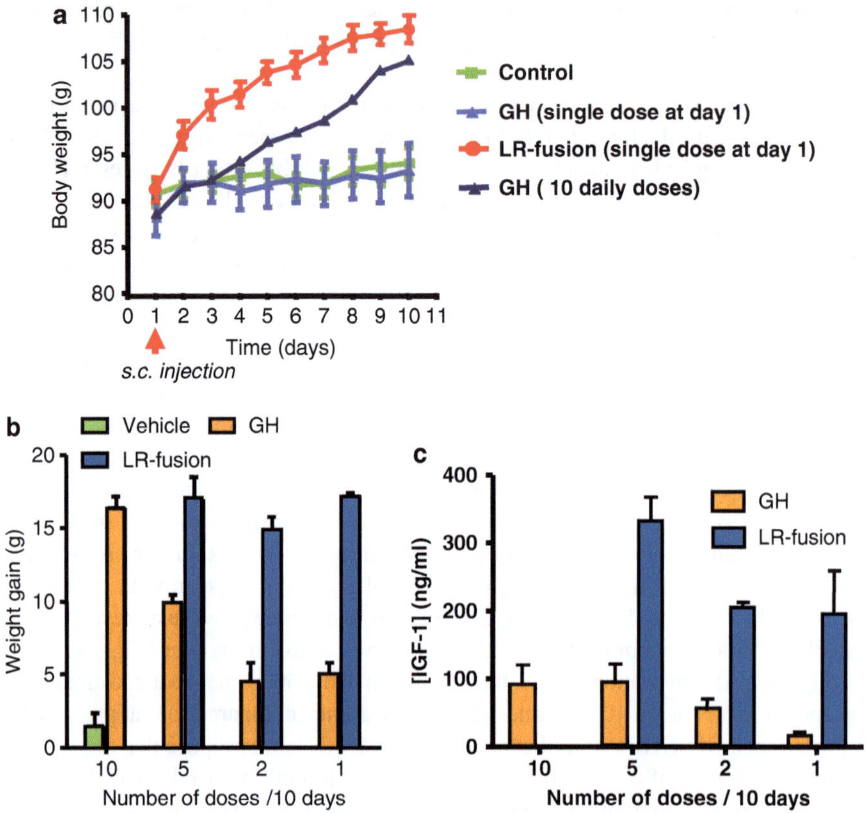

Fig. 18.2 (**a**) Graph showing the body weight of hypophysectomised mice over a 10-day period; the mice were injected s.c. with vehicle (daily injection), GH (single injection on day 1), GH (daily injection) and LR-fusion (single injection on day 1). (**b**) Graph showing weight gain of hypophysectomised mice injected, s.c. with vehicle, GH and LR-fusion, over a 10-day period with different dosing regimens. (**c**) Graph showing the terminal IGF-1 levels in hypophysectomised mice injected, s.c. with vehicle, GH and LR-fusion, over a 10-day period with different dosing regimens

Conclusions

It is intriguing that GH as one of the first biopharmaceuticals to be introduced into the clinic has proved to be the most challenging to generate a long-acting analogue. A number of long-acting GH technologies have been tested both in proof of concept animal models and human clinical studies; however to date, no long-acting GH agonist is in routine use in the clinic. Pegylation can delay clearance and prolong the biological action of GH; however, in GH-deficient patients, it caused lipoatrophy at the site of injection. Pegylation of the GH antagonist, Pegvisomant, results in a highly effective treatment for acromegaly and is currently the only marketed long-acting GH analogue. Albumin fusions of GH have been tested in animal models and proven effective; initially, studies were undertaken in man but currently no clinical studies are progressing. Ligand-receptor fusions are a novel approach to generating long-acting biopharmaceuticals and a ligand-receptor fusion of GH demonstrated the most delayed clearance of any technology to date with good biological activity. Clinical studies are now required to determine whether ligand-receptor fusions will translate into a new therapeutic approach in man.

References

1. Ross RJ, Rodda C, Tsagarakis S, Davies PS, Grossman A, Rees LH, et al. Treatment of growth-hormone deficiency with growth-hormone-releasing hormone. Lancet. 1987;1(8523):5–8.
2. Laursen T, Gravholt CH, Heickendorff L, Drustrup J, Kappelgaard AM, Jorgensen JO, et al. Long-term effects of continuous subcutaneous infusion versus daily subcutaneous injections of growth hormone (GH) on the insulin-like growth factor system, insulin sensitivity, body composition, and bone and lipoprotein metabolism in GH-deficient adults. J Clin Endocrinol Metab. 2001;86(3):1222–8.
3. Laursen T, Jorgensen JO, Jakobsen G, Hansen BL, Christiansen JS. Continuous infusion versus daily injections of growth hormone (GH) for 4 weeks in GH-deficient patients. J Clin Endocrinol Metab. 1995;80(8):2410–8.
4. Reiter EO, Attie KM, Moshang Jr T, Silverman BL, Kemp SF, Neuwirth RB, et al. A multicenter study of the efficacy and safety of sustained release GH in the treatment of naive pediatric patients with GH deficiency. J Clin Endocrinol Metab. 2001;86(10):4700–6.
5. Jostel A, Mukherjee A, Alenfall J, Smethurst L, Shalet SM. A new sustained-release preparation of human growth hormone and its pharmacokinetic, pharmacodynamic and safety profile. Clin Endocrinol (Oxf). 2005;62(5):623–7.
6. Cook DM, Biller BM, Vance ML, Hoffman AR, Phillips LS, Ford KM, et al. The pharmacokinetic and pharmacodynamic characteristics of a long-acting growth hormone (GH) preparation (nutropin depot) in GH-deficient adults. J Clin Endocrinol Metab. 2002;87(10):4508–14.
7. Abuchowski A, McCoy JR, Palczuk NC, van Es T, Davis FF. Effect of covalent attachment of polyethylene glycol on immunogenicity and circulating life of bovine liver catalase. J Biol Chem. 1977;252(11):3582–6.
8. Abuchowski A, van Es T, Palczuk NC, Davis FF. Alteration of immunological properties of bovine serum albumin by covalent attachment of polyethylene glycol. J Biol Chem. 1977; 252(11):3578–81.
9. Jain A, Jain SK. PEGylation: an approach for drug delivery. A review. Crit Rev Ther Drug Carrier Syst. 2008;25(5):403–47.
10. Gaberc-Porekar V, Zore I, Podobnik B, Menart V. Obstacles and pitfalls in the PEGylation of therapeutic proteins. Curr Opin Drug Discov Devel. 2008;11(2):242–50.
11. Bailon P, Won CY. PEG-modified biopharmaceuticals. Expert Opin Drug Deliv. 2009;6(1):1–16.
12. Veronese FM, Mero A. The impact of PEGylation on biological therapies. BioDrugs. 2008;22(5):315–29.
13. Doherty DH, Rosendahl MS, Smith DJ, Hughes JM, Chlipala EA, Cox GN. Site-specific PEGylation of engineered cysteine analogues of recombinant human granulocyte-macrophage colony-stimulating factor. Bioconjug Chem. 2005;16(5):1291–8.
14. Fishburn CS. The pharmacology of PEGylation: balancing PD with PK to generate novel therapeutics. J Pharm Sci. 2008;97(10):4167–83.
15. Sato AK, Viswanathan M, Kent RB, Wood CR. Therapeutic peptides: technological advances driving peptides into development. Curr Opin Biotechnol. 2006;17(6):638–42.

16. Clark R, Olson K, Fuh G, Marian M, Mortensen D, Teshima F, et al. Long-acting growth hormones produced by conjugation with polyethylene glycol. J Biol Chem. 1996;271(36):21969–77.
17. Maamra M, Kopchick JJ, Strasburger CJ, Ross RJ. Pegvisomant, a growth hormone-specific antagonist, undergoes cellular internalization. J Clin Endocrinol Metab. 2004;89(9):4532–7.
18. Bendele A, Seely J, Richey C, Sennello G, Shopp G. Short communication: renal tubular vacuolation in animals treated with polyethylene-glycol-conjugated proteins. Toxicol Sci. 1998;42(2):152–7.
19. Young MA, Malavalli A, Winslow N, Vandegriff KD, Winslow RM. Toxicity and hemodynamic effects after single dose administration of MalPEG-hemoglobin (MP4) in rhesus monkeys. Transl Res. 2007;149(6):333–42.
20. Cox GN, Rosendahl MS, Chlipala EA, Smith DJ, Carlson SJ, Doherty DH. A long-acting, mono-PEGylated human growth hormone analog is a potent stimulator of weight gain and bone growth in hypophysectomized rats. Endocrinology. 2007;148(4):1590–7.
21. Webster R, Xie R, Didier E, Finn R, Finnessy J, Edgington A, et al. PEGylation of somatropin (recombinant human growth hormone): impact on its clearance in humans. Xenobiotica. 2008;38(10):1340–51.
22. Touraine P, D'Souza GA, Kourides I, Abs R, Barclay P, Xie R, et al. Lipoatrophy in GH deficient patients treated with a long-acting pegylated GH. Eur J Endocrinol. 2009;161(4):533–40.
23. Chen WY, Wight DC, Wagner TE, Kopchick JJ. Expression of a mutated bovine growth hormone gene suppresses growth of transgenic mice. Proc Natl Acad Sci USA. 1990;87(13):5061–5.
24. Trainer PJ, Drake WM, Katznelson L, Freda PU, Herman-Bonert V, van der Lely AJ, et al. Treatment of acromegaly with the growth hormone-receptor antagonist pegvisomant. N Engl J Med. 2000;342(16):1171–7.
25. Pradhananga S, Wilkinson I, Ross RJ. Pegvisomant: structure and function. J Mol Endocrinol. 2002;29(1):11–4.
26. Ross RJM, Leung KC, Maamra M, Bennett W, Doyle N, Waters MJ, et al. Binding and functional studies with the growth hormone receptor antagonist, B2036-PEG (pegvisomant), reveal effects of pegylation and evidence that it binds to a receptor dimer. J Clin Endocrinol Metab. 2001;86:1716–23.
27. Higham CE, Trainer PJ. Growth hormone excess and the development of growth hormone receptor antagonists. Exp Physiol. 2008;93(11):1157–69.
28. Birzniece V, Sata A, Ho KK. Growth hormone receptor modulators. Rev Endocr Metab Disord. 2009;10(2):145–56.
29. Andersen JT, Sandlie I. The versatile MHC class I-related FcRn protects IgG and albumin from degradation: implications for development of new diagnostics and therapeutics. Drug Metab Pharmacokinet. 2009;24(4):318–32.
30. Dennis MS, Zhang M, Meng YG, Kadkhodayan M, Kirchhofer D, Combs D, et al. Albumin binding as a general strategy for improving the pharmacokinetics of proteins. J Biol Chem. 2002;277(38):35035–43.
31. Kratz F. Albumin as a drug carrier: design of prodrugs, drug conjugates and nanoparticles. J Control Release. 2008;132(3):171–83.
32. Schulte S. Use of albumin fusion technology to prolong the half-life of recombinant factor VIIa. Thromb Res. 2008;122:S14–9.
33. Weimer T, Wormsbacher W, Kronthaler U, Lang W, Liebing U, Schulte S. Prolonged in-vivo half-life of factor VIIa by fusion to albumin. Thromb Haemost. 2008;99(4):659–67.
34. Duttaroy A, Kanakaraj P, Osborn BL, Schneider H, Pickeral OK, Chen C, et al. Development of a long-acting insulin analog using albumin fusion technology. Diabetes. 2005;54(1):251–8.
35. Chuang VTG, Kragh-Hansen U, Otagiri M. Pharmaceutical strategies utilizing recombinant human serum albumin. Pharm Res. 2002;19(5):569–77.
36. Osborn BL, Sekut L, Corcoran M, Poortman C, Sturm B, Chen GX, et al. Albutropin: a growth hormone-albumin fusion with improved pharmacokinetics and pharmacodynamics in rats and monkeys. Eur J Pharmacol. 2002;456(1–3):149–58.

37. Muller-Newen G, Kohne C, Heinrich PC. Soluble receptors for cytokines and growth factors. Int Arch Allergy Immunol. 1996;111(2):99–106.
38. Baumann G, Amburn KD, Buchanan TA. The effect of circulating growth hormone-binding protein on metabolic clearance, distribution, and degradation of human growth hormone. J Clin Endocrinol Metab. 1987;64(4):657–60.
39. Baumann G. Growth hormone heterogeneity: genes, isohormones, variants, and binding proteins. Endocr Rev. 1991;12(4):424–49.
40. Baumann G, Shaw MA, Buchanan TA. In vivo kinetics of a covalent growth hormone-binding protein complex. Metabolism. 1989;38(4):330–3.
41. Clark RG, Mortensen DL, Carlsson LM, Spencer SA, McKay P, Mulkerrin M, et al. Recombinant human growth hormone (GH)-binding protein enhances the growth-promoting activity of human GH in the rat. Endocrinology. 1996;137(10):4308–15.
42. Baumann G. Growth hormone binding protein–errant receptor or active player? [editorial]. Endocrinology. 1995;136(2):377–8.
43. Ayling RM, Ross R, Towner P, Von Laue S, Finidori J, Moutoussamy S, et al. A dominant-negative mutation of the growth hormone receptor causes familial short stature [letter]. Nat Genet. 1997;16(1):13–4.
44. Ross RJ, Esposito N, Shen XY, Von Laue S, Chew SL, Dobson PR, et al. A short isoform of the human growth hormone receptor functions as a dominant negative inhibitor of the full-length receptor and generates large amounts of binding protein. Mol Endocrinol. 1997;11(3):265–73.
45. Wilkinson IR, Ferrandis E, Artymiuk PJ, Teillot M, Soulard C, Touvay C, et al. A ligand-receptor fusion of growth hormone forms a dimer and is a potent long-acting agonist. Nat Med. 2007;13(9):1108–13.
46. Cunningham BC, Ultsch M, de Vos AM, Mulkerrin MG, Clauser KR, Wells JA. Dimerization of the extracellular domain of the human growth hormone receptor by a single hormone molecule. Science. 1991;254(5033):821–5.
47. Huston JS, Tai MS, McCartney J, Keck P, Oppermann H. Antigen recognition and targeted delivery by the single-chain Fv. Cell Biophys. 1993;22(1–3):189–224.
48. Herington AC, Smith AI, Wallace C, Stevenson JL. Partial purification from human serum of a specific binding protein for human growth hormone. Mol Cell Endocrinol. 1987;53(3):203–9.
49. Frick GP, Tai LR, Baumbach WR, Goodman HM. Tissue distribution, turnover, and glycosylation of the long and short growth hormone receptor isoforms in rat tissues. Endocrinology. 1998;139(6):2824–30.
50. Mannor DA, Winer LM, Shaw MA, Baumann G. Plasma growth hormone (GH)-binding proteins: effect on GH binding to receptors and GH action. J Clin Endocrinol Metab. 1991;73(1):30–4.
51. Lim L, Spencer SA, McKay P, Waters MJ. Regulation of growth hormone (GH) bioactivity by a recombinant human GH-binding protein. Endocrinology. 1990;127(3):1287–91.
52. Keller A, Wu Z, Kratzsch J, Keller E, Blum WF, Kniess A, et al. Pharmacokinetics and pharmacodynamics of GH: dependence on route and dosage of administration. Eur J Endocrinol. 2007;156(6):647–53.
53. Roelfsema F, Biermasz NR, Pereira AM, Romijn J. Nanomedicines in the treatment of acromegaly: focus on pegvisomant. Int J Nanomedicine. 2006;1(4):385–98.

Chapter 19
Growth Hormone Supplementation in the Elderly

Ralf Nass and Jennifer Park

Abstract The expected demographic age shift in the world population over the next 3–5 decades will have a significant impact on society. Interventions with the potential to prevent or delay the loss of independence and to allow the majority of the older population to sustain strength, quality of life, and functionality are of increasing interest. Age-dependent muscle loss is recognized as a contributing factor of the age-dependent decline in strength and function. Circulating growth hormone (GH) levels show a significant decline with aging. Several age-dependent changes in body composition, such as muscle loss, have been associated with the age-dependent decline in GH in humans. The potential benefits and risks of restoring the GH-IGF-axis in the older population with GH or ghrelin-mimetics are discussed in this chapter.

Keywords Growth hormone • Ghrelin-mimetics • Sarcopenia • Aging

Introduction

In the developed world, people over the age of 80 are the fastest growing subset of the population [1]. In the United States alone, the proportion of the population over the age of 65 years is expected to increase from 38.9 million (12.4%) in 2008 to 71 million (19.4%) in 2030, and the number of persons aged > 80 years is projected to increase from 9.3 million in 2000 to 19.5 million in 2030 [2]. During the second half of the twentieth century, a 20-year increase in the average life span was observed,

R. Nass (✉)
Division of Endocrinology and Metabolism, University of Virginia,
450 Ray C. Hunt Drive, Charlottesville, VA 22902, USA
e-mail: rmn9a@virginia.edu

K. Ho (ed.), *Growth Hormone Related Diseases and Therapy: A Molecular and Physiological Perspective for the Clinician*, Contemporary Endocrinology,
DOI 10.1007/978-1-60761-317-6_19, © Springer Science+Business Media, LLC 2011

and the average life span worldwide is expected to increase by another 10 years by 2050 [3]. The worldwide population aged > 65 years is projected to increase between 2000 and 2030 from 420 million to 973 million. The largest increase in absolute numbers will occur in the developing countries where the population over the age of 65 will increase from about 249 million in 2000 to an estimated 690 million in 2030 [4]. The third national health and nutrition examination survey (NHANES III) data show that 23% of people aged 80 years and older are unable to prepare their own meals and 17% are unable to walk. Therefore, the major priority in the future of aging research should be to enhance "healthy aging" and thus reduce the number of years that the elderly spend in a diseased state with disabilities or with frailty. Maintaining independence is the number one priority. One of the major causes of dysfunction and disability in the elderly is loss of muscle mass and muscle strength – also called sarcopenia.

Features of GH Deficiency

Growth hormone deficiency (GHD) is a clinical syndrome associated with alterations in metabolism, mood, and quality of life. Specifically, there are changes in body composition, such as increased total fat mass which is principally manifest as central obesity [5, 6] as well as decreased lean body mass [7, 8]. Adults with GHD have significantly increased circulating leptin levels compared to age-matched controls [9]. Insulin resistance is also increased in GHD adults which may be related to the increased fat mass [10]. Given these clinical findings, there has been an association of GHD and increased prevalence of metabolic syndrome [11].

Growth hormone-deficient adults are at increased risk of cardiovascular and cerebrovascular diseases due to abnormal lipid profiles (i.e., increased total cholesterol, LDL cholesterol [12]) and increased coagulability, i.e., high fibrinogen and plasminogen activator inhibitor activity [13, 14]. These lead to premature atherosclerosis and an increased risk of death when compared to controls [15–18]. GHD adults have been reported to have impaired diastolic function [19, 20]. While some studies found preserved systolic function in GHD [20], others describe impaired left ventricular function [17, 21]. Colao et al. [17] report decreased left ventricular ejection fraction at peak exercise in elderly patients with hypopituitarism when compared with age-matched controls.

Another metabolic characteristic of GHD adults is decreased bone mass which is a marker for increased fracture rates [21–23]. In the European Vertebral Osteoporosis Study (EVOS), GHD patients aged >60 had a 2.66 times greater prevalence of fracture when compared to controls. These findings were statistically significant in men (p-value <0.05), but not in women (p-value = 0.06) [24].

GHD patients can have impaired quality of life and experience, decreased overall well-being, increased social isolation, and decreased energy [25, 26], although the validity of the applicability of quality of life questionnaires in GH deficiency (QoL-AGHDA) has been questioned in the past [27].

Changes in Body Composition in the Elderly

Several age-dependent changes in body composition [28–30] such as muscle loss have been associated with the age-dependent decline in GH in humans [31]. In elderly subjects, the 24-h integrated GH concentration is equal to levels observed in young patients with GH deficiency. Several authors have described a reduction in GH secretory parameters from 15 to 70% in men and women over 60 years of age [32, 33]. There is a change in the distribution of fat mass in the elderly with an increase in total body fat by about 9–18% in men and 12–13% in women [28, 29]. This age-dependent increase in body fat is associated with an excess of intraabdominal fat [34–36] rather than subcutaneous abdominal fat [28, 30, 37, 38]. In addition, a reduction of subcutaneous fat in the upper leg region has been described with aging [39]. Peak bone mass is usually reached by the third decade of life, followed by a significant gradual decline in bone mineral density [29, 40–43], which is associated with an increased incidence of fractures. Several studies show an age-dependent decrease in lean body mass (fat-free mass) [28, 44, 45], which predominantly represents muscle mass. Janssen et al. [46] report that the decrease of skeletal muscle mass starts in the third decade when expressed as relative muscle mass (see Fig. 19.1). However, absolute muscle mass remains stable until the age of 45 years and starts to decline afterwards (see Fig. 19.1).

Thus, the relative decrease in muscle mass before age 45 is mainly the result of an increase in body fat. This finding is in accordance with the studies showing that muscle fiber cross-sectional area (i.e., contractile muscle), body cell mass, and strength do not change substantially until around age 45 [47–52].

When reaching the eighth decade, men have lost about 7 kg of absolute muscle mass and women about 3.8 kg of muscle mass [46]. The age-related decrease in lower extremity strength is in the order of 20–40% in the seventh and eighth decade and exceeds 50% in the ninth decade [53].

Functional Implications of Loss of Muscle Mass

The age-related decline in muscle mass, which is also called sarcopenia, is a well-recognized phenomenon. However, a consensus clinical definition for the term sarcopenia is still lacking, and as a consequence of this, the criteria used in clinical studies to define and measure the age-dependent loss of muscle mass vary. The majority of published data show that the age-dependent loss of muscle mass and function is linked to (1) physical frailty [54], (2) increased risk of loss of independent living [55], and (3) increased risk of falling [56].

Janssen et al. [57] studied the relationship between sarcopenia and functional impairment in 4,504 adults aged 60 years and older using data of the Third National Health and Nutrition Examination Survey (NHANES III). Skeletal muscle mass was estimated using bioelectrical impedance and expressed as skeletal muscle mass index (SMI) (skeletal muscle mass/body mass × 100) and compared

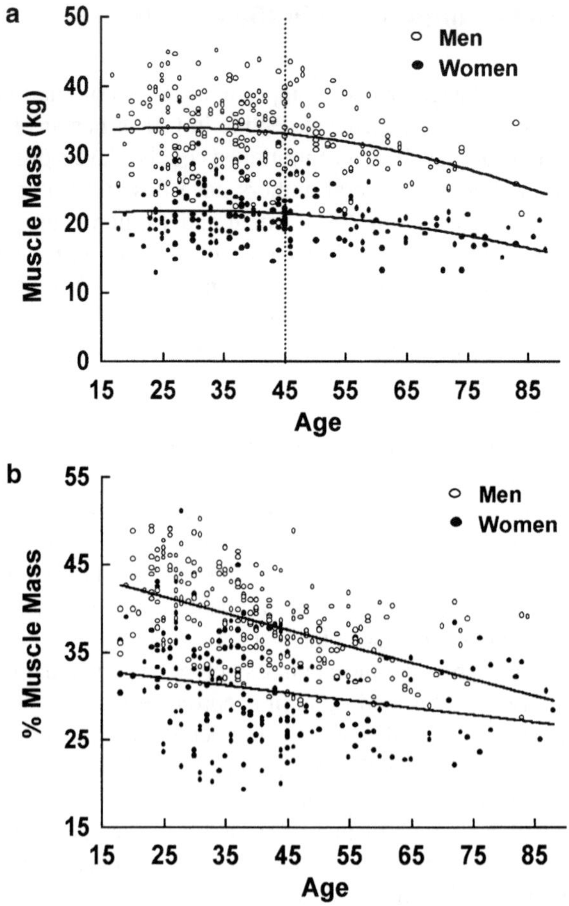

Fig. 19.1 (**a**) Shows the relationship between whole body skeletal muscle mass and age in men and women. *Solid lines*: regression lines. (**b**) Shows the relationship between the relative skeletal muscle mass (body mass/skeletal muscle mass) and age in men and women (from Janssen et al. [46]; with permission)

to young adults. The likelihood of functional impairment and physical disability was 2 times greater in older men and 3 times greater in older women with class II sarcopenia (defined as SMI of 2 SD below young adults). Interestingly, class I sarcopenia (defined as SMI within −1 to −2 SD of young adults), when adjusted for age, race, health behaviors, and comorbidity, was no longer associated with an increased likelihood of functional impairment and disability. This study suggests that modest reductions in skeletal muscle mass with aging do not cause functional impairment and disability. However, when skeletal muscle mass, when expressed relative to body weight, is 30% below the mean of young adults, there is an increased risk of functional impairment and disability. Melton et al. [58] show an association between sarcopenia and difficulties in walking and an increase in fractures in older men and women.

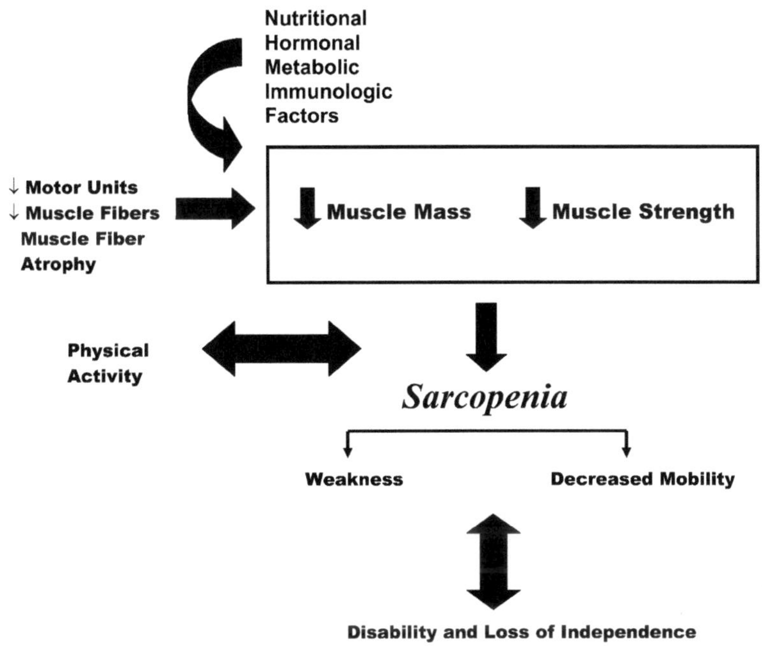

Fig. 19.2 Factors contributing to sarcopenia (from Doherty [53]; with permission)

Baumgartner et al. studied 808 older non-Hispanic Whites and Mexican-American men and women [56]. The authors report that sarcopenia is independently associated with disability and a history of falling. According to the guidelines of the American Geriatric Society for prevention of falls, sarcopenia is considered one of the main risk factors for falls in the older population [59]. One third of community-dwelling individuals older than 65 years fall every year [60]. Sixty-two percent of an estimated 2.7 million nonfatal injuries in 2001 were caused by falls in the age group 65 years and older and 5–10% of falls cause serious injuries such as major head trauma, major lacerations, or fracture [61]. Complications resulting from falls are the sixth leading cause for death in people over 65 years [62] and hospital admission rates from falls increase more than sixfold from between the ages of 65–69 years and to the over 85 years age group [63]. Finally, 75% of deaths due to falls in the U.S. occur in the population of age 65 years and over [64].

The etiology of the age-dependent muscle loss includes a wide variety of changes such as the loss of alpha-motor neurons [65], the reduction in dietary protein [66], a decreased level of physical activity [67], as well as an increase in catabolic cytokines such as interleukin-6 and interleukin-1 [68] and TNF-alpha [69] and CNTF (ciliary neutrophobic factor) [69, 70]. Most likely, quantitative and qualitative changes in Ca^{2+} and K^+ ion channels are also involved in the age-related decline in muscle force [69, 71]. The decrease in steroid hormones and growth hormone levels is also possibly involved in the age-dependent decrease in muscle mass [72].

Figure 19.2 summarizes the factors contributing to sarcopenia graphically.

Effects of GH Therapy in the Elderly

The expectation that GH treatment will have beneficial effects in the elderly is based on the observation that there is an age-dependent decline in GH and IGF-I levels [33] and the study results in patients with adulthood- or childhood-onset GH deficiency show that GH treatment increases lean body mass in patients with adulthood- as well as childhood-onset GH deficiency. Several studies also show that the alterations in body composition with aging are similar to those observed in adult GH deficiency [28–30, 32].

Some of the studies evaluating the outcomes of GH treatment in the elderly will be discussed here.

Treatment of healthy older adults, aged 60 year and older, for 1 week using different doses of GH resulted in a dose-dependent increase in IGF-I [73]. Rudman and coworkers showed that 6 months of treatment with a high dose of GH (0.03 mg/kg/week sc given 3 times per week) increased lean body mass and spinal bone density and decreased fat mass in men over the age of 60 year with low pretreatment IGF-I levels [74]. The GH dose used in this study was significantly higher than the current recommendation for the GH starting dose for GH-deficient adults. Similar results were found by Papadakis et al. [75] who studied 52 men over the age of 69 years. After 6 months of GH treatment (0.03 mg/kg sc given 3 times per week), fat mass decreased by 13.1% and lean body mass increased by 4.3%. However, these changes did not improve functional ability in this study population. Positive effects on body composition were also found in a 10-week study in 18 healthy elderly men when strength training was combined with GH treatment (0.02 mg/kg/day) [76]. In postmenopausal women, the combination of exercise, diet, and GH resulted in an enhanced loss of truncal fat rather than peripheral fat, when compared to placebo [77]. In another study, 26 weeks of GH treatment (0.02 mg/kg sc given 3 times per week) resulted in a decrease in visceral fat (compared to baseline) [78] in men 65 years of age or older; however, no change was found when compared to placebo. In a study of elderly malnourished patients [79], 3 weeks of GH treatment increased midarm muscle circumference. In a 12-week study with a small number of participants ($n = 17$), treatment with GH of healthy older men increased lean body mass by 3.2 kg [80].

The results of these studies show that GH administration results in a significant increase in IGF-I in healthy older adults and leads to an average reduction of fat mass by 2.08 kg and an average increase in lean body mass by 2.13 kg. Despite these beneficial changes in lean body mass, most of the available data on GH treatment in older adults show no beneficial effects on function or strength [81].

Additional hormones which have shown effects on skeletal muscle in the elderly are sex steroids [82]. Several studies suggest that both GH and testosterone have synergistic anabolic actions and small beneficial effects on strength in the elderly have been shown when GH was given in combination with testosterone. Blackman and colleagues found an increase in total body strength in men over 65 years of age who received GH and testosterone for 26 weeks. This was not the case for the group of women in the study who received GH and estrogen. In the same study, VO2 max increased (not significantly) in the group of men with 26 weeks of GH treatment. When GH and testosterone were administered simultaneously, VO2 max increased significantly when compared to placebo [83]. Ginnoulis et al. [84] compared GH, testosterone, and a combination of GH and Testosterone treatment in healthy older men between the age of 65 and 80 years. The GH dose was adjusted according to the IGF-I levels. GH treatment resulted in an increase in lean body mass which was more pronounced when GH and testosterone were given together. A mild increase in 1 of 6 strength measures (knee extension at 90°/s) was seen only when GH and testosterone were given combined.

Potential Risks of GH Replacement in the Elderly

The possible side effects of GH therapy in the elderly are similar to the side effects found in young adults. However, concerns have been voiced about the use of GH in the elderly. One concern is whether the risk of cancer is increased with GH therapy. Experimental data suggest that GH/IGF-I provides an anti-apoptotic environment, with IGF-I having powerful proliferative effects on almost all tissues. This could possibly favor the survival of genetically damaged cells and as a consequence increase the risk for developing cancer. This concern could be particularly relevant in the older population since there is a greater likelihood of the presence of genetically damaged cells. However, long-term data in children and adults treated with GH for 27,000 patient-years have shown no increased overall occurrence of de novo neoplasia or an increased rate of regrowth of primary pituitary tumors [85, 86]. A review of the current data suggests that GH replacement therapy in GH-deficient adults is safe and does not lead to tumor formation [87, 88]. However, these are patients who are GH-deficient and receive GH as a replacement therapy in order to restore age appropriate levels, whereas in the healthy elderly it is being given to restore "youthful" levels. Therefore, the currently available safety data from these surveillance studies in GH-deficient patients do not allow the conclusion that GH is safe to be given to healthy older adults. Another concern relates to the potential diabetogenic effects of growth hormone as it antagonizes insulin action. Only two cases of reversible diabetes were reported from a combined series of 400 treated GH-deficient adults [89] and insulin sensitivity may even normalize with GH treatment as shown in a 1-year study [90]. After 7 years of GH treatment in GH-deficient adults, no change in insulin sensitivity was found [91]. Nevertheless, the potential effects of GH on glucose metabolism, especially in the elderly, remain a concern and need to be carefully monitored. Other possible side effects include the development of edema, arthralgia, and carpal tunnel syndrome. The latter side effects are dose-dependent and reduction of the dose ameliorates these side effects. Another point to consider when discussing the use of GH treatment in the elderly are the data showing that GH knockout rodents live longer when compared to their GH intact controls [92, 93]. Whether these findings in rodents can be applied to a real-life situation is questionable [94].

Besson et al. [95] showed that patients with isolated childhood-onset GH deficiency caused by a GH-1 gene deletion who were untreated during childhood and adulthood have a reduction in life span when compared to controls.

Ghrelin-Mimetics as a Potential Therapeutic in the Elderly

Ghrelin mediates several effects which are potentially beneficial for the older population. Ghrelin increases GH and IGF-I through activation of the GHSR-1A (growth hormone secretagogue) receptor [96]and acyl-ghrelin has been shown to modulate GH release in the fed state. Ghrelin has also orexigenic effects [97]. Only a few groups have studied the effects of ghrelin-mimetics in the elderly. Optimal ghrelin-mimetic therapy requires the functional integrity of somatotroph cells as well as an intact hypothalamic-pituitary axis [98]. Both are present in the elderly. In addition, the pituitary GH-releasable pool in the elderly is comparable to young adults [99]. The use of ghrelin-mimetics also has the advantage that GH release is physiologic, i.e., pulsatile. Since the GH-IGF-I feedback loop remains intact when using a ghrelin-mimetic, supraphysiologic GH levels are unlikely to occur. This is not the case with GH treatment where high doses are associated with supraphysiologic IGF-I levels. Studies regarding the effects of ghrelin-mimetics on circulating IGF-I levels have shown conflicting results; this may be due to varying pharmacodynamic and pharmacokinetic properties. Rahim et al. [100] studied the effects of hexarelin on body composition in the elderly and could not show any effects after 16 weeks of treatment. In a double-blind placebo-controlled trial, 1-year treatment with the orally active ghrelin-mimetic MK677 led to an increase in FFM (fat-free mass) (measured by 4-compartment model and DXA) by 1.6 kg when compared to placebo. This gain in FFM was maintained in the second year of the study [101]. In the same study, the GH-IGF-I axis was restored into the normal range for young adults. Similar positive effects on body composition in the elderly with an orally active ghrelin-mimetic were described by White et al. [102].

In a randomized placebo-controlled study with 292 postmenopausal women, the treatment with MK677 alone or together with the antiresorptive agent alendronate resulted in an increase in bone mineral density at the femoral neck, described in the study by Murphy et al [103]. Another study reported the effects of 6 months' treatment with MK677 in healthy older adults recovering from hip fracture. The treatment with the ghrelin-mimetic resulted in improvements of some lower extremity functional measures. Overall, the compound was well tolerated and its use safe [104].

Summary

The age-dependent decline in muscle mass is one of the main features of the development of frailty. GH administration in the elderly has been shown to increase IGF-I levels, decrease fat mass, and increase lean body mass; however, improvement in muscle strength or function has not been shown consistently. Ghrelin-mimetics stimulate the GH/IGF-I axis, increase caloric intake, and have been shown to increase lean body mass in healthy older adults. Additional long-term studies, using validated outcome measures of function and the careful selection of a study populations, are necessary. Until the results of such studies are available, the use of GH and ghrelin-mimetics, albeit potentially beneficial, cannot be recommended in healthy older adults.

References

1. Butler RN. Population aging and health. BMJ. 1997;315(7115):1082–4.
2. U.S. Census Bureau. International database. Table 094. Midyear population, by age and sex. Available at http://www.census.gov/population/www/projections/natdet-D1A.html.
3. United Nations. Report of the Second World assembly on aging. Madrid: United Nations; 2002.
4. CDC. Public health and aging: trends in aging-United States and worldwide. JAMA. 2003;289:1371–3.
5. Toogood AA, Adams JE, O'Neill PA, Shalet SM. Body composition in growth hormone deficient adults over the age of 60 years. Clin Endocrinol (Oxf). 1996;45(4):399–405.
6. Barreto-Filho JA, Alcantara MR, Salvatori R, et al. Familial isolated growth hormone deficiency is associated with increased systolic blood pressure, central obesity, and dyslipidemia. J Clin Endocrinol Metab. 2002;87(5):2018–23.
7. Snel YE, Brummer RJ, Doerga ME, et al. Adipose tissue assessed by magnetic resonance imaging in growth hormone-deficient adults: the effect of growth hormone replacement and a comparison with control subjects. Am J Clin Nutr. 1995;61(6):1290–4.
8. Lonn L, Kvist H, Grangard U, Bengtsson BA, Sjostrom L. CT-determined body composition changes with recombinant human growth hormone treatment to adults with growth hormone deficiency. Basic Life Sci. 1993;60:229–31.
9. al-Shoumer KA, Anyaoku V, Richmond W, Johnston DG. Elevated leptin concentrations in growth hormone-deficient hypopituitary adults. Clin Endocrinol (Oxf). 1997;47(2):153–9.
10. Johansson JO, Fowelin J, Landin K, Lager I, Bengtsson BA. Growth hormone-deficient adults are insulin-resistant. Metabolism. 1995;44(9):1126–9.
11. Attanasio AF, Mo D, Erfurth EM, et al. Prevalence of metabolic syndrome in adult hypopituitary growth hormone (GH)-deficient patients before and after GH replacement. J Clin Endocrinol Metab. 2010;95(1):74–81.
12. Abdu TA, Neary R, Elhadd TA, Akber M, Clayton RN. Coronary risk in growth hormone deficient hypopituitary adults: increased predicted risk is due largely to lipid profile abnormalities. Clin Endocrinol (Oxf). 2001;55(2):209–16.
13. Johansson JO, Landin K, Tengborn L, Rosen T, Bengtsson BA. High fibrinogen and plasminogen activator inhibitor activity in growth hormone-deficient adults. Arterioscler Thromb. 1994;14(3):434–7.
14. Kvasnicka J, Marek J, Kvasnicka T, et al. Increase of adhesion molecules, fibrinogen, type-1 plasminogen activator inhibitor and orosomucoid in growth hormone (GH) deficient adults and their modulation by recombinant human GH replacement. Clin Endocrinol (Oxf). 2000;52(5):543–8.
15. Rosen T, Bengtsson BA. Premature mortality due to cardiovascular disease in hypopituitarism. Lancet. 1990;336(8710):285–8.
16. Bates AS, Van't Hoff W, Jones PJ, Clayton RN. The effect of hypopituitarism on life expectancy. J Clin Endocrinol Metab. 1996;81(3):1169–72.
17. Colao A, Cuocolo A, Di Somma C, et al. Impaired cardiac performance in elderly patients with growth hormone deficiency. J Clin Endocrinol Metab. 1999;84(11):3950–5.

18. Attanasio AF, Bates PC, Ho KK, et al. Human growth hormone replacement in adult hypopituitary patients: long-term effects on body composition and lipid status–3-year results from the HypoCCS Database. J Clin Endocrinol Metab. 2002;87(4):1600–6.
19. Sneppen SB, Steensgaard-Hansen F, Feldt-Rasmussen U. Cardiac effects of low-dose growth hormone replacement therapy in growth hormone-deficient adults. An 18-month randomised, placebo-controlled, double-blind study. Horm Res. 2002;58(1):21–9.
20. Shahi M, Beshyah SA, Hackett D, Sharp PS, Johnston DG, Foale RA. Myocardial dysfunction in treated adult hypopituitarism: a possible explanation for increased cardiovascular mortality. Br Heart J. 1992;67(1):92–6.
21. Amato G, Carella C, Fazio S, et al. Body composition, bone metabolism, and heart structure and function in growth hormone (GH)-deficient adults before and after GH replacement therapy at low doses. J Clin Endocrinol Metab. 1993;77(6):1671–6.
22. Bex M, Abs R, Maiter D, Beckers A, Lamberigts G, Bouillon R. The effects of growth hormone replacement therapy on bone metabolism in adult-onset growth hormone deficiency: a 2-year open randomized controlled multicenter trial. J Bone Miner Res. 2002;17(6):1081–94.
23. Rosen T, Wilhelmsen L, Landin-Wilhelmsen K, Lappas G, Bengtsson BA. Increased fracture frequency in adult patients with hypopituitarism and GH deficiency. Eur J Endocrinol. 1997;137(3):240–5.
24. Wuster C, Abs R, Bengtsson BA, et al. The influence of growth hormone deficiency, growth hormone replacement therapy, and other aspects of hypopituitarism on fracture rate and bone mineral density. J Bone Miner Res. 2001;16(2):398–405.
25. Rosen T, Johannson G, Hallgren P, Caidahl K, Bosaeus I, Bengtsson BA. Beneficial effects of 12 months replacement therapy with recombinant human growth hormone to growth hormone deficient adults. J Clin Endocrinol Metab. 1994;1:55–66.
26. Mahajan T, Crown A, Checkley S, Farmer A, Lightman S. Atypical depression in growth hormone deficient adults, and the beneficial effects of growth hormone treatment on depression and quality of life. Eur J Endocrinol. 2004;151(3):325–32.
27. Barkan AL. The "quality of life-assessment of growth hormone deficiency in adults" questionnaire: can it be used to assess quality of life in hypopituitarism? J Clin Endocrinol Metab. 2001;86(5):1905–7.
28. Novack L. Aging, total body potassium, fat-free mass, andcell mass in males and females between ages 18 and 85 years. J Gerontol. 1972;27:438–43.
29. Rudman D. Growth hormone, body composition, and aging. J Am Geriatr Soc. 1985;33(11): 800–7.
30. Clasey JL, Weltman A, Patrie J, et al. Abdominal visceral fat and fasting insulin are important predictors of 24-hour GH release independent of age, gender, and other physiological factors. J Clin Endocrinol Metab. 2001;86(8):3845–52.
31. Nass R, Thorner MO. Impact of the GH-cortisol ratio on the age-dependent changes in body composition. Growth Horm IGF Res. 2002;12(3):147–61.
32. Finkelstein JW, Roffwarg HP, Boyar RM, Kream J, Hellman L. Age-related change in the twenty-four-hour spontaneous secretion of growth hormone. J Clin Endocrinol Metab. 1972;35(5):665–70.
33. Zadik Z, Chalew SA, McCarter Jr RJ, Meistas M, Kowarski AA. The influence of age on the 24-hour integrated concentration of growth hormone in normal individuals. J Clin Endocrinol Metab. 1985;60(3):513–6.
34. Haarbo J, Marslew U, Gotfredsen A, Christiansen C. Postmenopausal hormone replacement therapy prevents central distribution of body fat after menopause. Metab Clin Exp. 1991; 40(12):1323–6.
35. Ashwell M, Cole TJ, Dixon AK. Obesity: new insight into the anthropometric classification of fat distribution shown by computed tomography. Br Med J (Clin Res Ed). 1985;290(6483): 1692–4.
36. DeNino WF, Tchernof A, Dionne IJ, et al. Contribution of abdominal adiposity to age-related differences in insulin sensitivity and plasma lipids in healthy nonobese women. Diabetes Care. 2001;24(5):925–32.

37. Enzi G, Gasparo M, Biondetti P, Fiore D, Semissa M, Zurlo F. Subcutaneous and visceral fat distribution according to sex, age and overweight, evaluated by computed tomography. Am J Clin Nutr. 1986;44:739–46.
38. Shimokata H, Tobin J, Muller D, Elahi D, Coon P, Andres R. Studies in the distribution of body fat. I. Effects of age, sex, and obesity. J Gerontol. 1989;44:67–73.
39. Borkan GA, Hults DE, Gerzof SG, Robbins AH, Silbert CK. Age changes in body composition revealed by computed tomography. J Gerontol. 1983;38(6):673–7.
40. Marcus R. Skeletal aging-understanding the functional and structural basis of osteoporosis. Trends Endocrinol Metab. 1991;2:53–8.
41. Hannan M, Felson D, Anderson J. Bone mineral density in elderly men and women: results from the Framingham Osteoporosis Study. J Bone Miner Res. 1992;7:547–53.
42. Stiegler C, Leb G. One year of replacement therapy in adults with growth hormone deficiency. J Clin Endocrinol Metab. 1994;1(Suppl A):37–42.
43. de Boer H, Blok GJ, van Lingen A, Teule GJ, Lips P, van der Veen EA. Consequences of childhood-onset growth hormone deficiency for adult bone mass. J Bone Miner Res. 1994;9(8):1319–26.
44. Thompson JL, Butterfield GE, Marcus R, et al. The effects of recombinant human insulin-like growth factor-I and growth hormone on body composition in elderly women. J Clin Endocrinol Metab. 1995;80(6):1845–52.
45. Forbes G, Reina J. Adult lean body mass declines with age: some longitudinal observations. Metabolism. 1970;19:653–63.
46. Janssen I, Heymsfield SB, Wang ZM, Ross R. Skeletal muscle mass and distribution in 468 men and women aged 18-88 yr. J Appl Physiol. 2000;89(1):81–8.
47. Tseng BS, Marsh DR, Hamilton MT, Booth FW. Strength and aerobic training attenuate muscle wasting and improve resistance to the development of disability with aging. J Gerontol A Biol Sci Med Sci. 1995;50 Spec No:113–9.
48. Bemben MG, Massey BH, Bemben DA, Misner JE, Boileau RA. Isometric muscle force production as a function of age in healthy 20- to 74-yr-old men. Med Sci Sports Exerc. 1991;23(11):1302–10.
49. Clement FJ. Longitudinal and cross-sectional assessments of age changes in physical strength as related to sex, social class, and mental ability. J Gerontol. 1974;29(4):423–9.
50. Hurley BF. Age, gender, and muscular strength. J Gerontol A Biol Sci Med Sci. 1995;50 Spec No:41–4.
51. Kehayias JJ, Fiatarone MA, Zhuang H, Roubenoff R. Total body potassium and body fat: relevance to aging. Am J Clin Nutr. 1997;66(4):904–10.
52. Lexell J, Downham D, Sjostrom M. Distribution of different fibre types in human skeletal muscles. Fibre type arrangement in m. vastus lateralis from three groups of healthy men between 15 and 83 years. J Neurol Sci. 1986;72(2–3):211–22.
53. Doherty TJ. Invited review: aging and sarcopenia. J Appl Physiol. 2003;95(4):1717–27.
54. Nass R, Johannsson G, Christiansen JS, Kopchick JJ, Thorner MO. The aging population–is there a role for endocrine interventions? Growth Horm IGF Res. 2009;19(2):89–100.
55. Walston J, Hadley EC, Ferrucci L, et al. Research agenda for frailty in older adults: toward a better understanding of physiology andetiology: summary from the American Geriatrics Society/National Institute on Aging Research Conference on Frailty in Older Adults. J Am Geriatr Soc. 2006;54(6):991–1001.
56. Baumgartner RN, Koehler KM, Gallagher D, et al. Epidemiology of sarcopenia among the elderly in New Mexico. Am J Epidemiol. 1998;147(8):755–63.
57. Janssen I, Heymsfield SB, Ross R. Low relative skeletal muscle mass (sarcopenia) in older persons is associated with functional impairment and physical disability. J Am Geriatr Soc. 2002;50(5):889–96.
58. Melton LJ 3rd, Khosla S, Crowson CS, O'Connor MK, O'Fallon WM, Riggs BL. Epidemiology of sarcopenia. J Am Geriatr Soc. 2000;48(6):625–30.
59. Society AG, Society G, Of AA, and On Falls Prevention OSP. Guideline for the prevention of falls in older persons. J Am Geriatr Soc. 2001;49:664–672.

60. Tinetti ME. Clinical practice. Preventing falls in elderly persons. N Engl J Med. 2003;-348(1):42–9.
61. Rubenstein LZ, Josephson KR. The epidemiology of falls and syncope. Clin Geriatr Med. 2002;18(2):141–58.
62. Tinetti ME, Baker DI, McAvay G, et al. A multifactorial intervention to reduce the risk of falling among elderly people living in the community. N Engl J Med. 1994;331(13):821–7.
63. Masud T, Morris RO. Epidemiology of falls. Age ageing. 2001;30 Suppl 4:3–7.
64. Josephson KR, Fabacher DA, Rubenstein LZ. Home safety and fall prevention. Clin Geriatr Med. 1991;7(4):707–31.
65. Brown WF. A method for estimating the number of motor units in thenar muscles and the changes in motor unit count with ageing. J Neurol Neurosurg Psychiatry. 1972;35(6):845–52.
66. Young VR. Amino acids and proteins in relation to the nutrition of elderly people. Age Ageing. 1990;19(4):S10–24.
67. Westerterp KR. Daily physical activity and ageing. Curr Opin Clin Nutr Metab Care. 2000; 3(6):485–8.
68. Roubenoff R, Harris TB, Abad LW, Wilson PW, Dallal GE, Dinarello CA. Monocyte cytokine production in an elderly population: effect of age and inflammation. J Gerontol A Biol Sci Med Sci. 1998;53(1):M20–6.
69. Argiles JM, Busquets S, Felipe A, Lopez-Soriano FJ. Molecular mechanisms involved in muscle wasting in cancer and ageing: cachexia versus sarcopenia. Int J Biochem Cell Biol. 2005;37(5):1084–104.
70. Henderson JT, Mullen BJ, Roder JC. Physiological effects of CNTF-induced wasting. Cytokine. 1996;8(10):784–93.
71. Delbono O. Molecular mechanisms and therapeutics of the deficit in specific force in ageing skeletal muscle. Biogerontology. 2002;3(5):265–70.
72. Labrie F, Belanger A, Luu-The V, et al. DHEA and the intracrine formation of androgens and estrogens in peripheral target tissues: its role during aging. Steroids. 1998;63(5–6):322–8.
73. Marcus R, Butterfield G, Holloway L, et al. Effects of short termadministration of recombinant human growth hormone to elderly people. J Clin Endocrinol Metab. 1990;70:519–27.
74. Rudman D, Feller A, Nagraj H, et al. Effects of human growth hormone in men over 60 years old. N Engl J Med. 1990;323:1–6.
75. Papadakis MA, Grady D, Black D, et al. Growth hormone replacement in healthy older men improves body composition but not functional ability. Ann Intern Med. 1996;124(8):708–16.
76. Taaffe DR, Pruitt L, Reim J, et al. Effect of recombinant human growth hormone on the muscle strength response to resistance exercise in elderly men. J Clin Endocrinol Metab. 1994;79(5):1361–6.
77. Taaffe DR, Thompson JL, Butterfield GE, Hoffman AR, Marcus R. Recombinant human growth hormone, but not insulin-like growth factor-I, enhances central fat loss in postmenopausal women undergoing a diet and exercise program. Horm Metab Res. 2001;33(3):156–62.
78. Muenzer T, Harman S, Hees P, et al. Effects of GH and/or sex steroid administration on abdominal subcutaneous and visceral fat in healthy aged women and men. J Clin Endocrinol Metab. 2001;86:3604–10.
79. Kaiser FE, Silver AJ, Morley JE. The effect of recombinant human growth hormone on malnourished older individuals. J Am Geriatr Soc. 1991;39(3):235–40.
80. Lange KH, Isaksson F, Rasmussen MH, Juul A, Bulow J, Kjaer M. GH administration and discontinuation in healthy elderly men: effects on body composition, GH-related serum markers, resting heart rate and resting oxygen uptake. Clin Endocrinol (Oxf). 2001;55(1):77–86.
81. Liu H, Bravata DM, Olkin I, et al. Systematic review: the safety and efficacy of growth hormone in the healthy elderly. Ann Intern Med. 2007;146(2):104–15.
82. Snyder PJ, Peachey H, Hannoush P, et al. Effect of testosterone treatment on body composition and muscle strength in men over 65 years of age. J Clin Endocrinol Metab. 1999; 84(8):2647–53.
83. Blackman MR, Sorkin JD, Munzer T, et al. Growth hormone and sex steroid administration in healthy aged women and men: a randomized controlled trial. JAMA. 2002;288(18):2282–92.

84. Giannoulis MG, Sonksen PH, Umpleby M, et al. The effects of growth hormone and/or testosterone in healthy elderly men: a randomized controlled trial. J Clin Endocrinol Metab. 2006;91(2):477–84.
85. Monson JP. Long-term experience with GH replacement therapy: efficacy and safety. Eur J Endocrinol. 2003;148 Suppl 2:S9–14.
86. Jenkins PJ, Mukherjee A, Shalet SM. Does growth hormone cause cancer? Clin Endocrinol (Oxf). 2006;64(2):115–21.
87. Orme SM, McNally RJ, Cartwright RA, Belchetz PE. Mortality and cancer incidence in acromegaly: a retrospective cohort study. United Kingdom Acromegaly Study Group. J Clin Endocrinol Metab. 1998;83(8):2730–4.
88. GHR Society Consensus. Critical evaluation of the safety of recombinant human growth hormone administration: Statement from the Growth Hormone Research Society. J Clin Endocrinol Metab. 2001;86:1868–70.
89. Chipman JJ, Attanasio AF, Birkett MA, Bates PC, Webb S, Lamberts SW. The safety profile of GH replacement therapy in adults. Clin Endocrinol (Oxf). 1997;46(4):473–81.
90. Hwu CM, Kwok CF, Lai TY, et al. Growth hormone (GH) replacement reduces total body fat and normalizes insulin sensitivity in GH-deficient adults: a report of one-year clinical experience. J Clin Endocrinol Metab. 1997;82(10):3285–92.
91. Svensson J, Fowelin J, Landin K, Bengtsson BA, Johansson JO. Effects of seven years of GH-replacement therapy on insulin sensitivity in GH-deficient adults. J Clin Endocrinol Metab. 2002;87(5):2121–7.
92. Bartke A. Long-lived Klotho mice: new insights into the roles of IGF-1 and insulin in aging. Trends Endocrinol Metab. 2006;17(2):33–5.
93. Berryman DE, Christiansen JS, Johannsson G, Thorner MO, Kopchick JJ. Role of the GH/IGF-1 axis in life span and health span: lessons from animal models. Growth Horm IGF Res. 2008;18(6):455–71.
94. Nass R, Thorner MO. Life extension versus improving quality of life. Best Pract Res Clin Endocrinol Metab. 2004;18(3):381–91.
95. Besson A, Salemi S, Gallati S, et al. Reduced longevity in untreated patients with isolated growth hormone deficiency. J Clin Endocrinol Metab. 2003;88(8):3664–7.
96. Smith RG. Development of growth hormone secretagogues. Endocr Rev. 2005;26(3):346–60.
97. Tong J, Pfluger PT, Tschop MH. Gastric O-acyl transferase activates hunger signal to the brain. Proc Natl Acad Sci U S A. 2008;105(17):6213–4.
98. Ghigo E, Arvat E, Aimaretti G, Broglio F, Giordano R, Camanni F. Diagnostic and therapeutic uses of growth hormone-releasing substances in adult and elderly subjects. Baillières Clin Endocrinol Metab. 1998;12(2):341–58.
99. Ghigo E, Arvat E, Gianotti L, et al. Hypothalamic growth hormone-insulin-like growth factor-I axis across the human life span. J Pediatr Endocrinol Metab. 2000;13 Suppl 6:1493–502.
100. Rahim A, O'Neill PA, Shalet SM. Growth hormone status during long-term hexarelin therapy. J Clin Endocrinol Metab. 1998;83(5):1644–9.
101. Nass R, Pezzoli SS, Oliveri MC, et al. Effects of an oral ghrelin mimetic on body composition and clinical outcomes in healthy older adults: a randomized trial. Ann Intern Med. 2008;149(9):601–11.
102. White HK, Petrie CD, Landschulz W, et al. Effects of an oral growth hormone secretagogue in older adults. J Clin Endocrinol Metab. 2009;94(4):1198–206.
103. Murphy MG, Weiss S, McClung M, et al. Effect of alendronate and MK-677 (a growth hormone secretagogue), individually and in combination, on markers of bone turnover and bone mineral density in postmenopausal osteoporotic women. J Clin Endocrinol Metab. 2001;86(3):1116–25.
104. Bach MA, Rockwood K, Zetterberg C, et al. The effects of MK-0677, an oral growth hormone secretagogue, in patients with hip fracture. J Am Geriatr Soc. 2004;52(4):516–23.

Chapter 20
Growth Hormone in Sports: Is There Evidence of Benefit?

Anne E. Nelson, Ken Ho, and Vita Birzniece

Abstract Despite being banned by the World Anti-Doping Agency, growth hormone (GH) is abused by athletes in sport, reportedly often in combination with anabolic androgenic steroids (AASs). There is evidence for clear benefits of GH treatment for growth hormone deficiency, including normalisation of body composition by increasing lean body mass (LBM) and reducing fat mass due to its lipolytic and anabolic actions, increase in muscle mass but not muscle strength with short-term treatment, and improved aerobic exercise capacity (VO_2max).

There is only limited evidence available for benefit of GH use in healthy young adults and athletes. GH decreases fat mass and significantly increases LBM in young, healthy physically fit adults. Fluid retention accounts for most of the increase in LBM induced by GH, rather than an increase in the muscle mass. There has been debate as to whether the improved body composition and metabolic changes induced by GH actually translate into improved performance. We have recently demonstrated in a randomised controlled study an increase in anaerobic sprint capacity following treatment with GH for 8 weeks. There is no evidence that GH enhances muscle strength or power, or aerobic exercise capacity in healthy young adults.

There is strong evidence of harm from GH abuse. Side effects of GH administration in healthy adults include swelling, arthralgias and paresthesias, fatigue, and changes in cardiac morphology and function. Side effects of GH such as myocardial hypertrophy are likely worsened by co-administration of AASs. The features of acromegaly, including cardiac complications, arthropathy, insulin resistance and increased risk of diabetes and malignancy, indicate potential health risks of chronic GH abuse.

A.E. Nelson (✉)
Pituitary Research Unit, Garvan Institute of Medical Research,
384 Victoria Street, Darlinghurst, NSW, Australia
e-mail: a.nelson@garvan.org.au

K. Ho (ed.), *Growth Hormone Related Diseases and Therapy: A Molecular and Physiological Perspective for the Clinician*, Contemporary Endocrinology,
DOI 10.1007/978-1-60761-317-6_20,

In conclusion, the evidence suggests that in healthy adults, GH may improve a selective aspect of performance, that of anaerobic exercise capacity; however, muscle strength, power and aerobic exercise capacity are not enhanced by GH administration. Athletes should be aware of the serious health risks from prolonged use of GH.

Keywords Growth hormone • Doping • Sport • Athletes • Physical performance • Body composition • Side effects

Introduction

Growth hormone (GH) is abused in sport by athletes to improve their performance, based on survey data, extensive anecdotal evidence including customs and police drugs seizures [1] and the number of website hits for GH supply. The abuse of GH by athletes likely arises from the immense pressure to perform in sport, reflected by a survey in which 98% of athletes said they would take a performance-enhancing substance that would guarantee an Olympic medal if they could not be caught [2]. When asked if they would take the drug with a guarantee they would not get caught and further that they won every competition for the next 5 years, even if they then died from its adverse effects, an amazing 50% also replied yes [2].

Doses of GH used by athletes are estimated to range from 5 to 10 times that of daily production rate, based on anecdotal evidence [1, 3]. Furthermore, GH is reportedly used in combination with anabolic steroids. A web-based survey reported use of GH (1–10 mg/day) and insulin together with anabolic androgenic steroids (AASs) by 25% of AAS users [4]. Another web-based survey of weightlifters and body builders reported use of GH together with anabolic steroids by 5% of the steroid users [5].

Abuse of performance-enhancing agents such as GH may start at young ages. An early survey of tenth grade boys in the US indicated that 5% had taken GH, with more than half using GH in conjunction with steroids [6]. In the National Collegiate Athletic Association survey of college athletes in the US, 1.2% of the athletes reported using GH in the past 12 months [7].

Abuse of GH in sport is banned by the World Anti-Doping Agency (WADA) and GH is listed in the 2010 Prohibited list (http://www.wada-ama.org/rtecontent/document/2010_Prohibited_List_FINAL_EN_Web.pdf) because of its theoretical potential to enhance sports performance, its violation of the spirit of sports and the health risks that it poses to athletes. However, development and implementation of a robust test for GH that would then deter its use has been challenging [8].

This chapter will review the available evidence for any beneficial effect of GH in sport, by first examining briefly the lessons from GH deficiency (GHD) and the response of these patients to GH replacement. The effects of GH in healthy adults including effects on body composition, protein metabolism, muscle strength, aerobic and anaerobic exercise capacity will be reviewed. The evidence of harm, including adverse effects of GH in healthy adults and the features of acromegaly, the state of GH excess, will also be discussed.

Growth Hormone Deficiency and GH Replacement

The state of GHD and the effects of GH replacement in GHD patients provide a good model for the role of GH in adult life.

There are body composition abnormalities in adults with GHD, which include increased fat mass and central abdominal obesity, reduction of lean body mass (LBM) and muscle atrophy [9]. GH replacement in adults with GHD has been shown to reverse these abnormalities of body composition over time, due to the anabolic and lipolytic effects of GH that act to optimise body composition [10]. Fat oxidation and lipolysis are stimulated by GH both directly and indirectly [10]. Adults with GHD have reduced protein synthesis compared with healthy controls [11]. Short-term GH replacement in GHD adults results in increased protein synthesis, whereas the anabolic effects on muscle associated with long-term replacement appear to be due to reduced protein proteolysis [9]. In adults with GHD, long-term GH replacement reverses muscle atrophy, reduces central and total body fat mass by up to 20% and increases LBM by about 3–7%, depending on the GH dose and the duration of treatment [12–14].

Adults with GHD have reduced muscle strength. The reduction in absolute maximal isometric, and possible isokinetic, muscle strength is at least partly due to reduced muscle mass and may possibly reflect reduced intrinsic muscle strength [9]. In adults with GHD, the evidence suggests that short-term (4–6 month) GH replacement increases muscle mass [9]. A recent meta-analysis of 9 patient cohorts, however, demonstrated no significant effect on isometric or isokinetic muscle strength of GH replacement over mean duration of 6.7 months in GHD patients [15], whereas long-term GH replacement results in increased muscle strength as well as muscle mass [10].

In addition to reduced muscle strength, aerobic exercise capacity, typically measured as VO_2max, is impaired in GH-deficient adults. Consistently reduced VO_2max (18–28% deficit) has been shown in adults with GHD, compared to that predicted for age, height and weight [9]. There has been some inconsistent evidence on the effect of GH replacement on VO_2max in adults with GHD; however, a meta-analysis of 11 randomised placebo-controlled studies showed that GH replacement improves VO_2max and maximal power output in GH-deficient subjects, with no association between GH dose and the degree of improvement [16].

In summary (Table 20.1), treatment of GHD adults with GH, due to its lipolytic and anabolic actions, normalises body composition by increasing LBM and reducing fat mass. GH replacement increases muscle mass, but not muscle strength with short-term treatment, and improves aerobic exercise capacity (VO_2max) in GHD adults.

Evidence for Benefit of GH: Effects of GH Use in Healthy Adults

While the extensive studies of GH replacement in GHD provide an understanding of the actions of GH, the effects of GH use must be demonstrated in healthy, fit adults in order to provide evidence of benefit of GH in sport. It should also be recognised

Table 20.1 Summary of effects of GH on body composition and physical performance in growth hormone deficiency (GHD) and in healthy young adults

	GH replacement in GHD adults	GH administration to healthy young adults
Body composition		
Lean body mass (LBM)	Increase	Increase
Extracellular water (ECW)	Increase	Increase
Muscle mass, or body cell mass	Increase	No evidence for significant change
Fat mass	Decrease	Decrease
Physical performance		
Muscle strength	Increased following long-term, but not short-term replacement	No evidence for significant change
Aerobic exercise	Increase	No evidence for significant change
Anaerobic exercise	Not known	Increase

The table summarises the evidence from reviews and meta-analyses [9, 10, 15, 16, 18], and from individual studies including our recent GH administration study in healthy young adults [17]; refer to text for details

that there are no published studies of GH administration in elite athletes, nor are these likely, since it is not ethical to administer banned agents such as GH to elite athletes.

Our group has recently studied the effects of 8 weeks' treatment with GH in a randomised, double-blind placebo-controlled study in 96 recreationally trained athletes (63 men and 33 women), which has provided evidence on the effects of GH in young, healthy, fit adults [17]. A recent systematic review has also identified a number of randomised controlled studies in physically fit, young individuals [18]. Meta-analyses were undertaken using 27 studies with a total of 404 predominantly male participants (mean age 27 years, SD: 3 years), of which 303 received GH; the mean daily dose of GH was 36 μg (micrograms)/kg. Seven of the studies evaluated the use of a single dose of GH and 20 studies evaluated GH for more than 1 day (mean treatment duration 20 days) [18].

Effect of GH in Healthy Adults on Basal Metabolism

Meta-analysis of seven studies by Liu et al. indicated that basal metabolic rate was significantly higher in GH-treated subjects than in those not treated with GH, by 141 kcal/day, 95% CI: 69–213 kcal/day [18]. Resting heart rate was significantly higher in GH-treated subjects by 3.8 beats/min, 95% CI: 0.2–7.4 beats/min. Resting respiratory exchange ratio or respiratory quotient was significantly lower in GH-treated subjects by −0.02, 95% CI: −0.03 to −0.01, which indicates increased use of lipids, rather than carbohydrates for fuel during rest [18].

Effect of GH in Healthy Adults on Protein Metabolism

There is evidence from several studies that GH induces a whole body protein anabolic effect in healthy young adults. Whole body protein synthesis increased more in young untrained men undertaking resistance training plus GH for 12 weeks, than in those undertaking resistance training plus placebo, although the quadriceps muscle protein synthesis rate did not increase more in the GH-treated group [19]. A net anabolic effect of GH on whole body protein metabolism at rest and during and after exercise has also been shown in a placebo-controlled study of young male endurance-trained athletes [20]. There is the suggestion, however, that GH may not increase whole body protein synthesis in some highly trained athletes, from a non-randomised study that showed short-term GH administration did not reduce the rate of whole body protein breakdown in weight lifters [21].

While there is evidence for a net anabolic effect of GH on whole body protein metabolism following GH administration, there is further recent evidence from molecular studies that GH may not stimulate muscle protein synthesis. Administration of GH for 14 days in a placebo-controlled cross-over study in healthy young men did not stimulate myofibrillar protein synthesis, assessed by mRNA expression in skeletal muscle biopsies and measurement of myofibrillar protein synthesis [22].

Effects of GH in Healthy Adults on Body Composition

There is strong evidence that GH has significant effects on body composition in young healthy subjects. Our recent study demonstrated a significant increase in LBM in recreationally trained athletes, in both men and women (Fig. 20.1). Treatment with GH (2 mg/day sc) for 8 weeks increased LBM in men by 2.9 kg (CI: 1.8–4.0, $P<0.005$) and in women by 2.5 kg (CI: 1.4–3.6, $P<0.005$), compared to placebo [17]. Meta-analysis by Liu et al. also indicated a significant increase in LBM of 2.1 kg, 95% CI: 1.3–2.9, $P=0.000$, in 11 randomised controlled studies undertaken in physically fit, young subjects [18].

The LBM, however, is heterogeneous, comprising an inert compartment of extracellular water (ECW) and a functional cellular compartment of mostly muscle, the body cell mass (BCM). Most measurement methods for LBM, such as dual-energy X-ray absorptiometry (DEXA), do not distinguish lean solid tissue from fluid. Therefore, a change in LBM measured by DEXA alone does not distinguish between changes in muscle mass and ECW. An estimate of muscle, or BCM, can be made by quantification of ECW using techniques such as bromide dilution [23, 24] and subtraction from the LBM. In our study in recreational athletes, ECW increased in men treated with GH for 8 weeks by 12% (absolute increase 2.4 kg, CI: 0.9–4.0, $P<0.005$) and in women by 8% (absolute increase 1.2 kg, CI: 0.1–2.3, $P<0.005$), compared to placebo [17]. These increases in ECW are consistent with the known

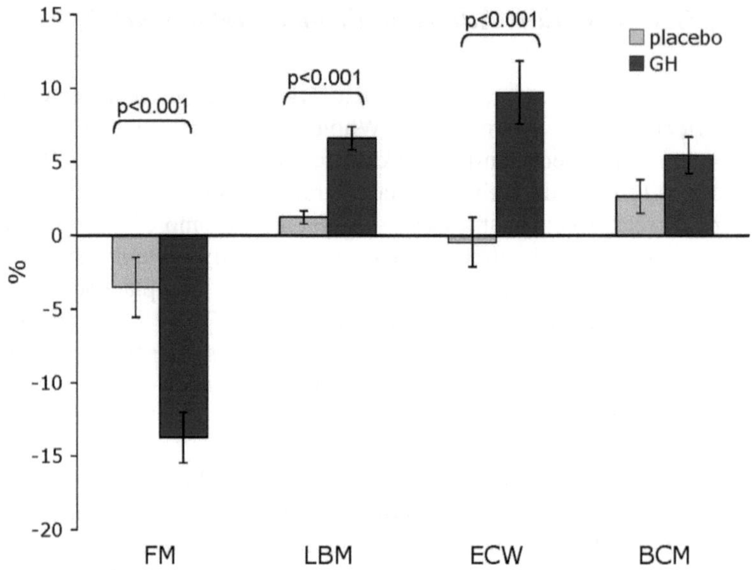

Fig. 20.1 Percentage changes in fat mass (FM), lean body mass (LBM), extracellular water (ECW), and body cell mass (BCM) in recreational athletes randomised to 8 weeks treatment with placebo or GH (2.0 mg a day) in a double-blinded study (adapted from Meinhardt et al. [17])

antinatriuretic properties of GH [25] and with the increases reported in ECW of 13% in men and 6% in women following 1 month's administration of GH in a randomised study in physically active young adults [26].

Following subtraction of ECW from LBM, there was no significant increase in our study in BCM following treatment with GH compared to placebo [17]. This indicates that the increase in LBM following GH was largely attributable to an increase in fluid. Following combined treatment of GH with testosterone in men, however, there was a significant increase in BCM of 6% (absolute increase 2.3 kg, CI: 0.7–3.8, $P<0.005$) compared to placebo [17].

GH also decreases fat mass in healthy adults. In our study, following 8 weeks' GH treatment, there was a non-significant decrease in fat mass in men of −0.5 kg, 95% CI: −1.6 to 0.6 and a significant decrease in women of −2.3 kg, 95% CI: −3.2 to −1.4, $P<0.005$, compared to placebo (Fig. 20.1). Meta-analysis by Liu et al. of ten randomised controlled studies undertaken in physically fit young subjects indicated a decrease in fat mass of −0.9 kg, 95% CI: −1.9 to −0.0 kg, $P=0.05$ [18].

In summary (Table 20.1), GH significantly increases LBM and decreases fat mass in young, healthy physically fit adults. The evidence suggests that fluid retention accounts for most of the increase in LBM induced by GH, rather than an increase in the muscle mass. There is evidence from our recent study of an increase in BCM following combined administration of GH and testosterone.

Effect of GH in Healthy Adults on Physical Performance

Physical performance can be assessed using a number of different measures. Our study in 96 young recreational athletes evaluated the effect of GH administration in men and women, and co-administration of GH and testosterone in men, on four measures of physical performance, namely strength, power, aerobic exercise capacity and anaerobic sprint capacity [17]. The meta-analysis by Liu et al. identified studies on strength outcomes and exercise capacity, but noted that limited evidence is available for the effects of GH on key athletic performance outcomes [18]. The results from our study and the meta-analysis are discussed in the following sections.

In addition, a single-blind study has recently shown that 6 days' administration of GH significantly increased strength (bench press and squat), peak power output in a 3-s ergometer test, and VO_2max [27]. This study, however, was undertaken in the specific model of dependent abstinent AAS-using weightlifters, who may have exhibited responses to GH specific to this state. It was suggested by the authors that these subjects may have been in a latent catabolic phase, which was ameliorated by the anabolic effect of GH administration.

Muscle Strength

In our recent study in recreational athletes, there was no significant increase in maximum strength assessed using an isometric deadlift dynamometer test, or in muscle power assessed using vertical jump height for maximal explosive power, following 8 weeks' GH administration compared to placebo (Fig. 20.2). Further, following co-administration of GH and testosterone, there was no significant increase compared to placebo in either deadlift (muscle strength) or in vertical jump (power) in our study [17].

Only two studies [19, 28] on the effect of GH on muscle strength were identified for meta-analysis by Liu et al., which found no significant change following GH with 1-repetition maximum voluntary strength testing in biceps strength (change −0.2 kg, 95% CI: −1.5 to 1.1) or in quadriceps strength (change −0.1 kg, 95% CI: −1.8 to 1.5) [18]. There were also no significant changes in maximum strength of seven other muscle groups following resistance training plus GH, compared to resistance training plus placebo for 12 weeks [19].

Aerobic Exercise Capacity

Aerobic exercise capacity is assessed by measuring VO_2max which depends not only on muscle function, but also on cardiorespiratory function and on motivation. In our study in recreational athletes, we found no significant increase in VO_2max following 8 weeks' administration of GH (2 mg/day) in men or women, or following combined administration of GH and testosterone in men, compared to placebo (Fig. 20.2) [17].

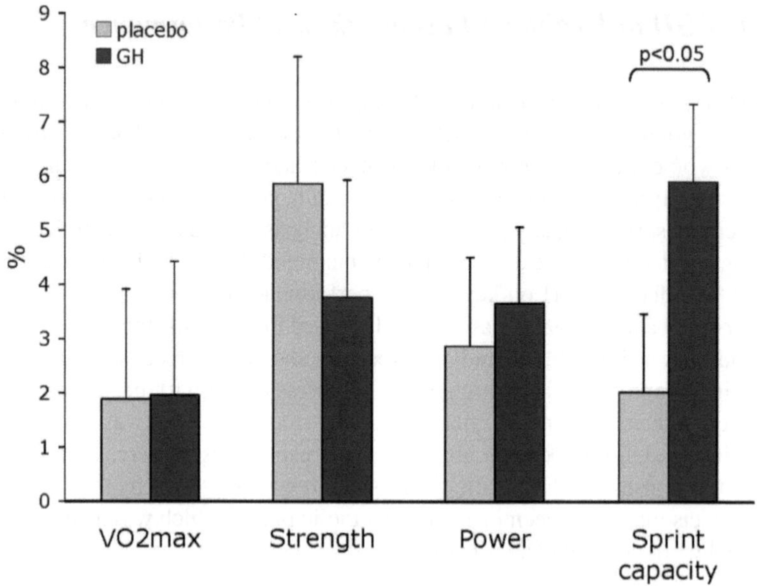

Fig. 20.2 Percentage changes in aerobic capacity (VO_2max), strength, power and sprint capacity in recreational athletes randomised to 8 weeks treatment with placebo or GH (2.0 mg a day) in a double-blinded study (adapted from Meinhardt et al. [17])

Liu et al. identified six studies that measured exercise capacity outcomes in adults [20, 27, 29–32]; however, the variability in exercise interventions precluded meta-analysis of the results [18]. Berggren et al. found no significant effect of GH on VO_2max in a placebo-controlled study of 30 healthy active young men and women administered 2 different doses of GH (approximately 5–10 times daily production rates) for 4 weeks [32]. No relationship was found in this study between changes in IGF-I and changes in oxygen uptake or maximum achieved power output. Similarly, we found no significant correlation between increases in IGF-I and VO_2max in our study.

An increase in lactate levels during exercise in GH-treated subjects following a single dose of GH has been shown in three studies [29–31]. This suggests that GH may in fact worsen exercise capacity, since increased exercising lactate levels are associated with decreased exercise stamina and physical exhaustion. In the study by Lange et al. [31], two of the seven subjects were unable to complete the 90 min bicycling protocol following GH, whereas they could complete the bicycling protocol following placebo.

Anaerobic Exercise Capacity

Anaerobic capacity assesses the ability to generate a relatively high power output of brief duration. It represents the capacity to exercise using predominately anaerobic sources of energy, derived from phosphocreatinine degradation and glycogenolysis, and is usually measured by the Wingate test, which is a 30 s all-out sprint capacity test.

We have recently provided the first evidence that growth hormone enhances anaerobic exercise capacity in our study of young recreational athletes following 8 weeks' administration of GH [17]. This is the first study we are aware of that has assessed the effect of GH on anaerobic capacity in healthy adults. Sprint capacity (total work using the 30 s maximal cycle ergometry Wingate test) increased significantly with GH in the group of men and women combined by 3.9% (95% CI: 0.0–7.7, $P=0.05$), and in men co-administered GH and testosterone by 8.3% (95% CI: 3.0–13.6, $P<0.005$), compared to placebo (Fig. 20.2). These increases were no longer present 6 weeks after treatments were discontinued.

We have suggested that the improvement in anaerobic exercise capacity following GH may be due to effects on muscle energy supply, rather than effects on muscle power. Both our study and previous studies have failed to show a beneficial effect of GH on muscle strength or power in athletes [17, 19, 28] and microarray studies in men with GHD have shown mixed effects of GH treatment on genes involved in protein synthesis and degradation in muscle, and on genes encoding myofibrillar proteins [33]. These gene expression studies showed that in muscle, GH replacement enhances use of glucose over fatty acids while suppressing oxidative mitochondrial energy production, suggesting regulation by GH through anaerobic metabolism. Increased anaerobic exercise capacity following GH treatment may involve improved ability to derive acute energy requirements from anaerobic metabolism.

Summary of Effects of GH on Physical Performance in Adults

In summary (Table 20.1), we have recently demonstrated in a randomised controlled study that GH supplementation increased sprint capacity (anaerobic exercise capacity) when administered alone and in combination with testosterone; however, there is no available evidence from studies in healthy young adults that GH enhances muscle strength or power, or aerobic exercise capacity. The available evidence from high-quality studies is limited, due to the small number of well-conducted studies, the relatively short duration of treatment and the small number of female participants in previous studies. The evidence suggests that, although GH may improve anaerobic sprint capacity, it does not seem to improve muscle strength, power or aerobic exercise capacity.

Evidence for Other Possible Benefits in Athletes

While there is no evidence from randomised controlled studies for significant effects of GH on physical performance in healthy, fit adults, apart from our recent data on anaerobic exercise capacity, the possibility remains of other potential benefits of GH in athletes.

There has been the suggestion that GH may accelerate recovery from soft tissue injury. Evidence has been provided by the recent demonstration of increased matrix collagen synthesis in skeletal muscle and tendon by up to sixfold in a placebo-controlled study of 14 days' GH administration in healthy young men [22]. The increased synthesis in muscle and tendon collagen, without any effect on myofibrillar protein synthesis, suggests that GH may be more important in strengthening the matrix tissue and stimulating the supporting connective tissue, than for muscle cell hypertrophy. The increases observed in muscle and tendon collagen synthesis are also supported by the observation of increased serum collagen peptides following administration of GH to healthy fit adults [34, 35]. There is also evidence for an effect of GH on healing of bone fractures from a randomised placebo-controlled study of 406 patients aged 18–64 years with tibial fractures. For patients with closed tibial fractures, there was a significantly shorter time to healing after treatment with 60 μg/kg/day GH, corresponding to approximately 26% decrease in healing time compared to placebo [36].

It is also possible that GH may have effects that are too small to be detected in randomised, controlled studies in healthy adults, which are usually statistically powered to detect much larger differences. Although difficult to detect in a controlled study, small differences may still be important in elite competition, where winning margins can be very small.

Evidence for Harm: Side Effects of GH Use and Acromegaly

High doses of GH may be used in real world of doping, with reports that athletes may be using GH in dosages ranging from approximately 10 to 25 IU/day 3–4 times a week [3]. These doses are approximately 5–10 times daily production rate. Athletes doping with GH are at risk of the side effects that have been observed in studies of GH administration in healthy adults. Due to the high doses reportedly used by athletes, they are likely at risk of side effects that potentially include features of acromegaly. The presentation of acromegaly, the human disease of GH excess, indicates the possible long-term effects of GH abuse.

Side Effects of GH in Healthy Adults

There are adverse effects of abuse of GH, broadly related to its known physiological effects on metabolism and growth. Most of the acute side effects, such as swelling, paresthesias ("pins and needles") and arthralgias, reported in clinical trials arise from fluid retention due to the antinatriuretic effect of GH (Fig. 20.3). In our study in recreational athletes following 8 weeks' treatment, swelling and joint pain were reported by significantly more GH-treated subjects than placebo, and men treated with GH reported more paresthesias [17]. In the meta-analysis of Liu et al., higher

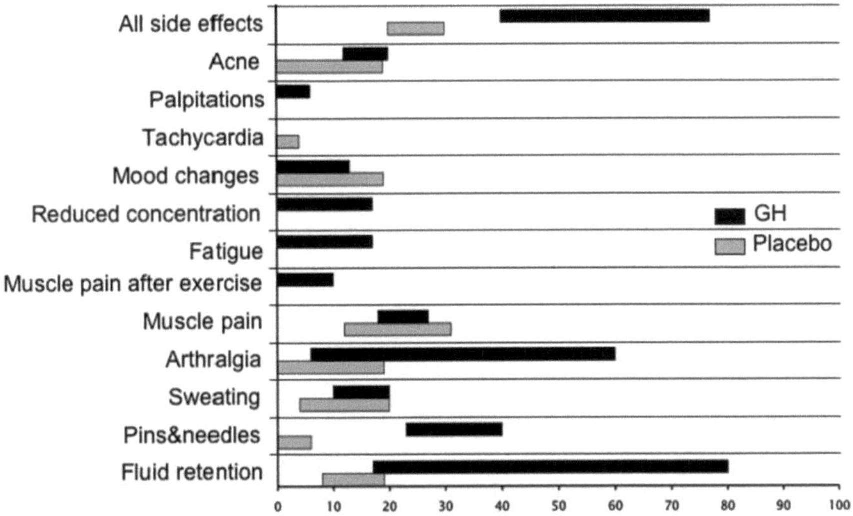

Fig. 20.3 Summary of side effects reported in 7 double-blind placebo-controlled trials in healthy subjects of GH administration for 4–12 weeks with median GH dose of 40 μg/kg/day [17, 19, 20, 26, 32, 34, 39]. Data are presented as a range of percent of the subjects reporting side effects after treatment with GH (*black bars*) and placebo (*grey bars*). Reprinted from Birzniece et al. [53] with permission from Elsevier

rates of adverse events were reported by GH-treated subjects than by those not treated with GH. GH-treated subjects reported more soft-tissue oedema (44 vs. 1%), more arthralgias (25 vs. 0%) and more carpal tunnel syndrome (15 vs. 0%), compared to those not treated [18].

Other side effects including sweating, fatigue and dizziness have been reported after GH administration in healthy subjects [18, 37]. Serious side effects including diabetes may also arise from the anti-insulin properties of GH which can cause insulin resistance and increases the risk of diabetes in susceptible individuals, and with high doses [38].

Excess levels of GH negatively affect cardiac function. In healthy young men and women treated with 33 or 67 μg/kg/day GH for 4 weeks, GH significantly affected cardiac morphology in the high-dose group compared to placebo. Cardiac output and left ventricular mass index were significantly increased and the accompanying increase in left ventricular wall thickness indicated that GH induced concentric left ventricular remodelling [39].

In addition, there are indirect adverse effects of GH that include the risk of transmission of prion diseases such as Creutzfelt-Jakob disease [40] from the use of cadaveric pituitary-derived GH that is still available on the black market because of the high cost of rhGH. There is also the risk of hepatitis or HIV contracted from shared needles.

The severity of the adverse effects of GH may be worsened by concurrent abuse of anabolic steroids. Many of these adverse effects are synergistic with those of

AASs, which potentially enhances their severity. GH and testosterone exert independent antinatriuretic effects, that in combination exacerbate fluid retention in hypopituitary men [24]. In our study in recreational athletes, combined administration of GH and testosterone in men resulted in a greater increase in ECW than in response to GH alone, compared to placebo (GH+T: 3.6 kg, 95%, CI 1.8–5.3 vs. GH: 2.4 kg, 95% CI 0.9–4.0). Men treated with combined GH and testosterone reported more swelling than men treated with GH alone (88 vs. 67%, respectively) [17]. In power athletes, potentiation by concomitant GH abuse of direct effects of AAS abuse on the myocardia resulting in myocardial hypertrophy has been reported [41, 42].

Acromegaly

Studies in acromegaly, the human disease of GH excess, indicate the potential health risks of long-term abuse of GH. The presentation of acromegaly reflects cardiac, metabolic and articular complications, which may include hypertension, cardiomyopathy, osteoarthritis and facial disfigurement [43].

In acromegaly, after many years of exposure to GH excess, muscle structure and function become impaired. It is generally accepted that GH excess results in muscles that appear larger, but are functionally weaker [9]. There is evidence for myopathy in acromegaly, which is associated with disease duration. Aerobic exercise capacity is significantly reduced in acromegalic patients [10]. Impaired aerobic function probably results from the underlying cardiomyopathy and impaired cardiac function in acromegaly. Morphological studies of the heart in acromegaly have reported ventricular hypertrophy with increased fibrosis and extracellular collagen, which coexists with myofibrillar derangement, myocyte necrosis, and lymphomononuclear infiltration, resembling a pattern of myocarditis [44, 45]. These features of acromegaly suggest that long-term GH excess is likely to be detrimental to exercise performance, although they appear to depend on disease duration [10].

Insulin action and hepatic and peripheral insulin sensitivity are impaired by GH, therefore prolonged GH excess conveys a state of insulin resistance, predisposing to the development of diabetes. Indeed, diabetes is found in up to 30% of patients with untreated acromegaly [46, 47]. GH excess induces dysregulated growth of cartilage causing arthropathy [48]. The changes in articular cartilage are irreversible, as exemplified by clinical and radiological evidence of osteoarthritis in high proportions of patients with long-term cure of acromegaly [49].

There is the possibility that prolonged GH excess may be associated with increased risk of neoplasms, with evidence for increased risk of colorectal cancer in acromegaly [50, 51]. Life expectancy is reduced in acromegaly which is normalised by achieving disease control. Overall standardised mortality rates are approximately two times higher for untreated acromegalic patients than in the general population, relating to an average reduction in life expectancy of around 10 years [52].

In summary, athletes doping with GH are at risk of the well-described acute side effects of GH administration in healthy adults, which include swelling, arthralgias, paresthesias, fatigue, and changes in cardiac morphology and function. GH is reportedly often abused by athletes in combination with AASs, and there is evidence that the side effects of GH are likely worsened by co-administration of AASs, such as testosterone. The features of acromegaly, the disease of GH excess, indicate the potential health risks of chronic abuse of GH, which include cardiac complications, arthropathy, insulin resistance and increased risk of diabetes and malignancy. Finally, there is evidence in acromegaly for weaker muscles and myopathy, and reduced aerobic exercise capacity.

Summary and Conclusions

There is evidence for clear benefits of GH in treatment of GHD, including normalisation of body composition by increasing LBM and reducing fat mass due to its lipolytic and anabolic actions, increase in muscle mass but not muscle strength with short-term treatment, and improved aerobic exercise capacity (VO_2max). There is only limited evidence available, however, for benefit of GH use in healthy young adults and in athletes, and it should be recognised that there are no studies of GH administration in elite athletes since it is not ethical to administer banned agents to elite athletes.

In young, healthy adults, GH treatment increases basal metabolic rate and heart rate and increases use of lipids, rather than carbohydrates, for fuel during rest. There is evidence for a net anabolic effect of GH on whole body protein metabolism following GH administration; however, GH may not increase muscle protein synthesis. There is strong evidence that GH decreases fat mass and significantly increases LBM in young, healthy physically fit adults; however, fluid retention accounts for most of the increase in LBM induced by GH, rather than an increase in the muscle mass. Our recent study demonstrated an increase in BCM, an estimate of muscle mass, following combined administration of GH and testosterone.

There has been debate as to whether the improved body composition and metabolic changes induced by GH actually translate into improved performance. We have recently demonstrated in a randomised controlled study an increase in anaerobic sprint capacity following treatment with GH for 8 weeks. There is no evidence that GH enhances muscle strength or power, or aerobic exercise capacity in healthy young adults. There may be other potential benefits of GH, such as accelerated recovery from soft tissue injury.

There is strong evidence of harm from GH abuse. Side effects of GH administration in healthy adults include swelling, arthralgias and paresthesias, fatigue, and changes in cardiac morphology and function. The side effects of GH such as myocardial hypertrophy are likely worsened by co-administration of AASs. Potential health risks of chronic abuse of GH, indicated by the features of acromegaly, include cardiac complications, arthropathy, insulin resistance and increased risk of diabetes and malignancy. There is also evidence for weaker muscles and myopathy, and reduced aerobic exercise capacity in acromegaly.

In conclusion, contrary to improvements in exercise capacity by GH replacement in GHD adults, the evidence suggests that in healthy adults, muscle strength, power, and aerobic exercise capacity are not enhanced by GH administration. Our recent data indicate that GH may improve a selective aspect of performance, that of anaerobic exercise capacity. More research is needed to determine the effects of GH on performance measures in well-designed, controlled studies and to confirm our findings. There are serious health risks, however, from prolonged use of GH in healthy adults due to its known adverse effects.

Acknowledgements Dr Anne E. Nelson was supported by the World Anti-Doping Agency and by the Australian Government through the Anti-Doping Research Program and the Department of Communications, Information Technology and the Arts. Dr Vita Birzniece was supported by the National Health and Medical Research Council of Australia.

References

1. Holt RI, Sonksen PH. Growth hormone, IGF-I and insulin and their abuse in sport. Br J Pharmacol. 2008;154(3):542–56.
2. Bamberger M, Yaeger D. Over the edge: special report. Sports Illustrated. 1997;86:64.
3. Saugy M, Robinson N, Saudan C, Baume N, Avois L, Mangin P. Human growth hormone doping in sport. Br J Sports Med. 2006;40 Suppl 1:i35–9.
4. Parkinson AB, Evans NA. Anabolic androgenic steroids: a survey of 500 users. Med Sci Sports Exerc. 2006;38(4):644–51.
5. Perry PJ, Lund BC, Deninger MJ, Kutscher EC, Schneider J. Anabolic steroid use in weightlifters and bodybuilders: an internet survey of drug utilization. Clin J Sport Med. 2005;15(5):326–30.
6. Rickert VI, Pawlak-Morello C, Sheppard V, Jay MS. Human growth hormone: a new substance of abuse among adolescents? Clin Pediatr (Phila). 1992;31(12):723–6.
7. National Collegiate Athletic Association. NCAA Study of Substance Use of College Student-Athletes. 2010. www.ncaa.org/library/research/substance_use_habits/2006/2006_substance_use_report.pdf. Accessed February 2010.
8. Nelson AE, Ho KK. A robust test for growth hormone doping–present status and future prospects. Asian J Androl. 2008;10(3):416–25.
9. Woodhouse LJ, Mukherjee A, Shalet SM, Ezzat S. The influence of growth hormone status on physical impairments, functional limitations, and health-related quality of life in adults. Endocr Rev. 2006;27(3):287–317.
10. Gibney J, Healy ML, Sonksen PH. The growth hormone/insulin-like growth factor-I axis in exercise and sport. Endocr Rev. 2007;28(6):603–24.
11. Hoffman DM, Pallasser R, Duncan M, Nguyen TV, Ho KK. How is whole body protein turnover perturbed in growth hormone-deficient adults? J Clin Endocrinol Metab. 1998;83(12):4344–9.
12. Burt MG, Gibney J, Hoffman DM, Umpleby AM, Ho KK. Relationship between GH-induced metabolic changes and changes in body composition: a dose and time course study in GH-deficient adults. Growth Horm IGF Res. 2008;18(1):55–64.
13. Gotherstrom G, Bengtsson BA, Bosaeus I, Johannsson G, Svensson J. A 10-year, prospective study of the metabolic effects of growth hormone replacement in adults. J Clin Endocrinol Metab. 2007;92(4):1442–5.

14. Attanasio AF, Bates PC, Ho KK, et al. Human growth hormone replacement in adult hypopituitary patients: long-term effects on body composition and lipid status–3-year results from the HypoCCS Database. J Clin Endocrinol Metab. 2002;87(4):1600–6.
15. Widdowson WM, Gibney J. The effect of growth hormone (GH) replacement on muscle strength in patients with GH-deficiency: a meta-analysis. Clin Endocrinol (Oxf). 2010;72:787–92.
16. Widdowson WM, Gibney J. The effect of growth hormone replacement on exercise capacity in patients with GH deficiency: a metaanalysis. J Clin Endocrinol Metab. 2008;93(11):4413–7.
17. Meinhardt U, Nelson AE, Hansen JL, et al. The effect of growth hormone on body composition and physical performance in recreational athletes. Ann Intern Med. 2010;152(9):568–79.
18. Liu H, Bravata DM, Olkin I, et al. Systematic review: the effects of growth hormone on athletic performance. Ann Intern Med. 2008;148(10):747–58.
19. Yarasheski KE, Campbell JA, Smith K, Rennie MJ, Holloszy JO, Bier DM. Effect of growth hormone and resistance exercise on muscle growth in young men. Am J Physiol. 1992;262(3 Pt 1):E261–7.
20. Healy ML, Gibney J, Russell-Jones DL, et al. High dose growth hormone exerts an anabolic effect at rest and during exercise in endurance-trained athletes. J Clin Endocrinol Metab. 2003;88(11):5221–6.
21. Yarasheski KE, Zachweija JJ, Angelopoulos TJ, Bier DM. Short-term growth hormone treatment does not increase muscle protein synthesis in experienced weight lifters. J Appl Physiol. 1993;74(6):3073–6.
22. Doessing S, Heinemeier KM, Holm L, et al. Growth hormone stimulates the collagen synthesis in human tendon and skeletal muscle without affecting myofibrillar protein synthesis. J Physiol. 2010;588(Pt 2):341–51.
23. O'Sullivan AJ, Kelly JJ, Hoffman DM, Freund J, Ho KK. Body composition and energy expenditure in acromegaly. J Clin Endocrinol Metab. 1994;78(2):381–6.
24. Johannsson G, Gibney J, Wolthers T, Leung KC, Ho KK. Independent and combined effects of testosterone and growth hormone on extracellular water in hypopituitary men. J Clin Endocrinol Metab. 2005;90(7):3989–94.
25. Ho KY, Weissberger AJ. The antinatriuretic action of biosynthetic human growth hormone in man involves activation of the renin-angiotensin system. Metabolism. 1990;39(2):133–7.
26. Ehrnborg C, Ellegard L, Bosaeus I, Bengtsson BA, Rosen T. Supraphysiological growth hormone: less fat, more extracellular fluid but uncertain effects on muscles in healthy, active young adults. Clin Endocrinol (Oxf). 2005;62(4):449–57.
27. Graham MR, Baker JS, Evans P, et al. Physical effects of short-term recombinant human growth hormone administration in abstinent steroid dependency. Horm Res. 2008; 69(6):343–54.
28. Deyssig R, Frisch H, Blum WF, Waldhor T. Effect of growth hormone treatment on hormonal parameters, body composition and strength in athletes. Acta Endocrinol (Copenh). 1993;128(4):313–8.
29. Hansen M, Morthorst R, Larsson B, et al. No effect of growth hormone administration on substrate oxidation during exercise in young, lean men. J Physiol. 2005;567(Pt 3):1035–45.
30. Irving BA, Patrie JT, Anderson SM, et al. The effects of time following acute growth hormone administration on metabolic and power output measures during acute exercise. J Clin Endocrinol Metab. 2004;89(9):4298–305.
31. Lange KH, Larsson B, Flyvbjerg A, et al. Acute growth hormone administration causes exaggerated increases in plasma lactate and glycerol during moderate to high intensity bicycling in trained young men. J Clin Endocrinol Metab. 2002;87(11):4966–75.
32. Berggren A, Ehrnborg C, Rosen T, Ellegard L, Bengtsson BA, Caidahl K. Short-term administration of supraphysiological recombinant human growth hormone (GH) does not increase maximum endurance exercise capacity in healthy, active young men and women with normal GH-insulin-like growth factor I axes. J Clin Endocrinol Metab. 2005;90(6):3268–73.
33. Sjogren K, Leung KC, Kaplan W, Gardiner-Garden M, Gibney J, Ho KK. Growth hormone regulation of metabolic gene expression in muscle: a microarray study in hypopituitary men. Am J Physiol Endocrinol Metab. 2007;293(1):E364–71.

34. Longobardi S, Keay N, Ehrnborg C, et al. Growth hormone (GH) effects on bone and collagen turnover in healthy adults and its potential as a marker of GH abuse in sports: a double blind, placebo-controlled study. The GH-2000 Study Group. J Clin Endocrinol Metab. 2000;85(4):1505–12.
35. Nelson AE, Meinhardt U, Hansen JL, et al. Pharmacodynamics of growth hormone abuse biomarkers and the influence of gender and testosterone: a randomized double-blind placebo-controlled study in young recreational athletes. J Clin Endocrinol Metab. 2008;93(6):2213–22.
36. Raschke M, Rasmussen MH, Govender S, Segal D, Suntum M, Christiansen JS. Effects of growth hormone in patients with tibial fracture: a randomised, double-blind, placebo-controlled clinical trial. Eur J Endocrinol. 2007;156(3):341–51.
37. Keller A, Wu Z, Kratzsch J, et al. Pharmacokinetics and pharmacodynamics of GH: dependence on route and dosage of administration. Eur J Endocrinol. 2007;156(6):647–53.
38. Young J, Anwar A. Strong diabetes. Br J Sports Med. 2007;41(5):335–6; discussion 336.
39. Cittadini A, Berggren A, Longobardi S, et al. Supraphysiological doses of GH induce rapid changes in cardiac morphology and function. J Clin Endocrinol Metab. 2002;87(4):1654–9.
40. Brown P, Preece M, Brandel JP, et al. Iatrogenic Creutzfeldt-Jakob disease at the millennium. Neurology. 2000;55(8):1075–81.
41. Karila TA, Karjalainen JE, Mantysaari MJ, Viitasalo MT, Seppala TA. Anabolic androgenic steroids produce dose-dependant increase in left ventricular mass in power atheletes, and this effect is potentiated by concomitant use of growth hormone. Int J Sports Med. 2003;24(5):337–43.
42. Mark PB, Watkins S, Dargie HJ. Cardiomyopathy induced by performance enhancing drugs in a competitive bodybuilder. Heart. 2005;91(7):888.
43. Melmed S. Acromegaly pathogenesis and treatment. J Clin Invest. 2009;119(11):3189–202.
44. Colao A, Marzullo P, Di Somma C, Lombardi G. Growth hormone and the heart. Clin Endocrinol (Oxf). 2001;54(2):137–54.
45. Meyers DE, Cuneo RC. Controversies regarding the effects of growth hormone on the heart. Mayo Clin Proc. 2003;78(12):1521–6.
46. Ezzat S, Forster MJ, Berchtold P, Redelmeier DA, Boerlin V, Harris AG. Acromegaly. Clinical and biochemical features in 500 patients. Medicine (Baltimore). 1994;73(5):233–40.
47. Colao A, Baldelli R, Marzullo P, et al. Systemic hypertension and impaired glucose tolerance are independently correlated to the severity of the acromegalic cardiomyopathy. J Clin Endocrinol Metab. 2000;85(1):193–9.
48. Colao A, Pivonello R, Scarpa R, Vallone G, Ruosi C, Lombardi G. The acromegalic arthropathy. J Endocrinol Invest. 2005;28(8 Suppl):24–31.
49. Wassenaar MJ, Biermasz NR, van Duinen N, et al. High prevalence of arthropathy, according to the definitions of radiological and clinical osteoarthritis, in patients with long-term cure of acromegaly: a case-control study. Eur J Endocrinol. 2009;160(3):357–65.
50. Jenkins PJ, Mukherjee A, Shalet SM. Does growth hormone cause cancer? Clin Endocrinol (Oxf). 2006;64(2):115–21.
51. Perry JK, Emerald BS, Mertani HC, Lobie PE. The oncogenic potential of growth hormone. Growth Horm IGF Res. 2006;16(5–6):277–89.
52. Ayuk J, Sheppard MC. Does acromegaly enhance mortality? Rev Endocr Metab Disord. 2008;9(1):33–9.
53. Birzniece V, Nelson AE, Ho KK. Growth hormone administration: is it safe and effective for athletic performance. Endocrinol Metab Clin North Am. 2010;39(1):11–23, vii.

Index

K. Ho (ed.), *Growth Hormone Related Diseases and Therapy: A Molecular and Physiological Perspective for the Clinician*, Contemporary Endocrinology, DOI 10.1007/978-1-60761-317-6, © Springer Science+Business Media, LLC 2011

MIX
Papier aus verantwortungsvollen Quellen
Paper from responsible sources
FSC® C105338

If you have any concerns about our products,
you can contact us on
ProductSafety@springernature.com

In case Publisher is established outside the EU,
the EU authorized representative is:
Springer Nature Customer Service Center GmbH
Europaplatz 3, 69115 Heidelberg, Germany

Printed by Libri Plureos GmbH
in Hamburg, Germany